Multi-Sensor Image Fusion
and
Its Applications

Signal Processing and Communications

Multi-Sensor Image Fusion and Its Applications

edited by

Rick S. Blum and Zheng Liu

Taylor & Francis

Taylor & Francis Group

Boca Raton   London   New York   Singapore

A CRC title, part of the Taylor & Francis imprint, a member of the
Taylor & Francis Group, the academic division of T&F Informa plc.

FIRST INDIAN REPRINT, 2015

Published in 2006 by
CRC Press
Taylor & Francis Group
6000 Broken Sound Parkway NW, Suite 300
Boca Raton, FL 33487-2742

International Standard Book Number-10: 0-8493-3417-9 (Hardcover)
International Standard Book Number-13: 978-0-8493-3417-7
Library of Congress Card Number 2005041824

Library of Congress Cataloging-in-Publication Data

Multi-sensor image fusion and its applications / Rick S. Blum and Zheng Liu [editors].
       p. cm. – (Signal processing and communications : 26)
  Includes bibliographical references and index.
  ISBN 0-8493-3417-9 (alk. paper)
  1. Intelligent control systems. 2. Multisensor data fusion. I. Blum, Rick S. II. Liu, Zheng, 1969 July 7- III. Series.

TJ217.5.M86 2005
621.36'7—dc22   2005041824

Taylor & Francis Group
is the Academic Division of T&F Informa plc.

Visit the Taylor & Francis Web site at
http://www.taylorandfrancis.com

and the CRC Press Web site at
http://www.crcpress.com

Printed and bound in India by Nutech Print Services

FOR SALE IN SOUTH ASIA ONLY.

Preface

Imaging sensors are an important resource in today's world. Further, a single sensor cannot provide a complete view of the environment in many applications. Just as humans use all five senses to great advantage, computers use image fusion algorithms to gain similar advantages. The resulting fused image, if suitably obtained from a set of source sensor images, can provide a better view of the true scene than the view provided by any of the individual source images. In particular, image sharpening, feature enhancement, better object detection, and improved classification can result. This book explores both the theoretical and practical aspects of image fusion techniques. The approaches include, but are not limited to, statistical methods, color related techniques, model-based methods, and visual information display strategies. Applications of image fusion techniques are also highlighted in this book. The particular applications discussed include medical diagnosis, surveillance systems, biometric systems, remote sensing, nondestructive evaluation, blurred image restoration, and image quality assessment. Each chapter contains a description of research progress on image fusion and highlights methods for solving practical problems with proposed techniques. Due to the intimate relationship between registration and fusion, registering multisensor images is another topic featured. In particular, an in-depth discussion of multisensor image registration can be found in several chapters. In total, this book consists of 16 chapters. Each of these contributions will now be described in more detail.

In the first chapter, Blum and co-authors review the state-of-the-art in image fusion techniques. An in-depth overview of fusion algorithms and fusion performance evaluation is given, along with some introductory discussion on multisensor image registration. The authors present a detailed description of multiscale-transform-based approaches, along with a discussion of nonmultiscale-transform-based approaches, including pixel averaging methods, nonlinear methods, estimation-based methods, color-based fusion methods, and artificial neural network-based methods. In the (multifocus) digital camera and the concealed weapon detection applications, fusion results were accessed by different evaluation metrics, with or without a perfect reference image. Finally, an image registration scheme that takes advantage of both intensity- and feature-based components is also presented.

In the next five chapters, the authors consider registration and fusion of multimodal medical images. Chen and Varshney introduce a mutual information-based registration algorithm proposed for registration of a computed tomography (CT) brain image with a magnetic resonance image (MRI) of the same human brain. The method employs higher order B-spline kernels for joint histogram estimation, and overcomes the interpolation-induced artefacts, a known challenge for mutual information based registration. The proposed generalized partial volume estimation algorithm successfully suppresses the artefacts and improves registration accuracy considerably. The registration and fusion of human retinal images is investigated by Laliberté and Gagnon. The retinal images are of different modalities (color, fluorescein angiogram), different resolutions, and taken at different times (from a few minutes during an angiography examination to several years between two examinations). The authors review and describe the principal methods for retinal image registration and fusion. Extensive and comparative tests are carried out. Both the registration and fusion results, which are presented using a set of images covering a large ensemble of retinal diseases, are assessed qualitatively and quantitatively. Pennec et al. present an image-based technique to register intraoperative 3-D ultrasound images with preoperative magnetic resonance data. An automatic rigid registration is achieved by maximizing a similarity measure, and brain deformations are tracked in the 3-D ultrasound time sequence using an innovative algorithm. Experimental results show that a registration accuracy of the magnetic resonance voxel size is achieved for the rigid part, and a qualitative accuracy of a few millimeters could be obtained for the complete tracking system. Based on the fact that a scene boundary structure is shared in different image modalities, Weisenseel et al. present a unified multimodality fusion formulation which combines the reconstruction, enhancement, and alignment processes. An objective energy function consisting of four terms is minimized by an optimization approach. The terms include the sensor observation model term, the noise suppression term, the alignment term, and the boundary term. Fusion results obtained for examples of simulated and real vascular imaging with multidetector computed tomography (MDCT) and volume computed tomography (VCT) are presented. In the last chapter of this group, Neemuchwala and Hero introduce a technique to extend image registration to a higher dimension feature space. The authors use independent component analysis and multi-dimensional wavelets to represent local image features. Classes of entropic-graph methods are used to approximate similarity measures between multisensor images. Experiments on registering multiband satellite images and dual-modality MRI images of a human brain demonstrate the efficiency and sensitivity of the proposed approaches, while correlation-based registration methods are not effective.

Surveillance and security systems that employ multiple image modalities continue to receive great attention, and it is becoming clear that fusion can significantly improve the overall system performance. Three contributions in this book consider surveillance and security system applications. In the first contribution, Toet provides experiments that indicate that a fused image provides a better representation of the spatial layout of the scene. Therefore, graylevel and color image fusion schemes are tested for fusing intensified visual and thermal imagery for global scene recognition and perception. The appropriate use of color information will greatly increase the operator's situation awareness for a given condition and task. Yang and Blum present a statistical approach for fusing visual and nonvisual images for concealed weapon detection and night-vision applications. The newly developed algorithm employs a hidden Markov tree model to relate the wavelet coefficients across different scales. A mathematical model employs a Gaussian mixture distortion distribution to relate the sensor observations to the true scene. The model acknowledges that certain objects are only visible to certain sensors. The parameters of the mathematical model are estimated using the expectation-maximization algorithm. In the concealed weapon detection application, a visual image is fused with a millimeter wave image. Information on the identification of the people in the scene as well as the concealed weapon is available in the fusion result. The proposed method also shows a superior performance in night vision applications. In the next contribution from Ross and Jain, a multimodal biometric system is implemented to meet the stringent requirements imposed by high-security applications. Through fusing multiple evidence, the multimodal biometric system outperforms the unimodal system in the presence of noisy data, intraclass variations, spoof attacks, and so forth. The fusion scenarios and system design issues are also discussed in this chapter.

The chapter presented by Jouan et al. deals with a remote sensing application for a surveillance task. A prototype is designed to fuse the evidence of contextual attributes extracted from space-borne imagery and ancillary data that includes incomplete and uncertain data. The changes in a set of multidate RADARSAT-1 scenes are detected and classified to locate or highlight regions of potential concerns. The issues specific to registering remotely sensed earth data are discussed in the chapter contributed by Le Moigne and Eastman. The topics include: the need for subpixel accurate image registration, geo-registration, operational registration systems, and methods for characterizing the registration components in the framework of remote sensing.

In this book, two chapters are dedicated to studies of fusing nondestructive testing images. The system- and model-based approaches for fusing eddy current and ultrasonic data are presented by Upda et al. The eddy current relies on diffusion for propagating energy while a wave propagation equation governs the ultrasonic method. The ultrasonic sensor produces a measurement with high resolution, but the measurement is subject to the surface roughness of the specimen and the grain structure of materials. In contrast, eddy current images have a relatively high signal-to-noise ratio and contain depth information of cracks. In the system-based method, a linear minimum mean square error filter is applied to the eddy current and ultrasonic images to achieve an improved signal-to-noise ratio, while morphological processing can reduce the noise in the ultrasonic image and integrate the depth information from the eddy current testing. The so-called Q-transform is a model-based method that can map the data provided by the eddy current and ultrasonic sensors onto a common format, thus providing a preprocessing step for multisensor data fusion. In the next chapter, Liu et al. investigate the detection and quantification of hidden corrosion in the multilayered lap joint structure of aircraft by fusing the images of multifrequency eddy current testing and the lift-off-intersection scan from pulsed eddy current testing. The fusion process is implemented in three different ways; that is, pixel-level fusion, classification-based fusion, and a generalized additive model-based fusion. The fusion of the multiple images achieves an optimal solution to reveal the corrosion at different layers with the two nondestructive testing techniques.

In the last part of the book, Šroubek and Flusser propose a generic solution to the scenario where multichannel inputs are blurred, impaired by noise, and possibly geometrically distorted. The fusion algorithm makes use of multichannel blind deconvolution theory and is highly capable for several types of degradations, including uniform blurring and misregistration. The experimental results verify the reliable performance of the proposed fusion algorithm. A gaze-contingent multi-modality system implemented with OpenGL along with texture mapping techniques, proposed by Nikolov et al. integrates the visualization of multimodal images. The system promises great potential and flexibility for the human visual system to fuse/perceive multisource information through the synthetic visualization of 2-D medical, surveillance, and remote sensing images, and 3-D volumetric images and multilayer information maps. The final chapter in this book by Seshadrinathan et al., concerns another very important aspect of the studies on image fusion, that is, image quality assessment. The objective assessment of image quality provides guidance for efficient and optimal fusion of multisensor images, when the fusion algorithm is under development. This chapter describes recent innovations in image quality assessment. A top-down approach is developed by Seshadrinathan et al. to derive models that quantify the structural distortions in an image. Using these models, image quality may be extracted.

Editors

Rick S. Blum, Ph.D., received a B.S. in electrical engineering from the Pennsylvania State University in 1984 and an M.S. and Ph.D. in electrical engineering from the University of Pennsylvania in 1987 and 1991.

From 1984 to 1991 he was a member of the technical staff at General Electric Aerospace in Valley Forge, Pennsylvania and he graduated from GE's Advanced Course in Engineering. Since 1991, he has been with the Electrical and Computer Engineering Department at Lehigh University in Bethlehem, Pennsylvania where he is currently a Professor and holds the Robert W. Wieseman Chair in Electrical Engineering. His research interests include signal detection and estimation and related topics in the areas of signal processing and communications. He has been an associate editor for the *IEEE Transactions on Signal Processing* and for the *IEEE Communications Letters* and for the *Journal of Advances in Information Fusion*. He was a member of the Signal Processing for Communications Technical Committee of the IEEE Signal Processing Society.

Dr. Blum, an IEEE Fellow, is a member of Eta Kappa Nu and Sigma Xi, and holds a patent for a parallel signal and image processor architecture. He was awarded an ONR Young Investigator Award in 1997 and an NSF Research Initiation Award in 1992.

Zheng Liu, Ph.D., completed his B.E. degree in mechanical and automation engineering from the Beijing Institute of Chemical Fibre Technology (P.R. China) in 1991 and obtained an M.E. degree in automatic instrumentation from the Beijing University of Chemical Technology (P.R. China) in 1996. He was awarded the Peace and Friendship Scholarship by Japanese Association of International Education and studied in Kyoto University for one year as part of the masters program. He earned a doctorate in engineering from Kyoto University in 2000 (Japan).

From 1991 to 1993, he worked as an engineer in the Fujian TV Factory, which belongs to Fujian Hitachi Group. From 2000 to 2001, he worked as a research fellow for the Control and Instrument Division of the School of Electrical and Electronic Engineering at the Nanyang Technological University (Singapore). After that, he joined the Institute for Aerospace Research, National Research Council Canada as a government laboratory visiting fellow selected by NSERC. Now he is an assistant research officer.

Dr. Liu is a member of IEEE and the Canadian Image Processing and Pattern Recognition Society.

Contributors

Yannick Allard
Lockheed Martin
Montreal, Quebec, Canada

Nicholas Ayache
EPIDAURE, INRIA Sophia-Antipolis
Sophia–Antipolis, France

Rick S. Blum
ECE Department
Lehigh University
Bethlehem, Pennsylvania

Alan C. Bovik
Laboratory for Image and
  Video Engineering
Department of Electrical and
  Computer Engineering
The University of Texas at Austin
Austin, Texas

David Bull
Signal Processing Group
Centre for Communications Research
University of Bristol
Bristol, UK

Nishan Canagarajah
Signal Processing Group
Centre for Communications Research
University of Bristol
Bristol, UK

Pascal Cathier
CAD Solutions Group
Siemens Medical Solutions
Malvern, Pennsylvania

Raymond C. Chan
Department of Radiology
Massachusetts General Hospital
Harvard Medical School
Cambridge, Massachusetts

Hua-Mei Chen
Department of Computer Science and
  Engineering
The University of Texas at Arlington
Arlington, Texas

Roger D. Eastman
Loyola College of Maryland
Baltimore, Maryland

Jan Flusser
Institute of Information Theory and
  Automation
Academy of Sciences of the
  Czech Republic
Prague, Czech Republic

David S. Forsyth
Structures, Materials Performance
  Laboratory
Institute for Aerospace Research
National Research Council
  Canada
Ottawa, Ontario, Canada

Langis Gagnon
R&D Department
Computer Research Institute of
  Montreal (CRIM)
Montreal, Quebec, Canada

Iain Gilchrist
Department of Experimental Psychology
University of Bristol
Bristol, UK

Alfred Hero
Departments of Electrical Engineering
  and Computer Science
Biomedical Engineering and Statistics
University of Michigan
Ann Arbor, Michigan

Anil K. Jain
Department of Computer Science
  and Engineering
Michigan State University
East Lansing, Michigan

Michael Jones
School of Information Technology
Griffith University Gold Coast
Queensland, Australia

Alexandre Jouan
Space Optronics Section at Defense
  Research and Development
Val-Belair, Quebec, Canada

William C. Karl
Departments of Electrical and
  Computer Engineering, and
  Biomedical Engineering
Boston University
Boston, Massachusetts

Jerzy P. Komorowski
Structures, Materials Performance
  Laboratory
Institute for Aerospace Research
National Research Council Canada
Ottawa, Ontario, Canada

France Laliberté
R&D Department
Computer Research Institute of
  Montreal (CRIM)
Montreal, Quebec, Canada

Zheng Liu
Structures, Materials Performance
  Laboratory
Institute for Aerospace Research
National Research Council
  Canada
Ottawa, Ontario, Canada

Yves Marcoz
Centre de Recherches Mathematiques
Universite de Montreal
Montreal, Quebec, Canada

Jacqueline Le Moigne
NASA Goddard Space Flight Center
Applied Information Sciences Branch
Greenbelt, Maryland

Huzefa Neemuchwala
Departments of Biomedical
  Engineering, Electrical Engineering
  and Computer Science and Radiology
University of Michigan
Ann Arbor, Michigan

Stavri Nikolov
Signal Processing Group
Centre for Communications Research
University of Bristol
Bristol, UK

Xavier Pennec
EPIDAURE, INRIA Sophia-Antipolis
Sophia–Antipolis, France

Alexis Roche
Service Hospitalier Frédéric Joliot
Orsay, France

Arun Ross
Department of Computer Science
  and Electrical Engineering
West Virginia University
Morgantown, West Virginia

Kalpana Seshadrinathan
Laboratory for Image and
  Video Engineering
Department of Electrical and
  Computer Engineering
The University of Texas at Austin
Austin, Texas

Hamid R. Sheikh
Laboratory for Image and
  Video Engineering
Department of Electrical and
  Computer Engineering
The University of Texas at Austin
Austin, Texas

Filip Šroubek
Institute of Information Theory and
  Automation
Academy of Sciences of the
  Czech Republic
Prague, Czech Republic

Antonello Tamburrino
Associazione EURATOM/ENEA/
  CREATE, DAEIMI
Università di Cassino
Cassino, Italy

Alexander Toet
TNO Human Factors Research Institute
Soesterberg, The Netherlands

Lalita Udpa
Department of Electrical and
  Computer Engineering
Michigan State University
East Lansing, Michigan

Satish Udpa
Department of Electrical and
  Computer Engineering
Michigan State University
East Lansing, Michigan

Pramod K. Varshney
Department of Electrical Engineering
  and Computer Sciences
Syracuse University
Syracuse, New York

Zhou Wang
Laboratory for Computational
  Vision (LCV)
New York University
New York, New York

Robert A. Weisenseel
Departments of Electrical and
  Computer Engineering, and
  Biomedical Engineering
Boston University
Boston, Massachusetts

Zhiyun Xue
ECE Department
Lehigh University
Bethlehem, Pennsylvania

Jinzhong Yang
ECE Department
Lehigh University
Bethlehem, Pennsylvania

Zhong Zhang
ObjectVideo
Reston, Virginia

Table of Contents

Dedication

Rick S. Blum would like to dedicate this book to his family: Karleigh, Kyle, Kirstyn and Karen. He would like to thank them for their love and for graciously allowing him the time to develop his research and this book project.

He would also like to thank the United States Army Research Office for supporting his research under grant number DAAD19-00-1-0431. The content of the information does not necessarily reflect the position or the policy of the federal government, and no official endorsement should be inferred.

Zheng Liu would like to dedicate this book to those who support him with great patience and love.

1 An Overview of Image Fusion*

Rick S. Blum, Zhiyun Xue, and Zhong Zhang

CONTENTS

* This material is based on work supported by the U. S. Army Research Office under grant number DAAD19-00-1-0431. The content of the information does not necessarily reflect the position or the policy of the federal government, and no official endorsement should be inferred.

I. INTRODUCTION TO IMAGE FUSION

The information science research associated with the development of sensory systems focuses mainly on how information about the world can be extracted from sensory data. The sensing process can be interpreted as a mapping of the state of the world into a set of much lower dimensionality. The mapping is many-to-one which means that there are typically many possible configurations of the world that could give rise to the measured sensory data. Thus in many cases, a single sensor is not sufficient to provide an accurate perception of the real world.

There has been growing interest in the use of multiple sensors to increase the capabilities of intelligent machines and systems. As a result, multisensor fusion has become an area of intense research and development activity in the past few years.[1-11] Multisensor fusion refers to the synergistic combination of different sources of sensory information into a single representational format. The information to be fused may come from multiple sensory devices monitored over a common period of time, or from a single sensory device monitored over an extended time period. Multisensor fusion is a very broad topic that involves contributions from many different groups of people. These groups include academic researchers in mathematics, physics, and engineering. These groups also include defense agencies, defense laboratories, corporate agencies and corporate laboratories.

Multisensor fusion can occur at the signal, image, feature, or symbol level of representation. Signal-level fusion refers to the direct combination of several signals in order to provide a signal that has the same general format as the source signals. Image-level fusion (also called pixel-level fusion in some literature[6]) generates a fused image in which each pixel is determined from a set of pixels in each source image. Clearly, image-level fusion, or image fusion, is closely related to signal-level fusion since an image can be considered a two-dimensional (2D) signal. We make a distinction since we focus on image fusion here. Feature-level fusion first employs feature extraction on the source data so that features from each source can be jointly employed for some purposes. A common type of feature-level fusion involves fusion of edge maps. Symbol-level fusion allows the information from multiple sensors to be effectively combined at the highest level of abstraction. The symbols used for the fusion can be originated either from processing only the information provided or through a symbolic reasoning process that may include *a priori* information. A common type of symbol-level fusion is decision fusion. Most common sensors provide data that can be fused at one or more of these levels. The different levels of multisensor fusion can be used to provide information to a system that can be used for a variety of purposes.

One should recognize that image fusion means different things to different people. In the rest of this chapter, image fusion is defined as a procedure for generating a fused image in which each pixel is determined from a set of pixels in each source image. Other researchers define image fusion as any form of fusion involving images. This includes, for example, cases of decision fusion using images, a specific example of which is shown in Figure 1.1.

FIGURE 1.1 Decision fusion using images. Images from November 1993 Fort Carson Data Collection Final Report, J. Ross Beveridge, Steve Hennessy, Durga Panda, Bill Hoff, and Theodore Yachik, Colorado State Tech. report CS-94-118.

Many of today's most advanced sensors produce images. Examples include optical cameras, millimeter wave (MMW) cameras, infrared (IR) cameras, x-ray imagers, and radar imagers. In this chapter, we focus on fusing images from such sensors using techniques we call image fusion. The purpose of image fusion is to generate a single image which contains a more accurate description of the scene than any of the individual source images. This fused image should be more useful for human visual or machine perception. The different images to be fused can come from different sensors of the same basic type, or they may come from different types of sensors.

The sensors used for image fusion need to be accurately aligned so that their images will be in spatial registration. This might be achieved through locating the sensors on the same platform and then using a system of optical lenses to approximately register the images for a prespecified viewing point. If this is not possible, then registration algorithms are the next choice. The improvement in quality associated with image fusion can be assessed through the improvements noted in the performance of image processing tasks applied following fusion.

This chapter is organized as follows. In the rest of Section I, we give an introduction to the advantages of image fusion and illustrate some applications of this technique. In Section II, several typical image fusion methods are described.

In Section III, two important issues, image evaluation and image registration, are discussed. Section IV gives conclusions.

A. POTENTIAL ADVANTAGES OF IMAGE FUSION

The workload of a human operator increases significantly with the number of images that need simultaneous monitoring. Moreover, a human observer cannot reliably combine visual information by viewing multiple images separately. Further, the integration of information across multiple human observers is often unreliable. Thus, a fusion system that can provide one single fused image with more accurate and reliable information than any source image is of great practical value.

The potential advantages of image fusion are that information can be obtained more accurately, as well as in less time and at a lower cost. Further, image fusion enables features to be distinguished that are impossible to perceive with any individual sensor. These advantages correspond to the notions of *redundant*, *complementary*, *more timely*, and *less costly* information.

Redundant information is provided by a group of sensors (or a single sensor over time) when each sensor is perceiving, possibly with a different fidelity, the same features in the environment. The fusion of redundant information can reduce overall uncertainty and thus serve to increase accuracy. Multiple sensors providing redundant information can also serve to increase reliability in the case of sensor error or failure.

Complementary information from multiple sensors allows features in the environment to be perceived that are impossible to perceive using just the information from each individual sensor operating separately. If the features to be perceived are considered dimensions in a space of features, then complementary information is provided when each sensor is only able to present information concerning a subset of the feature space.

More timely information may result when multiple sensors are employed, since the overall system may be able to reach a conclusion about some property of the environment more rapidly.

Less costly information may be obtained from a system using multisensor fusion. For example, a set of ordinary sensors can be used to obtain performance that could only otherwise be achieved using a very expensive single sensor.

B. APPLICATIONS OF IMAGE FUSION

In recent years, image fusion has been attracting a large amount of attention in a wide variety of applications such as concealed weapon detection (CWD),[12–17] remote sensing,[18–21] medical diagnosis,[22–25] defect inspection,[26–28] and military surveillance.[29–33] Some example applications of image fusion are shown in Figures 1.2–1.6.

The first example illustrates a CWD application.[15] CWD is an increasingly important topic in the general area of law enforcement and it appears to be a

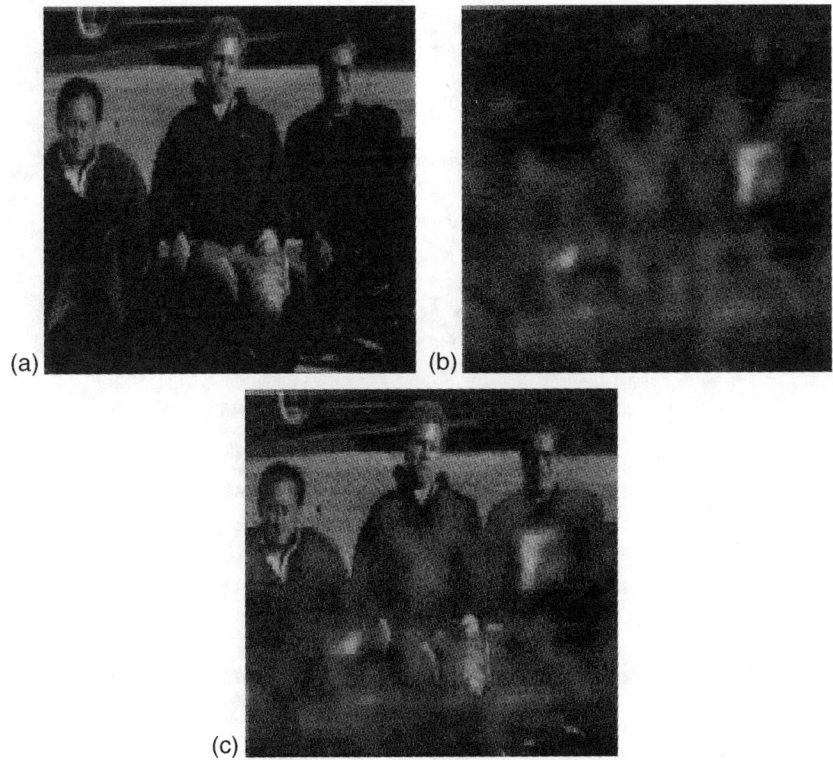

FIGURE 1.2 Example application of image fusion — CWD. (a) Visual image; (b) MMW image; (c) fused image.

critical technology for dealing with terrorism, which appears to be the most significant law enforcement problem for the next decade. Since no single sensor technology can provide acceptable performance in CWD applications, image fusion has been identified as a key technology to achieve improved CWD procedures. Figure 1.2 shows a pair of visual and 94 GHz MMW images[*]. The visual image provides the outline and the appearance of the people while the MMW image shows the existence of a gun. From the fused image, there is considerable evidence to suspect that the person on the right has a concealed gun underneath his clothes.

The second example illustrates a multifocus image fusion application[34] which is useful to enhance digital camera images. Inexpensive cameras may have difficulty in obtaining images which are in-focus everywhere in some situations due to the limited depth-of-focus of the employed lenses.[35] To overcome this problem, we can take several images with different focus points and combine them together into a single composite image using image fusion.

---

[*] The source images were obtained from Thermotex Corporation.

FIGURE 1.3 Example application of image fusion — multifocus images. (a) Focus on the left; (b) focus on the right; (c) fused image.

Figure 1.3 shows a pair of digital camera images. In one image, the focus is on the Pepsi can. In the other image, the focus is on the testing card. In the fused image, the Pepsi can, the table, and the testing card are all in focus.

The third example illustrates a head-tracked vision system for night vision applications. The multiple imaging sensors employed can enhance a driver's overall situational awareness.[36] Figure 1.4 shows a scene captured by the head-tracked vision system during a field exercise. This scene includes a person, a road, a house, grass, and trees. Figure 1.4(a) shows the image-intensified CCD (IICCD) sensor image of the scene, and Figure 1.4(b) shows the corresponding thermal imaging forward-looking-infrared (FLIR) sensor image of the same scene. These images contain complementary features as illustrated by the fused image shown in Figure 1.4(c).

The fourth example illustrates nondestructive testing (NDT) defect detection. NDT of composite materials is difficult and more than one method is usually required to provide a more complete assessment of the current state of the material. Figure 1.5 shows one example of improving defect detection through the fusion of images from multiple NDT sources gathered during the inspection

FIGURE 1.4 Example application of image fusion — night vision. (a) IICCD image; (b) FLIR image; (c) fused image.

of composite material damaged by impact.[37] Figure 1.5(a) shows the image obtained from the eddy current inspection and Figure 1.5(b) shows the image obtained from infrared thermographic inspections. These two images are fused to generate a single image shown in Figure 1.5(c) which provides comprehensive information on the extent of damage.

The fifth example demonstrates an approach for improving the spectral and spatial resolution of remotely sensed imagery.[38] Figure 1.6(a) shows a high spatial resolution panchromatic image. Figure 1.6(b) shows a low spatial resolution multispectral image of the same scene. Using image fusion, we can obtain a high-resolution multispectral image which combines the spectral characteristic of the low-resolution data with the spatial resolution the panchromatic image, as shown in Figure 1.6(c).

II. METHODS OF IMAGE FUSION

A variety of image fusion techniques have been developed. While it would be impossible to discuss each and every approach, we describe some common example approaches. They can be roughly divided into two groups, multiscale-decomposition-based fusion methods, and nonmultiscale-decomposition-based (NMDB) fusion methods.

A. MULTISCALE-DECOMPOSITION-BASED FUSION METHODS

In recent years, many researchers have recognized that multiscale transforms (MST) are very useful for analyzing the information content of images for the

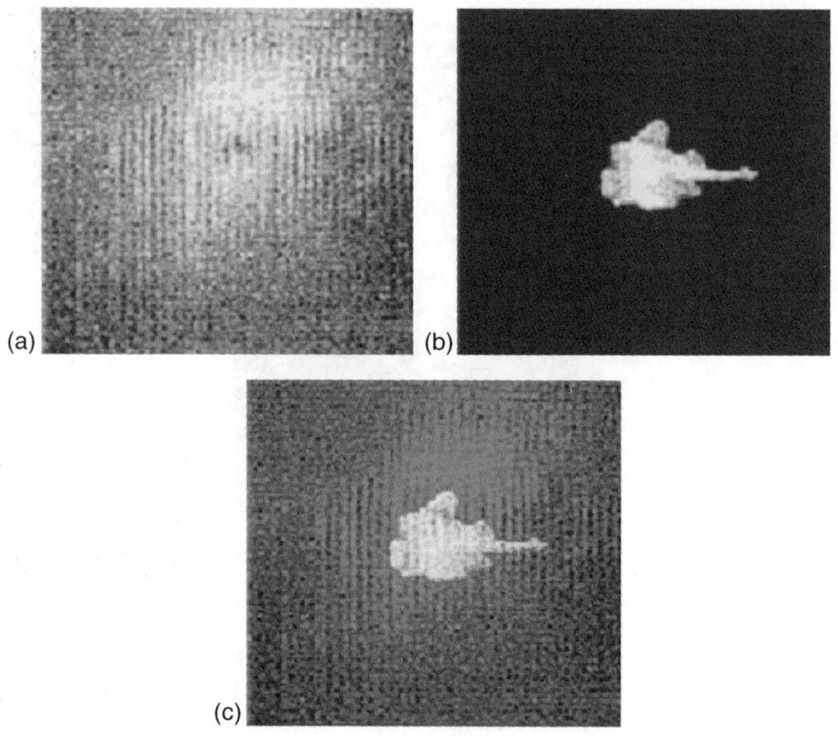

FIGURE 1.5 Example application of image fusion — defect detection. (a) Eddy current inspection (image courtesy of X. Gros); (b) thermograph (image courtesy of X. Gros); (c) fused image (image courtesy of X. Gros).

purpose of fusion. Multiscale signal representation was first studied by Rosenfeld,[39] Witkin,[40] and others. Researchers such as Marr,[41] Burt and Adelson,[42] and Lindeberg[43] established that multiscale information can be useful in a number of image processing applications. More recently, wavelet theory has emerged as a well developed, yet rapidly expanding, mathematical foundation for a class of multiscale representations. At the same time, several sophisticated image fusion approaches based on multiscale representations began to emerge and receive increased attention. Most of these approaches were based on combining the multiscale decompositions (MSDs) of the source images. Figure 1.7 illustrates the block diagram of a generic image fusion scheme based on multiscale analysis. The basic idea is to perform a MST on each source image, then construct a composite multiscale representation from these. The fused image is obtained by taking an inverse multiscale transform (IMST).

In Ref. 34, the authors proposed a generic framework for multiscale image fusion schemes as illustrated in Figure 1.8. This framework describes a family of methods for using the MSD representations of the source images to construct the MSD representation of the fused image. These methods combine the source images by consulting a quantity we call the activity level measurement.

FIGURE 1.6 (See color insert following page 236) Example application of image fusion — remote sensing. (a) Panchromatic image (courtesy of John R. Schott); (b) multi-spectral image (courtesy of John R. Schott); (c) fused image (courtesy of John R. Schott).

The activity level measurement attempts to determine the quality of each source image. Grouping and combining methods are also used to obtain the composite MSD representation of the fused image. A consistency verification procedure is then performed which incorporates the idea that a composite MSD coefficient is unlikely to be generated in a completely different manner from all its neighbors. There are multiple alternatives in the procedures noted by the dashed boxes in Figure 1.8. Different combinations of these alternatives lead to different fusion schemes. We will next discuss each procedure in Figure 1.8 in more detail.

1. Multiscale Decomposition

The three most commonly employed MSD methods appear to be the pyramid transform (PT), the discrete wavelet transform (DWT), and the discrete wavelet frame (DWF). A pyramid transform of a source image produces a set of coefficients organized in a pyramid structure. A pyramid structure is an efficient organization methodology for implementing multiscale representation and

(b)

FIGURE 1.6 (Continued).

computation. A pyramid structure can be described as a collection of images at different scales which together represent the original image. One of the most frequently studied versions of the pyramid transform is the Laplacian pyramid (LPT).[42,44] Other typical pyramids include, the filter subtract decimate pyramid,[45] the contrast pyramid,[46] the gradient pyramid,[47] the morphological pyramid,[48,49] and the steerable pyramid.[50] Each level of the LPT is recursively constructed from its lower level by the following four basic procedures: blurring (low-pass filtering), subsampling (reduce size), interpolation (expand in size), and differencing (to subtract two images pixel-by-pixel) in the order we have given. In fact, a blurred and subsampled image is produced by the first two procedures at each decomposition level, and these partial results, taken from different decomposition levels, can be used to construct a pyramid known as the Gaussian pyramid. In both the Laplacian and Gaussian pyramids, the lowest level of the pyramid is constructed from the original image.

In computing the Laplacian and Gaussian pyramids, the blurring is achieved using a convolution mask ω which should obey certain constraints.[51] Let G_k be the kth level of the Gaussian pyramid for the image I. Then $G_0 \equiv I$ and for $k > 0$,

$$G_k = [\omega * G_{k-1}]_{\downarrow 2} \qquad (1.1)$$

FIGURE 1.6 (Continued).

where $[\cdot]_{\downarrow 2}$ denotes downsampling of the signal by 2, which means to keep one sample out of two. This downsampling operation is performed in both horizontal and vertical directions. The kth level of the LPT is defined as the weighted difference between successive levels of the Gaussian pyramid

$$L_k = G_k - 4\omega * [G_{k+1}]_{\uparrow 2} \tag{1.2}$$

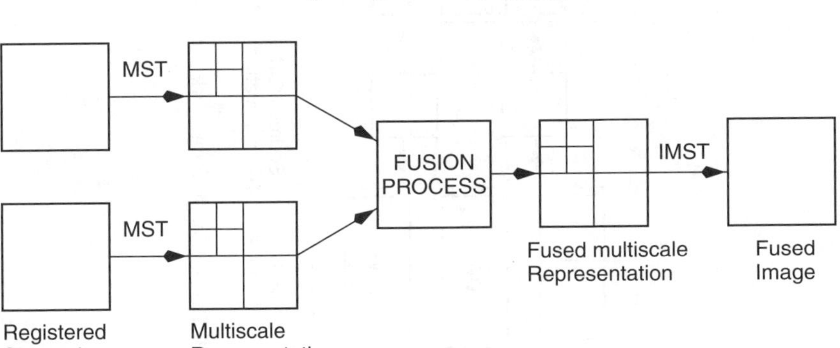

FIGURE 1.7 Block diagram of a generic image fusion scheme.

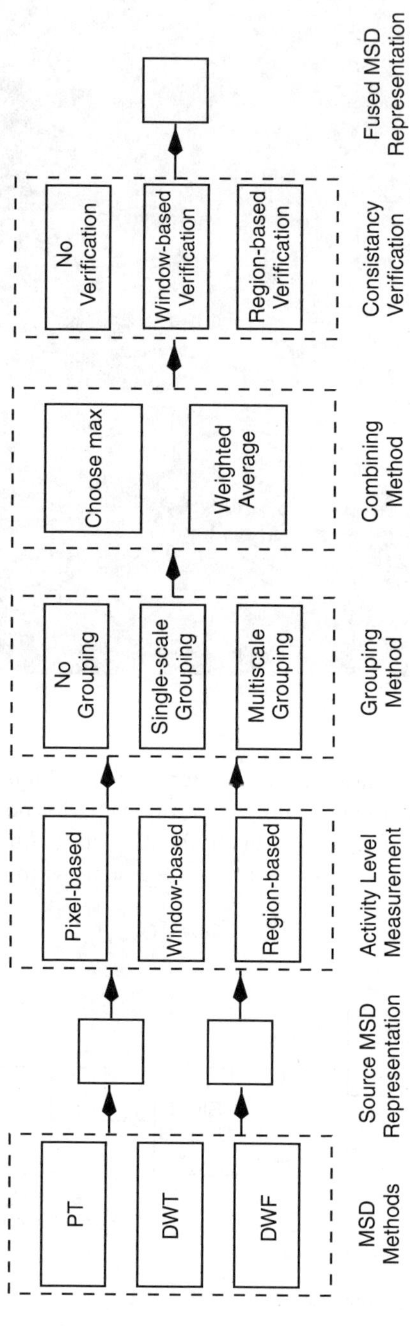

FIGURE 1.8 A generic framework for image fusion.

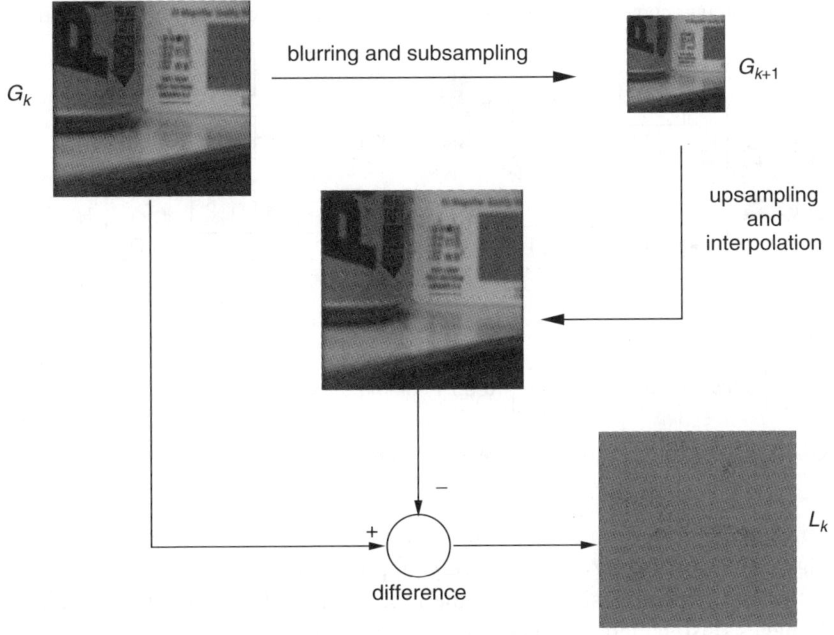

FIGURE 1.9 One stage of a Laplacian pyramid transform decomposition.

where $[\cdot]_{\uparrow 2}$ denotes upsampling, which means to insert a zero between every two samples in the signal. This upsampling operation is also performed in both horizontal and vertical directions. Here, convolution by ω has the effect of interpolating the inserted zero samples. Figure 1.9 illustrates one stage of the decomposition of the LPT.

An image can be reconstructed by the reverse procedure. Let $\hat{G}$ be the recovered Gaussian pyramid. Reconstruction requires all levels of the LPT, as well as the top level of the Gaussian pyramid $\hat{G}_N$. Thus the procedure is to set $\hat{G}_N = G_N$, and for $k < N$,

$$\hat{G}_k = L_k + 4\omega * [\hat{G}_{k+1}]_{\uparrow 2} \tag{1.3}$$

which can be used to compute $\hat{G}_0$, the reconstructed version of the original image G_0.

The wavelet representation introduced by Mallat[52–54] suggests that, for efficiency reasons, successive layers of the pyramid should include only the additional details, which are not already available at preceding levels. Taking this approach leads to using a set of filters to perform the wavelet transform. The DWT will have a pyramid hierarchy that can be understood by considering the frequency bands passed by these filters. The sizes of frequency bands will decrease as the decomposition moves from one level to the next. Figure 1.10 illustrates the $i + 1$th level of 2D DWT,[52] where the particular pair of analysis

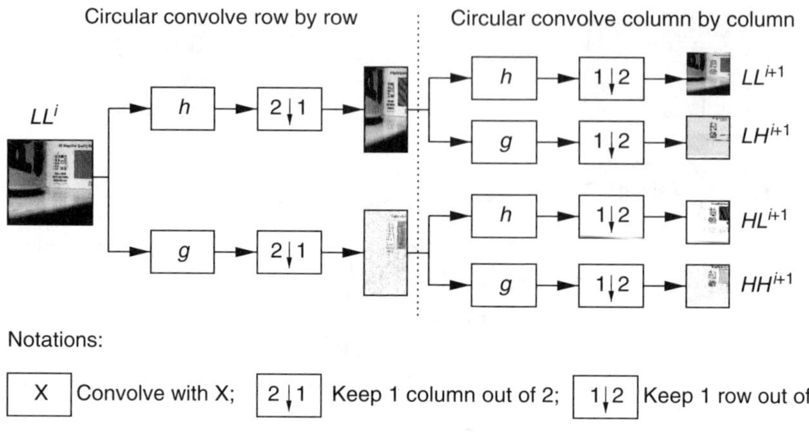

FIGURE 1.10 One stage of a 2D DWT decomposition.

filters h and g correspond to a particular type of wavelet used. LL^0 is the original image. The processing is recursively applied for each decomposition level. For the $i + 1$ decomposition level, the input image LL^i is decomposed into four subimages (also called frequency bands), low–low LL^{i+1} (an image at a coarser resolution level), low–high LH^{i+1} (a image containing horizontal edge information), high–low HL^{i+1} (a image containing vertical edge information), and high–high HH^{i+1} (a image containing diagonal edge information), by the following two steps. Firstly, each row of the image LL^i is separately convolved with the low pass filter h and the corresponding mirror high pass filter g followed by downsampling as illustrated in Figure 1.10. Secondly, each column of the two resulting images is convolved with h and g followed by downsampling. The procedure is then repeated until the desired level is reached. Thus, a DWT with N decomposition levels will have $M = 3 \times N + 1$ frequency bands. The inverse DWT follows a similar pattern.

It is well known that the DWT yields a shift variant signal representation. By this we mean that a simple integer shift of the input signal will usually result in a nontrivial modification of the DWT coefficients. This shift-variant property is introduced by the subsampling process. Thus, an image fusion scheme based on the DWT will also be shift dependent, which is undesirable in practice (especially considering misregistration problems). One approach to solve the problem is given in Ref. 55 by employing the concept of discrete wavelet frames (DWF).[56,57] In contrast to the standard DWT, in each decomposition stage the DWF uses dilated analysis filters and drops the downsampling process. Consequently, the subimages (frequency bands) in the DWF will have the same size for all the levels, while the size of the subimages in the DWT changes. Thus, for the same number of levels, the DWF employs more coefficients which can be exploited to obtain a more robust image fusion method. One stage of the 2D DWF decomposition is illustrated in Figure 1.11.

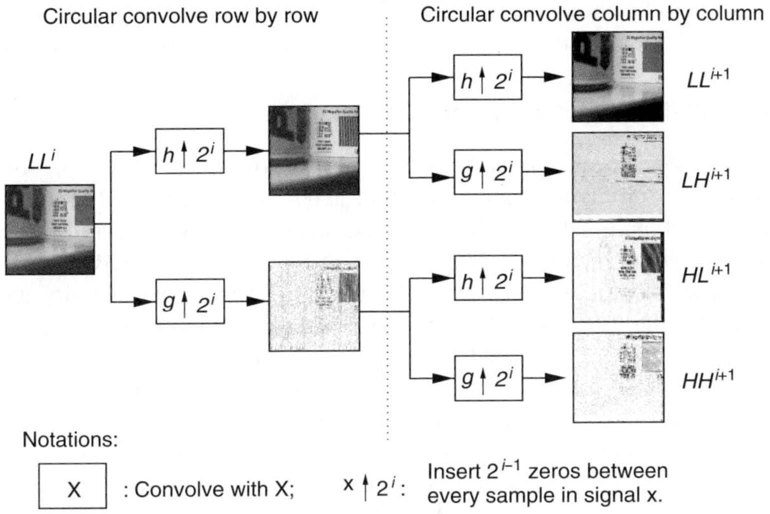

FIGURE 1.11 One stage of a 2D DWF decomposition.

Figure 1.12 shows the 2D structures of the LPT, DWT, and DWF multiscale transforms with two decomposition levels. The dark blocks in different frequency bands correspond to coefficients whose computation depends on the same group of pixels in the original image, which indicates the spatial localization of the transform. For a transform with K levels of decomposition, there is always only one low frequency band (G^K or LL^K in Figure 1.12), the rest of bands are high frequency bands, which provide detailed image information at different scales.[52]

Besides the MSD analysis, another key issue in MSD-based image fusion is how to form the fused MSD representation from the MSD representations of the source images. We call the processing to achieve this goal a fusion rule. Some general alternatives for constructing a fusion rule are illustrated in Figure 1.8.

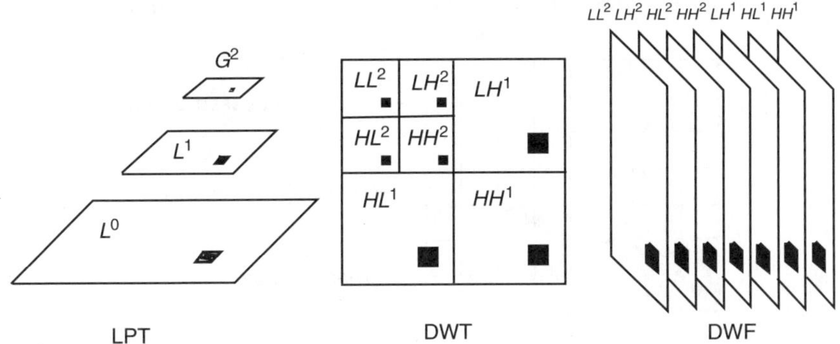

FIGURE 1.12 MSD structures.

These include the choice of an activity level measurement, coefficient grouping method, coefficient combining method, and consistency verification method.

2. Activity-Level Measurement

The activity level measurement is used to judge the quality of a given part of each source image. Activity level measurement can be categorized into three classes as illustrated in Figure 1.13: coefficient-based measures, window-based measures, and region-based measures. The coefficient-based activity (CBA) measures consider each coefficient separately. In its simplest form, the activity level is described by the absolute value or square of corresponding coefficient in the MSD representation. The window-based activity (WBA) measures employ a small (typically 3×3 or 5×5) window centered at the current coefficient position. When a WBA measure is employed, several alternatives exist. One option is the weighted average method (WA-WBA).[58] Nonlinear operators are also possible. An example is the rank filter method (RF-WBA).[59] The regions used in region-based activity (RBA) measurement are similar to windows with odd shapes. The regions are typically defined using the low–low band, which is a subsampled version of the original image. One RBA method was introduced in Ref. 15, where each region corresponds to an object or part of an object. More complicated activity-level measurement approaches, including those that use *a priori* information to identify high frequency noise and distortion, can be employed to further improve performance.

3. Coefficient Grouping Method

When constructing the fused MSD representation from two source images, one simple method is to set each MSD coefficient in the fused image to be the corresponding MSD coefficient in that source image with the larger activity level. We notice that each MSD coefficient will have a set of related coefficients in other frequency bands and other decomposition levels, as illustrated in Figure 1.12 by the dark squares. These coefficients relate to the same pixel position in the low-pass filtered source image or the low frequency band of

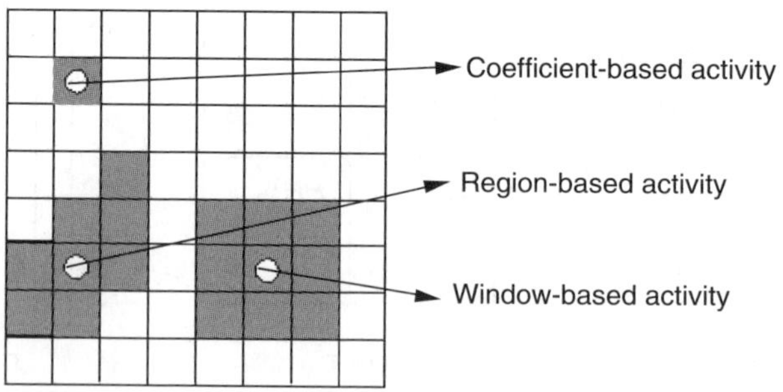

FIGURE 1.13 Categorization of activity level measurement.

the MSD. For the standard image fusion approaches, when determining the composite MSD representation, these coefficients are not associated with each other. We call these schemes no-grouping (NG) schemes. If the corresponding coefficients in the same decomposition scale are jointly constrained to be obtained using the same fusion approach, we call this a single-scale grouping (SG) scheme. This is a more restrictive case. The most restrictive case is to consider all the corresponding MSD samples together and ensure that they are all obtained using the same fusion approach. We call this a multiscale grouping (MG) scheme. A simple way to accomplish this is to compute activity by jointly considering all the related coefficients and then simply selecting the coefficients to come from the source image with higher activity. Of course a linear or nonlinear combination of these coefficients is also possible.

4. Coefficient Combining Method

When combining the source MSD representations to produce the composite MSD representation, there are several alternatives. One is the choose-max (CM) scheme, which means picking the coefficient with larger activity level and discarding the other. Another combining scheme is the weighted average (WA) scheme. At each coefficient position $\vec{p}$, the composite MSD D_Z can be obtained from the source MSDs D_X and D_Y by

$$D_Z(\vec{p}) = W_X(\vec{p})D_X(\vec{p}) + W_Y(\vec{p})D_Y(\vec{p}) \tag{1.4}$$

The weights W_X and W_Y may depend on the activity levels of the source MSD coefficients and the similarity between the source images at the current position. One popular way to determine W_X and W_Y can be found in Ref. 58. Clearly other nonlinear combination methods, possibly employing *a priori* information, can also be adopted.

5. Consistency Verification

Consistency verification attempts to ensure neighboring coefficients in the composite MSD are obtained in a similar manner. This is often applied with the CM combining scheme for example. When constructing the composite MSD representation using CM combining, it is unlikely that a given MSD coefficient is better described by a different source image from all its neighbors. Thus consistency verification ensures that a composite MSD coefficient does not come from a different source image from all its neighbors. Li[59] applied consistency verification using a majority filter. Specifically, if the center composite MSD coefficient comes from image X while the majority of the surrounding coefficients come from image Y, the center sample is then changed to come from image Y.

B. NONMULTISCALE-DECOMPOSITION-BASED METHODS

There are also many other image fusion schemes which are not based on the multiscale transforms. We classify all the other methods into five general classes and we acknowledge that some existing algorithms fall into several of these classes. Furthermore, some of these methods can also be applied in the transform domain.

1. Pixel-Level Weighted Averaging

A straightforward approach to image fusion is to take each pixel in the fused image as the weighted average of the pixel intensity of the corresponding pixels of the two source images. Two representative methods are the principal component analysis (PCA) method[60] and adaptive weight averaging (AWA) method.[61] In the PCA method, the weightings for each source image are obtained from the eigenvector corresponding to the largest eigenvalue of the covariance matrix of each source image.[62] In the AWA method,[61] a scheme for fusing thermal and visual images in a surveillance scenario, the weighting algorithm for the IR image will assign larger weights to either the warmer and cooler pixels, while the weighting algorithm for the visual image will assign larger weights to those pixels whose intensities are much different from its neighbors (based on the local variance of the pixels' intensity).

2. Nonlinear Method

Nonlinear fusion methods are also possible. The nonlinear method developed by Therrien and colleagues[63] is a scheme for the enhancement and fusion of low-light visible and thermal infrared images. The method separates the source image into low and high pass components. The method first adaptively modifies the low pass component of each source image to enhance the local luminance mean, and then fuses the low pass components by a nonlinear mapping. The method then adaptively modifies the high pass component of each source image to enhance the local contrast, and then fuses the high pass components by weighted averaging. Finally, the fused high and low pass images are added to produce the final enhanced fused image.

3. Estimation Theory Based Methods

Estimation theory[64,65] has also been applied to the image fusion problem.[17,66,67] Common estimation procedures include the maximum *a posteriori* (MAP) estimate and the maximum likelihood (ML) estimate. Let a denote the set of sensor images and let s denote the underlying true scene to be estimated. Further assume $p(s|a)$ and $p(a|s)$ denote conditional probabilities while $p(s)$ and $p(a)$ denote the corresponding marginal probabilities. The MAP estimate chooses to maximize $p(s|a)$. The ML estimate chooses to maximize $p(a|s)$. Bayes formula

$$p(s|a) = \frac{p(a|s)p(s)}{p(a)} \tag{1.5}$$

indicates that the MAP and the ML estimates are closely related. Estimation theory based image fusion methods can often also be described using two primary components called the image formation model and the prior model. The image formation model represents the relationship between the sensor images and the underlying true scene. In Ref. 66, a distorted, locally affine transformation whose parameters are allowed to vary across the image is employed. The prior model represents the knowledge we have about the true scene or the distortion. After specific forms are chosen for the image formation model and the prior model, different possible solutions are assigned probabilities of being the true scene. A generalization to the approach in Ref. 66 which allows the use of non-Gaussian distributions was introduced in Ref. 68. This approach adaptively estimates the distributions by employing a Gaussian mixture model for the distributions and a powerful iterative estimation approach. One very interesting approach which uses estimation theory is the Markov Random Field (MRF) method.[69–71]

4. Color Composite Fusion

Since the human visual system (HVS) is very sensitive to colors, one class of image fusion approaches involves combining the input images (or the processed ones) in a color space to obtain a false color representation of the fused image. In many cases, this technique is applied in combination with other image fusion procedures, such as principle component analysis, intensity–hue–saturation techniques, or neural network approaches. Such a procedure is sometimes called color composite fusion. Color composite fusion methods have been employed in the field of remote sensing.[20,72,73] For example, Nunez et al.,[20] use color to fuse a high-resolution panchromatic image with a low-resolution multispectral image by adding the wavelet coefficients of the high-resolution image to the intensity component of the multispectral image. Another technique utilizing color is based on opponent-color processing which maps opponent-sensors to human opponent colors (red vs. green, blue vs. yellow). This technique has been used in the field of night time surveillance.[74–78] For example, Waxman, Aguilar et al.,[74,75] use a neural network to fuse a low-light visible image and a thermal IR image to generate a three-channel false color image used for night operation. In addition, Aguilar[79,80] extended their work to fuse multimodality volumetric medical imagery based on techniques originally developed for night time surveillance.

5. Artificial Neural Networks

Artificial neural networks have been employed in the image fusion process.[81–84] They are sometimes physiologically motivated by the fusion of different sensor signals in biological systems. Multilayer perceptron neural networks (MLP) and pulse-coupled neural networks (PCNN) are two types of neural networks that have been frequently utilized for image fusion. In order to combine the natural appearance of a low light level television (LLLTV) image with the prominent thermal information from a FLIR image, Fechner and Godlewski[81] use the LLLTV image as a base image. On this base image they superimpose the important details of the FLIR image by training a MLP neural network to generate a binary mask indicating regions of interest in the FLIR image which should appear in the composite image. In Ref. 84, Broussard and coauthors designed a fusion network based on PCNNs to improve object detection accuracy. First, filtering techniques are used to extract target features. Then, the original image and the filtered versions are taken as inputs to the proposed PCNN fusion network. The PCNNs are used to segment and fuse the images. The output of the fusion network is a single image in which the desired objects are supposed to be the brightest so they can be easily detected.

III. BEYOND FUSION ALGORITHMS: PERFORMANCE EVALUATION AND REGISTRATION

Some general requirements for fusion schemes are: (1) the fusion algorithm should be able to extract the complimentary information from the input images for use in the fused image; (2) the fusion algorithm must not introduce any artifacts or inconsistencies which would distract the human observer or following processing tasks; and (3) the fusion algorithm should be reliable, robust, and have the capability to tolerate imperfections such as misregistration. These requirements are often very difficult to achieve. One of the most critical problems is that there is still a lack of reliable and efficient methods to validate and evaluate different fusion schemes. In image fusion, it is very difficult to access the truth, making evaluations often subjective. Dealing with unregistered images further complicates these problems. Most fusion schemes usually assume that the source images to be fused are perfectly registered. In practice, this requirement may be difficult to achieve. Thus we believe, that, after the consideration of the fusion algorithms themselves, the two most important issues for image fusion are how to judge performance, and how to register the images. In this section, we discuss these two issues.

A. PERFORMANCE EVALUATION

The best method for assessing the quality of the fused image generally depends on the application domain. In many applications, a human observer is the end user of the fused image. Therefore, the human perception and interpretation of the fused image is very important. Consequently, one way to assess the fused images is to use subjective tests. In these tests, human observers are asked to view a series of fused images and rate them. Although the subjective tests are typically accurate if performed correctly, they are inconvenient, expensive, and time consuming. Hence, an objective performance measure that can accurately predict human perception would be a valuable complementary method. However, it is difficult to find a good, easy to calculate, objective evaluation criterion which matches favorably with visual inspection and is suitable for a variety of different application requirements. In the literature, there are two broad classes of objective performance measures. One class requires a reference image, while the other does not.

1.   Objective Evaluation Measures Requiring a Reference Image

For certain applications, it is possible to generate an ideal fused image. This ideal fused image is then used as a reference to compare with the experimental fused results. Some typical quality metrics which are used for these comparisons are listed:

1. The root mean square error (RMSE)

$$\text{RMSE} = \sqrt{\frac{1}{NM} \sum_{i=1}^{N} \sum_{j=1}^{M} |R(i,j) - F(i,j)|^2} \qquad (1.6)$$

   where R denotes the reference image, F denotes the fused image, (i, j) denotes a given pixel, and $N \times M$ is the size of the image.

2. The correlation (CORR)

$$\text{CORR} = \frac{2R_{R,F}}{R_R + R_F} \qquad (1.7)$$

   where $R_R = \sum_{i=1}^{N} \sum_{j=1}^{M} R(i,j)^2$, $R_F = \sum_{i=1}^{N} \sum_{j=1}^{M} F(i,j)^2$, and $R_{R,F} = \sum_{i=1}^{N} \sum_{j=1}^{M} R(i,j)F(i,j)$

3. The peak signal to noise ratio (PSNR)

$$\text{PSNR} = 10 \log_{10} \left(\frac{L^2}{\frac{1}{NM} \sum_{i=1}^{N} \sum_{j=1}^{M} |R(i,j) - F(i,j)|^2} \right) \qquad (1.8)$$

   where L is the number of graylevels in the image.

4. The mutual information (MI)

$$\text{MI} = \sum_{i_1=1}^{L} \sum_{i_2=1}^{L} h_{R,F}(i_1, i_2) \log_2 \frac{h_{R,F}(i_1, i_2)}{h_R(i_1) h_F(i_2)} \qquad (1.9)$$

   where $h_{R,F}$ denotes the normalized joint graylevel histogram of images R and F while h_R, h_F are the normalized marginal histograms of the two images.

5. The universal quality index (QI)[85]

$$Q = \frac{4\sigma_{xy}\bar{x}\bar{y}}{\left(\sigma_x^2 + \sigma_y^2\right)[(\bar{x})^2 + (\bar{y})^2]} = \frac{\sigma_{xy}}{\sigma_x \sigma_y} \frac{2\bar{x}\bar{y}}{(\bar{x})^2 + (\bar{y})^2} \frac{2\sigma_x \sigma_y}{\sigma_x^2 + \sigma_y^2} \qquad (1.10)$$

where $x = \{x_i | i = 1,2,\ldots,N\}$ and $y = \{y_i | i = 1,2,\ldots,N\}$ denote the reference and the test images, respectively, and

$$\bar{x} = \frac{1}{N} \sum_{i=1}^{N} x_i, \quad \bar{y} = \frac{1}{N} \sum_{i=1}^{N} y_i, \quad \sigma_x^2 = \frac{1}{N-1} \sum_{i=1}^{N} (x_i - \bar{x})^2,$$

$$\sigma_y^2 = \frac{1}{N-1} \sum_{i=1}^{N} (y_i - \bar{y})^2, \quad \sigma_{xy} = \frac{1}{N-1} \sum_{i=1}^{N} (x_i - \bar{x})(y_i - \bar{y})$$

$$(1.11)$$

This index[85] can be rewritten as a product of three components which model the loss of correlation, the luminance distortion, and the contrast distortion, respectively.

The first four objective image quality measures, RMSE, PSNR, CORR, MI, are widely employed due to their simplicity. However, they must be used carefully because they have been found to sometimes not correlate well with human evaluation when sensors of different types are considered. Here we give two examples of some of our own studies on comparing different image fusion approaches using the standard evaluation measures just described. One is for a digital camera (multifocus) application.[86] The other is for a CWD application.[87] In Ref. 86, we studied and compared several MSD-based image fusion schemes from within the generic image fusion framework of Figure 1.8 for a digital camera application. The study was focused on how to use the MSD data of the source images to produce a fused MSD representation which should be more informative to the observer. In Ref. 87, we studied and tested some of the MSD-based and NMSD-based image fusion schemes described in Section 2 for a CWD problem in which a visual image of the scene is fused with a nonvisual image (MMW or IR) that shows a hidden object, typically a weapon.

For the digital camera application where the source images come from the same type of sensors, it is possible to generate pairs of distorted source images from a known test image in a controlled setting. From a test image, two out-of-focus images are created by radial blurring. We take the original image as the reference image. For the CWD application where the source images come from different type of sensors, the reference image, which should be an ideal composition of the two sensor (for example, IR and visual) images, is normally unknown. The reference image used in our tests was created by extracting the gun from the IR image and pasting it into the visual image. We justify this approach by noticing that, for our CWD application, the ideal fusion result should preserve the high-resolution information of the visual image and combine the important complementary information of the IR image (the object of interest, a gun in our examples). However, a more practical consideration shows that while a true ideal image may be impossible to determine (this is equivalent to having a perfect fusion algorithm), this approach will yield some useful information. Here, we provide only one of a large set of results given in

TABLE 1.1
Some Typical Image Fusion Schemes and their Performance for Digital Camera Applications

Scheme	MSD	Activity	Grouping	Combining	Verification	$\overline{\text{RMSE}}$	$\overline{\text{MI}}$	Reference
1	LPT	CBA	NG	CM	NV	4.09	3.79	88, 89
2	DWT	CBA	NG	CM	NV	5.04	3.33	90–92
3	DWT	CBA	NG	CM	WBV	4.83	3.45	93, 94
4	DWT	WA-WBA	NG	WA	NV	4.77	3.46	13, 95
5	DWT	WA-WBA	SG	CM	NV	4.65	3.52	96
6	DWT	RF-WBA	NG	CM	WBV	4.81	3.43	59
7	DWT	RBA	MG	CM	WBV	2.72	5.02	15
8	LPT	RF-WBA	MG	CM	RBV	2.74	5.18	86
9	DWT	RF-WBA	MG	CM	RBV	2.69	5.25	86
10	DWF	RF-WBA	MG	CM	RBV	2.58	5.22	86

Refs. 86 and 87 to illustrate the nature of these studies. Table 1.1 shows representative results for the digital camera application. Table 1.1 also provides average fusion performance, where the average is taken over 16 test images. Table 1.2 shows representative results for the CWD application. Here the averages are taken over nine test images.

TABLE 1.2
Some Typical Image Fusion Schemes and their Performance for CWD Applications

	Scheme	$\overline{\text{RMSE}}$	$\overline{\text{CORR}}$	$\overline{\text{PSNR}}$	Reference
1	PCA	0.2891	0.8780	14.8487	60
2	AWA	0.3753	0.8191	12.8190	61
3	MAX	0.5251	0.7645	9.9581	87
4	NONLINEAR	0.2999	0.8663	14.5672	63
5	LPT + CBA + NG + CM + NV	0.3351	0.8379	13.6195	88
6	FSD + CBA + NG + CM + NV	0.3438	0.8425	13.4118	45
7	CONT + CBA + NG + CM + NV	0.3577	0.7954	12.9649	46
8	GRAD + CBA + NG + CM + NV	0.3434	0.8434	13.4235	47
9	MORPH + CBA + NG + CM + NV	0.3730	0.7870	12.6671	48
10	DWT + CBA + SG + CM + NV	0.3941	0.7994	12.5073	34
11	DWT + RF-WBA + NG + CM + NV	0.3332	0.8465	13.6775	34
12	DWT + WA-WBA + NG + CM + NV	0.3333	0.8465	13.6756	34
13	DWT + WA-WBA + NG + WA + NV	0.3330	0.8468	13.6848	34
14	DWT + RF-WBA + NG + CM + WBV	0.3332	0.8465	13.6776	34
15	DWF + WA-WBA + NG + CM + WBV	0.5225	0.7665	10.0050	87

Comparing the evaluation results for the two applications, we found that the performance of a given image fusion algorithm can be highly application specific. For example, the DWT image fusion method with single-scale grouping is very good when applied to multifocus visual digital camera images, but it is not suitable for detecting a concealed weapon. The reason for this phenomenon is due to the significant differences between visual image and IR image that makes using grouping unsuitable. For the multifocus digital camera application, we found the evaluation result from using quantitative quality metrics was consistent with the visual evaluation. However, for CWD application, the quantitative results differ greatly from the visual comparisons. The differences are so extreme that we feel the use of the quantitative measures we have considered is questionable. At the very least, these measures must be used with caution. Some methods that result in poor visual evaluation results may have relatively good quantitative evaluation results. This is partly due to the global property of the quantitative evaluation algorithm. For CWD, we pay more attention to the face and the weapon, while these quantitative evaluation algorithms do not put more weight on these aspects. Clearly, improved quantitative evaluation methods are needed. In fact, we believe this is a topic that should receive significant attention in the future. Automated evaluation methods would allow larger studies and the possibility of adapting the fusion processing during operation to improve performance. As a hint to guide future research, we note that the idea of using a reference image must be employed carefully. We note that we have found that the exact method used to compose the reference image will generally have great influence on the statistical evaluation results.

At present, the standard simple mathematical measures such as PSNR and RMSE are still used widely in spite of their questionable performance. The QI measure given in Equation 1.10 is a new image quality index recently developed by Wang and Bovik[85] attempting to address some of the flaws of PSNR and RMSE.

Evaluating a fused image for cases where an ideal image is available is similar to evaluating a compressed image when the original (uncompressed) image is available. The image quality assessment has been widely studied in the field of image compression. A comprehensive discussion of the image quality metrics for assessing the results of image compression is given in Ref. 97. In Ref. 97, the image quality metrics are categorized into six groups according to the type of information they are using: pixel difference-based measures, correlation-based measures, edge-based measures, spectral distance-based measures, context-based measures, and HVS-based measures. Some of these metrics can be revised or applied directly for the application of image fusion.

2.   Objective Evaluation Measures Not Requiring a Reference Image

As stated previously, quantitatively accessing the performance in practical applications is a complicated issue because it is usually very difficult to access the truth (i.e., the ideal composite images is normally unknown). Several simple quantitative evaluation methods which do not require a reference image are listed below.

1. The standard deviation (SD)

$$\sigma = \sqrt{\sum_{i=0}^{L} (i - \bar{\imath})^2 h(i)}, \qquad \bar{\imath} = \sum_{i=0}^{L} i h(i) \qquad (1.12)$$

where h is the normalized histogram of image.

2. The entropy (H)

$$H = -\sum_{i=0}^{L} h(i) \log_2 h(i) \qquad (1.13)$$

where h is the normalized histogram of image.

3. The overall cross entropy (CE) of the source images X, Y and the fused image F

$$CE(X, Y; F) = \frac{CE(X; F) + CE(Y; F)}{2} \qquad (1.14)$$

where $CE(X; F)$ $(CE(Y; F))$ is the cross entropy of the source image X (Y) and the fused image F

$$CE(X; F) = \sum_{i=0}^{L} h_X(i) \log_2 \frac{h_X(i)}{h_F(i)} \qquad (1.15)$$

Other objective image fusion performance evaluation measures which do not require the availability of a reference image have been proposed in the literature. Based on the Wang–Bovik image quality index described previously, Piella and Heijmans[98] defined an objective quality index for image fusion which does not require a reference image. This quality index gives an indication of how much of the salient information contained in each of the input images has been transferred into the fused image. Xydeas and Petrovic[99] also proposed an objective performance metric which measures the amount of information that is transferred from the input images into the fused image. Their approach is based on the assumption that important visual information is related with edge information. Thus, weighted edge information is used to form a measure that quantifies fusion performance.[99]

In Ref. 100, we proposed a technique to blindly estimate the quality of an image of a natural scene. The method involves forming histograms of an edge intensity image. The characteristics of the histograms give information about the amount of noise added to the image. The edge intensity image is obtained from a Canny edge detection operation[101] by first convolving the image with a one-dimensional mask to find the edge contributions in both the horizontal and vertical directions. We model the histograms of the edge contributions in each direction using a mixture of Gaussian probability density functions (pdfs), a model which is known to be useful for modeling non-Gaussian distributions. The edge intensity is computed by the square root of the sum of the squares of the edge contributions in the two directions. Since this operation converts Gaussian random variables to Rayleigh random variables, we model the histograms of the edge intensity images by a mixture of Rayleigh pdfs. The free parameters of the Rayleigh mixture model correspond to the variances and probabilities of a sample being generated from a given mixture term in the original Gaussian mixture model. Approximate maximum likelihood estimates of these parameters are obtained directly from the overall edge intensity image using the EM algorithm.[102,103] By studying the effects of noise on these parameters, they motivate techniques to estimate the amount of noise and image quality (see Ref. 100 for further details). Evaluation of the fused image is one of the most critical problems in the area of image fusion. It is undoubtedly true that there is great need for improved objective performance assessment of image fusion algorithms. Another important topic in image fusion is image registration, introduced next.

B. IMAGE REGISTRATION

Image registration will be addressed in greater detail in later chapters. Our purpose here is to introduce the topic. Image registration is an important prerequisite for image fusion. In the majority of previous image fusion research, it is assumed that the source images are perfectly aligned. In fact, this is difficult to achieve in many practical situations. The images to be fused may have relative translation, rotation, scale, and other geometric transformations in relation to each other. The task of image registration is to align the source images to one another. Over the years, a broad range of techniques have been developed for various types of sensors and applications, resulting in a large body of research.[104–116]

There are two general types of differences between the images to be registered. The first type is due to changes in acquisition, which cause the images be spatially misaligned. In the second type, the difference can not be modeled by a spatial transform alone. The differences which are not due to spatial misalignment can be attributed to factors such as lighting changes, using different type of sensors, using similar sensors but with different parameters, object movements, or scene changes. The differences which are not due to spatial misalignment will not be removed by registration, but they make the registration more difficult as there is no longer an exact match between two images, even after spatial transformation.

Existing image registration techniques can be generally classified into two categories: the intensity-based methods, and the feature-based methods. In the intensity-based methods, the images are essentially registered by selecting a number of windows in high-variance areas of one image, locating the corresponding windows in the other image, and using the window geometric centers or mass centers as control points to determine the registration parameters. Feature-based methods extract and match the common features from the source images. Frequently used features include edges, corners, and contours. The feature-based approach has received more attention for the purpose of multisensor image registration.

There has been extensive study on registration. As a concrete example, we describe a particular approach we previously developed. In Ref. 117, we proposed a hybrid image registration algorithm for a multifocus digital camera application. In our study, we found that neither the intensity-based or feature-based image registration approaches, without modification, is well suited to the problem of multifocus image registration. For example, if the two images were not obtained using the same focus, a strong feature in one image may appear much weaker in the other image. Thus, it is not always true that a high-variance area in one image will correspond to a high-variance area in the other image. This leads to difficulties in finding control points for the intensity-based methods. Similarly, a common feature may appear to look quite different in the two images, which makes feature matching nearly impossible.

In Ref. 117, we proposed a hierarchical scheme, which uses both feature-based and intensity-based methods. We apply the idea of robust estimation of optical flow, formally used to compress videos, and supplement this with a coarse-to-fine multiresolution approach and feature-based registration to overcome some of the limitations of the intensity-based scheme. Figure 1.14 illustrates the basic framework of the scheme. First, a multiscale decomposition is applied to the source images to obtain two sets of Gaussian pyramids.

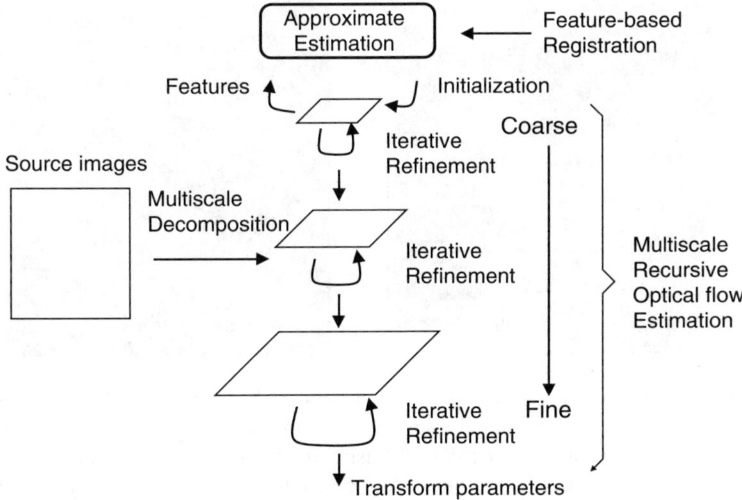

FIGURE 1.14 Block diagram of a hybrid image registration scheme.

Canny edge detection is then performed on the coarsest level of the pyramids to produce a pair of edge images. The edges are used as the major image feature for the initial matching. The results of this matching produce the initial parameters used in the optical flow estimation. This is followed by several iterative refinement steps which use optical flow estimation on this coarsest decomposition level. The updated matching parameters are passed on to the next decomposition level (finer resolution) and the same iterative refinement is repeated for this level. The process will continue until the finest decomposition level is reached, where we obtain the final and the most accurate matching parameters. A representative example is given in Figure 1.15 to illustrate the procedures of the hybrid image registration approach. Figure 1.15(a) and (b) show the two unregistered images. Figure 1.15(c) and (d) are the edge intensity images on the coarsest level of the pyramids after the Canny edge detection, hysteresis

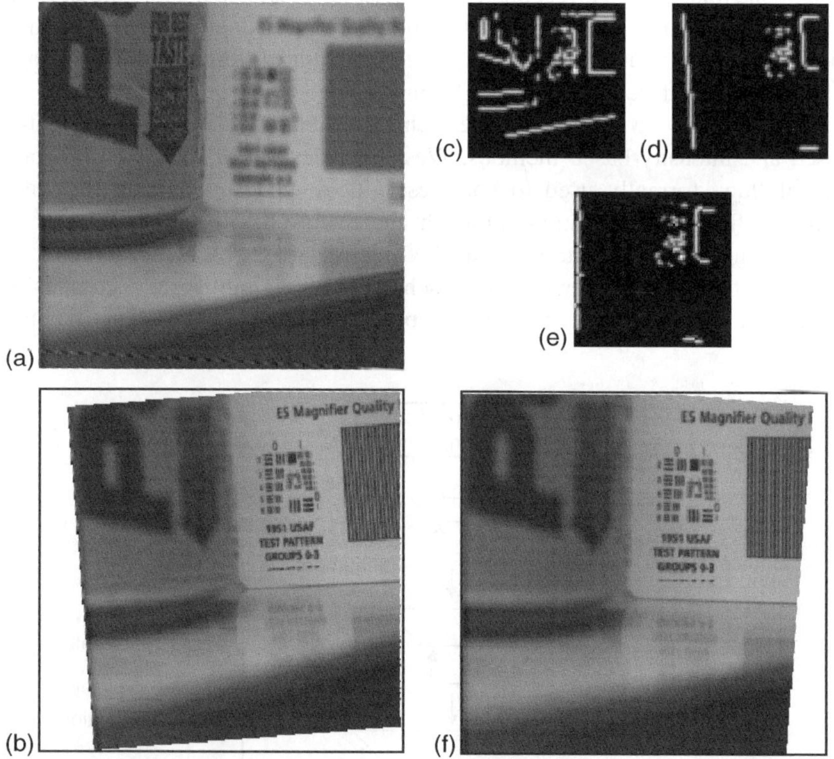

FIGURE 1.15 Illustration of the image registration in Ref. 117 (a) and (b): two source images; (c) and (d): modified edge map for matching; (e) warped version of (d) using initial parameters; (f) warped version of (b) using final parameters. (From Zhang, Z., and Blum, R. S., Image registration for multi-focus image fusion. *SPIE AeroSense, Conference on Battlefield Digitization and Network Centric Warfare* (4396-39), Orlando, FL, 2001, April. With permission.)

thresholding, connectivity checking and smoothing. These two images will be used for the feature-based initial matching. Figure 1.15(e) is the warped edge image of Figure 1.15(d) using the transformation matrix obtained by the initial matching. If we compare Figure 1.15(e) with Figure 1.15(c), we see that the initial estimation is reasonably accurate. This initial estimation is then used as the initial value for the iterative refinement from the coarsest scale to the finest scale. Figure 1.15(f) shows the warped original image from Figure 1.15(b) using the final estimated transformation parameters. Further details are provided in.[117]

IV. CONCLUSIONS

Image fusion is a process of combining information from different sources in order to get a more accurate understanding of the observed scene. It has already been successfully applied for many applications. In this chapter, we gave an overview of image fusion techniques, the advantages and limitations of image fusion, the applications of the image fusion, the typical methods of image fusion, and examples of image registration and fusion performance evaluation.

REFERENCES

1. Clark, J. J., and Yuille, A. L., *Data Fusion for Sensory Information Processing System*, Kluwer, Norwell, MA, 1990.
2. Hall, D. L., *Mathematical Techniques in Multisensor Data Fusion*, Artech House, Boston–London, 1992.
3. Abidi, M. A., and Gonzalez, R. C., *Data Fusion in Robotics and Machine Intelligence*, Academic Press, Boston, 1992.
4. Aggarwal, J. K., *Multisensor Fusion for Computer Vision*, Springer, Berlin, 1993.
5. Klein, L. A., Sensor and data fusion concepts and applications, *SPIE*, 1993.
6. Luo, R. C., and Kay, M. G., *Multisensor Integration and Fusion for Intelligent Machines and Systems*, Ablex Publishing Corporation, Norwood, NJ, 1995.
7. Hall, D. L., and Llinas, J., An introduction to multisensor data fusion, *Proc. IEEE*, 85(1), 6–23, 1997.
8. Varshney, P. K., Multisensor data fusion, *Electron. Commun. Eng. J.*, 9(6), 245–253, 1997, December.
9. Brooks, R. R., and Iyengar, S. S., *Multi-Sensor Fusion: Fundamentals and Applications with Software*, Prince-Hall, Upper Saddle River, NJ, 1998.
10. Viswanathan, R., and Varshney, P. K., Distributed detection with multiple sensors: part I — fundamentals, *Proc. IEEE*, 85(1), 54–63, 1997, January.
11. Blum, R. S., Kassam, S. A., and Poor, H. V., Distributed detection with multiple sensors: part II — advanced topics, *Proc. IEEE*, 85(1), 64–79, 1997, January.
12. Varshney, P. K., Chen, H., and Ramac, L. C., Registration and fusion of infrared and millimeter wave images for concealed weapon detection, pp. 532–536. In *Proceedings of International Conference on Image Processing*, Vol. 3. Japan, 1999.
13. Uner, M. K., Ramac, L. C., and Varshney, P. K., Concealed weapon detection: an image fusion approach, pp. 123–132. In *Proceedings of SPIE*, Vol. 2942, 1997.

14. Ramac, L. C., Uner, M. K., Varshney, P. K., Alford, M., and Ferris, D., Morphological filters and wavelet based image fusion for concealed weapons detection, pp. 110–119. In *Proceedings of the SPIE — The International Society for Optical Engineering*, Vol. 3376, 1998.

15. Zhang, Z., and Blum, R. S., A region-based image fusion scheme for concealed weapon detection, pp. 168–173. In *Proceedings of 31st Annual Conference on Information Sciences and Systems*. Baltimore, MD, 1997.

16. Xue, Z., and Blum, R. S., Concealed weapon detection using color image fusion, pp. 622–627. In *The Sixth International Conference on Image Fusion*. Queensland, Australia, 2003.

17. Yang, J., and Blum, R. S., A statistical signal processing approach to image fusion for concealed weapon detection, pp. 513–516. In *IEEE International Conference on Image Processing*. Rochester, NY, 2002.

18. Tu, T. M., Su, S. C., Shyu, H. C., and Huang, P. S., Efficient intensity–hue–saturation-based image fusion with saturation compensation, *Opt. Eng.*, 40(5), 720–728, 2001.

19. Pohl, C., and Van Genderen, J. L., Multisensor image fusion in remote sensing: concepts, methods and applications, *Int. J. Remote Sens.*, 19(5), 823–854, 1998.

20. Nunez, J., Otazu, X., Fors, O., Prades, A., Pala, V., and Arbiol, R., Multiresolution-based image fusion with additive wavelet decomposition, *IEEE Trans. Geosci. Remote Sens.*, 37(3), 1204–1211, 1999.

21. Simone, G., Farina, A., Morabito, F. C., Serpico, S. B., and Bruzzone, L., Image fusion techniques for remote sensing applications, *Inf. Fusion*, 3(1), 3–15, 2002.

22. Laliberte, F., Gagnon, L., and Sheng, Y., Registration and fusion of retinal images-an evaluation study, *IEEE Trans. Med. Imaging*, 22(5), 661–673, 2003.

23. Zhang, Z., Yao, J., Bajwa, S., and Gudas, T., Automatic multimodal medical image fusion, pp. 42–49. In *Proceedings of the 16th IEEE Symposium on Computer-based Medical Systems*. New York, 2003.

24. Qu, G., Zhang, D., and Yan, P., Medical image fusion using two dimensional discrete wavelet transform, *Proc. SPIE*, 4556, 86–95, 2001.

25. Pattichis, C. S., Pattichis, M. S., and Micheli-Tzanakou, E., Medical imaging fusion applications: an overview, pp. 1263–1267. In *The 35th Asilomar Conference on Signals, Systems and Computers*, Vol. 2. Pacific Grove, CA, 2001.

26. Reed, J. M., and Hutchinson, S., Image fusion and subpixel parameter estimation for automated optical inspection of electronic components, *IEEE Trans. Ind. Electron.*, 43(3), 346–354, 1996.

27. Simone, G., and Morabito, F. C., NDT image fusion using eddy current and ultrasonic data, pp. 857–868. In *The Sixth International Workshop on Optimization and Inverse Problems in Electromagnetism (OPIE 2000)*, Vol. 20. Italy, 2000.

28. Leon, F. P., and Kammel, S., Image fusion techniques for robust inspection of specular surfaces, *Proc. SPIE — Int. Soc. Opt. Eng.*, 5099, 77–86, 2003.

29. Dobeck, G. J., Fusing sonar images for mine detection and classification, *Proc. SPIE*, 3710, 602–614, 1999.

30. McDaniel, R., Scribner, D., Krebs, W., Warren, P., Ockman, N., and McCarley, J., Image fusion for tactical applications, *Proc. SPIE*, 3436, 685–695, 1998.

31. Smith, M. I., and Rood, G., Image fusion of II and IR data for helicopter pilotage. In *Proceedings of the Night Vision Conference*. Washington, DC, 1999.

32. Murphy, R. R., Sensor and information fusion for improved vision-based vehicle guidance, *IEEE Intell. Syst.*, 13(6), 49–56, 1998.
33. Reese, C. E., and Bender, E. J., Multi-spectral/image fused head tracked vision system for driving applications, *Proc. SPIE*, 4361, 1–11, 2001.
34. Zhang, Z., and Blum, R. S., A categorization of multiscale-decomposition-based image fusion schemes with a performance study for a digital camera application, *Proc. IEEE*, 87(8), 1315–1326, 1999.
35. Seales, W. B., and Dutta, S., Everywhere-in-focus image fusion using controllable cameras, *Proc. SPIE*, 2905, 227–234, 1996.
36. Reese, C. E., Bender, E. J., and Reed, R. D., Advancements of the head-tracked vision system, *Proc. SPIE*, 4711, 105–116, 2002.
37. Gros, X. E., Liu, Z., Tsukada, K., and Hanasaki, K., Experimenting with pixel-level NDT data fusion techniques, *IEEE Trans. Instrum. Meas.*, 49(5), 1083–1090, 2000.
38. Gross, H. N., and Schott, J. R., Application of spectral mixing to image fusion. In *The 26th International Symposium on Remote Sensing of Environment*. Vancouver, BC, 1996, February.
39. Rosenfeld, A., and Thurston, M., Edge and curve detection for visual scene analysis, *IEEE Trans. Comput.*, C-20, 562–569, 1971.
40. Witkin, A. P., and Tenenbaum, J. M., On the role of structure in vision, In *Human and Machine Vision*, Beck, J., Hope, B., and Rosenfeld, A., Eds., Academic Press, New York, 1983, pp. 481–544.
41. Marr, D., *Vision*, WH Freeman, San Francisco, CA, 1982.
42. Burt, P. J., and Adelson, E. H., The Laplacian pyramid as a compact image code, *IEEE Trans. Commun.*, 31(4), 532–540, 1983.
43. Lindeberg, T., *Scale-Space Theory in Computer Vision*, Kluwer Academic Publisher, Dordrecht, 1994.
44. Burt, P. J., and Adelson, E. H., Merging images through pattern decomposition, *Proc. SPIE*, 575, 173–182, 1985.
45. Anderson, C. H., A filter-subtract-decimate hierarchical pyramid signal analyzing and synthesizing technique, US Patent 4,718,104, Washington, DC, 1987.
46. Toet, A., van Ruyven, L. J., and Valeton, J. M., Merging thermal and visual images by a contrast pyramid, *Opt. Eng.*, 28(7), 789–792, 1989.
47. Burt, P. J., A gradient pyramid basis for pattern-selective image fusion, *SID Tech. Digest*, 16, 467–470, 1992.
48. Toet, A., A morphological pyramidal image decomposition, *Pattern Recognit. Lett.*, 9, 255–261, 1989.
49. Matsopoulos, G. K., Marshall, S., and Brunt, J., Multiresolution morphological fusion of MR and CT images of human brain, *Proc. IEE Vision Image Signal Process.*, 141(3), 137–142, 1994.
50. Liu, Z., Tsukada, K., Hanasaki, K., Ho, Y. K., and Dai, Y. P., Image fusion by using steerable pyramid, *Pattern Recognit. Lett.*, 22(9), 929–939, 2001.
51. Wechsler, H., *Computational Vision*, Academic Press, New York, 1990.
52. Mallat, S. G., A theory for multiresolution signal decomposition: the wavelet representation, *IEEE Trans. Pattern Anal. Machine Intell.*, 11, 674–693, 1989.
53. Vetterli, M., and Herley, C., Wavelets and filter banks: theory, and design, *IEEE Trans. Signal Process.*, 40, 2207–2232, 1992.
54. Daubechies, I., Orthonormal bases of compactly supported wavelets, *Commun. Pure Appl. Math.*, 41, 909–996, 1988.

55. Unser, M., Texture classification and segmentation using wavelet frames, *IEEE Trans. Image Process.*, 4(11), 1549–1560, 1995.

56. Young, R. M., *An Introduction to Nonharmonic Fourier Series*, Academic Press, New York, 1980.

57. Daubechies, I., *Ten Lectures on Wavelets*, Society for Industrial and Applied Mathematics, Philadelphia, PA, 1992.

58. Burt, P. J., and Kolczynski, R. J., Enhanced image capture through fusion, pp. 173–182. In *Proceedings of the Fourth International Conference on Computer Vision*. Berlin, Germany, 1993.

59. Li, H., Manjunath, B., and Mitra, S., Multisensor image fusion using the wavelet transform, *Graphical Models Image Process.*, 57, 235–245, 1995.

60. Rockinger, O., and Fechner, T., Pixel-level image fusion: the case of image sequences, *Proc. SPIE*, 3374, 378–388, 1998.

61. Lallier, E., and Farooq, M., A real time pixel-level based image fusion via adaptive weight averaging, pp. WeC3_3–WeC3_13. In *The Third International Conference on Information Fusion*. Paris, France, 2000.

62. Gonzalez, R. C., and Woods, R. E., *Digital Image Processing*, 2nd ed., Prentice-Hall, New Jersey, 2002.

63. Therrien, C. W., and Krebs, W. K., An adaptive technique for the enhanced fusion of low-light visible with uncooled thermal infrared imagery, pp. 405–408. In *IEEE International Conference on Image Processing*. Santa Barbara, CA, 1997.

64. Poor, H. V., *An Introduction to Signal Detection and Estimation*, 2nd ed., Springer, Berlin, 1994.

65. Clark, J. J., and Yuille, A. L., *Data Fusion for Sensory Information Processing Systems*, Kluwer, Boston, 1990.

66. Sharma, R. K., Leen, T. K., and Pavel, M., Bayesian sensor image fusion using local linear generative models, *Opt. Eng.*, 40(7), 1364–1376, 2001.

67. Ma, B., Lakshmanan, S., and Hero, A. O., Simultaneous detection of lane and pavement boundaries using model-based multisensor fusion, *IEEE Trans. Intell. Trans. Syst.*, 1(3), 135–147, 2000.

68. Blum, R. S., and Yang, J., Image fusion using the expectation-maximization algorithm and a Gaussian mixture model, In *Advanced Video-Based Surveillance Systems*, Foresti, G. L., Regazzoni, C. S., and Varshney, P. K., Eds., Kluwer, Boston, 2003.

69. Wright, W. A., and Bristol, F., Quick Markov random field image fusion, *Proc. SPIE*, 3374, 302–308, 1998.

70. Kundur, D., Hatzinakos, D., and Leung, H., Robust classification of blurred imagery, *IEEE Trans. Image Process.*, 9(2), 243–255, 2000.

71. Azencott, R., Chalmond, B., and Coldefy, F., Markov fusion of a pair of noisy images to detect intensity valleys, *Int. J. Comput. Vision*, 16(2), 135–145, 1995.

72. Yocky, D. A., Multiresolution wavelet decomposition image merger of landsat thematic mapper and spot panchromatic data, *Photogrammetric Engineering and Remote Sensing*, 62, 1067–1074, 1996.

73. Carper, W. J., Lillesand, T. M., and Kiefer, R. W., The use of intensity-hue-saturation transformations for merging spot panchromatic and multi-spectral image data, *Photogrammetric Engineering Remote Sensing*, 56, 459–467, 1990.

74. Waxman, A. M., Aguilar, M., Baxter, R. A., Fay, D. A., Ireland, D. B., Racamato, J. P., and Ross, W. D., Opponent-color fusion of multi-sensor imagery: visible, IR and SAR, *Proc. IRIS Passive Sens.*, 1, 43–61, 1998.

75. Aguilar, M., Fay, D. A., Ross, W. D., Waxman, A. M., Ireland, D. B., and Racamato, J. P., Real-time fusion of low-light CCD and uncooled IR imagery for color night vision, *Proc. SPIE*, 3364, 124–135, 1998.

76. Xue Z., and Blum R. S., Concealed weapon detection using color image fusion. In *The Sixth International Conference on Image Fusion*. Queensland, Australia, 2003, July.

77. Toet, A., and Walraven, J., New false color mapping for image fusion, *Opt. Eng.*, 35(3), 650–658, 1996.

78. Essock, E. A., Sinai, M. J., McCarley, J. S., Krebs, W. K., and DeFord, J. K., Perceptual ability with real-world nighttime scenes: image-intensified, infrared, and fused-color imagery, *Hum. Factors*, 41(3), 438–452, 1999.

79. Aguilar, M., and Garret, A. L., Biologically based sensor fusion for medical imaging, In *Sensor Fusion: Architectures, Algorithms, and Applications V*, Dasarathy, B. V., Ed., The International Society for Optical Engineering, Bellingham, WA, 2001, pp. 149–158.

80. Aguilar, M., and New, J. R., Fusion of multi-modality volumetric medical imagery, pp. 1206–1212. In *Proceedings of The Fifth International Conference of Information Fusion*. Annapolis, MD, 2002.

81. Fechner, T., and Godlewski, G., Optimal fusion of TV and infrared images using artificial neural networks, *Proc. SPIE*, 2492, 919–925, 1995.

82. Kinser, J. M., Pulse-coupled image fusion, *Opt. Eng.*, 36(3), 737–742, 1997.

83. Johnson, J. L., Schamschula, M. P., Inguva, R., and Caulfield, H. J., Pulse coupled neural network sensor fusion, *Proc. SPIE*, 3376, 219–226, 1998.

84. Broussard, R. P., Rogers, S. K., Oxley, M. E., and Tarr, G. L., Physiologically motivated image fusion for object detection using a pulse coupled neural network, *IEEE Trans. Neural Netw.*, 10(3), 554–563, 1999.

85. Wang, Z., and Bovik, A. C., A Universal image quality index, *IEEE Signal Process. Lett.*, 9(3), 81–84, 2002.

86. Zhang, Z., and Blum, R. S., Image fusion for a digital camera application, pp. 603–607. In *The proceedings of the 32nd Asilomar Conference on Signals, Systems, and Computers*. Monterey, CA, 1998.

87. Xue, Z., Blum, R. S., and Li, Y., Fusion of visual and IR images for concealed weapon detection. In *Proceedings of the Fifth International Conference on Information Fusion*. Annapolis, MD, 2002.

88. Burt, P. J., The pyramid as structure for efficient computation, *Multiresolution Image Processing and Analysis*, Springer, Berlin, 1984.

89. Akerman, A. III, Pyramid techniques for multisensor fusion, *Proc. SPIE*, 1828, 124–131, 1992.

90. Huntsberger, T., and Jawerth, B., Wavelet based sensor fusion, *Proc. SPIE*, 2059, 488–498, 1993.

91. Ranchin, T., Wald, L., and Mangolini, M., Efficient data fusion using wavelet transform: the case of spot satellite images, *Proc. SPIE*, 2934, 171–178, 1993.

92. Koren, I., Laine, A., and Taylor, F., Image fusion using steerable dyadic wavelet transform, *Proc. IEEE Int. Conf. Image Process.*, 3, 232–235, 1995.

93. Chipman, L. J., Orr, T. M., and Graham, L. N., Wavelets and image fusion, *Proc. SPIE*, 2569, 208–219, 1995.

94. Peytavin, L., Cross-sensor resolution enhancement of hyperspectral images using wavelet decomposition, *Proc. SPIE*, 2758, 193–197, 1996.

95. Wilson, T. A., Rogers, S. K., and Myers, L. R., Perceptual-based hyperspectral image fusion using multiresolution analysis, *Opt. Eng.*, 34(11), 3154–3164, 1995.

96. Jiang, X., Zhou, L., and Gao, Z., Multispectral image fusion using wavelet transform, *Proc. SPIE*, 2898, 35–42, 1996.

97. Avcıbas, I., Sankur, B., and Sayood, K., Statistical evaluation of image quality measures, *J. Electron. Imaging*, 11, 206–223, 2002.

98. Piella, G., and Heijmans, H., A new quality metric for image fusion. In *Proceedings of International Conference on Image Processing*. Barcelona, 2003.

99. Xydeas, C., and Petrovic, V., Objective pixel-level image fusion performance measure, *Proc. SPIE*, 4051, 88–99, 2000.

100. Zhang, Z., and Blum, R. S., On estimating the quality of noisy images, pp. 2897–2901. In *IEEE International Conference on Acoustics, Speech and Signal Processing*. Seattle, WA, 1998.

101. Canny, J., A computational approach to edge detection, *IEEE Trans. Pattern Anal. Machine Intelligence*, 8(6), 679–698, 1986.

102. Redner, R. A., and Walker, H. F., Mixture densities, maximum likelihood and the EM algorithm, *SIAM Rev.*, 26, 195–239, 1984.

103. Dempster, A. P., Laird, N. M., and Rubin, D. B., Maximum likelihood from incomplete data via the EM algorithm, *J. R. Stat. Soc.*, 39(1), 1–38, 1977.

104. Li, H., Manjunath, B. S., and Mitra, S. K., A contour-based approach to multisensor image registration, *IEEE Trans. Image Process.*, 4, 320–334, 1995.

105. Likar, B., and Pernus, F., Automatic extraction of corresponding points for the registration of medical images, *Med. Phys.*, 26, 1678–1686, 1999.

106. Chen, H., and Varshney, P. K., Automatic registration of infrared and millimeter wave images for concealed weapons detection, *Proceedings of SPIE Conference on Sensor Fusion: Architectures, Algorithms, and Applications III*, 3719, 152–160, 1999.

107. Chen, H., and Varshney, P. K., Cooperative search algorithm for mutual-information-based image registration, *Proc. SPIE*, 4385, 117–128, 2001.

108. Chen, Y., Brooks, R. R., Iyengar, S. S., Rao, N. S. V., and Barhen, J., Efficient global optimization for image registration, *IEEE Trans. Knowledge Data Eng.*, 14(1), 79–92, 2002.

109. Fonseca, L., Hewer, G. A., Kenney, C. S., and Manjunath, B. S., Registration and fusion of multispectral images using a new control point assessment method derived from optical flow ideas, *Proc. SPIE*, 3717, 104–111, 1999.

110. Le Moigne, J., and Cromp, R. F., Wavelets for remote sensing image registration and fusion, *Proc. SPIE*, 2762, 535–544, 1996.

111. Mostafa, M. G., Farag, A. A., and Essock, E., Multimodality image registration and fusion using neural network, *Inf. Fusion*, 2, 3–9, 2000.

112. Remagnino, P., and Jones, G. A., Automated registration of surveillance data for multi-camera fusion, *Inf. Fusion*, 2, 1190–1197, 2002.

113. Turcajova, R., and Kautsky, J., Hierarchical multiresolution technique for image registration, *Proc. SPIE*, 2825, 686–696, 1996.

114. Nevel, A. J. V., Image registration: a key element for information processing, *Proc. SPIE*, 4471, 190–200, 2001.
115. Brown, L. G., A survey of image registration techniques, *ACM Computing Survey*, 24(4), 325–376, 1992.
116. Lester, H., and Arridge, S. R., A survey of hierarchical non-linear medical image registration, *Pattern Recogn.*, 32, 129–149, 1999.
117. Zhang, Z., and Blum, R. S., Image registration for multi-focus image fusion. *SPIE AeroSense, Conference on Battlefield Digitization and Network Centric Warfare* (4396-39), Orlando, FL, 2001, April.

2 Mutual Information Based Image Registration with Application to 3D Medical Brain Imagery

Hua-Mei Chen and Pramod K. Varshney

CONTENTS

I. INTRODUCTION

An image fusion algorithm accepts two or more images of the same region and produces an image with higher information content. The first step toward image fusion is a precise alignment of the images involved, such that the corresponding pixels/voxels in the two images/volumes represent the same physical point of the common region. This task is usually referred to as *image registration* in the literature. When the two images are acquired from different types of imaging sensors, this process is called *multimodality* image registration. Multimodality image registration has become an important research topic because of its great value in a variety of applications. For medical image analysis, an image showing functional and metabolic activity such as single photon emission computed tomography (SPECT), positron emission tomography (PET), and magnetic resonance spectroscopy (MRS), is often registered to an image which shows

"

anatomical structures such as magnetic resonance image (MRI), computed tomography (CT), and ultrasound. These registered multimodality images are fused which, in turn, lead to improved diagnosis, better surgical planning, more accurate radiation therapy, and many other medical benefits.[1] Over the years, many image registration techniques have been developed for different applications. For a recent review of image registration techniques, see Ref. 2. Existing image registration techniques can be broadly classified into two categories: feature-based and intensity-based methods.[2,3] A feature-based method requires the extraction of features common in both images. Obviously, this method is data dependent. Since different image data may have different features, feature extraction algorithms adopted in a feature-based image registration algorithm are expected to be different for diverse applications. In contrast, intensity-based image registration techniques are free from this limitation because they do not deal with the identification of geometrical landmarks. The general design criterion of an intensity-based image registration technique can be expressed as:

$$\alpha^* = \arg \mathrm{opt}(S(F(\tilde{x}), R(T_\alpha(\tilde{x})))) \tag{2.1}$$

where F and R are the images to be registered. F is referred to as the *floating image*, whose pixel co-ordinates ($\tilde{x}$) are to be mapped to new co-ordinates on the *reference image R*, which are to be resampled according to the positions defined by the new co-ordinates $T_\alpha(\tilde{x})$, where T denotes the transformation model and the dependence of T on its associated parameters α is indicated by the use of notation T_α. S is an intensity-based similarity measure calculated over the region of overlap of the two images. The above criterion says that the two images F and R are registered through T_{α^*} when α^* optimizes the selected similarity measure S. Among a variety of existing similarity measures, mutual information (MI) has received substantial attention recently because of its ability to measure the similarity between images from different modalities, especially in, but not limited to, medical imaging applications.[3-8] A general treatment of MI based registration techniques with applications to medical imagery can be found in two recently published survey papers.[9,10] In this chapter, in addition to the introduction of MI based image registration techniques, we focus our discussion on a phenomenon known as interpolation induced artifacts which has been reported to limit the performance of MI based registration techniques in many practical applications like three-dimensional (3D) brain image registration in medical imaging[11,12] and multitemporal image registration in remote sensing.[6] Figure 2.1 shows the typical artifact patterns encountered in the MI registration function in one dimension: (1) through linear interpolation,[8] and (2) through partial volume interpolation (PVI).[15] These patterns have at least two consequences: (1) they hamper the global optimization process because of the introduction of periodical local extrema and (2) they influence registration accuracy[11,12] because the true global optimum is now buried in the artifact pattern. We want to point out that this influence is not always negative, that is, registration accuracy does not always

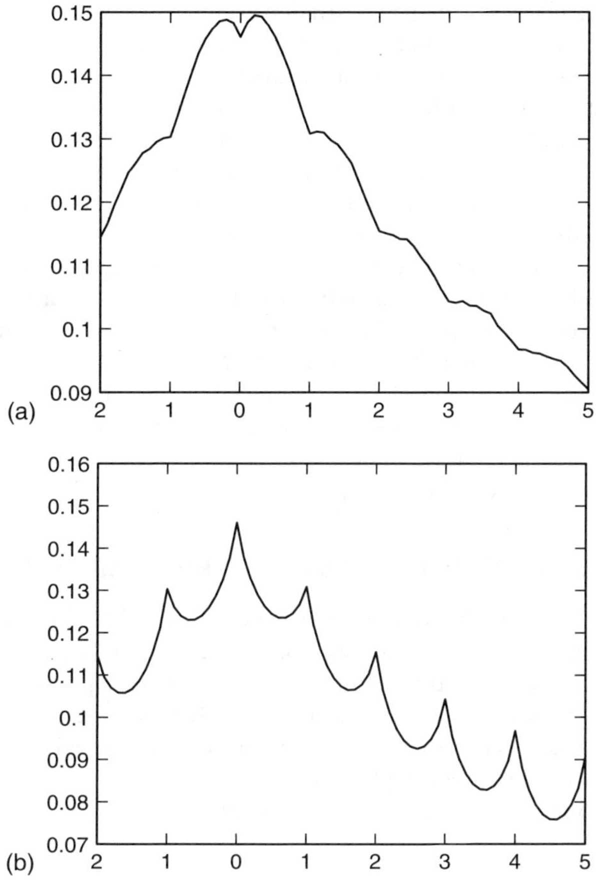

FIGURE 2.1 Typical interpolation-induced artifact patterns for a MI based registration function. In both cases, the vertical axis is the MI based measure, and the horizontal axis can be displacement in x- or y- or z-dimension. (a) artifact pattern resulting from linear interpolation. (b) artifact pattern resulting from PVI algorithm.

become worse. It depends on the position of the highest peak of the artifact pattern. To facilitate the explanation, let α_{ideal} denote the pose parameters (which is a vector in a multi-dimensional search space) that result in perfect registration, $\alpha_{\text{arti-free}}$ the pose parameters found using an artifact-free MI based registration algorithm, and α_{arti} the pose parameters resulting from a MI based registration algorithm that is known to suffer from artifacts due to the use of an algorithm such as the PVI algorithm. Clearly, α_{arti} is determined by the position of the highest peak of the artifact pattern. If this position happens to be closer than $\alpha_{\text{arti-free}}$ to α_{ideal}, the influence of the artifacts on the registration accuracy is positive, that is, the registration accuracy is improved. On the other hand, if the distance between α_{arti} and α_{ideal} is larger than that between $\alpha_{\text{arti-free}}$ and α_{ideal}, the influence is negative and the registration accuracy becomes worse. Therefore, assuming perfect optimization, registration accuracy would depend on the position of the highest peak of the artifact pattern. Although the influence is not always negative, it is still desirable to develop an artifact-free MI based registration algorithm. There are two main reasons: (1) an artifact-free MI based registration algorithm facilitates the global optimization process, and (2) when the influence of the artifacts on the registration accuracy is negative, artifact-free MI based registration can improve registration accuracy.

Many ideas have been suggested to suppress the interpolation artifacts,[12–14] however, whether those ideas lead to improved registration accuracy has not been conclusively shown except for Ref. 12. In this chapter, we introduce the joint histogram estimation algorithm presented in Ref. 12, which has been shown to lead to improved registration accuracy degraded by interpolation artifacts for 3D medical brain computed tomography magnetic resonance (CT-MR) image registration. Experimental results using the image data furnished by Vanderbilt University[16] are included in this chapter.

II. MUTUAL INFORMATION AS A GENERIC SIMILARITY MEASURE

Mutual information has its roots in information theory.[17] It was developed to set fundamental limits on the performance of communication systems. However, it has made vital contributions to many different disciplines like physics, mathematics, economics, and computer science. In this subsection, we introduce the use of MI as a generic similarity measure for image registration problems along with two widely used implementations.

MI of two random variables A and B is defined by

$$I(A, B) = \sum_{a,b} P_{A,B}(a, b) \log \frac{P_{A,B}(a, b)}{P_A(a) \cdot P_B(b)} \tag{2.2}$$

where $P_A(a)$ and $P_B(b)$ are the marginal probabilities, and $P_{A,B}(a, b)$ is the joint probability. MI measures the degree of dependence of A and B by measuring the distance between the joint probability $P_{A,B}(a, b)$ and the probability associated with the case of complete independence $P_A(a) \cdot P_B(b)$, by means of the relative entropy or the Kullback–Leibler measure.[17] MI is related to entropies by the relationship

$$I(A, B) = H(A) + H(B) - H(A, B) \tag{2.3}$$

where $H(A)$ and $H(B)$ are the entropies of A and B and $H(A, B)$ is their joint entropy. Considering A and B as two images, *floating image* (F) and *reference image* (R), respectively, the MI based image registration criterion postulates that the images shall be registered when $I(F, R)$ is maximum. The entropies and joint entropy can be computed from,

$$H(F) = \sum_f - P_F(f) \log P_F(f) \tag{2.4}$$

$$H(R) = \sum_r - P_R(r) \log P_R(r) \tag{2.5}$$

$$H(F, R) = \sum_{f,r} - P_{F,R}(f, r) \log P_{F,R}(f, r) \tag{2.6}$$

where $P_F(f)$ and $P_R(r)$ are the marginal probability mass functions, and $P_{F,R}(f, r)$ is the joint probability mass function of the two images F and R. The probability mass functions can be obtained from

$$P_{F,R}(f, r) = \frac{h(f, r)}{\displaystyle\sum_{f,r} h(f, r)} \tag{2.7}$$

$$P_F(f) = \sum_r P_{F,R}(f, r) \tag{2.8}$$

$$P_R(r) = \sum_f P_{F,R}(f, r) \tag{2.9}$$

where h is the joint histogram of the image pair. It is a two-dimensional (2D) matrix given by

$$h = \begin{bmatrix} h(0,0) & h(0,1) & \cdots & h(0, N-1) \\ h(1,0) & h(1,1) & \cdots & h(1, N-1) \\ \vdots & \vdots & \ddots & \vdots \\ h(M-1,0) & h(M-1,1) & \cdots & h(M-1, N-1) \end{bmatrix} \tag{2.10}$$

The value $h(f, r)$, $f \in [0, M-1]$, $r \in [0, N-1]$ is the number of pixel pairs with intensity value f in the first image (i.e., F) and intensity value r in the second image (i.e., R). M and N are the number of gray levels used in the images F and R, respectively. Typically, $M = N = 256$ for 8-bit images. It can thus be seen from Equation 2.3 to Equation 2.9 that the joint histogram estimate is sufficient to determine the MI between two images.

To interpret Equation 2.3 in the context of image registration with random variables F and R, let us first assume that both $H(F)$ and $H(R)$ are constant. Under this assumption, maximization of mutual information in Equation 2.3 is equivalent to the minimization of joint entropy. However, this assumption is usually violated in most of the applications. When the assumption of constant entropies $H(F)$ and $H(R)$ is not satisfied, we may employ the notion of conditional entropy to interpret the MI criterion. We rewrite MI in terms of conditional entropy as

$$I(F, R) = H(F) - H(F|R) \qquad (2.11a)$$

$$= H(R) - H(R|F) \qquad (2.11b)$$

where the conditional entropies $H(F|R)$ and $H(R|F)$ are defined as:

$$H(F|R) = - \sum_{f,r} P_{F,R}(f, r) \log P_{F|R}(f|r) \qquad (2.12)$$

$$H(R|F) = - \sum_{f,r} P_{F,R}(f, r) \log P_{R|F}(r|f) \qquad (2.13)$$

Conditional entropy $H(F|R)$ represents the entropy (uncertainty) of F when knowing R. Similarly, $H(R|F)$ is the entropy (uncertainty) of R given F. Therefore, mutual information can be realized as *the reduction of uncertainty (entropy) of one image by the knowledge of another image*. In other words, the MI criterion implies that when two images are registered, one gains most knowledge about one image by observing the other one. For example, based on Equation 2.11a, the uncertainty of F without the knowledge of R is $H(F)$ and its uncertainty when knowing R is $H(F|R)$. Therefore, the reduction of the uncertainty $H(F) - H(F|R)$ defines the MI between F and R and its maximization implies the most reduction in the uncertainty. Since its introduction, MI has been used widely for many different applications because of its high accuracy and robustness.

From Equation 2.3 to Equation 2.9, it is clear that the main task involved in the determination of the mutual information between two images is to estimate the joint histogram of the two images. The joint histogram of the overlap of F and R can be obtained by counting the intensity pairs $(F(\tilde{x}), R(T_\alpha(\tilde{x})))$ over the overlap of the two images. To facilitate the "counting" process both $F(\tilde{x})$ and $R(T_\alpha(\tilde{x}))$ need to be integers. However, the co-ordinates of a transformed grid point $T_\alpha(\tilde{x})$ are, in general, nonintegers and the image intensity value at a nongrid point is not defined. To overcome this problem, two methods are widely used in the literature. The first one is to obtain the intensity value at a nongrid point through linear interpolation, followed by a rounding operation to assure that the interpolated intensity value is an integer. This procedure can be explained with the help of the graphic shown in Figure 2.2 in the 2D case. In Figure 2.2, $\tilde{x}$ represents the co-ordinates of a pixel in the floating image F and $\tilde{y}$ the transformed co-ordinates of that pixel in R, i.e., $\tilde{y} = T_\alpha(\tilde{x})$. In Figure 2.2, $\tilde{y}$ splits the cell $\tilde{y}_1, \tilde{y}_2, \tilde{y}_3, \tilde{y}_4$ into four subcells each with area $\omega_1, \omega_2, \omega_3,$ and ω_4, respectively. The sum of $\omega_1, \omega_2, \omega_3,$ and ω_4 is restricted to be 1. For linear interpolation, the intensity value of R at the position $\tilde{y}$ is obtained by

$$R(\tilde{y}) = \sum_{i=1}^{4} \omega_i R(\tilde{y}_i) \qquad (2.14)$$

Since the interpolated value is not an integer in general, a rounding operation is

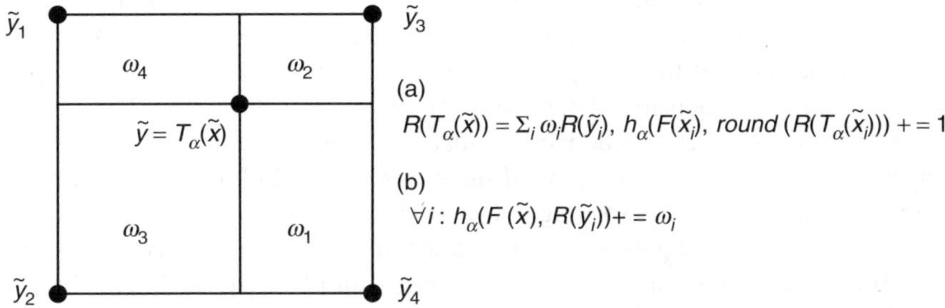

FIGURE 2.2 Graphical illustration of (a) linear interpolation, and (b) partial volume interpolation (PVI) in two dimensions.

used when updating the joint histogram by

$$h_\alpha(F(\tilde{x}), \, round(R(\tilde{y}))) + = 1 \tag{2.15}$$

This procedure is repeated for every pixel in F over the overlap region.

Another widely used method for joint histogram estimation is PVI. The idea of PVI is the following. Instead of interpolating the intensity value at the co-ordinates $\tilde{y} = T_\alpha(\tilde{x})$ through the fractions of the intensities of the surrounding grids ($\tilde{y}_1 \sim \tilde{y}_4$ in Figure 2.2), PVI estimates the joint histogram by updating multiple intensity pairs defined by $F(\tilde{x})$ and the intensities of the surrounding grids (for instance, $R(\tilde{y}_1)$, $R(\tilde{y}_2)$, $R(\tilde{y}_3)$, $R(\tilde{y}_4)$ in Figure 2.2) by the same fractions. This is mathematically expressed by

$$\forall i : h_\alpha(F(\tilde{x}), R(\tilde{y}_i)) + = \omega_i \tag{2.16}$$

One advantage of PVI over linear interpolation is that the resulting registration function is much smoother.[15] This facilitates the optimization process. Nevertheless, it is pointed out in Ref. 11 that the mutual information measure obtained through either linear interpolation or PVI suffers from a phenomenon known as interpolation induced artifacts when both images have equal sample spacing along at least one dimension. In the next section, we discuss this phenomenon in some detail.

III. INTERPOLATION INDUCED ARTIFACTS

One major limitation of current implementations of the MI criterion through either linear interpolation or PVI is the occurrence of periodical extrema in the registration function under certain conditions. This phenomenon, known as interpolation induced artifacts, was first discussed in detail by Pluim et al. in Ref. 11. In the paper, it was pointed out that when two images have equal sample spacing in one or more dimensions, existing joint histogram estimation algorithms like PVI and linear interpolation may result in certain types of artifact patterns in a MI based registration function as shown in Figure 2.1.

More precisely, the artifacts will occur when the ratio of the two sample spacings along a certain dimension is a simple rational number. The reason is that under this situation, many of the grid planes (3D case) or grid lines (2D case) may be aligned along that dimension under certain geometric transformations. Therefore, fewer interpolations are needed to estimate the joint histogram of these two images than in the case where none of the grid planes/grid lines are aligned. For example, when the sample spacings of the two image volumes along the z-axis are 5 and 3 mm, assuming the same sample spacing along the other two axes, then by shifting one of the image volumes, the grids on planes 1, 4, 7,…of the first image volume (the one with sample spacing 5 mm along the z-axis) can be made to coincide with the grids on planes 1, 6, 11,…of the second image volume. In this case, the contribution of the coincident grids to the joint histogram can be directly counted without the need of any estimation. But if one of the image volumes is further shifted along the z-axis, then none of the grids of the two image volumes will be coincident with each other and the joint histogram has to be estimated completely. In this case, this abrupt change between "much less" estimation and "substantially more" estimation causes the artifacts. In addition, the artifacts are expected to be periodic in 1 mm along the z-direction because in this case, there are certain grid planes that will be aligned for every 1 mm shift in the z-direction. One such example is presented in Figure 2.3. Simulated brain MR T1 image with voxel size $1 \times 1 \times 5$ mm and MR T2 image with voxel size $1 \times 1 \times 3$ mm from BrainWeb[18,19] were used to produce the artifact patterns. In Figure 2.3, the curve in solid line was the pattern resulting from PVI while the dashed line resulted from linear interpolation. The curves have been shifted vertically for visualization purposes. Notice that the artifacts occur for every 1 mm spacing as expected.

As indicated above, the underlying mechanism causing the artifacts in the registration function obtained through linear interpolation and PVI were

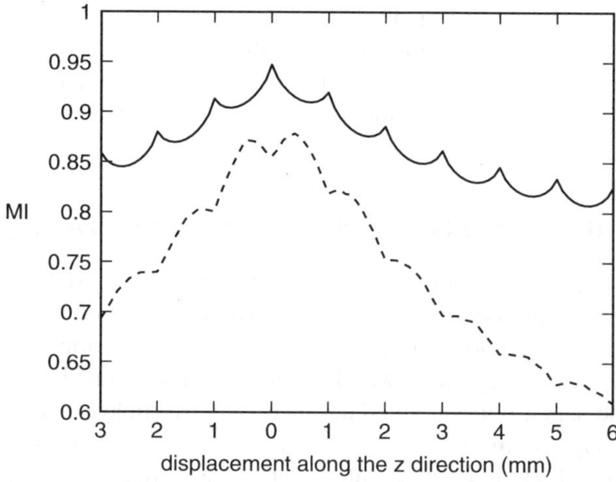

FIGURE 2.3 Artifact patterns in the case where the ratio of sample spacings equals 5/3.

thoroughly discussed in Ref. 11. In short, the blurring nature of linear interpolation was concluded as the cause of the convex artifact pattern and the abrupt changes in the joint histogram dispersion resulting from the PVI method for grid-aligning transformations cause the concave artifact pattern. Since most existing intensity interpolation algorithms involve blurring to some extent, the chance of developing a new joint histogram estimation scheme through intensity interpolation is remote. In Ref. 20, it was shown that similar artifact patterns were still present when more sophisticated intensity interpolation algorithms like cubic convolution interpolation[21] and B-spline interpolation[22,23] were used. As another approach for reducing the artifacts, one may think of devising a new joint histogram estimation scheme similar to the PVI method. In the next section, we describe such an algorithm by generalizing the PVI algorithm.

IV. GENERALIZED PARTIAL VOLUME ESTIMATION OF JOINT HISTOGRAM

Based on the idea of PVI, which corresponds to the use of a linear function as the interpolation kernel, a more general scheme allowing the use of kernels of larger support was developed in Ref. 12. This scheme is known as generalized partial volume estimation (GPVE). To better understand the structure of the GPVE algorithm, we first rewrite the PVI algorithm proposed by Maes and Collignon[15] as shown in Equation 2.16 in the 2D case. This sets the stage for the algorithm by providing some of the terminology used.

Refer to Figure 2.4 and let F and R be the floating image and reference image, respectively, which can be considered as two mappings:

$$F : \tilde{x} \rightarrow F(\tilde{x}), \quad \tilde{x} \in X$$
$$R : \tilde{y} \rightarrow R(\tilde{y}), \quad \tilde{y} \in Y$$

$$(2.17)$$

where X is the discrete domain of the floating image F and Y is the discrete

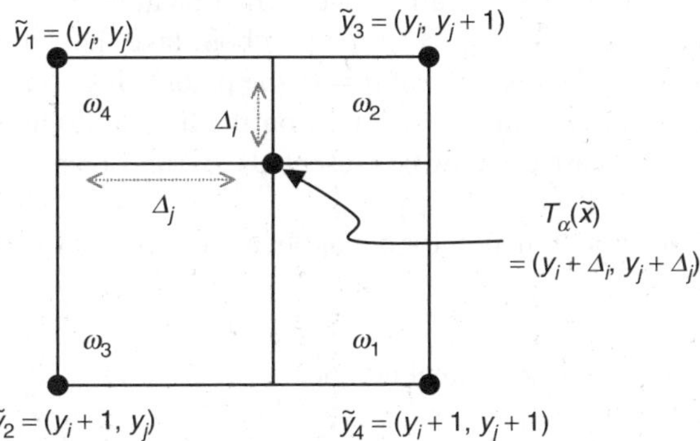

FIGURE 2.4 Another graphical illustration of the PVI algorithm in two-dimensional space.

domain of the reference image R. The intensity of the floating image at the grid point with co-ordinates $\tilde{x} = (x_i, x_j)$ is represented by $F(\tilde{x})$. Also, we represent the transformation characterized by the parameter set α that is applied to the grid points of F as T_α and assume that T_α maps the grid point (x_i, x_j) in image F into the point with co-ordinates, in terms of sample spacing, $(y_i + \Delta_i, y_j + \Delta_j)$ in the image R, where (y_i, y_j) is a grid point in R and $0 \leq \Delta_i, \Delta_j < 1$. Recall that in Figure 2.4, $\tilde{y}_1, \tilde{y}_2, \tilde{y}_3, \tilde{y}_4$ are the grid points on the reference image R that are closest to the transformed grid point $T_\alpha(\tilde{x})$ and $T_\alpha(\tilde{x})$ splits the cell $\tilde{y}_1, \tilde{y}_2, \tilde{y}_3, \tilde{y}_4$ into four subcells. The subcells having areas $\omega_1, \omega_2, \omega_3,$ and ω_4 as shown satisfy the constraint

$$\sum_{l=1}^{4} \omega_l(T_\alpha(\mathbf{x})) = 1$$

The PVI algorithm obtains the joint histogram using the following updating procedure:

$$h(F(\tilde{x}), R(\tilde{y}_l)) + = \omega_l \ \text{ for } \ l = 1, \ldots, 4 \tag{2.18}$$

Now if f is a triangular function defined by

$$f(t) = \begin{cases} 1 + t; & \text{if } -1 \leq t < 0 \\ 1 - t; & \text{if } 0 \leq t \leq 1 \\ 0; & \text{otherwise} \end{cases} \tag{2.19}$$

then we can rewrite the updating procedure Equation 2.18 as

$$h(F(x_i, x_j), R(y_i + p, y_j + q)) + = f(p - \Delta_i)f(q - \Delta_j) \ \ \forall p, q \in Z \tag{2.20}$$

where p and q are used to specify the pixels involved in the histogram updating procedure, and Z is the set of all integers. In Equation 2.20, notice that the increments are all zeros except when p and q both take the values from the set $\{0,1\}$. For example, when $p = 0$ and $q = 0$, the pixel being considered is $\tilde{y}_1$ and the corresponding increment is ω_1. Based on the above discussion, the GPVE algorithm for the 3D case can now be presented in terms of a more general kernel function.

To be a valid kernel function for our application, two conditions are imposed. They are:

1. $f(x) \geq 0,$ where x is a real number $\tag{2.21}$

2. $\displaystyle\sum_{n=-\infty}^{\infty} f(n - \Delta) = 1,$ where n is an integer, $0 \leq \Delta < 1$ $\tag{2.22}$

For a valid kernel function f, the joint histogram h is updated in the following manner:

$$h(F(x_i, x_j, x_k), R(y_i + p, y_j + q, y_k + r)) +$$
$$= f(p - \Delta_i)f(q - \Delta_j)f(r - \Delta_k) \quad \forall p, q, r \in Z \qquad (2.23)$$

where f is referred to as the kernel function of GPVE. Notice that the contribution of each voxel involved toward the joint histogram is now represented in terms of the kernel functions along each direction. The first condition on f ensures that the contributions are nonnegative. Otherwise, it is possible to obtain a histogram with negative entries. The purpose of imposing the second condition is to make the sum of the updated amounts equal to one for each corresponding pair of points (x_i, x_j, x_k) in F and $(y_i + \Delta_i, y_j + \Delta_j, y_k + \Delta_k)$ in R.

From this generalization, we notice that the PVI algorithm proposed by Maes et al. is a special case of the GPVE algorithm. That is, if we elect to use a triangular function defined by Equation 2.19 as the kernel function f, Equation 2.23 reduces to Equation 2.16 in the 2D case. In Ref. 12, B-spline functions were suggested as the kernel functions f because they satisfy both the conditions in Equation 2.21 and Equation 2.22 and furthermore, they have finite support. Figure 2.5 shows the shapes of the first, second, and third-order B-splines. From the shapes of the B-spline functions shown in Figure 2.5, it should be understood that the kernel function introduced in Equation 2.21 and Equation 2.22 is an approximator rather than an interpolator because an interpolator has zero crossings at positions $0, \pm 1, \pm 2, \ldots$. The kernel function is introduced to assign a value to the contribution of each voxel involved in updating the joint histogram. For more details on B-spline functions, interested readers are referred to Refs. 22,23. It is interesting to note that the triangular function defined in Equation 2.19 is identical to the first-order B-spline function. In this case, the PVI algorithm is, in fact, equivalent to the first-order GPVE algorithm (using the first-order B-spline function as the kernel). In the GPVE algorithm, the kernel functions can be different along different directions. That is, we can rewrite Equation 2.23 as

$$h(F(x_i, x_j, x_k), R(y_i + p, y_j + q, y_k + r)) +$$
$$= f_1(p - \Delta_i)f_2(q - \Delta_j)f_3(r - \Delta_k) \quad \forall p, q, r \in Z \qquad (2.24)$$

where f_1, f_2, and f_3 can be different kernels. For example, if we know that the artifact is going to appear in the z-direction only, then we can choose both f_1 and f_2 as the first-order B-spline but choose f_3 as the third-order B-spline. This strategy was validated experimentally in Ref. 12.

V. OPTIMIZATION

Once we estimate the joint histogram that allows the computation of mutual information, the next step is to find the pose parameters that result in the maximal value of the mutual information measure. While there have been some efforts

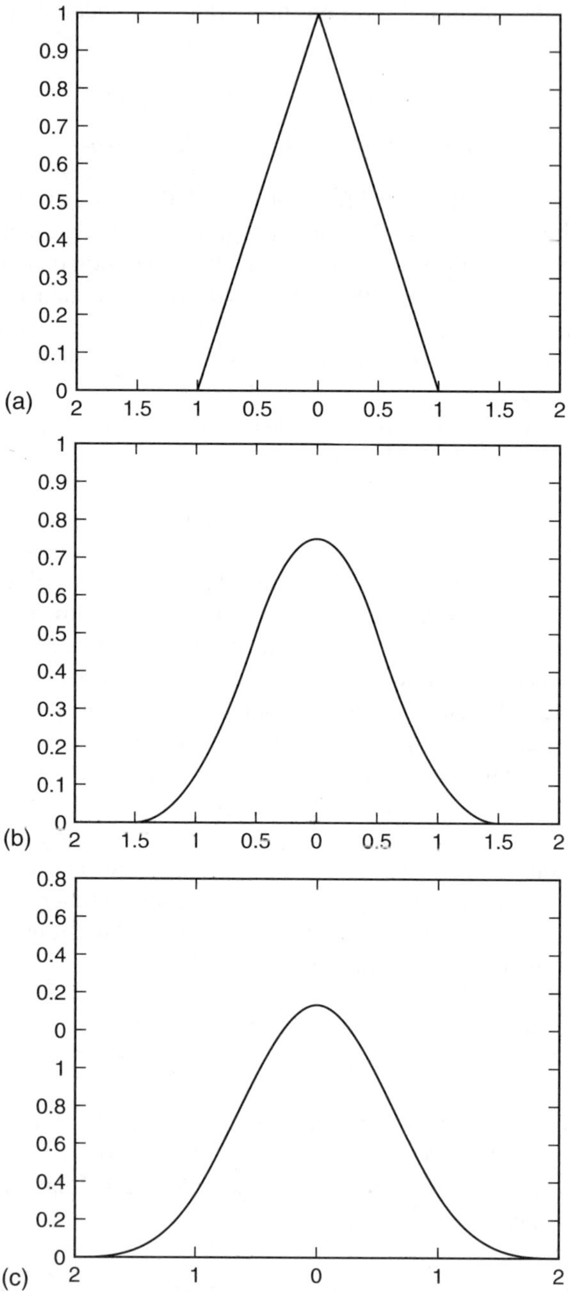

FIGURE 2.5 B-splines (a) first-order, (b) second-order, (c) third-order.

using global optimization algorithms like simulated annealing (SA)[26] and genetic algorithms (GA)[27] for MI based image registration, most of the work reported in the literature uses local optimization algorithms to maximize the MI measure of two images, with the assumption that the initial misregistration is not too large, which is usually true in most of the medical brain imaging applications. In Refs. 28–30, Thévenaz and Unser use the optimization method of Jeeves and a modified Marquardt–Levenberg optimizer to maximize the MI measure. In Ref. 31, Pluim et al. use Powell's method and in Ref. 24, a locally exhaustive search algorithm was used. In this section, we review another popularly used local optimization algorithm known as the simplex search algorithm[25] along with a widely used optimization strategy called multiresolution optimization. The simplex search algorithm together with multiresolution optimization is adopted in the experiments in this chapter.

A. SIMPLEX SEARCH ALGORITHM

The simplex method is a method for minimizing a function of n variables. A simplex is a geometrical figure which is formed by $N + 1$ points in an N dimensional space. An example of the simplex in three dimensions is given in Figure 2.6. The first step of this method uses a set of points to form an initial simplex. At each subsequent step in the algorithm, a new simplex is created according to three basic operations: reflection, contraction, and expansion. The definitions of these three operations are

$$P_{\text{ref}} = \bar{P} + \alpha(\bar{P} - P_h), \ \alpha > 0 \tag{2.25}$$

$$P_{\text{con}} = \bar{P} - \beta(\bar{P} - P_h), \ 1 > \beta > 0 \tag{2.26}$$

$$P_{\text{exp}} = \bar{P} + \gamma(\bar{P} - P_h), \ \gamma > \alpha \tag{2.27}$$

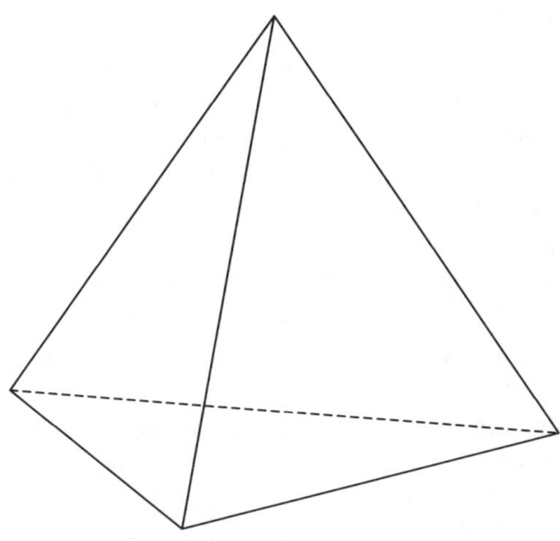

FIGURE 2.6 A simplex in three-dimensional space.

The constants α, β, and γ are referred to as the reflection coefficient, contraction coefficient, and expansion coefficient, respectively. P_h is the vertex of interest and $\bar{P}$ is the centroid of the vertices except for the vertex P_h. With these three basic operations, the simplex method can successfully reach the local optimum without resorting to the derivatives.

Let us consider the minimization of a function of n variables. Let $P_0, P_1, \ldots, P_N$ be the $(N + 1)$ points in an N dimensional space defining the initial simplex. Let y_i denote the function value at P_i, and we define $y_h = \max_i (y_i)$ and $y_l = \min_i (y_i)$. So P_h is the point resulting in the largest function value and P_l is the point resulting in the smallest function value within the $(N + 1)$ points. Further $\bar{P}$ is defined as the centroid of the points with $i \neq h$. At each step in the process, P_h is replaced by a new point. If y_{ref} is between y_h and y_l then P_h is replaced by P_{ref}. If $y_{\mathrm{ref}} < y_l$, that is reflection has produced a new minimum, then we check whether $y_{\mathrm{exp}} < y_l$. If $y_{\mathrm{exp}} < y_l$, we replace P_h by P_{exp}, otherwise, we replace P_h by P_{ref}. If $y_{\mathrm{ref}} > y_i$ for all $i \neq h$, that is, replacing P_h by P_{ref} leaves P_{ref} the maximum, then P_h is replaced by P_{con}. Finally, if $y_{\mathrm{con}} > y_h$ then we replace all P_i's by $(P_i + P_l)/2$ and restart the process. The whole process is terminated when $(y_h - y_l)$ is less than a predetermined threshold or the final simplex is smaller than a predetermined size.

B. MULTIRESOLUTION OPTIMIZATION

A local optimizer is often used in conjunction with a coarse-to-fine/multiresolution optimization strategy. This coarse-to-fine strategy has been used to improve the efficiency of many image-processing tasks,[32] including image registration.[28–31] There are at least two advantages of employing this scheme. The first one is the acceleration of the optimization process,[31] and the second one is the increase in the capture range.[28] Here the capture range is defined as the range of transformations within which a specified similarity measure is a monotonic function of misregistration.[7] The basic idea behind the multiresolution optimization strategy is that, first pose parameter estimation is performed at the coarsest resolution level of image pyramids constructed from the two images/volumes to be registered. After this first estimation, we switch to the next finer resolution level and use the previously obtained pose parameters as initial conditions to carry out the optimization at this new level. The same procedure is repeated until the finest level is reached. Care needs to be taken when employing the pose parameters obtained from the previous level as the initial conditions at the current resolution level in that the displacements must be doubled if the pixel sizes are used as the units for displacements. The reason is that the pixel sizes are reduced by a factor of two when we switch to the next finer resolution level.

The core of this multiresolution optimization approach is the construction of the image pyramid, which is a series of images of different resolutions. There are many image pyramid construction algorithms in the literature. Existing algorithms include Burt's Laplacian pyramid,[33] wavelet based pyramids,[34] cubic spline pyramids,[32] and the pyramid obtained by down sampling the image.[31]

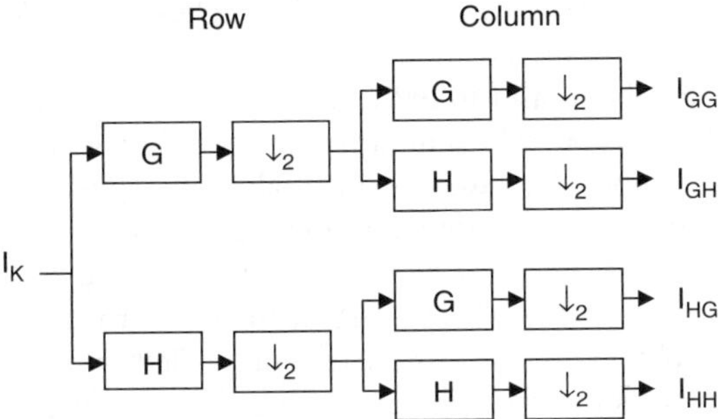

FIGURE 2.7 One level wavelet decomposition from resolution K to $K + 1$. I_{K+1} equals I_{HH}.

In our experiment, we employ the Haar wavelet decomposition approach to construct an image pyramid for optimization because of its simplicity. General 2D wavelet based multiresolution pyramid decomposition is performed using filters H (low pass) and G (high pass), which are quadrature mirror filters. The filtering is performed first by convolving the input image with H and G in the horizontal direction. This is followed by down sampling each output along the rows. Then the two resulting images are further processed along the vertical direction followed by down sampling along the columns. At the output, the source image at resolution level K is decomposed into four subimages: an image at coarser resolution level $K + 1$, a horizontally oriented image, a vertically oriented image, and a diagonally oriented image. The filtering can be repeated by using the coarser resolution image as the input source image at the next level until the desired level is reached. For our application, only the coarser resolution subimage is needed. In this case, we can repeatedly apply the low pass filter H horizontally and vertically to obtain the whole pyramid. Figure 2.7 summarizes this procedure to obtain the image at a coarser level. In Figure 2.7, I_{HH} is the image of the next coarser resolution level.

VI. APPLICATION TO 3D BRAIN IMAGE REGISTRATION

The performances of the first, second, and third-order GPVE algorithms for MI based image registration were evaluated and compared through the 3D brain CT to MR registration using clinical image data furnished by Vanderbilt University in Ref. 12 as a part of the Retrospective Image Registration Evaluation Project.[16] It was concluded in Ref. 12 that in cases where the registration accuracy is clearly affected by the interpolation artifacts, the use of a higher-order GPVE algorithm can effectively improve the registration accuracy. Otherwise, the performances of the higher-order (second-order or third-order) GPVE are about the same as the first-order GPVE, which is actually the PVI algorithm. To determine whether interpolation artifacts affect the registration accuracy, a measure called peak-shift (PS) was utilized in Ref. 12. It was defined as the distance between the position of the highest peak determined in the presence of artifacts and the position of the maximum value determined by the interpolated curve. This is illustrated in Figure 2.8. Curves in Figure 2.8 were obtained by plotting the MI similarity measure (calculated through PVI algorithm) as a function of the displacement along the axis in which the artifact pattern was expected. For example, if the sample spacings for both image volumes are the same along the z-axis, then the curves are obtained by plotting the MI measure as a function of the z-displacement after registration. Observing the curves thus obtained, we can determine whether the position of the maximum MI value is the result of the artifacts or not. In Figure 2.8, the first curve indicates that the position of the maximum is irrelevant to the artifacts because no artifact pattern is observed. However, Figure 2.8(b) and (c) clearly show the artifact patterns. In this case, we can draw a smooth curve through interpolation and the distance between the positions of the peaks of the interpolated smooth curve, and the original curve is defined as the PS measure. The value of the PS measure is used as a quantitative indicator to determine the influence. Figure 2.8(b) shows the case where this influence is insignificant, while in Figure 2.8(c) this influence is significant. In this manner, 4 out of 21 cases were identified as cases whose registration results were adversely affected by artifacts resulting from the PVI algorithm using the Vanderbilt furnished CT and MR image data in Ref. 12. Table 2.1 shows the registration errors using the PVI, the second-order GPVE, and the third-order GPVE algorithms for these cases. The improvement in registration accuracy is clearly demonstrated in Table 2.1. Figure 2.9 shows some registered image slices for CT-MR T2 and PET-MR T1 pairs using the second-order GPVE algorithm introduced in this chapter. More extensive experimental results are available in Ref. 12.

VII. SUMMARY

Mutual information based image registration has become an important tool for medical imaging applications. In this chapter, we have introduced the rationale of the maximization of the mutual information criterion and reviewed two widely used joint histogram estimation methods for MI calculation, namely linear interpolation and PVI, as well as a simple optimization method to actually maximize the MI measure. In many instances of brain image registration tasks, the two joint histogram estimation methods suffer from a phenomenon called interpolation induced artifacts that hampers the optimization process and influence registration accuracy. For this reason, we included a new joint histogram estimation scheme that was specifically designed for use with the MI based registration algorithm to overcome the artifact problem in this chapter. This method employs higher-order B-spline kernels during the estimation of the joint histogram. The objective is to reduce the abrupt change of the joint histogram dispersion caused by the PVI method. It is also the reason that it is sufficient to apply a higher-order kernel along the z-direction only for the

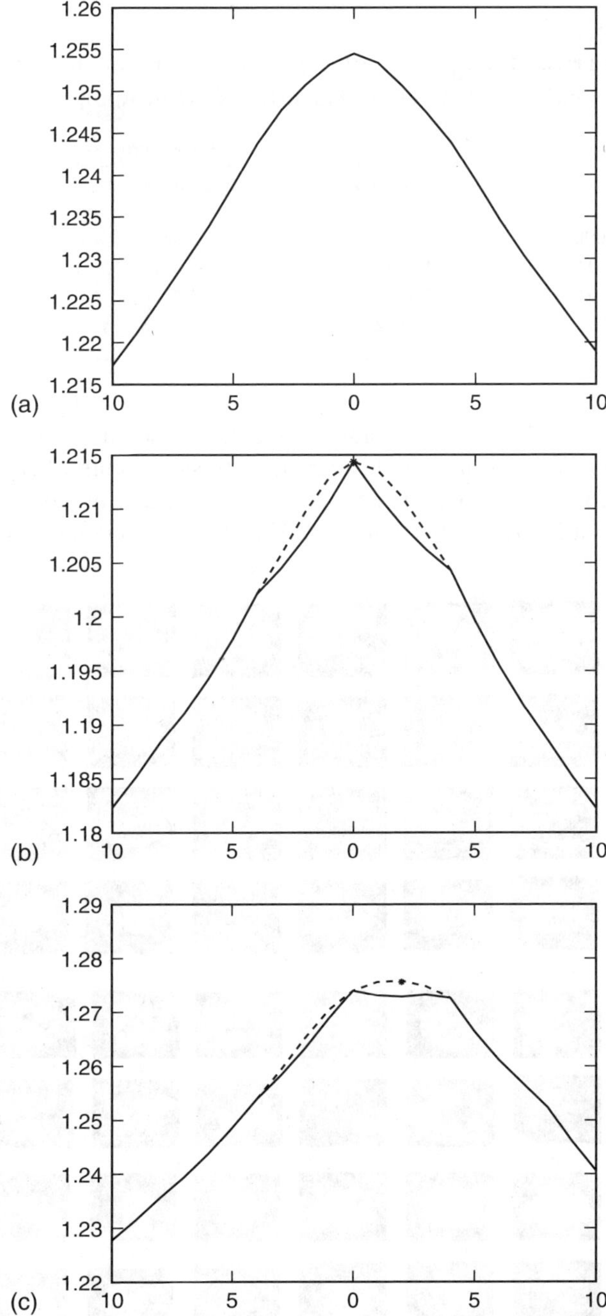

FIGURE 2.8 Illustrations of the PS measure utilized in the experiments (a) no PS measure is defined because of the absence of artifacts, (b) insignificant PS measure, and (c) significant PS measure.

TABLE 2.1
Registration Errors Using GPVE of Different Orders for the Identified Cases from Vanderbilt University Furnished CT-MR Image Data

CT-PD	Error (mm)		
	First-Order	Second-Order	Third-Order
CT-PD patient_006	1.5194	0.5897	0.5851
CT-T1 patient_002	1.2745	0.8215	0.8475
CT-T1 patient_004	2.9410	1.7740	1.8078
CT-T2 patient_006	1.2842	0.6738	0.6729

Vanderbilt University furnished brain image data, since it is the z-direction in which abrupt change of the joint histogram dispersion occurs when PVI is used. The effectiveness of the GPVE approach was evaluated by applying it to the brain image data provided by Vanderbilt University for brain CT to MR image

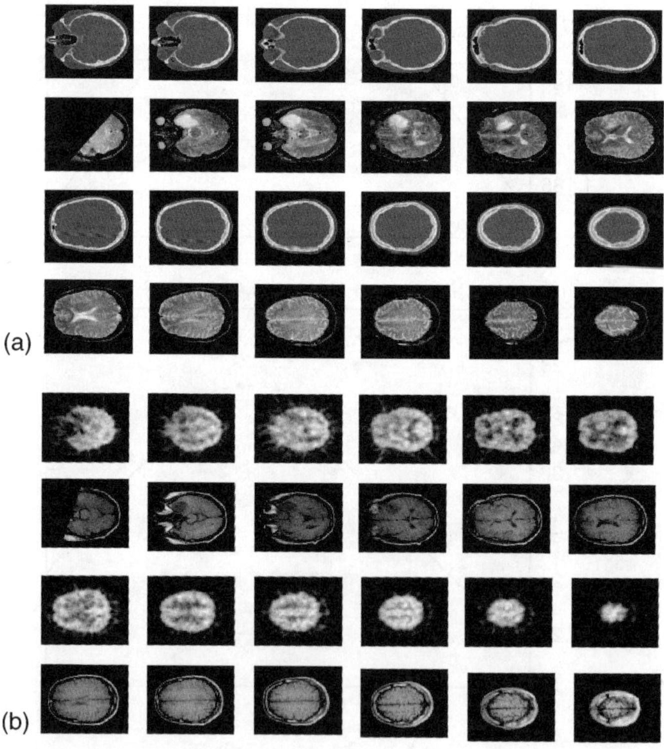

FIGURE 2.9 Examples of registered image slices (a) CT-MR T2 registration, (b) PET-MR T1 registration.

registration application. Experimental results in terms of registered image slices for CT-MR and PET-MR registrations were also provided in this chapter.

REFERENCES

1. Maguire, G. Q. Jr., Noz, M. E., Rusinek, H., Jaeger, J., Kramer, E. L., Sanger, J. J., and Smith, G., Graphics applied to medical image registration, *IEEE Comput. Graph. Appl.*, 11(2), 20–28, 1991, March.
2. Zitová, B., and Flusser, J., Image registration methods: a survey, *Image Vis. Comput.*, 21, 977–1000, 2003.
3. Brown, L. G., A survey of image registration techniques, *ACM Comput. Surv.*, 24(4), 325–376, 1992, December.
4. Chen, H., and Varshney, P. K., Automatic two stage IR and MMW image registration algorithm for concealed weapon detection, *IEE Proc. Vis. Image Signal Process.*, 148(4), 209–216, 2001, August.
5. Chen, H., Varshney, P. K., and Arora, M. K., Mutual information based image registration for remote sensing data, *Int. J. Remote Sen.*, 24(18), 3701–3706, 2003.
6. Chen, H., Varshney, P. K., and Arora, M. K., Performance of mutual information similarity measure for registration of multitemporal remote sensing images, *IEEE Trans. Geosci. Remote Sens.*, 41(11), 2445–2454, 2003.
7. Studholme, C., Hill, D. L. G., and Hawkes, D. J., Automated three-dimensional registration of magnetic resonance and positron emission tomography brain images by multiresolution optimization of voxel similarity measures, *Med. Phys.*, 24(1), 25–35, 1997, January.
8. Holden, M., Hill, D. L. G., Denton, E. R. E., Jarosz, J. M., Cox, T. C. S., Rohlfing, T., Goodey, J., and Hawkes, D. J., Voxel similarity measures for 3-D serial MR brain image registration, *IEEE Trans. Med. Imaging*, 19(2), 94–102, 2000.
9. Pluim, J. P. W., Maintz, J. B. A., and Viergever, M. A., Mutual-information-based registration of medical images: a survey, *IEEE Trans. Med. Imaging*, 22(8), 986–1004, 2003, August.
10. Maes, F., Vandermeulen, D., and Suetens, P., Medical image registration using mutual information, *Proc. IEEE*, 91(10), 1699–1722, 2003, October.
11. Pluim, J. P. W., Maintz, J. B. A., and Viergever, M. A., Interpolation artefacts in mutual information-based image registration, *Comput. Vis. Image Understanding*, 77, 211–232, 2000.
12. Chen, H., and Varshney, P. K., Mutual information based CT-MR brain image registration using generalized partial volume joint histogram estimation, *IEEE Trans. Med. Imaging*, 22(9), 1111–1119, 2003.
13. Taso, J., Interpolation artifacts in multimodality image registration based on maximization of mutual information, *IEEE Trans. Med. Imaging*, 22(7), 854–864, 2003.
14. Ji, J., Pan, H., and Liang, Z., Further analysis of interpolation effects in mutual information-based image registration, *IEEE Trans. Med. Imaging*, 22(9), 1131–1140, 2003.
15. Maes, F., Collignon, A., Vandermeulen, D., Marchal, G., and Suetens, P., Multimodality image registration by maximization of mutual information, *IEEE Trans. Med. Imaging*, 16(2), 187–198, 1997.

16. http://www.vuse.vanderbilt.edu/~image/registration/.

17. Cover, T. M., and Thomas, J. A., *Elements of Information Theory*, Wiley, New York, 1991.

18. http://www.bic.mni.mcgill.ca/brainweb/.

19. Collins, D. L., Zijdenbos, A. P., Kollokian, V., Sled, J. G., Kabani, N. J., Holms, C. J., and Evans, A. C., Design and construction of a realistic digital brain phantom, *IEEE Trans. Med. Imaging*, 17(3), 463–468, 1998.

20. Chen, H., and Varshney, P. K., Registration of multimodal brain images: some experimental results, pp. 122–133. In *Proceedings of SPIE Conference on Sensor Fusion: Architectures, Algorithms, and Applications VI*, Vol. 4731. Orlando, FL, 2002, April.

21. Keys, R. G., Cubic convolution interpolation for digital image processing, *IEEE Trans. Acoustics, Speech Signal Process.*, Assp-29(6), 1153–1160, 1981, December.

22. Unser, M., Aldroubi, A., and Eden, M., B-spline signal processing: part I-theory, *IEEE Trans. Signal Process.*, 41(2), 821–833, 1993.

23. Unser, M., Aldroubi, A., and Eden, M., B-spline signal processing: part II-efficient design, *IEEE Trans. Signal Process.*, 41(2), 834–848, 1993.

24. Studholme, C., Hill, D. L. G., and Hawkes, D. J., An overlap invariant entropy measure of 3D medical image alignment, *Pattern Recognit.*, 32, 71–86, 1999.

25. Nelder, J. A., and Mead, R., A simplex method for function minimization, *Comput. J.*, 7(4), 308–313, 1965.

26. Farsaii, B., and Sablauer, A., Global cost optimization in image registration using simulated annealing, pp. 117–125. In *Proceedings of SPIE Conference on Mathematical Modeling and Estimation Techniques in Computer Vision*, Vol. 3457. San Diego, California, 1998, July.

27. Matsopoulos, G. K., Mouravliansky, N. A., Delibasis, K. K., and Nikita, K. S., Automatic retinal image registration scheme using global optimization techniques, *IEEE Trans. Inf. Technol. Biomed.*, 3(1), 47–60, 1999, March.

28. Thévenaz, P., and Unser, M., A pyramid approach to sub-pixel image fusion based on mutual information, pp. 265–268. In *Proceedings of IEEE International Conference on Image Processing*, Vol. 1. Lausanne, Switzerland, 1996.

29. Thévenaz, P., and Unser, M., An efficient mutual information optimizer for multi-resolution image registration, pp. 833–837. In *Proceedings of IEEE International Conference on Image Processing*, Vol. I. Chicago IL, USA, 1998.

30. Thévenaz, P., and Unser, M., Optimization of mutual information for multi-resolution image registration, *IEEE Trans. Image Process.*, 9(12), 2083–2098, 2000.

31. Pluim, J. P. W., Maintz, J. B. A., and Viergever, M. A., A multiscale approach to mutual information matching, pp. 1334–1344. In *Proceedings of SPIE Conference on Image Processing*, Vol. 3338. San Diego, California, 1998.

32. Unser, M., Aldroubi, A., and Eden, M., The L2 polynomial spline pyramid, *IEEE Trans. Pattern Anal. Machine Intell.*, 15(4), 364–379, 1993, April.

33. Burt, P. J., and Adelson, E. H., The Laplacian pyramid as a compact code, *IEEE Trans. Commun.*, COM-31, 337–345, 1983, April.

34. Mallat, S. G., A theory of multiresolution signal decomposition: the wavelet representation, *IEEE Trans. Pattern Anal. Machine Intell.*, 11, 674–693, 1989.

3 Studies on Registration and Fusion of Retinal Images

France Laliberté and Langis Gagnon

CONTENTS

I. INTRODUCTION

This chapter aims to review recent work on the registration and fusion of human retinal images. While registration is an active topic, fusion of ophthalmic images is a relatively new application domain. There are two main reasons for the emerging interest. First, image acquisition technology is becoming fully digital with the commercial availability of nonmydriatic fundus cameras. This enables the development of software that can manipulate, enhance, and process images on the fly. Second, ophthalmology necessitates the manipulation of an increasing number of images due to the follow-up requirements of mass diseases like diabetic retinopathy. These follow-ups involve the manipulation of images acquired through different modalities (color and angiograms) over many years, and forms a large part of the ophthalmologist's workload. Automatic image manipulation procedures help to reduce that workload. For instance, registration of images help in following disease evolution or identifying the types of retinal lesions, while multimodal fusion allows the combination and enhancement of pathological information to facilitate diagnoses.

The chapter is divided into two main sections: registration and fusion. The principal retinal image registration methods used over the last 10 years are first reviewed. These are mainly correlation and point-matching methods. Some registration results are then presented. Secondly, works in retinal image fusion are reviewed. These mainly involve pixel-level fusion and can be classified in four types: linear, nonlinear, image pyramid, and wavelet transform. We describe these and present several comparative tests. Results are presented using a set of images covering a large ensemble of retinal diseases such as diabetic retinopathy, age-related macular degeneration, branch retinal vein occlusion, cytomegalo-virus retinitis, anterior ischemic optic neuropathy, choroidal neovascular membrane, cystoid macular edema, histoplasmosis retinitis/choroiditis, telan-giectasia, and so forth.

II. IMAGING MODALITIES

The eye (Figure 3.1) can be imaged under many modalities with various sensors. We focus on the imaging conditions of the eye fundus (Figure 3.2). There are two main diseases associated to the retina and choroid: diabetic retinopathy and age-related macular degeneration. Diabetes can cause a weakening of blood vessel bodies, in particular within the retina. The capillaries can leak and become potential hosts for microaneurisms (small bulges that develop in the capillary walls and are the first indicator of diabetic retinopathy). The development of new weak capillaries (neovascularization), which leak easily, can also occur. An edema appears when fluids accumulate in the retina. These fluids are also responsible for exudates, which are metabolic waste products. Closure of capillaries is another possible change that may lead

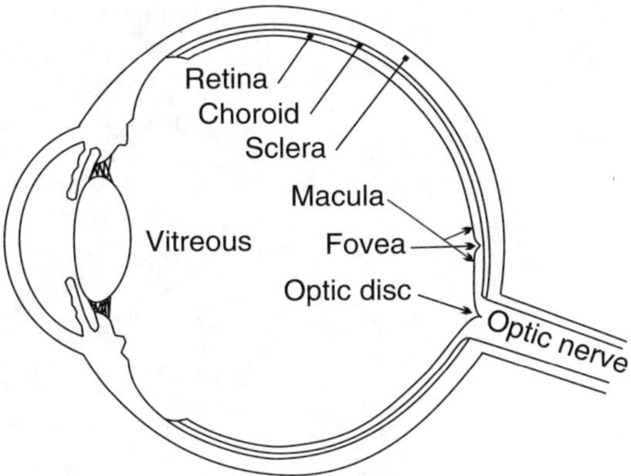

FIGURE 3.1 Diagram of a human eye.

to a lack of oxygen in the retina (ischemia). The cotton wool spots indicate the areas of oxygen-starved retina. Figure 3.3 shows the fundus of a normal eye with the principal anatomical components identified. Figure 3.4 shows the fundus of an eye with diabetic retinopathy on which some common lesions are identified. Choroidal neovascularization associated with age-related macular degeneration leads to irreversible damage in retinal tissue. Early and complete

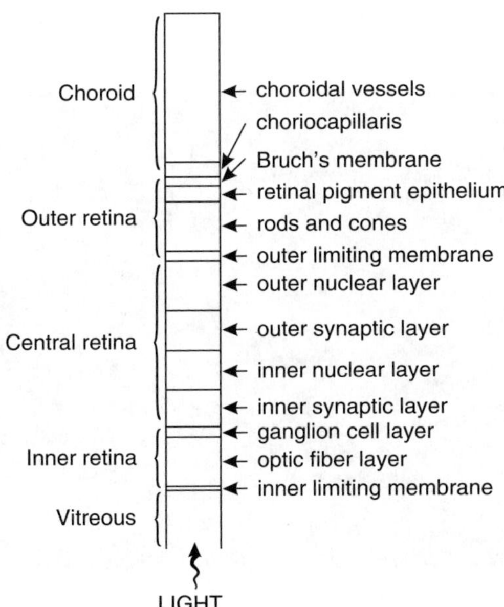

FIGURE 3.2 Diagram of the layers of the eye fundus (retina and choroid).

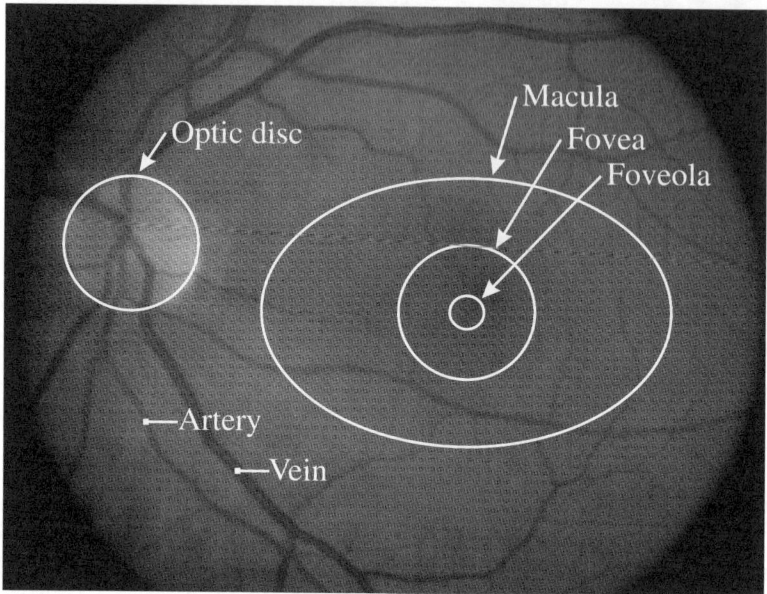

FIGURE 3.3 (See color insert following page 236) Image of a normal eye (color photography) with the principal anatomical components identified.

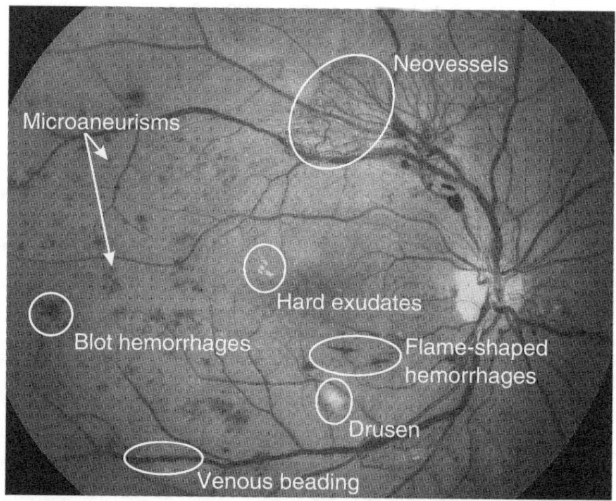

FIGURE 3.4 Image of an eye with diabetic retinopathy on which some common lesions are identified.

photocoagulation of the affected areas is the only treatment for delaying or preventing decreases in visual acuity.

A fundus camera is used to take photographs (color and red-free) and angiograms (fluorescein sodium and indocyanine green). A fundus camera can be mydriatic or nonmydriatic (i.e., one that does not require the use of mydriatic eye drops to dilate the pupil). Fundus cameras can be digital or film-based.

Photographs are taken under natural illumination with or without the use of a green filter (color or red-free photography). Red-free images enhance the visibility of the vessels, hemorrhages, drusens, and exudates, while color images are useful in detecting exudates and hemorrhages associated with diabetic retinopathy.

Angiograms are taken after the injection of a dye in the patient's arm to enhance the appearance of blood vessels, under a specific illumination and filter. The dye has an absorption peak, which determines the illumination light, and an emission peak, which determines the filter that retains only the fluorescent contribution. Fluorescein has absorption and emission peaks at 493 nm (blue) and 525 nm (green), respectively, and is used to image the retina. Images are acquired from 10 sec to 15 min after the injection while the dye circulates through the retinal arteries, capillaries, veins, and is eliminated from the vasculature, while staining in the optic disc and lesions. Indocyanine-green, whose peaks are at 805 nm (infrared) and 835 nm (infrared), is used to image the choroid. Fluorescein angiograms are very different from the green band of color images: they have a better contrast and some features (vessels, microaneurisms) are intensity reversed. Fluorescein angiograms are useful in detecting leaks and other abnormalities in arteries and veins. They allow the detection of occluded and leaking capillaries, microaneurisms, macular edema, and neovascularization. Acquiring angiograms at different phases is useful in following the dye progression and staining, which indicate the vasculature and lesion state. Angiograms taken at different phases are also very different: vessels go from black to white and to black again eventually.

The scanning laser ophthalmoscope (SLO) is used to take fluorescein and indocyanine-green angiograms, measure scotomas, topography and tomography of structures, and so forth. It is a very low-light imaging device compared to the fundus camera because it uses a focused laser beam as illumination source, allowing the remaining pupil area to collect the light.[1] It provides a better resolution than the fundus camera. It is useful, in its infrared mode, to detect the choroidal neovessels.

Certain substances in the eye act naturally like a dye: they absorb and emit light. Autofluorescence is documented in the case of optic nerve head drusen and astrocytic hamartoma.[1] Autofluorescence images can be obtained with a fundus camera or a SLO. They are useful in seeing subretinal deposits, which cannot be seen by another method. Many retinal diseases are believed to be associated with an accumulation or absence of such deposits like lipofuscin, an age pigment.[2]

The optical coherence tomography scanner (OCT) is used to acquire cross-sectional images of the retina using the optical backscattering of light. It allows the measurement of the retina thickness and the differentiation of its anatomical layers. OCT images allow the documentation of diseases like diabetic retinopathy, macular holes, cystoid macula edema, central serous choroidopathy, and optic disc pits.[3]

III. REGISTRATION

Image registration is a necessary step in (1) producing mosaics that display a large area of the retina, (2) combining segmented features on one image, or (3) performing pixel-level image fusion. Registration allows compensation for image distortion due to changes in viewpoint. The type of distortion determines the type of transformation that must be used to register the images.[4]

A. TRANSFORMATION TYPES

Retinal image distortions come from different sources: change in patient sitting position (large horizontal translation), change in chin cup position (smaller vertical translation), head tilting and ocular torsion (rotation), distance change between the eye and the camera (scaling), three-dimensional retinal surface (spherical distortion), and inherent aberrations of the eye and camera optical systems. Seven different transformation types are used in the literature to register retinal images (see Table 3.1):

1. x and y translations,
2. Rigid body (x and y translations and rotation),
3. Similarity (x and y translations, rotation, and uniform scaling),
4. Affine,
5. Bilinear,
6. Projective,
7. Second-order polynomial.

The similarity transformation has the following form

$$\begin{pmatrix} x \\ y \end{pmatrix} = s \begin{pmatrix} \cos\theta & -\sin\theta \\ \sin\theta & \cos\theta \end{pmatrix} \begin{pmatrix} x_0 \\ y_0 \end{pmatrix} + \begin{pmatrix} t_x \\ t_y \end{pmatrix} \tag{3.1}$$

where s, θ, t_x, t_y are the scaling, rotation, and x and y translation parameters. The original point co-ordinates are identified by the zero subscript. Since there are four parameters, two pairs of matches are required to solve the system. Unlike the other transformation types, the parameters of the similarity transformation

TABLE 3.1
Registration Methods, Which are Used in Ophthalmology, Defined in Terms of Feature Space, Similarity Measure, Search Space, and Search Strategy

Authors	Feature Space	Similarity Measure	Search Space	Search Strategy	Image Modality
Peli[7]	Vessels in a template	Sum of the normalized absolute values of the differences	x and y translations	Exhaustive hierarchical	Temporal (3 years) C; temporal (2 s) FS; SLO video
Lee[16]	Power spectrum and power cepstrum	Sum of the absolute values of the differences	Rigid body	Hierarchical	Not mentioned *optic disc*
Yu[8]	Vessels in a template	Sum of the absolute values of the differences	x and y translations	Hierarchical	Red-free
Jasiobedzki[24]	Vessels represented as active contours and nodes	Minimization of the contour energy	Not mentioned	Active contours	SLO FS
Hart[25]	End of vessel segments	Correlation coefficient and empirical criteria based on the knowledge of the transformation	Affine		Temporal C
Mendonca[26]	Edges represented as two matrices by image	Number of edge points with a correspondent	x and y translations	Spiral scan	Temporal FS

Continued

TABLE 3.1
Continued

Authors	Feature Space	Similarity Measure	Search Space	Search Strategy	Image Modality
Cideciyan[6]	Fourier amplitude in the log–polar domain	Cross-correlation	Similarity	Exhaustive	Not mentioned
Berger[20]	Edges	Partial Hausdorff distance	Similarity	Exhaustive	C–FS; C–C
Park[12]	Vessel contour points	Hough transform + sum of the absolute values of the differences	Rigid body	Hierarchical	Video
Pinz[19]	Monopixel binary retinal network	Number of overlap after the transformation	Similarity	Exhaustive	Many modalities of SLO
Matsopoulos[22]	Binary retinal network	Objective function (measure of match)	Affine; bilinear; projective	Simulated annealing; genetic algorithms	FS–red-free; ICG–red-free
Ritter[10]	Intensity	Mutual information	Similarity	Simulated annealing	Temporal (3 years) C; stereo C *optic disc*
Zana[27]	Bifurcation points with vessel orientations	Angle-based invariant	Affine	Bayesian Hough transform	C–FS; temporal (2 years) FS
Lloret[13]	Vessels	Cross-power spectrum	Similarity	Simplex algorithm	Stereo FS; temporal (1 year) FS; red-free–FS; SLO sequence

Corona[17]	Vessels	Fourier spectrum + power cepstrum + correlation coefficient	Rigid body	Hierarchical	FS video
Heneghan[5]	Bifurcation points		Bilinear; second order polynomial	Expectation maximization algorithm	C–FS; temporal (13 days) C
Rosin[18]	Intensity	Windowed mutual information	Similarity	Exhaustive multiscale	C–SLO *optic disc*
Laliberté[32]	Bifurcation points with vessel orientation	Distance, number and angle of surrounding vessels	Similarity; affine; second order polynomial	Relaxation	C–FS; NMC–C; NMC–FS; FS sequence
Raman[9]	Power spectrum and power cepstrum	Sum of the absolute values of the differences	Rigid body		FS video; C–FS
Raman[9]	Intensity	Correlation coefficient	Not mentioned	Exhaustive	FS video
Shen [29]	Bifurcation points with vessel orientations		Similarity; affine; second order polynomial		Not mentioned

The last column indicates the imaging modalities used (FS: fluorescein sodium angiogram, ICG: indocyanine green angiogram, C: color photography, NM: nonmydriatic, SLO: scanning laser ophthalmoscope).

cannot be determined by least-mean-squares. They are usually obtained by computing all the possible transformations from two pairs of matches, given by $0.5(M^2 - M)(N^2 - N)$ where M and N are, respectively, the number of points in image one and two, and taking the median or the mean of the parameters.

The affine transformation is the same as the similarity transformation but it allows a different scaling in the x and y directions and a shearing. This transformation maps straight lines to straight lines, preserves parallelism between lines, and is defined by

$$\begin{pmatrix} x \\ y \end{pmatrix} = \begin{pmatrix} a_0 & a_1 & a_2 \\ b_0 & b_1 & b_2 \end{pmatrix} \begin{pmatrix} 1 \\ x_0 \\ y_0 \end{pmatrix} \tag{3.2}$$

The bilinear transformation maps straight lines to curves. It is the simplest polynomial transformation and is given by

$$\begin{pmatrix} x \\ y \end{pmatrix} = \begin{pmatrix} a_0 & a_1 & a_2 & a_3 \\ b_0 & b_1 & b_2 & b_3 \end{pmatrix} \begin{pmatrix} 1 \\ x_0 \\ y_0 \\ x_0 y_0 \end{pmatrix} \tag{3.3}$$

The projective transformation maps straight lines to straight lines but does not preserve parallelism. It is defined by

$$\begin{pmatrix} u \\ v \\ h \end{pmatrix} = \begin{pmatrix} a_0 & a_1 & a_2 \\ b_0 & b_1 & b_2 \\ c_0 & c_1 & c_2 \end{pmatrix} \begin{pmatrix} x_0 \\ y_0 \\ 1 \end{pmatrix} \tag{3.4}$$

where $x = u/h$ and $y = v/h$.

The second-order polynomial transformation is given by

$$\begin{pmatrix} x \\ y \end{pmatrix} = \begin{pmatrix} a_0 & a_1 & a_2 & a_3 & a_4 & a_5 \\ b_0 & b_1 & b_2 & b_3 & b_4 & b_5 \end{pmatrix} \begin{pmatrix} 1 \\ x_0 \\ y_0 \\ x_0 y_0 \\ x_0^2 \\ y_0^2 \end{pmatrix} \tag{3.5}$$

The higher the order of the transformation type, the more flexible the mapping provided, but at the expense of an increased sensitivity to bad parameter estimation.[5]

B. REGISTRATION METHODS

Registration methods are characterized by a feature space, a similarity metric, a search space, and a search strategy.[4] The feature space and similarity metric are, respectively, the type of image information and the comparison criterion used for matching. The search space is the transformation type that can align the images. The search strategy is the procedure used to find the transformation parameters.

There are many registration methods. The ones used in ophthalmology are summarized in Table 3.1 and presented in more detail below. Since ophthalmic image deformations are mainly global, elastic model registration methods are unnecessary and have not been used.

In Ref. 6, fundus image registration methods are classified in two groups: local and global. In local methods the registration is performed on extracted features, while in global methods all the pixel intensities are used to find the best transformation. Unlike global methods, local ones are difficult to generalize in order to deal with different problems because they are based on application-dependent rules for the feature identification and correspondence. This classification is not the one used here; rather we organize the methods according to the classification of registration methods presented in Ref. 4.

1. Correlation and Sequential Methods

Registration methods using similarity metrics like the cross-correlation metric and the correlation coefficient metric, or metrics defined by the sum of the absolute values of the differences or the normalized sum of the absolute values of the differences, have been tested for unimodal ophthalmic image registration.[7-9] They are not affected by white noise but suffer from some limitations. First, they cannot deal with local intensity reversal in multimodal registration.[4,10] Second, if the correlation surface is smooth or the image is noisy, the correlation peak can be difficult to detect. Third, the computation cost increases with the window and search area sizes as well as registration accuracy. Finally, since it is impossible to distinguish between local and global maxima, there is no guarantee of finding the right solution.

a. Correlation Methods

The normalized cross-correlation metric measures the intensity similarity between a template T_1 extracted from an image I_1 and different parts of an image I_2 according to

$$C(t_x, t_y) = \frac{\sum_k \sum_l T_1(k, l) I_2(k + t_x, l + t_y)}{\sqrt{\sum_k \sum_l I_2^2(k + t_x, l + t_y)}} \tag{3.6}$$

where (t_x, t_y) are the translation parameters and (k, l) are pixel co-ordinates. The transformation with maximum cross-correlation indicates how the template must be registered to the image.

The correlation coefficient metric is defined by

$$D(t_x, t_y) = \frac{\sum_k \sum_l (T_1(k, l) - \mu_{T_1})(I_2(k + t_x, l + t_y) - \mu_{I_2})}{\sqrt{\sum_k \sum_l (I_2(k + t_x, l + t_y) - \mu_{I_2})^2 \sum_k \sum_l (T_1(k, l) - \mu_{T_1})^2}} \tag{3.7}$$

where μ_{I_2} is the mean of the window being analyzed in I_2 and μ_{T_1} is the mean of the template. This registration method was used in Ref. 9 to register a fluorescein angiogram sequence and compared to the Fourier registration method (Section III.B.2). The authors found that the latter performs better (less misalignment compared to a manual registration), and thus used it to register fluorescein angiograms and color images. In addition, it is not a good idea to use correlation coefficient to register fluorescein angiograms and color images since the modalities are different.

b. Sequential Methods

In Ref. 11, a more computationally efficient algorithm than correlation methods is introduced: the sequential similarity detection algorithm. It combines a new similarity measure and search strategy. The similarity measure is given by

$$S(t_x, t_y) = \sum_k \sum_l |I_2(k + t_x, l + t_y) - T_1(k, l)| \tag{3.8}$$

In its normalized version, Equation 3.8 becomes

$$S(t_x, t_y) = \sum_k \sum_l |(I_2(k + t_x, l + t_y) - \mu_{I_2}) - (T_1(k, l) - \mu_{T_1})| \tag{3.9}$$

The search strategy first consists in choosing a maximum threshold for S. Then, for each window of the search area, the number of pixels that are processed before reaching the threshold is recorded. The window for which the largest number of pixels is processed represents the optimal transformation. According to Ref. 10, this method is not robust to large intensity changes and white noise.

In Ref. 7, the similarity measure (Equation 3.9) is applied at carefully chosen feature points rather than to all pixels in the template. This is to reduce the computation time and limit the effect of intensity variation on the sequential similarity detection algorithm. A template in the image to be registered and a search area in the reference image are chosen manually. Adaptive thresholding is then used to detect vessels in the template. Some of these template vessel points are chosen manually. The similarity measure is applied at the position of the template vessel points. Also, a two-stage approach is used for template matching to further reduce the computation time. It consists in using two different threshold levels. The first threshold level is chosen low enough in order to reject the unlikely template positions in the search area (those for which the number of feature points that are processed, before reaching the threshold, is lower than a percentage of the total number of feature points). The value of the first threshold level is multiplied by a constant, which gives the value of the second threshold level. The feature-based sequential similarity detection algorithm is applied to further process the likely template positions.

In Ref. 8, the optic disc is first detected in the source images (the method used is not mentioned). The image to be registered is then divided into a number of subimages. The blood vessel bifurcation points are detected in the subimages until a subimage with a sufficient number of bifurcation points is found. This subimage is chosen as the template. The relative position between the optic disc and the bifurcation points is used to define the search area. Equation 3.8 is used as a similarity measure. The method was only tested on artificial image pairs (one of the images is transformed to create the other image of the pair).

In Ref. 12, the similarity measure (Equation 3.8) is applied at specific feature points like in Ref. 7. Mathematical morphology is used to detect the retinal vessels and their contours. First, the original image is opened and closed with a small square structuring element to eliminate noise and small features. The original image is also closed with a large square structuring element. Second, the noise free image is subtracted from the closed image to obtain an image where the intensity of a pixel indicates a degree of belonging to a retinal vessel. Third, the resultant image is thresholded and edge detected to extract the vessel contours. Finally, the vessel contour points are subsampled to further reduce the number of points to be processed by the sequential similarity detection algorithm. A computationally efficient template matching approach is proposed: a Hough-based matching method is used to obtain a coarse registration, followed by a feature-based sequential similarity detection algorithm to refine the registration. The parameter space used for the Hough-based matching method is composed of rotation and translations and is quantized. Scaling can be added, but it was not required for this work since the images to register came from a video sequence where they were separated by 1/60 sec. For the translation, the same quantization as for the subsampling of the feature points is used. For the rotation, the parameter space ranges from $-8°$ to $8°$ and is quantized at every degree. The rotation is the first transformation parameter to be estimated between the two feature point sets P and P'. The points in P are rotated and the translation parameters are estimated for each angle. The translations for all possible point pairs between P and P' are computed and the corresponding cell in the parameter space is increased by one to estimate the translation parameters. The parameter space cell with the maximum value provides a rough estimate of the transformation parameters $(\theta_e, t_{xe}, t_{ye})$. The feature-based sequential similarity detection algorithm is applied over a small search area around the rough estimate of the transformation parameters $(\theta_e \pm 1°, t_{xe} \pm 5, t_{ye} \pm 5)$ in order to refine the registration. The advantages of using the Hough transform, besides the gain in computation time, are its robustness and resistance to noise. Unfortunately, the effectiveness of the algorithm in detecting the rotation transformation parameter has only been tested on an artificial image sequence.

2. Fourier Methods

Fourier methods register images by exploiting the fact that translation, rotation, and scale have corresponding representations in the frequency domain. Fourier methods are appropriate for images having intensity variations due to illumination change. They also achieve better robustness, accuracy, and computation time than correlation and sequential methods. However, since they rely on their invariant properties, they are restricted to certain transformation types such as translation, rotation, and scaling.[4]

Phase correlation is a Fourier-based technique that can be used to register shifted images. Two shifted images have the same Fourier amplitude and a phase difference proportional to the translation. The cross-power spectrum (Fourier transform of the cross-correlation function) of the two images is computed, and its phase is transformed back in the spatial domain to obtain the translation. Rotation and scale can be taken into account but only at the expense of computation time or robustness.

The phase correlation technique is used in Ref. 13. First, the vessels are extracted using the level set extrinsic curvature (MLSEC-ST) operator,[14] which detects ridges and valleys. The ridges and valleys are thresholded to reduce the number of pixels to be transformed in the image to be registered. Then, one of the images is rotated and the best translation, which is found with the phase correlation technique, for each rotation is kept. These rotation and translation parameters provide initial estimates to be optimized by the iterative Simplex algorithm.[15] Scaling is added in a final additional step.

Another spectral-based method is developed in Ref. 16 for the early detection of glaucoma and is tested on one image pair of the optic disc. The registration is realized using the power cepstrum (power spectrum of the logarithm of the power spectrum of an image) technique. The rotation is obtained from the angle that minimizes the sum of the absolute values of the difference between the power spectra (i.e., the Fourier transforms of the autocorrelation functions) of the images. In absence of rotation and for identical images, this difference is zero. The translation is obtained from the power cepstrum, which is given by

$$P[J(x,y)] = |F(\ln\{|F[J(x,y)]|^2\})|^2 = P[I(x,y)] + A\delta(x,y)$$
$$+ B\delta(x \pm x_0, y \pm y_0) + C\delta(x \pm 2x_0, y \pm 2y_0) + \cdots \tag{3.10}$$

where F represents the Fourier transform operation, $I(x,y)$ the reference image, $I(x + x_0, y + y_0)$ the image to be registered, $J(x,y)$ the sum of these two images, and A, B, and C are constants. The cepstrum of the reference image is subtracted from the one of the image to be registered. The resulting cepstrum is composed of deltas (see Equation 3.10) from which a subset is chosen. These deltas represent the translation (or a multiple integer of it) for each corresponding region in the images. This registration method is also used in Ref. 9.

In Ref. 17, the following procedure is used to extract vessels before performing the registration. First, the histograms of the images are computed and the histogram with the lowest brightness is matched against the other histogram to ensure that similar features have similar contrast in the image pair. Second, each image is subtracted from its Gaussian blurred version to obtain an edge enhanced grayscale image. Third, these images are binarized using the maximum value of the histogram as threshold. Finally, the images are median filtered to eliminate impulse noise. The Fourier spectra of these extracted vessel images are used to find the rotation parameter. The spectrum of the image to be registered is rotated, with steps smaller than $1°$, and cross-correlated with the spectrum of the reference image. The search for the rotation parameter uses a coarse-to-fine strategy. In the coarse stage, the rotated image is obtained through an inverse transformation with each output pixel being the nearest neighbor of the computed pixel. In the fine stage, the image is rotated around the angle corresponding to the highest correlation coefficient obtained in the coarse stage and each output pixel is obtained through bilinear interpolation. The angle that gives the highest correlation coefficient corresponds to the rotation parameter. The power cepstrum technique[16] is used to find the translation parameters, with the following extension to find the right delta among those of the cepstrum (see Equation 3.10). The correlation coefficient is computed between the reference image and the image to be registered, shifted by the translation corresponding to each delta. The delta giving the highest correlation coefficient corresponds to the translation parameters.

In Ref. 6, images are assumed to be related by a similarity transformation and the cross-correlation of a triple invariant image descriptor is used. This descriptor is the log−polar transform (polar mapping followed by a logarithmic transformation of the r-axis) of the Fourier amplitudes. It removes translation and converts rotation and scaling into independent shifts in orthogonal directions. The rotation and scaling are determined by computing the cross-correlation of the descriptor. The location of the peak gives the difference in rotation and scale between the source images. These parameters are applied to the image to be registered and the translation is found by cross-correlation.

This algorithm was tested on three images/patient for three patients. The image modality and disease are not mentioned.

3. Feature-Based Methods

In feature-based methods, transformation parameters are found from correspondences between features extracted from the source images (as opposed to point mapping methods where only isolated points are used to establish the correspondences). A drawback of these methods is that good and reliable feature extraction is required, which is often difficult to achieve, particularly in medical imaging.[18] However, since features are used for the correspondence, these methods can cope with intensity variations between images of different modalities or different stage angiograms.

In Ref. 19, a feature-based registration algorithm using extracted vessel segments from SLO images is presented. The vessels are modeled as dark or bright tubes (single or triple). They are modeled as triple tubes, by combining three parallel single tubes, when the vessels have a visible vascular wall. A visible vascular wall results in an intensity profile going from bright to dark to bright again (on angiograms where vessels are usually bright). First, the significant edgels (edge elements) along the vessel boundary are extracted to identify the vessels. Second, they are paired in cross-sections according to specified criteria. Third, the triple tubes are identified and the cross-sections modified accordingly. Fourth, the cross-sections are joined to form the vessel centerline. Finally, short segments of the vessel centerline are connected to long ones. The reference image is the infrared reflection image and the other three modalities (argon-blue reflection, fluorescein, and static scotometry images) are registered to it using a similarity transformation. The algorithm has three steps:

1. Sample points are selected from the extracted vessels of the image to be registered to reduce the computation cost,
2. A distance image $d(x, y)$ is computed, from the extracted vessels of the reference image, using an exponentially decreasing distance function,
3. The transformation parameters are computed by optimizing an objective function from different seed points,

$$E(\vec{P}) = \sum_{i=1}^{N} d(x_i(\vec{P}), y_i(\vec{P})) \qquad (3.11)$$

where $\vec{P} = (s, \theta, t_x, t_y)$, N is the number of sample points, $(x_i(\vec{P}), y_i(\vec{P}))$ are the co-ordinates of the transformed sample points, and $d(x_i(\vec{P}), y_i(\vec{P}))$ is the pixel value at position $(x_i(\vec{P}), y_i(\vec{P}))$ on the distance image. The vessels have to be well distributed over the retina to provide a good global maximum. The algorithm has two parameters: the number of sample points given by

$$N = \frac{\text{TubeLength}_{\text{reference}} + \text{TubeLength}_{\text{other}}}{10} \qquad (3.12)$$

where TubeLength$_{reference}$ and TubeLength$_{other}$ are the total lengths of the tubes in the reference image and the one to be registered, and the number of seed points, which is set empirically. The advantage of this method is that vessel extraction takes into account the fact that vessels can have visible vascular walls. However, the method has three drawbacks. First, the range of possible transformation parameters is limited by the method used and must be provided by the user and implemented as penalty terms in the objective function evaluation. Second, the image to be registered and the reference image must be sorted in a way that ensures that $s < 1$. Third, the seed point strategy for the optimization is not guaranteed to converge to a global optimum.

In Ref. 20, the goal is to construct a mosaic from partially overlapping images and to overlay this mosaic on real time captured images. The images are smoothed and the edges are detected with the Canny edge detector to create the mosaic. The edges are then thresholded to obtain a binary image. An extension of the partial Hausdorff distance[21] is used to register the images. Once the images are registered, the overlapping parts are "fused" to create the mosaic. Since fundus images usually have better brightness and contrast in the central region, the combination favors the pixels that are near the image center. The pixels are combined according to

$$I(x, y) = \frac{\sum_{i=1}^{N} (d_i^2(x, y) \times I_i(x, y))}{\sum_{i=1}^{N} d_i^2(x, y)} \tag{3.13}$$

where $I(x, y)$ is the mosaic, N the number of images, $d_i(x, y)$ the distance between the pixel $(x, y) \in I_i(x, y)$ and the nearest border of the ith image. With the use of the Hausdorff distance, the method can cope with outliers and can be used for any transformation type.

In Ref. 22, three transformation types (affine, bilinear, and projective) and two global optimization methods (simulated annealing and genetic algorithms) are compared in terms of accuracy and efficiency. Motivation for this work comes from the following. During laser treatment, surgeons observe red-free images. Because sensitive anatomical areas (fovea) and lesions are more visible on fluorescein or indocyanine green angiograms, registration of these modalities with red-free images allows boundaries traced by the ophthalmologist on angiograms to be automatically superimposed on the red-free images used during the operation. A sample of 26 pairs of red-free/fluorescein and red-free/ indocyanine green images were registered, with 18 of these pairs taken from normal eyes. The vessels are extracted and used for the registration. Red-free images are first intensity reversed so that their vessels appear bright as in angiograms. Second, the image is opened with a disk-shaped structuring element of diameter larger than the maximum vessel width to obtain a background image. Third, this opened image is subtracted from the original image to obtain an image where the background is suppressed and the vessels enhanced. Fourth, matched filtering[23] is used to extract vessels as a binary image. The similarity metric is the objective function measure of match (MOM) given by

$$\mathrm{MOM} = \frac{1}{n} \sum_{(x_0,y_0):I_2(x_0,y_0)\neq 0} I_1(x,y) \tag{3.14}$$

where I_1 and I_2 are, respectively, the reference image and the image to be registered, (x_0, y_0) and (x, y) are, respectively, the original and transformed pixel co-ordinates, and n is the number of nonzero (vessel) pixels in I_2. The maximum possible value of MOM corresponds to the unlikely scenario where the images are identical and perfectly registered. The authors use stochastic global optimization methods because the transformation types have many independent parameters covering a large range of values, and the objective function to be optimized has a narrow global maximum (which corresponds to the solution) and many local maxima. Ten independent executions of the global optimization methods are performed and the mean solution yields the transformation parameters. Furthermore, the objective function is evaluated a certain number of times during each execution to study the convergence. Results show that (1) genetic algorithms not only perform better than simulated annealing in terms of registration accuracy but also require a smaller number of objective function evaluations to converge, and (2) the transformation type influences the performance (i.e., genetic algorithms perform the best for a bilinear transformation while simulated annealing performs the best for an affine transformation). One limitation of this method is that the matched filtering assumes a Gaussian vessel intensity profile, which is not always the case. Also, the method is computationally intensive. Unfortunately, the dependency between best transformation type vs. optimization method is not analyzed.

In Ref. 18, five different methods are tested in registering digital color and SLO images. The registration of color and SLO images is useful for glaucoma diagnosis. SLO images indicate the degree of reflected light, which is determined by the surface topography. A low reflectivity is obtained from areas of high slope change like the edge of the optic nerve. Because photographs are less dependent on surface topography, they are used to delineate the border of the optic disc. Complementary information from these modalities allows a better correlation of topographic and visible optic nerve head damage. The first two registration methods are feature-based and will be described here (the three others will be discussed in Section III.B.6). For both methods, the color and SLO images are first processed to extract the retinal vessels as monopixel binary networks using the following procedure. Color images are corrected for the nonuniform illumination, closed with morphological filtering to eliminate the blood vessels, the result is then subtracted from the original image, thresholded, median filtered to remove noise, and thinned. Ridges and valleys are detected and thresholded on SLO images. For the first method the distance transform is computed on the network of the color image. The images are matched by transforming the SLO retinal network to maximize the sum of the distance transform values lying under it. In the second method, the only difference is the fact that the distance transform values are weighted by the vessel width in order to increase the influence of wider vessels.

4. Active Contours Methods

Active contour registration methods can be classified as feature-based methods. However, because features are detected only in one source image, we think that they can constitute a class by themselves.

In Ref. 24, the image to be registered is morphologically smoothed using a structuring element smaller than the smallest feature of interest in the image. Then a watershed transform is applied to separate the image into primary regions (closed plateaus separated by monopixel vessels). The regions, curves (vessels), and nodes are represented as a dual region adjacency graph to be classified and simplified. The curves are then converted to active contours constrained by different forces. Registration is performed by overlaying active contours on the reference image. Some nodes are fixed to the image, like at the image border, while others are replaced by springs and are allowed to adapt. The contours connected to two fixed nodes are not allowed to move. The other contours move and change shape according to an energy minimization criterion. The registration obtained in this way gives only vessel locations in the reference image. Interpolation is used to register the rest of the image. This method can be useful in characterizing local deformations like venous beading. However, a drawback of the classical active contour approach is its sensitivity to initialization. Since active contours are only attracted by the neighboring structures, this method is prone to fail for a large transformation between the images. Level set-based active contours should be exploited.

5. Point Mapping Methods

Point mapping methods refer to methods that use sparse control points detected in the images to establish a correspondence that leads to the image transformation. They are useful when the transformation type is unknown. However, they highly depend on the quality of the extraction step. An insufficient number of detected points or matches can cause the method to fail.

a. Vessel Segment Ends

In Ref. 25, vessel segments are detected using the matched filters presented in Ref. 23. Images are thresholded and thinned to obtain binary monopixel retinal networks. The control points are the vessel segment ends. The procedure to match the control points is not mentioned by the authors. The following steps are used to eliminate outliers:

1. The matches composed of points separated by a distance larger than a given threshold are eliminated,
2. The matches are ordered according to the correlation coefficient response in a window centered on each point and the first 20% are kept,
3. A scale estimate ($\tilde{s}$) is obtained from the highest correlation match and a random match at a distance larger than a given threshold; a match is eliminated if the difference between its scale (with an accepted match) and $\tilde{s}$ is greater than a given threshold or if it is not far enough from an accepted match.

The least-mean-square method is then used to find the affine transformation parameters, which are refined by iteratively removing the match that contributes the most to the RMS error of the affine transformation. This is done until the error is less than a threshold or when there are only four matches left.

b. Highest Intensity Edges

In Ref. 26, the edges are first detected. A template is chosen in the image to be registered and represented as two matrices. Each row of the first matrix is associated to an image column and contains the vertical co-ordinates of the highest intensity edges. Each row of the second matrix is associated to an image row and contains the horizontal co-ordinates of the highest intensity edges. Two matrices are also obtained for each area of the search region in the reference image. The similarity measure is the number of co-ordinate matches between the matrices. The search strategy is a spiral scan instead of a sequential one. The drawback of this method is that it works only for translation transformations. The advantage is that the similarity measure is not designed specifically for ophthalmic images and can therefore be used in other domains.

c. Blood Vessel Bifurcations

A good point matching registration process requires a sufficient number of corresponding control points uniformly distributed in both images. For this reason, blood vessel bifurcation points are a natural choice.

A point matching method for temporal and multimodal registration of mydriatic color images and fluorescein angiograms of the retina is proposed in Ref. 27. A binary image of the vessel is obtained using morphological operators. The vessel bifurcation points are then detected with a supremum of openings with revolving T-shaped structuring elements. The bifurcation points are identified by the orientations of the vessels that surround them. The use of the vessel orientations has two advantages: first, the selection of candidate matches without having to use a neighborhood, which contributes to speeding up the algorithm; and second, a Bayesian generalization of the Hough transform, which allows a more progressive evaluation of the transformation. An angle-based invariant is computed to obtain the probability that two points form a match. The matching process and search for the best transformation is based on the Hough transform.

This method seems to give good results, but it requires the assumption of a Gaussian shape vessel intensity profile, which is not always satisfied.[28] Furthermore, the extracted bifurcation points and their orientations are not all valid because false edges are created by the retinal network detection procedure.

In Ref. 5, the expectation maximization (EM) algorithm is used to identify the parameters of the optimal transformation. This algorithm iteratively finds the maximum likelihood estimate of the parameters of a system from incomplete data. The blood vessel bifurcations are selected manually. For control point matching, the authors use the similarity transformation as an approximation to the optimal transformation. The similarity transformation is computed for all possible combinations of two control point pairs. The number of similarity transformations for the good matches is $J = 0.5(K^2 - K)$, where K is the number of good matches. Finding a tight cluster of data points in the transformation space identifies good matches, which are used to compute the final transformation. An exhaustive search is used to find this cluster.

The work presented in Ref. 29 and references therein aims to build a mosaic on which the physician manually delineates a laser treatment area. During surgery, an online image is registered indicating the striking point of the laser on the retina. This registration indicates if the laser is inside the treatment area drawn by the physician. This work is justified by the fact that laser treatment only has a 50% success rate partly due to a lack of spatial information during surgery. The energy delivered at each position on the retina is not quantified and there is no safety shut-off when the eye moves. First, the mosaic must be constructed from diagnostic images. The vessel centerlines and bifurcation points are extracted using an iterative tracing algorithm that starts with automatically detected seed points. Next, a second-order polynomial transformation between each image pair is found. Then, pairs and triples of bifurcation points are determined and invariants (under the similarity transformation) are computed. These invariants are based on the position of the points and the orientation of the surrounding vessels. Once the mosaic is ready, an online image can be registered to it. The vessels and bifurcation points are extracted (with the same algorithm used for the mosaic but optimized for speed) and invariants are computed. Invariant indexing (rapid online look-up) allows establishing initial matches and finding a similarity transformation between the bifurcation points of the mosaic and those of the online image. This initial similarity transformation is then checked, extended to an affine one followed by a second-order polynomial transformation, and refined. The authors tested the algorithm on groups of images. Because they did not have online images, they used an image of each group. Each image of each group is chosen once as the online image. A 96.5% success rate is achieved, and the average registration error is 1.21 pixels. An important contribution of this work is that the registration algorithm, unlike previous tracking algorithms,[30,31] determines the position of the online image without knowledge of the previous position. This is important in order to cope with saccadic eye and/or head movement, blinking, and glare. However, the algorithms are only tested on healthy images of the retina.

Another registration method based on global point mapping is proposed in Ref. 32. It uses blood vessel bifurcations as control points and a search for control point matches based on the local structural information of the retinal network. Three transformation types (similarity, affine, and second-order polynomial) are evaluated on each image pair. The method is designed to cope with the presence of various types of lesions associated with retinal diseases as well as the high visual quality variation between images. The method is applicable to temporal and multimodal registration of ophthalmic images. Unlike the method presented in Ref. 27, it does not assume a Gaussian intensity profile for the vessels and it is much easier to implement. For the color images, only the green band is used because it has the best contrast. The algorithm has ten parameters of which seven depend on the image resolution only. The control point detection involves two steps: retinal vessel centerline detection followed by bifurcation point detection. The angles of the vessels surrounding each bifurcation point are obtained by computing the intersection between the vessels and a circle of fixed diameter centered on each point.[27] There is no constraint regarding which of the images is to be used as the reference; the matching procedure in Ref. 32 gives results that are independent of the choice of the reference image. Each bifurcation point in image one is linked to the bifurcation points in image two that are located within a given distance d, have the same number of surrounding vessels and with a difference between the angles smaller than a given threshold o. Bifurcation points without matches are eliminated. A relaxation method is then used to eliminate outliers and estimate the transformation parameters. The affine transformation is used for the outlier rejection process but once the final good matches are obtained, they can be used to compute any type of registration transformation. Similarity, affine and second-order polynomial transformations are tested. The relaxation process for outlier rejection follows five steps:

1. Computation of the affine transformation parameters in the two directions (to register image one to image two and *vice versa*),
2. Transformation of the control points according to these parameters,
3. Computation of a root-mean-square (RMS) error between the transformed and reference points for each image,
4. Computation of the average RMS,
5. Successive elimination of matches that contribute the most to the RMS error when used for estimating the parameters of the affine transformation, until the average RMS is smaller than a threshold or only three matches remain.

If there are bifurcation points with more than one match, the best match is kept. If there are more matches than needed to find an exact solution, the least-mean-square method is used to estimate the transformation parameters. This algorithm was tested on 70 image pairs and succeeded in registering 45 of them. The image pairs that could not be registered were of very bad quality

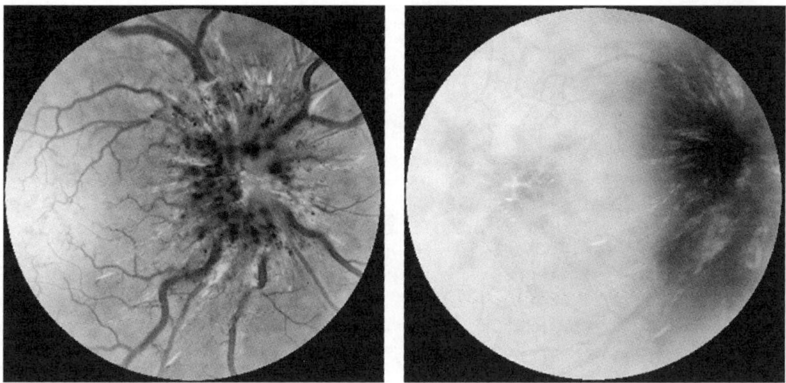

FIGURE 3.5 Image pair that the algorithm presented in Ref. 32 failed to register because the vessels are deteriorated by disease and one of the images is a late stage angiogram (for which vessels are hardly visible because of the absence of dye).

(very few vessels due to illumination problems, zoom on the macula, or vessel deterioration caused by disease), or late stage angiograms (for which vessels are hardly visible because of the absence of dye). Figure 3.5 shows an example of a late stage angiogram with bad image quality. On average for each image pair, 90 ± 21 control points are detected leading to 14 ± 7 good matches. Figure 3.6 shows an example of an image pair. Figure 3.7 and Figure 3.8 show the registration result of this image pair, under two formats: retinal network and checkerboard image. The average registration performance is given in Table 3.2. On average, the three transformation types are equally good; although there are a significant number of cases (19%) for which a transformation type is better than another one (see Table 3.3). This registration algorithm can be used for any global transformation type. It has been pointed out in Ref. 32 that some issues like robustness of the image registration algorithm remain open and should be addressed in the future. The quality of retinal images is highly variable regarding color, texture, and lesions. Human pigmentation, age, diseases, eye movement,

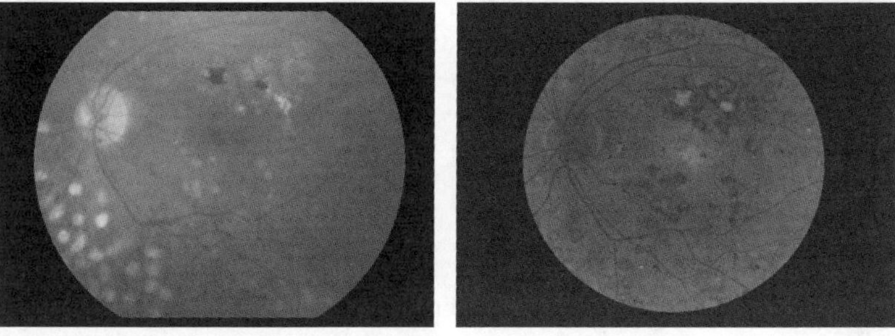

FIGURE 3.6 (See color insert) Example of color mydriatic image and positive angiogram.

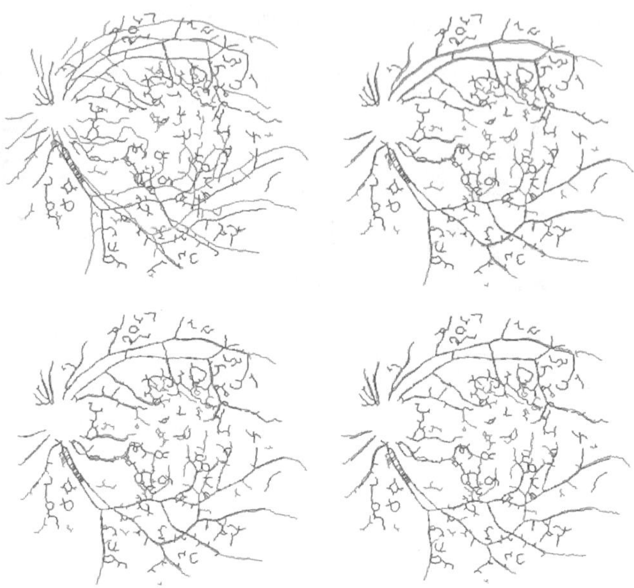

FIGURE 3.7 (See color insert) Registered retinal networks of Figure 3.6 color image (in red) and angiogram (in black) with no (top-left), similarity (top-right), affine (bottom-left), and second-order polynomial transformation (bottom-right).

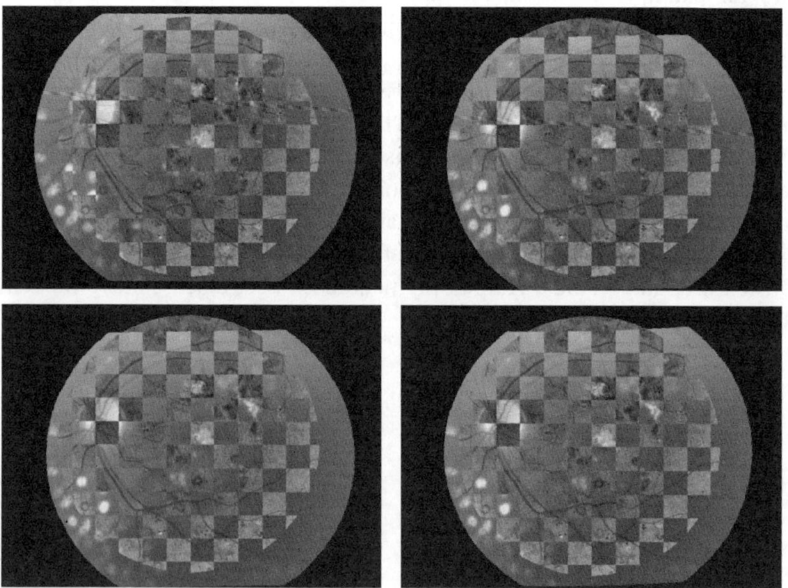

FIGURE 3.8 Checkerboard images showing the registration of Figure 3.6 with no (top-left), similarity (top-right), affine (bottom-left), and second-order polynomial transformation (bottom-right).

TABLE 3.2
**Registration Performance for Different Transformation Types and
Alignment Improvement Brought by the Different Transformation Types[32]**

Transformation Type	Registration Performance (%)	Registration Improvement (%)
No	20 ± 7	—
Similarity	52 ± 19	32 ± 18
Affine	54 ± 19	34 ± 18
Second order polynomial	54 ± 19	35 ± 18

camera calibration, illumination, and film defects can contribute to this variability. Despite that, clinicians can still use bad quality images for diagnosis. But bad image quality affects the ability of the algorithm to reliably extract and match control points provided by the retinal network, so alternative control point extraction strategies should be explored.

6. Mutual Information Methods

Mutual information methods are based on the statistical dependence or information redundancy between the image intensities.[13] Mutual information can cope with substantial image variations between different modalities. It is defined as

$$\text{MI} = \sum_{i=1}^{m} \sum_{j=1}^{n} P(g_i, g_j) \ln\left(\frac{P(g_i, g_j)}{P(g_i)P(g_j)} \right) \tag{3.15}$$

and is computed only for the overlapping part of the images. In Equation 3.15, m and n are the number of gray levels in image I and J, respectively, $P(g_i)$ and $P(g_j)$ represent the number of pixels in image I and J with values g_i and g_j over the total

TABLE 3.3
**Cases where a Transformation Type is Significantly Better (Difference
Greater than 1 σ) than Another[32]**

Worse\Better	Similarity (%)	Affine (%)	Second Order Polynomial (%)
Similarity	—	10	9
Affine	7	—	9
Second order polynomial	11	10	—

number of pixels, and $P(g_i, g_j)$ is the number of pixel pairs with value (g_i, g_j) over the total number of pixels.

In Ref. 10, simulated annealing is used to minimize the negative of Equation 3.15. The algorithm is tested on 49 stereo image pairs and 48 temporal image pairs centered on the optic disc and covering only this area. A drawback is that the simulated annealing algorithm depends on many hidden parameters that are nearly impossible to tune in order to cope with changes in the image constraints.[13] However, simulated annealing is able to deal with local minima, which are numerous for the mutual information function, and does not assume a given shape for the function being minimized.

Three of the five registration methods in Ref. 18 use mutual information. See Section III.B.3 for the first two. In the third method, the vessels are extracted as previously but they are dilated to serve as a mask to limit the computation of the mutual information. This is done to overcome the difficulty of getting a higher match response from a homogeneous noisy background than from the sparse features of interest.[18] In the fourth method, the image intensity values are used along with a variance-weighted version of the mutual information. This version assumes a constraint based on the fact that a given intensity is mapped to a small range of intensities given by

$$(MI)_\sigma = \sum_{i=1}^{m} \sum_{j=1}^{n} \frac{P(g_i, g_j)}{\sigma_i + \sigma_j + 1} \ln\left(\frac{P(g_i, g_j)}{P(g_i)P(g_j)} \right) \tag{3.16}$$

where σ_i is the variance of all the mapped values of g_i in image J. In the fifth method, the image to be registered is divided into windows and the mutual information is computed for each. This is to take into account the fact that some image areas contain more features than others do. The mutual information outputs of each window are then combined by taking the mean, the median, and the mean weighted by the intensity variance of the window, and the right transformation is the one that maximizes the combined mutual information results. Among the five registration methods in Ref. 18, the one that uses the mean to combine the mutual information output in each window provides the best registration. The authors thus conclude that windowed mutual information offers better performance than standard mutual information, which performs very poorly probably because the differences between the modalities are too big.

Registration using mutual information has the advantage of requiring neither image preprocessing nor knowledge about the image content. However it seems to work only for, at most, similarity transformations.[10] Furthermore, standard mutual information (as opposed to windowed mutual information) ignores the fact that pixel values are continuous (although quantified). In other words, if the mapping $a \rightarrow b$ is frequent, the similar mapping $a + 1 \rightarrow b$, which happens less frequently, will be considered as an important change.[18]

C. REGISTRATION PERFORMANCE ASSESSMENT

1. Qualitative

In the literature, a visual performance assessment is often done by creating a checkerboard of the reference and transformed images,[10] or by superposing the transformed extracted vessels on the reference image,[20] or on its extracted vessels.[13] In Ref. 18, the registration is graded on a four-point scale by a clinician looking at an SLO image overlaid on a color image.

Visual evaluation is useful but may be inconsistent due to inter and intrareader variability, or too demanding when comparing registration performance for various transformation types.

2. Quantitative

A technique called cross-validation is defined in Ref. 6. It requires more than two images of a given scene. One image is chosen as the reference and the others are registered to it. The transformed (registered) images are then registered together. The magnitude of this last transformation measures the registration internal consistency.

In Ref. 19, the quality of the transformation parameters is estimated because the registration algorithm proposed is not guaranteed to find the global maximum. The criterion is the percentage of overlap given by

$$\text{Overlap} = \frac{\text{Max}_{\text{computed}}}{\text{Max}_{\text{possible}}} = \frac{\text{Score}_{\text{max}}}{N \cdot \text{DistValue}_{\text{max}}} \qquad (3.17)$$

where $\text{Score}_{\text{max}}$ is the maximum value of the objective function (Equation 3.11), N is the number of sample points randomly selected from the extracted tubes of the image to be registered, and $\text{DistValue}_{\text{max}}$ is the maximum value of the distance function. The authors found that a minimum of 40% overlap is necessary for a 90% probability of finding the global maximum.

An overlap-based criterion to assess registration performance is also used in Ref. 32. The retinal network having the smallest number of pixels is used as the reference. Overlap is defined as the presence of another pixel in a fixed size window centered on a reference pixel. A window of size unity corresponds to the true overlap definition. The larger the window is, the larger the accepted registration error. The overlap percentage is the registration performance evaluation criterion. A 100% overlap cannot be reached in practice because that would require perfectly registered, almost identical retinal networks.

In Ref. 22, a comparison between automatic and manual registration is carried out. For manual registration, three ophthalmologists choose at least three bifurcation points in each image pair. Then bilinear transformation parameters are estimated and the objective function (Equation 3.14) is evaluated. The results show that automatic registration is better (more vessels are superimposed), and more consistent (smaller standard deviation) than the manual registration performed by the ophthalmologists.

IV. FUSION

A fused image must contain all the pertinent information of the source images without fusion artifacts. Image fusion exploits complementary information from different image modalities. It allows the establishment of the relationship between lesions visible in only one modality, or between a lesion visible in a modality and an anatomical area that is more visible in another one. Such relationships cannot or, at least, not without difficulty be established from a single modality or by viewing sequentially different modalities while trying to mentally integrate the information provided by each modality.

An example of the usefulness of fusion in ophthalmology is in evaluating sight-threatening changes in diabetic retinopathy. Laser treatment reduces the risk of visual loss from clinically significant diabetic macular edema. Clinical significance is defined by the presence of hard exudates within 500 μm of the macula center combined with a thickening of the adjacent retina.[33] Hard exudates being clearly visible on photographs and macula being more visible on angiograms, the fusion of these two modalities would allow the distance measurement required.[9]

At the time of writing this chapter, we knew of three studies regarding retinal image fusion. Although not specific to pixel-level fusion, the work of Ref. 19 addresses a kind of information fusion that can be described as a combination of segmented features by graphical superposition. The coarse contours of anatomical and pathological features (vessels, fovea, optic disc, scotoma, and subretinal leakage) are extracted from scanning ophthalmoscope images, registered, and superposed on the same image to create retinal maps. Such maps are required for diagnosis and treatment (scotoma-based photocoagulation) of age-related macular degeneration. The authors developed and tested software to generate these retinal maps, which previously had to be drawn manually.

To our knowledge, the first published works that really address pixel-level fusion in the ophthalmology domain are Refs. 32,34. Fourteen methods are tested to fuse images of different modalities (color and fluorescein angiogram) and resolutions, images taken at different times (separated by a few minutes during an angiography examination to several years between two examinations), and images presenting different pathologies (diabetic retinopathy, age-related macular degeneration, retinal vein occlusion, cytomegalovirus retinitis, anterior ischemic optic neuropathy, choroidal neovascular membrane, cystoid macular edema, histoplasmosis retinitis/choroiditis, telangiectasia, etc.). The authors evaluate and classify the grayscale fusion methods based on four quantitative criteria. They also visually compare the color fusion methods. In this study, 45 image pairs are used. Each pair is registered with the transformation type (similarity, affine, or second-order polynomial) that gives the best registration performance for this particular pair.

Another work,[9] which makes use of the wavelet transform, aims to demonstrate that the information content of retinal pathologies is increased in a multimodal fused image. In particular, the authors seek an improved visualization of pathological changes at an earlier stage of the disease and/or an enhancement of particular patterns within the retina (and/or choroid), which could improve our visual sensitivity to abnormal physiological occurrences. The authors fused three pairs of *angiogram videos — color images* presenting different pathologies (diabetic retinopathy, cystoid macular edema, and branch retinal vein occlusion). Only the green band of the color image is used for the fusion.

A. COMBINATION BY GRAPHICAL SUPERPOSITION

The work presented in Ref. 19 is described in more detail in this section. The authors used four different image modalities of the SLO:

1. Infrared reflection image taken with a 2 mm circular aperture and an infrared diode source,
2. Argon-blue reflection image taken with a 2 mm circular aperture,
3. Fluorescein angiogram,
4. Static scotometry.

The infrared reflection image shows the surface of the fundus. It allows the extraction of blood vessels and optic disc. The vessels, which are needed for registration, are also extracted from the three other image modalities. The optic disc is needed, as a landmark, for the extraction of the fovea, scotoma, and subretinal leakage.

Argon-blue wavelengths are used because a yellow pigment (xanthophyll) present in the fovea absorbs them, letting the corresponding area appear as a dark blob. The fovea and foveola, which are not visible in any image, can be located by extracting this dark blob and by using their anatomical relationship (position and size) with the optic disc. A procedure in two steps is used to extract the dark blob:

1. The macula, defined as a $3d_{OD} \times 2d_{OD}$ ($d_{OD} \equiv$ optic disc diameter) area centered at 4 mm temporal and 0.8 mm below the optic disc center, is smoothed and thresholded to identify the darkest region (candidate dark blob),
2. The fovea, defined as a $d_{OD} \times d_{OD}$ area centered on the centroid of the candidate dark blob, is thresholded to extract the blob.

The fovea and foveola are the areas, of size $d_{OD} \times d_{OD}$ and $0.23d_{OD} \times 0.23d_{OD}$, respectively, centered on the dark blob. Two parameters are used to determine if the extracted dark blob is pathological

$$\text{XanthoSize}_{\text{rel}} = \frac{4\text{XanthoSize}_{\text{abs}}}{d_{\text{Fovea}}^2 \pi},$$

$$\text{ShapeError} = \frac{\text{XanthoPeri} - \text{CHullPeri}}{\text{XanthoPeri}}$$

$$(3.18)$$

where $\text{XanthoSize}_{\text{abs}}$ and XanthoPeri are the absolute size and perimeter of the xanthophyll area, d_{Fovea} is the fovea diameter, and CHullPeri is the perimeter of the convex hull of the xanthophyll area in pixels. Points outside a certain area on the graph ShapeError vs. $\text{XanthoSize}_{\text{rel}}$ indicate an abnormal fovea.

Position and time behavior of subretinal leakage can be assessed from fluorescein angiograms on which they appear as bright blobs. Unfortunately, the authors were unable to develop an automatic and robust method to extract subretinal leakage.

Static scotometry aims to determine the boundary of the group of points in the macula that do not respond to visual stimulus: the scotoma.

Images are digitized from a 760×430 SLO video taken with a $40°$ field of view. Image resolution is approximately 8 to 10 μm. The static scotometry results, which are points, are already in a digital form. The authors used a data set of 59 images from 11 patients of three categories (undefined, subfoveal, and extrafoveal neovascularization). The extracted features are registered to the infrared reflection image using the computed transformation parameters (see Section III.B.3).

The retinal maps and the occurrence and location of choroidal neovascularization are used to classify patients in three categories. Then, a one-year clinical follow-up allowed the authors to conclude that, unlike the recommendations of previous clinical studies, only the patients of the third category should receive the scotoma-based photocoagulation treatment. Furthermore, the treatment should be applied only where a dense scotoma covers the choroidal neovascularization in order to minimize the resulting laser scar. The authors found that (1) a choroidal neovascularization visible on an angiogram is always covered by a scotoma with coinciding boundaries, (2) the scotoma-based photocoagulation fails in case of vascularized pigment epithelium detachment, and (3) the existence of a choroidal neovascularization cannot be confirmed if it is not detected in both an angiogram and static scotometry.

Therefore, the recommendations of the authors for the diagnosis of age-related macular degeneration are:

1. The use of a SLO,
2. Automatic scotometry acquisition, and
3. Retinal map generation.

B. Pixel-Level Fusion Methods

1. Grayscale Fusion Methods

Twelve of the 14 methods studied in Ref. 32 are classical grayscale image fusion methods implemented in the *fusetool* MATLAB™ toolbox.[35] These methods can be classified in three groups: linear, nonlinear, and multiresolution.

a. Linear Methods

Linear methods such as average and principal component analysis (PCA) are based on a weighted superposition of the source images.

The average method is simply the average of the two source images. Its use is appropriate when fusing images from the same modality (no local intensity reversal), without complementary features, and with additive Gaussian noise (for N images with equal noise variance, the noise variance of the fused image is reduced by a factor N).

In the PCA method, the optimal coefficients (in terms of information content and redundancy elimination) are calculated using a Karhunen–Loeve transform of the intensities. The coefficients for each source image are obtained from the normalized eigenvector associated with the largest eigenvalue of the covariance matrix of the two source images.

b. Nonlinear Methods

In nonlinear methods, such as select minimum and select maximum, nonlinear operators are applied, which select the minimum or maximum intensity pixel between the source images. These methods are appropriate if the features of interest are, respectively, dark and bright.

c. Multiresolution Methods

Multiresolution methods can be divided in two subgroups: image pyramid and wavelet transform methods. Each source image is decomposed into a pyramid representing the edge information at different resolutions. Fusing images then amounts to combine their pyramids and apply the inverse transform to the resulting pyramid. Since the fusion takes place at each resolution level, it keeps the salient features at each resolution.

The salient features of an image coincide with the sharper intensity changes, thus with the larger transform coefficients. Therefore, the high pass coefficients can be combined by choosing the maximum coefficient at each pixel between the images. This method is implemented in *fusetool* along with two other methods: salience/match measure[36] and consistency check.[37]

For the salience/match measure, the high pass coefficients are combined according to

$$D_C(m, n, k) = w_A(m, n, k)D_A(m, n, k) + w_B(m, n, k)D_B(m, n, k) \qquad (3.19)$$

where (m, n) are pixel co-ordinates, k is the pyramid level, D_A and D_B are the first and second pyramids, D_C is the fused pyramid, and w_A and w_B are weights. The salience must first be computed to obtain the salience/match measure

$$S_A(m, n, k) = \sum_{m',n'} p(m', n')D_A^2(m + m', n + n', k),$$

$$S_B(m, n, k) = \sum_{m',n'} p(m', n')D_B^2(m + m', n + n', k) \tag{3.20}$$

where p is the neighborhood. If $S_A > S_B$, $w_A = w_{\max}$ and $w_B = w_{\min}$, otherwise $w_A = w_{\min}$ and $w_B = w_{\max}$. Then, the match has to be computed according to

$$M_{AB}(m,n,k) = \frac{2\sum_{m',n'} p(m',n')D_A(m+m',n+n',k)D_B(m+m',n+n',k)}{S_A(m,n,k)+S_B(m,n,k)} \tag{3.21}$$

A threshold α is chosen to determine the amount of selection and averaging for the coefficients. A larger α implies more selection than averaging. If the match is smaller than α, the pixel with the largest salience is picked, otherwise a weighted average of the two pixels is used according to

$$M_{AB} < \alpha \Rightarrow w_{\min} = 0, \; w_{\max} = 1,$$

$$M_{AB} \geq \alpha \Rightarrow w_{\min} = \frac{1}{2} - \frac{1}{2}\left(\frac{1 - M_{AB}}{1 - \alpha}\right), \; w_{\max} = 1 - w_{\min} \tag{3.22}$$

The consistency check method consists of the following steps:

1. The maximum absolute value in a window determines which coefficient is picked,
2. A binary decision image is created,
3. The consistency is verified in this image (if a pixel comes from image A and the majority of its neighbors are from B, it is replaced by the corresponding pixel from B).

An area is used for the high pass coefficient combination since the significant features are always larger than a single pixel. Finally, for the low pass coefficient combination one can select a low pass source image or the average of both.

Best results were obtained with seven decomposition levels, salience/match measure and an area of 3×3 for high pass coefficients combination, and the average of the source images for the low pass coefficient combination.[32]

i. *Image Pyramids.* Image pyramid methods can be divided in six subgroups: Laplacian, filter–subtract–decimate, ratio, contrast, gradient, and morphological methods.

In the Laplacian method,[38] each level of the pyramid is recursively constructed from its lower level by the following four steps: low pass filtering, subsampling, interpolating, and differencing. The filter–subtract–decimate method[39] is an alternative to the Laplacian pyramid where the subsampling and interpolating steps are skipped. The process to obtain the ratio pyramid is identical to the one used for the Laplacian pyramid but a division replaces the differencing. It results in a nonlinear pyramid. The contrast method[40] is different from the ratio pyramid in that there is a subtraction by one after the division. This method preserves the local intensity contrast and it selects features with maximum contrast instead of maximum magnitude. It is designed to make the fused image more appropriate for visual analysis by humans. In the gradient method,[36] the image is decomposed into its directional edges using directional derivative filters. In the morphological method,[41] the linear filters are replaced by nonlinear morphological filters. Each level of the pyramid is constructed from the preceding level by the following four steps: opening, closing, subsampling–oversampling dilation, and differencing.

ii.   Wavelet Transform. Discrete wavelet transform methods are similar to image pyramid methods but provide nonredundant image representations. They are supposed to give better results than the Laplacian pyramid[37] because of their:

1. Compactness (the wavelet transform size is equal to the image size while the Laplacian pyramid size is 4/3 of the image size),
2. Orthogonality (which eliminates the redundancy between the resolutions),
3. Directional information (there is no selection based on spatial orientation in the Laplacian pyramid).

The first wavelet transform method implemented in *fusetool* and tested in ophthalmology[32] is a discrete wavelet transform with Daubechies wavelets.

In Ref. 9, Symlet wavelets of lengths 4 to 20 are tested and the best results are obtained with the largest length. Using the consistency check technique for the high pass coefficient combination, the authors fused three pairs of images. The first fused image, which is a case of diabetic retinopathy, emphasizes the spatial relationship between the exudates, which are only visible in the color image, and the small focal areas of hypofluorescence visible on the fluorescein angiogram, which are indicative of capillary closure or early ischemia. This could allow the determination of the position of the exudates with respect to the fovea, which is relevant in deciding if the patient should undergo laser treatment. For the second case (cystoid macular edema), some subtle changes in the retinal pigment epithelium can be seen in the color image. The fluorescein angiogram shows swelling in the intraretinal tissue, which appears as a petaloid. The fused image allows the positioning of the swelling in the green band. For the third case (branch retinal vein occlusion), the color image shows a vein that appears thicker and tortuous. Hemorrhages are also visible near the macula and in the upper half of the retina. On the fluorescein angiogram, the abnormal blood return produces leakages of dye, which are scattered through the retina, and swollen pockets of tissues above the macula. The fused image emphasizes the relationship between the leakage sources and the hemorrhages.

A drawback of the discrete wavelet transform is that it is not shift-invariant, which means that a translation of the original signal leads to different transform coefficients. The second wavelet transform method implemented in *fusetool* is a shift-invariant extension of the discrete wavelet transform.[35]

Figure 3.9 shows the results of the 12 grayscale fusion methods implemented in *fusetool* on the image pair of Figure 3.6.

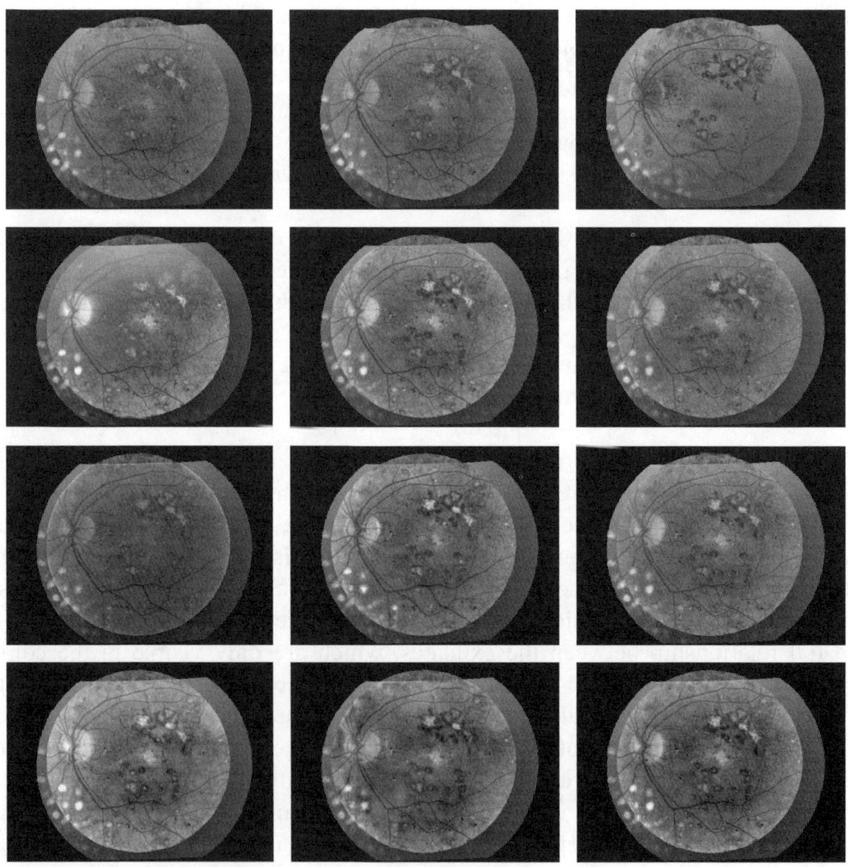

FIGURE 3.9 Fused images of Figure 3.6 with (from top to bottom and left to right) average, PCA, select minimum, maximum, laplacian, filter–subtract–decimate, ratio, contrast, gradient, and morphological pyramid, discrete wavelet transform (dwt), and shift-invariant dwt methods.

2. Color Fusion Methods

Color fusion provides color contrast information in addition to the brightness contrast information.

a. *True-Color Method*

Color information is an important feature for the diagnosis of some diseases (e.g., diabetic retinopathy). When an image pair containing at least one color image is fused with a grayscale fusion method (Section IV.B.1), the color information content is lost. A fusion method is proposed in Ref. 32 that adds the R and B bands of the color image (with the best resolution if the two source images are color images) to the grayscale fused image to overcome this. It is found that the addition of R and B bands to a grayscale fused image provides additional information of potential interest for medical diagnosis.

In grayscale fusion (with or without adding color bands), there is no way to retrieve the source image information from the fused image. This is why two false-color fusion methods, proposed by the Netherlands Organization for Applied Scientific Research (TNO) and Massachusetts Institute of Technology (MIT) are studied in Ref. 32.

b. *TNO False-Color Method*

The TNO method[42] is a false color mapping of the unique component of each source image given by

$$\begin{pmatrix} R \\ G \\ B \end{pmatrix} = \begin{pmatrix} I_2 - I_1^* \\ I_1 - I_2^* \\ I_1^* - I_2^* \end{pmatrix} \qquad (3.23)$$

where

$$I_1^* = I_1 - \min\{I_1(i,j), I_2(i,j)\} \qquad (3.24)$$

is the unique component of the first source image, that is, the source image minus the common component of both source images.

c. *MIT False-Color Method*

The MIT method[43] is based on a color perception model of the human visual system. Many different mappings have been proposed in the past according to the application. The one proposed in Ref. 44, which is illustrated in Figure 3.10, is used in Ref. 32. This mapping is appropriate when fusion involves one image with a better contrast or a more natural appearance, as is the case for temporal and multimodal fusion. As a result, the "best" image is preserved by its mapping to the green band. Step 1 in Figure 3.10 is a within-band contrast enhancement

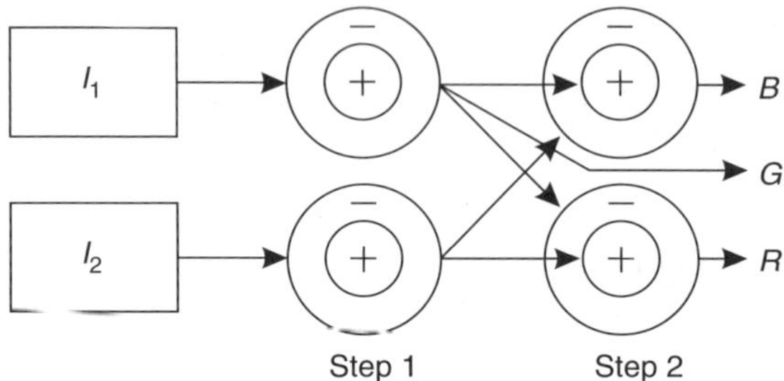

FIGURE 3.10 Fusion architecture of the MIT false-color fusion method. Step 1 is a within-band contrast enhancement and normalization. Step 2 is a between-band fusion.

and normalization. Step 2 is a between-band fusion. The concentric circles in Figure 3.10 represent a convolution operator defined by

$$x_i = \frac{(BCG_{\text{in}} \otimes I_{\text{c}} - DG_{\text{out}} \otimes I_{\text{s}})}{A + (CG_{\text{in}} \otimes I_{\text{c}} + G_{\text{out}} \otimes I_{\text{s}})} \tag{3.25}$$

where x_i is the value of the ith pixel of the processed image, A, B, C, and D are constants, and G_{in} and G_{out} are Gaussian weighted averaging masks. The other terms are described below.

In Figure 3.10, the small circles with the plus sign represent the excitatory center (I_{c}) while the large circles with the minus sign represent the inhibitory surround (I_{s}). Arrows indicate which image feeds the center and which image feeds the surround of the operator. The upper-left operator in Figure 3.10 is characterized by $I_{\text{c}} = I_{\text{s}} = I_1$, the lower-left one by $I_{\text{c}} = I_{\text{s}} = I_2$, the upper-right one by $I_{\text{c}} = I_1'$ and $I_{\text{s}} = I_2'$ (I_1' and I_2' being the output images of the upper-left and lower-left operator, respectively), and the lower-right one by $I_{\text{c}} = I_2'$ and $I_{\text{s}} = I_1'$. The parameter values suggested in Ref. 43 are used in Ref. 32; they are $A = \overline{CG_{\text{in}} \otimes I_{\text{c}} + G_{\text{out}} \otimes I_{\text{s}}}$, $B = 1$, $C = 2$, $D = 1$, $G_{\text{in}} = 1$, and

$$G_{\text{out}} = \frac{1}{256}\begin{bmatrix} 1 & 7 & 26 & 57 & 74 & 57 & 26 & 7 & 1 \end{bmatrix}$$

Figure 3.11 shows the fusion of the image pair of Figure 3.6 with the shift-invariant discrete wavelet transform (SIDWT) method of *fusetool* (R and B bands added), the TNO, and the MIT methods. The fused images combine the exudates of the color image and the microaneurisms of the angiogram. The addition of the R and B bands restores the yellow color of the optic disc and exudates. In the TNO-fused image, the contribution of the color image is shown in red, the contribution of the angiogram is shown in green, and the difference between the two images is shown in blue. In the MIT-fused image, the contribution of the color image is shown in red too, the contribution of the angiogram is shown in blue, and the normalized and contrast enhanced angiogram is shown

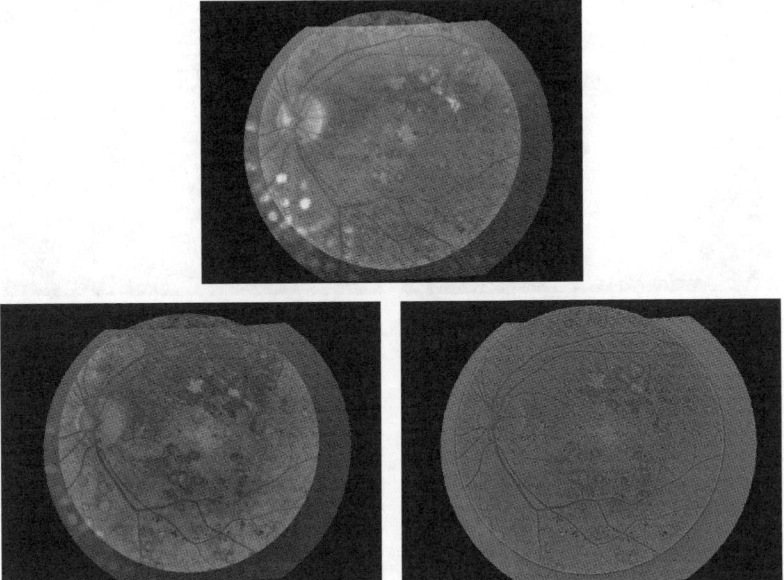

FIGURE 3.11 (See color insert) Fused images of Figure 3.6 with SIDWT R and B bands added (top), TNO (bottom-left), and MIT (bottom-right) methods.

in green. A visual evaluation by an ophthalmologist has raised the following points:

1. The use of the SIDWT + R + B method highlights the hypovascularization of the macula,
2. The macula is the most visible in the fused image obtained with the TNO method,
3. The microaneurisms are best seen in the fused image obtained with the MIT method,
4. The TNO and MIT methods show the neovessels very well, which are only visible in the angiogram.

In general, the ophthalmologist has shown a preference for the MIT over the TNO method because the lesions appear sharper without a halo around them.

Figure 3.12 shows the result of the fusion of a color image and an angiogram for an eye with histoplasmosis retinitis/choroiditis. The fused image combines the blocking effects in the macula and above it, which are seen only in the angiogram, and the details of the lesions, which are visible only in the color image.

Figure 3.13 shows the fused results of a color image and an angiogram for an eye with cytomegalovirus retinitis. The fused images allow establishing the relationship between the blocking effect seen in the angiogram, and the more extended diseases visible in the color image. The fused image with the

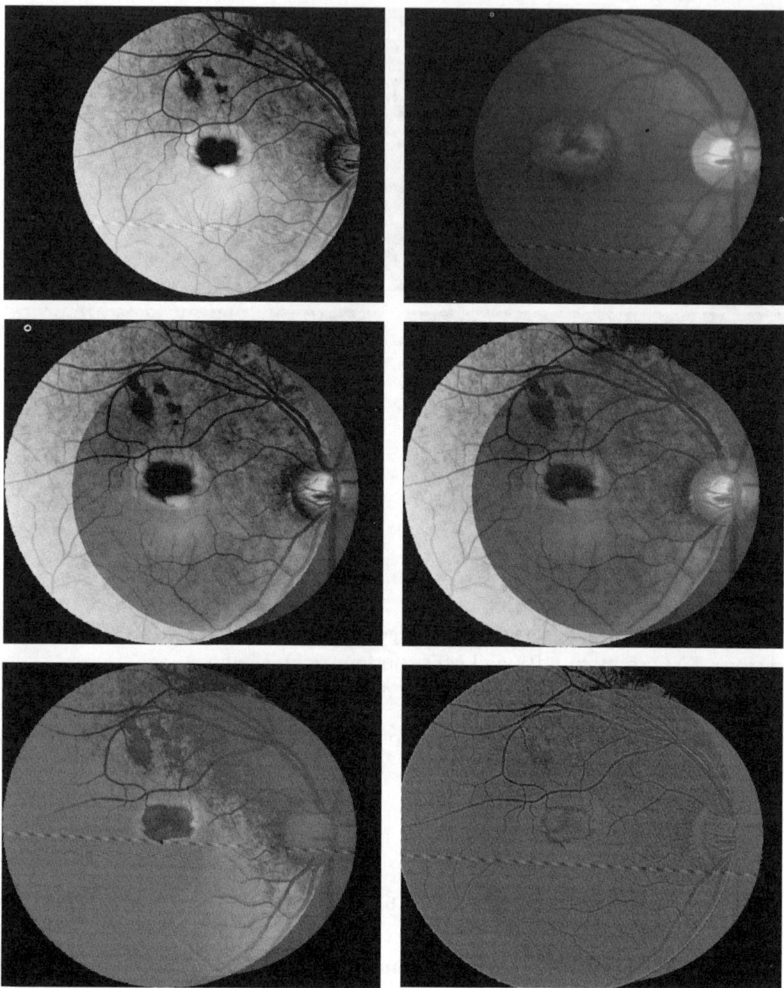

FIGURE 3.12 (See color insert) Angiogram (top-left), color image (top-right), fused images with SIDWT (middle-left), SIDWT + R + B (middle-right), TNO (bottom-left), and MIT (bottom-right) methods for an eye with histoplasmosis retinitis/choroiditis.

SIDWT + R + B method is really interesting. Usually, when an angiogram and a color image are fused with the MIT method, the angiogram is chosen as the image to be used for the green band because it has the most details. Here is a counterexample: the color image shows much more details than the angiogram and should have been used for the green band. For these ambiguous modalities, maybe a more neutral mapping, which does not favor one image over the other, should be used like the one shown in Figure 3.14.

Finally, Figure 3.15 shows the fused results of two angiograms at different stages for an eye with anterior ischemic optic neuropathy.

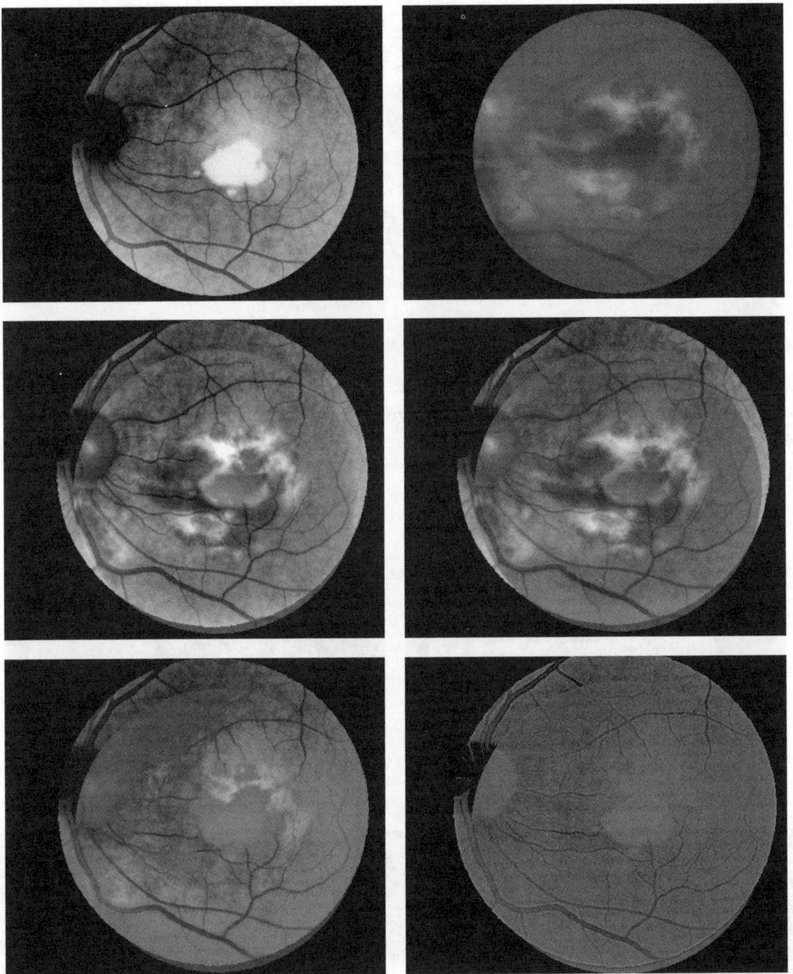

FIGURE 3.13 (See color insert) Angiogram (top-left), color image (top-right), fused images with SIDWT (middle-left), SIDWT + R + B (middle-right), TNO (bottom-left), and MIT (bottom-right) methods for an eye with cytomegalovirus retinitis.

C. FUSION PERFORMANCE ASSESSMENT

1. Qualitative

In the literature, almost all image fusion evaluation is done qualitatively because of the user perception factor. This is particularly obvious for color image fusion since it is not easy to find a criterion to evaluate color information.

In Ref. 19, the retinal maps generated automatically received the comment "excellent" by the ophthalmologists who evaluated them visually. In Ref. 9,

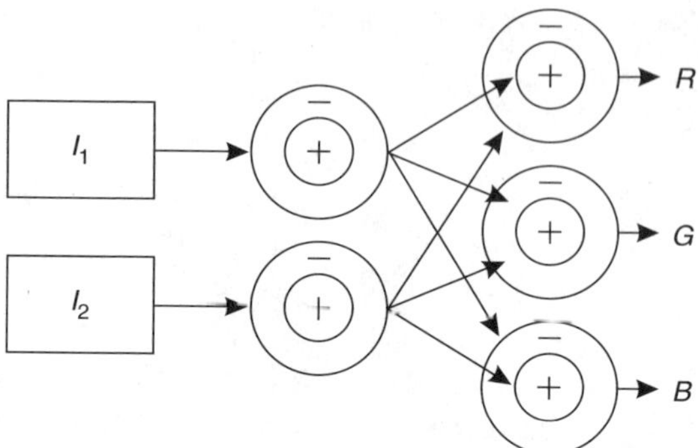

FIGURE 3.14 Different fusion architecture of the MIT false-color fusion method. The two contrast enhanced and normalized images are mapped in the green band instead than only one.

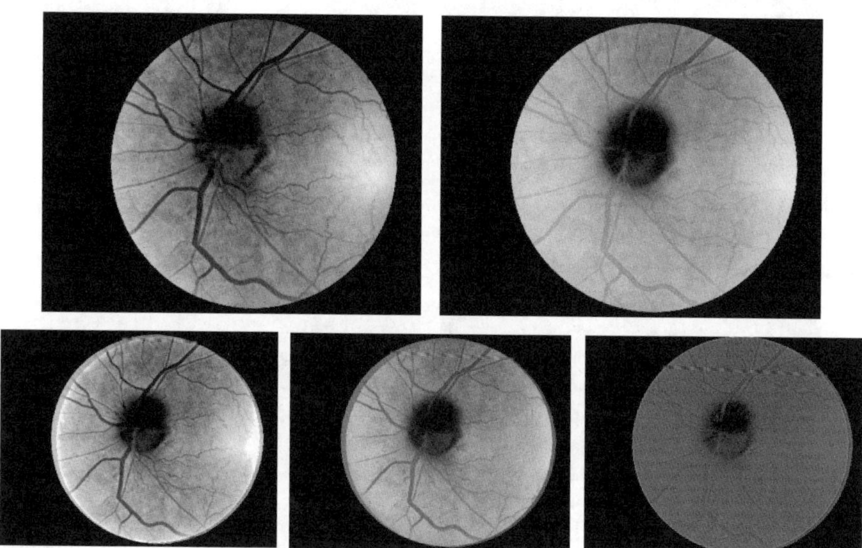

FIGURE 3.15 (See color insert) Angiograms taken at different stages and fused images with SIDWT (bottom-left), TNO (bottom-middle), and MIT (bottom-right) methods for an eye with anterior ischemic optic neuropathy.

the authors found that the contrast is better for the wavelet transform fusion method than for the average fusion method, a result which is hardly surprising. We believe it would have been a good idea to compare the wavelet transform fusion method with a better fusion method. Furthermore, the change in contrast between the source images and the fused image should have been evaluated and not only the change in contrast between the two fusion methods (average and wavelet transform). The false-color fusion methods are evaluated in Ref. 32 on the basis of the brightness contrast, the presence of complementary features without attenuation, the enhancement of common features, and the color contrast.

All fusion methods have qualitative advantages and drawbacks depending on the underlying medical context (application, disease type, subjective color appreciation, etc.).

2. Quantitative

Some quantitative criteria (standard deviation, entropy, cross-entropy, image noise index, spatial frequency) have been identified to objectively evaluate the fusion performance.[45–46]

a. Standard Deviation

The standard deviation criterion measures the contrast in an image; an image with a high contrast will have a high standard deviation. It is defined as

$$\sigma = \sqrt{\sum_{g=0}^{L-1} (g - \bar{g})^2 p(g)} \tag{3.26}$$

where L is the number of gray levels in the image, g is the gray level value, $\bar{g} = \sum_{g=0}^{L-1} g \cdot p(g)$ is the average gray value, and

$$p(g) = \frac{\text{number of pixels with value } g}{\text{total number of pixels}}$$

is the probability that a pixel has a value g.

b. Entropy

The entropy criterion measures the information content in an image and is defined as

$$H = -\sum_{g=0}^{L-1} p(g)\log_2 p(g) \text{ bits/pixel} \tag{3.27}$$

An image with high information content will have a high entropy.

c. Cross-Entropy

Cross-entropy evaluates the similarity in information content between images; images containing approximately the same information will have a low cross-entropy. For two images, it is defined as

$$\text{cen}(A : F) = \sum_{g=0}^{L-1} p(g)\log_2 \frac{p(g)}{q(g)} \tag{3.28}$$

where $p(g)$ and $q(g)$ are the probabilities that a pixel has a value g in images A and F, respectively; A being the first source images and F the fused image. The overall cross-entropy is defined as

$$\text{cen}_\alpha = \frac{\text{cen}(A:F) + \text{cen}(B:F)}{2},$$

$$\text{cen}_\beta = \sqrt{\frac{\text{cen}^2(A:F) + \text{cen}^2(B:F)}{2}} \tag{3.29}$$

This criterion is used in Ref. 32 in the form

$$\text{cen} = \frac{\text{cen}_\alpha + \text{cen}_\beta}{2} \tag{3.30}$$

d.   Image Noise Index

A drawback of the entropy is its incapacity to distinguish between noise and relevant information in an image since its value reflects both.[46] This distinction can be made using the image noise index, which is based on entropy difference. This criterion evaluates whether the fused image contains more relevant information than either of the source images. It is computed by first restoring the fused image to one of its source images through the reverse fusion process. If the forward and reverse fusion processes are perfect, the entropies of the fused and restored images should be identical. If not, this means the fusion process adds noise at either stage or both. Given the entropy of one of the source images H_I and of the fused and restored images H_F and H_R, the image noise index is defined as

$$\text{INI} = \frac{H_F - H_I}{|H_I - H_R|} - 1 \tag{3.31}$$

The entropy difference between the fused and one of the source images, $H_F - H_I$, represents the change in information content, which can be noise or relevant information. The entropy difference between one of the source images and the restored image, $|H_I - H_R|$, represents the noise. Thus, the addition of relevant information by the fusion process is $H_F - H_I - |H_I - H_R|$. The ratio between the relevant information ($H_F - H_I - |H_I - H_R|$) and the noise ($|H_I - H_R|$) is the image noise index. If $H_F - H_I > 0$, there is more information content in the fused image than in the source image I. But this increase in information content is an improvement only if $\text{INI} > 0$.

This criterion has two weaknesses. First, the $|H_I - H_R|$ term takes into account the noise coming from the reverse fusion process, which is applied to obtain the restored image R. This noise being not present in the fused image, the noise content is overestimated and the image noise index underestimated. Thus, the improvement brought by the fusion is underestimated. Second, only one of the source images is considered.

This criterion is used in Ref. 9. The authors obtained a positive image noise index, which indicates that the fusion increases the information content and that this increase is an improvement.

e. Spatial Frequency

The spatial frequency criterion SF is defined as

$$SF = \sqrt{(RF)^2 + (CF)^2} \tag{3.32}$$

where

$$RF = \sqrt{\frac{1}{MN} \sum_{m=0}^{M-1} \sum_{n=1}^{N-1} [F(m,n) - F(m, n-1)]^2}$$

$$\tag{3.33}$$

$$CF = \sqrt{\frac{1}{MN} \sum_{n=0}^{N-1} \sum_{m=1}^{M-1} [F(m,n) - F(m-1, n)]^2}$$

are, respectively, the row and column frequencies of the image, M and N the number of rows and columns and $F(m,n)$ the value of the pixel (m,n) of the fused image.

TABLE 3.4

Grayscale Fusion Methods, Implemented in the *fusetool* MATLAB™ Toolbox, Classified in Decreasing Performance Order According to Four Quantitative Performance Assessment Criteria and to the Combination of these Criteria[32]

Standard Deviation	Entropy	Cross-Entropy	Spatial Frequency	All Criteria
Maximum	Laplacian	SIDWT	Morphological	Laplacian
Morphological	SIDWT	Laplacian	DWT	SIDWT
Laplacian	Contrast	Contrast	Contrast	Contrast
Sidwt	Morphological	PCA	Laplacian	Morphological
Contrast	PCA	FSD	SIDWT	DWT
Dwt	DWT	Minimum	FSD	PCA
Pca	FSD	DWT	Gradient	FSD
Minimum	Average	Morphological	Maximum	Maximum
Ratio	Maximum	Average	Ratio	Minimum
Fsd	Minimum	Maximum	Minimum	Ratio
Gradient	Ratio	Gradient	PCA	Gradient
Average	Gradient	Ratio	Average	Average

The 12 grayscale fusion methods implemented in fusetool are evaluated in Ref. 32 with the standard deviation, entropy, cross-entropy, and spatial frequency criteria. Table 3.4 shows the classification of the methods, in decreasing order of performance, for each criterion. It shows also their classification according to the sum of their order of performance for the four criteria. For example, for the 45 image pairs fused with the SIDWT method, the contrast is improved by 4 ± 4, the information content by 0.7 ± 0.3 bit/pixel, and the spatial frequency by 4 ± 2. Select maximum is the best method according to the standard deviation criterion, but the fused images do not have a natural appearance and important dark features (microaneurisms, hemorrhages) are missing. This emphasizes the need for caution in using a single criterion to assess fusion performance.

V. CONCLUSION

Among all the papers published about registration and fusion of fundus images, some deserve particular attention. For example, the use of the orientations of the vessels surrounding a bifurcation was first introduced in Ref. 27. Using local structural information to facilitate the matching process was an important step in fundus image registration. This local structural information was used afterwards in Refs. 29,32,34. Another important paper is Ref. 10, who published an exhaustive review of the fundus registration literature. A first work addressing a kind of information fusion of fundus images is Ref. 19. Previously, ophthalmologists had to (1) draw manually a retinal map from a printout of each modality, (2) overlay them manually, and (3) find which area should be treated. Also, during the treatment, the image provided to the ophthalmologist only displayed the vessels. Finally, the work presented in Refs. 32,34 is the first to address the pixel-level fusion of fundus images and its quantitative performance evaluation.

This chapter aimed to review works in the field of registration and fusion of human retinal images. Unlike registration, fusion of ophthalmic images has not yet been thoroughly studied. The emerging interest in registration and fusion of fundus images can be explained by (1) the commercial availability of digital fundus cameras, which stimulates software development for image manipulation, enhancement and processing, and (2) the increasing workload of the ophthalmologists.

This chapter was divided into two main sections: registration and fusion. The main registration and fusion methods tested up to now on retinal images have been reviewed. The results presented cover a large range of diseases (diabetic retinopathy, age-related macular degeneration, branch retinal vein occlusion, cytomegalovirus retinitis, anterior ischemic optic neuropathy, choroidal neovascular membrane, cystoid macular edema, histoplasmosis retinitis/choroiditis, telangiectasia, etc.)

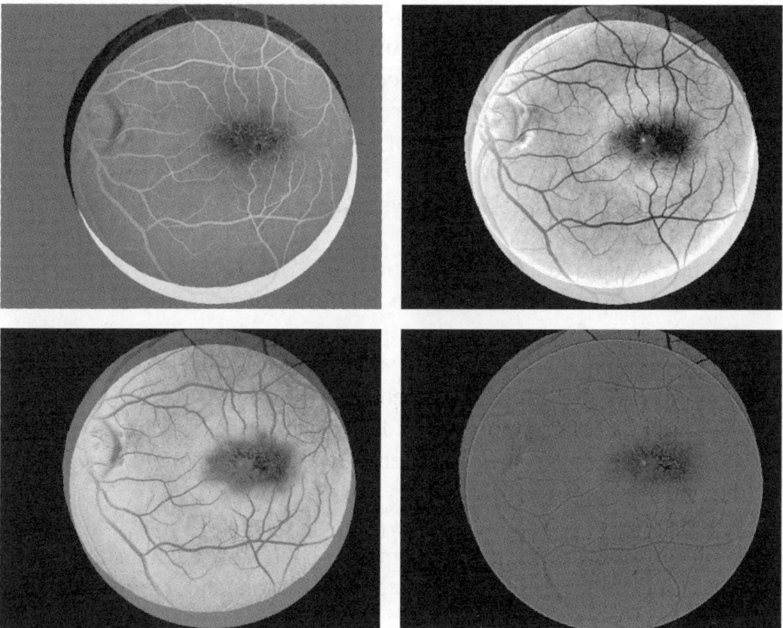

FIGURE 3.16 (See color insert) Comparison of image fusion and simple subtraction. Subtraction of the source images (top-left), fused image with the SIDWT (top-right), TNO (bottom-left), and MIT (bottom-right) methods.

Finally, let us mention a few issues we consider important for the future of this field:

- Image fusion is a promising tool to help in retinal diagnosis. However, simple alternatives should not be put aside. When evaluating fused images, ophthalmologists may be more interested in a simple subtraction of the source images than a fancy fused result. An example is shown in Figure 3.16 where the subtracted image shows all the capillaries in the macula area, unlike the fused images.
- When a user has to register fundus images, his natural choice is to use the anatomical features (vessels, optic disc, and macula) to align the images. After the review of the registration methods presented here, we think that the methods that use such a strategy have a clear advantage over the others. However, feature extraction still has to be improved. A strong co-operation with physicians is essential for this. Good feature extraction will have to rely on knowledge of these features, which may have been ignored until now. It is generally accepted that a registration algorithm is application specific so it is not beneficial to attempt to design an algorithm that will register any **type** of image. It is a better time investment to characterize an image type (human fundus

images in the case at hand) and to design a robust and efficient application-specific registration algorithm, ideally capable of providing a confidence level or deciding the success of a registration.

- Registration tools have to be refined to allow improved operation planning, real-time tracking of a fundus mosaic during manual photocoagulation treatment, and ultimately computer-controlled treatment. The reliable tracking of eye movements would also constitute a major improvement.

- Besides image fusion, registration is a necessary step for change detection. Ophthalmologists are interested, for example, in change detection in fluorescein angiogram sequences. Two factors are important: the fluorescein flow and intensity change in time. This kind of issue should be addressed in the future.

- Although image fusion is a standard topic in medical imaging, much work has still to be done in order to evaluate the potential benefits of the technique in ophthalmology.

- It would be very useful to build a publicly available database of graded images by experienced technicians or physicians. Such a standardized database would greatly facilitate registration and fusion algorithm evaluation and comparison.

ACKNOWLEDGMENTS

This work was financially supported by the Natural Science and Engineering Research Council (NSERC) of Canada, the Canadian Foundation for Innovation (CFI), the Fonds Québécois de la Recherche sur la Nature et les Technologies (FQRNT) and the Ministère du Développement Économique, de I'Innovation et de l'Exportation (MDEIE). The authors also thank Marie-Carole Boucher M.D. from Maisonneuve-Rosemont hospital (Montreal, QC) for her medical inputs.

REFERENCES

1. Saine, P. J., and Tyler, M. E., *Ophthalmic photography: retinal photography, angiography, and electronic imaging*, 2nd ed., Butterworth-Heinemann, Boston, 2002.

2. Marchand, D., and Wade, A., PC-based imaging technology helps advance research in ophthalmology, 1997; http://www.matrox.com/imaging/news_events/feature/archives/bio.cfm.

3. New York Eye and Ear Infirmary, Clinical services and specialty facilities, Optical coherence tomography clinical database; http://www.nyee.edu/page_deliv.html?page_no=118&origin=95.

4. Brown, L. G., A survey of image registration techniques, *ACM Comput. Surveys*, 24(4), 325–376, 1992.

5. Heneghan, C., Maguire, P., de Chazal, P., and Ryan, N., Retinal image registration using control points, pp. 349–352. In *Proceedings of International Symposium on Biomedical Imaging*, Washington, 2002.

6. Cideciyan, A. V., Registration of ocular fundus images by cross-correlation of triple invariant image descriptors, *IEEE Eng. Med. Biol.*, 14, 52–58, 1995.

7. Peli, E., Augliere, R. A., and Timberlake, G. T., Feature-based registration of retinal images, *IEEE Transactions on Medical Imaging*, 6(3), 272–278, 1987.

8. Yu, J. J. H., Hung, B. N., and Liou, C. L., Fast algorithm for digital retinal image alignment, pp. 374–375. In *Proceedings of International Conference of the IEEE Engineering in Medicine and Biology Society*, 1989.

9. Raman, B., Nemeth, S. C., Wilson, M. P., and Soliz, P., Clinical and quantitative assessment of multimodal retinal image fusion. In *Proceedings of SPIE Medical Imaging*, San Diego, CA, 2003.

10. Ritter, N., Owens, R., Cooper, J., Eikelboom, R. H., and van Saarloos, P. P., Registration of stereo and temporal images of the retina, *IEEE Trans. Med. Imaging*, 18(5), 404–418, 1999.

11. Barnea, D. I., and Silverman, H. F., A class of algorithms for fast digital registration, *IEEE Trans. Comput.*, C-21, 179–186, 1972.

12. Park, J., Keller, J. M., Gader, P. D., and Schuchard, R. A., Hough-based registration of retinal images, pp. 4550–4555. In *Proceedings of IEEE International Conference on Systems, Man and Cybernetics*, San Diego, CA, 1998.

13. Lloret, D., Serrat, J., López, A. M., Sole, A., and Villanueva, J. J., Retinal image registration using creases as anatomical landmarks, pp. 203–206. In *Proceedings of International Conference on Pattern Recognition*, Barcelona, 2000.

14. López, A. M., Lloret, D., Serrat, J., and Villanueva, J. J., Multilocal creaseness based on the level set extrinsic curvature, *Comput. Vis. Image Understand.*, 77, 111–144, 2000.

15. Dantzig, G. B., *Linear Programming and Extensions*, Princeton University Press, Princeton, 1963.

16. Lee, D. J., Krile, T. F., and Mitra, S., Power cepstrum and spectrum techniques applied to image registration, *Appl. Opt.*, 27, 1099–1106, 1988.

17. Corona, E., Mitra, S., and Wilson, M., A fast algorithm for registration of individual frames and information recovery in fluorescein angiography video image analysis, pp. 265–268. In *Proceedings of Southwest Symposium on Image Analysis and Interpretation*, Santa Fe, New Mexico, 2002.

18. Rosin, P. L., Marshall, D., and Morgan, J. E., Multimodal retinal imaging: new strategies for the detection of glaucoma, pp. 137–140. In *Proceedings of International Conference on Image Processing*, Rochester, New York, 2002.

19. Pinz, A., Bernogger, S., and Kruger, A., Mapping the human retina, *IEEE Trans. Med. Imaging*, 17(4), 606–619, 1998.

20. Berger, J. W., Leventon, M. E., Hata, N., Wells, W. M., and Kikinis R., Design considerations for a computer-vision-enabled ophthalmic augmented reality environment. In *Proceedings of First Joint Conference Computer Vision, Virtual Reality and Robotics in Medicine and Medial Robotics and Computer-Assisted Surgery*, Grenoble, 1997.

21. Huttenlocher, D. P., Klanderman, G. A., and Rucklidge, W. J., Comparing images using the Hausdorff distance, *IEEE Trans. Pattern Anal. Mach. Intell.*, 15(9), 850–863, 1993.

22. Matsopoulos, G. K., Mouravliansky, N. A., Delibasis, K. K., and Nikita, K. S., Automatic retinal image registration scheme using global optimization techniques, *IEEE Trans. Inf. Technol. Biomed.*, 3(1), 47–60, 1999.

23. Chauduri, S., Chatterjee, S., Katz, N., Nelson, M., and Goldbaum, M., Detection of blood vessels in retinal images using two-dimensional matched filters, *IEEE Trans. Med. Imaging*, 8(3), 263–269, 1989.

24. Jasiobedzki, P., Registration of retinal images using adaptive adjacency graphs, pp. 40–45. In *Proceedings of IEEE Symposium on Computer-Based Medical Systems*, Ann Arbor, Michigan, 1993.

25. Hart, W. E., and Goldbaum, M. H., Registering retinal images using automatically selected control point pairs, pp. 576–580. In *Proceedings of International Conference on Image Processing*, Austin, Texas, 1994.

26. Mendonca, A. M., Campilho, A. J. C., and Nunes, J. M. R., A new similarity criterion for retinal image registration, pp. 696–700. In *Proceedings of International Conference on Image Processing*, Los Alamitos, CA, 1994.

27. Zana, F., and Klein, J. C., A multimodal registration algorithm of eye fundus images using vessels detection and Hough transform, *IEEE Trans. Med. Imaging*, 18(5), 419–428, 1999.

28. Lalonde, M., Gagnon, L., and Boucher, M. C., Non-recursive paired tracking for vessel extraction from retinal images, pp. 61–68. In *Proceedings of Vision Interface*, Montreal, 2000.

29. Shen, H., Stewart, C. V., Roysam, B., Lin, G., and Tanenbaum, H. L., Frame-rate spatial referencing based on invariant indexing and alignment with application to online retinal image registration, *IEEE Trans. Pattern Anal. Mach. Intell.*, 25(3), 379–384, 2003.

30. Markow, M. S., Rylander, H. G., and Welch, A. J., Real-time algorithm for retinal tracking, *IEEE Trans. Biomed. Eng.*, 40(12), 1269–1281, 1993.

31. Barrett, S. F., Jerath, M. R., Rylander, H. G., and Welch, A. J., Digital tracking and control of retinal images, *Opt. Eng.*, 33(1), 150–159, 1994.

32. Laliberté, F., Gagnon, L., and Sheng, Y., Registration and fusion of retinal images: an evaluation study, *IEEE Trans. Med. Imaging*, 25(5), 661–673, 2003.

33. Early Treatment Diabetic Retinopathy Study (ETDRS), http://www.nei.nih.gov/neitrials/static/study53.htm.

34. Laliberté, F., Gagnon, L., and Sheng, Y., Registration and fusion of retinal images: a comparative study, pp. 715–718. In *Proceedings of International Conference on Pattern Recognition*, Quebec, 2002.

35. Rockinger, O., and Fechner T., Pixel-level image fusion: the case of image sequences, pp. 378–388. In *Proceedings of SPIE Signal Processing, Sensor Fusion, and Target Recognition VII*, San Diego, CA, 1998.

36. Burt, P. J., and Kolczynski, R. J., Enhanced image capture through fusion. pp. 173–182. In *Proceedings of Fourth International Conference on Computer Vision*, Berlin, 1993.

37. Li, H., Manjunath, B. S., and Mitra, S. K., Multisensor image fusion using the wavelet transform, *Graph. Models Image Process.*, 57(3), 235–245, 1995.

38. Burt, P. J., and Adelson, E. H., The Laplacian pyramid as a compact image code, *IEEE Trans. Commun.*, 31(4), 532–540, 1983.

39. Anderson, C. H., 1984. An Alternative to the Burt Pyramid Algorithm. RCA correspondence.

40. Toet, A., Image fusion by a ratio of low-pass pyramid, *Pattern Recognit. Lett.*, 9(4), 245–253, 1989.

41. Toet, A., A morphological pyramid image decomposition, *Pattern Recognit. Lett.*, 9(4), 255–261, 1989.

42. Toet, A., and Walraven, J., New false color mapping for image fusion, *Opt. Eng.*, 35(3), 650–658, 1996.
43. Waxman, A. M., Fay, D. A., Gove, A. N., Seibert, M. C., and Racamato, J. P., Method and apparatus for generating a synthetic image by the fusion of signals representative of different views of the same scene. U.S. Patent, 5,555,324, 1996.
44. Aguilar, M., and Garrett, A. L., Neurophysiologically-motivated sensor fusion for visualization and characterization of medical imagery. In *Proceedings of International Conference on Information Fusion*, Montreal, 2001.
45. Wang, Y., and Lohmann, B., Multisensor Image Fusion: Concept, Method and Applications, Technical report, University of Bremen, Germany, 2000.
46. Leung, L. W., King, B., and Vohora, V., Comparison of image data fusion techniques using entropy and INI. pp. 152–157. In *Proceedings of Asian Conference on Remote Sensing*, Singapore, 2001.

FURTHER READING

Becker, D. E., Can, A., Turner, J. N., Tanenbaum, H. L., and Roysam, B., Image processing algorithms for retinal montage synthesis, mapping, and real-time location determination, *IEEE Trans. Biomed. Eng.*, 45(1), 105–118, 1998.

Can, A., Stewart, C. V., Roysam, B., and Tanenbaum, H. L., A feature-based, robust, hierarchical algorithm for registering pairs of images of the curved human retina, *IEEE Trans. Pattern Anal. Mach. Intell.*, 24(3), 347–364, 2002.

Can, A., Stewart, C. V., Roysam, B., and Tanenbaum, H. L., A feature-based technique for joint, linear estimation of high-order image-to-mosaic transformations: mosaicing the curved human retina, *IEEE Trans. Pattern Anal. Mach. Intell.*, 24(3), 412–419, 2002.

Can, A., Stewart, C. V., Roysam, B., and Tanenbaum, H. L., A feature-based technique for joint, linear estimation of high-order image-to-mosaic transformations: application to mosaicing the curved human retina, pp. 2585–2591. In *Proceedings of Computer Vision and Pattern Recognition*, Hilton Head Island, South Carolina, 2000.

Can, A., Stewart, C. V., and Roysam, B., Robust hierarchical algorithm for constructing a mosaic from images of the curved human retina, pp. 286–292. In *Proceedings of Computer Vision and Pattern Recognition*, Fort Collins, CO, 1999.

Domingo, J., Ayala, G., Simo, A., de Ves, E., Martinez-Costa, L., and Marco, P., Irregular motion recovery in fluorescein angiograms, *Pattern Recognit. Lett.*, 18(8), 805–821, 1997.

Ege, B. M., Dahl, T., Sondergaard, T., Bek, T., Hejlesen, O. K., and Larsen, O. V., Automatic registration of ocular fundus images. In *Proceedings of Computer Assisted Fundus Image Analysis*, Herlev, 2000.

Ellis, E. L., Software tools for registration of digital ocular fundus imagery. In *Proceedings of Khoros Symposium*, Albuquerque, NM, 1997.

Ellis, E. L., Registration of Ocular Fundus Images, M.Sc. dissertation, University of New Mexico, Microelectronics Research Center, 1997.

Glazer, F., Reynolds, G., and Annanan, P., Scene matching by hierarchical correlation, pp. 432–441. In *Proceedings of Computer Vision and Pattern Recognition*, 1983.

Goldbaum, M. H., Kouznetsova, V., Cote, B., Hart, W. E., and Nelson, M., Automated registration of digital ocular fundus images for comparison of lesions, pp. 94–99. In *Proceedings of SPIE Ophthalmic Technologies III*, Los Angeles, CA, 1993.

Jagoe, R., Arnold, J., Blauth, C., Smith, P. L. C., Taylor, K. M., and Wootton, R., Retinal vessel circulation patterns visualized from a sequence of computer-aligned angiograms, *Investig. Ophthalmol Visual Sci.*, 34, 2881–2887, 1993.

Laliberté, F., Gagnon, L., and Sheng, Y., Registration and fusion of retinal images: a comparative study, pp. 715–718. In *Proceedings of International Conference on Pattern Recognition*, Quebec, 2002.

Lee, D. J., Krile, T. F., and Mitra, S., Digital registration techniques for sequential fundus images, pp. 293–300. In *Proceedings of SPIE Applications of Digital Image Processing X*, 1987.

Ling, A., Krile, T., Mitra, S., and Shihab, Z., Early detection of glaucoma using digital image processing, *Investig. Ophthalmol. Visual Sci.*, 27, 160, 1986.

Lloret, D., Marino, C., Serrat, J., Lopez, A. M., and Villanueva, J. J., Landmark-based registration of full SLO video sequences, pp. 189–194. In *Proceedings of Spanish National Symposium on Pattern Recognition and Image Analysis*, Castellon, Spain, 2001.

López, A. M., Lloret, D., Serrat, J., and Villanueva, J. J., Multilocal creaseness based on the level set extrinsic curvature, *Comput. Vis. Image Understand.*, 77, 111–144, 2000.

Markow, M. S., Rylander, H. G., and Welch, A. J., Real-time algorithm for retinal tracking, *IEEE Trans. Biomed. Eng.*, 40(12), 1269–1281, 1993.

Mendonca, A. M., Campilho, A. J. C., Restivo, F., and Nunes, J. M. R., Image registration of eye fundus angiograms, In *Signal Processing V: Theories and Applications*, Torres, L., Masgrau, E., Lagunas, M. A., Eds., Elsevier Science Publishers, pp. 943–946, 1990.

Nagin, P., Schwartz, B., and Reynolds, G., Measurement of fluorescein angiograms of the optic disc and retina using computerized image analysis, *Ophthalmology*, 92, 547–552, 1985.

Parker, J. A., Kenyon, R. V., and Young, L. R., Measurement of torsion from multitemporal images of the eye using digital signal processing techniques, *IEEE Trans. Biomed. Eng.*, 32, 28–36, 1985.

Pinz, A., Ganster, H., and Prantl, M., Mapping the retina by information fusion of multiple medical datasets, pp. 321–332. In *Proceedings of SPIE Human Vision, Visual Processing, and Digital Display VI*, San Jose, CA, 1995.

Pinz, A., Prantl, M., and Datlinger, P., Mapping the human retina. pp. 189–198. In *Proceedings of Scandinavian Conference on Image Analysis*, Uppsala, 1995.

Pinz, A., Prantl, M., and Ganster, H., A robust affine matching algorithm using an exponentially decreasing distance function, *J. Universal Comput. Sci.*, 1(8) 1995.

Pinz, A., Prantl, M., and Ganster, H., Affine matching of intermediate symbolic representations, pp. 359–367. In *Proceedings of Computer Analysis of Images and Patterns*, Prague, 1995.

Ritter, N., Owens, R., Yogesan, K., and van Saarloos, P. P., The application of mutual information to the registration of stereo and temporal images of the retina, pp. 67–76. In *Proceedings of Australian Joint Conference on Artificial Intelligence*, Perth, Australia, 1997.

Ryan, N., and Heneghan, C., Image registration techniques for digital ophthalmic images. In *Proceedings of Irish Signals and Systems Conference*, Galway, 1999.

Stewart, C. V., Tsai, C. L., and Roysam, B., The Dual-Bootstrap Iterative Closest Point Algorithm with Application to Retinal Image Registration. Technical Report, Rensselaer Polytechnic Institute, Troy, NY, 2002.

Walter, T., Klein, J. C., Massin, P., and Zana, F., Automatic segmentation and registration of retinal fluorescein angiographies. Application to diabetic retinopathy. In *Proceedings of Computer Assisted Fundus Image Analysis*, Herlev, 2000.

4 Nonrigid MR/US Registration for Tracking Brain Deformations

Xavier Pennec, Nicholas Ayache, Alexis Roche, and Pascal Cathier

CONTENTS

I. INTRODUCTION

The use of stereotactic systems is now a standard procedure for neurosurgery. However, these systems assume that the brain is in fixed relation to the skull during surgery. In practice, relative motion of the brain with respect to the skull (also called brain shift) occurs, mainly due to tumor resection, cerebrospinal fluid drainage, hemorrhage, or even the use of diuretics. Furthermore, this motion is likely to increase with the size of the skull opening and the duration of the operation.

Recently, the development of real time three-dimensional (3D) ultrasound (US) imaging has revealed a number of potential applications in image-guided surgery as an alternative approach to open MR and intrainterventional computed tomography (CT). The major advantages of 3D US over existing intraoperative imaging techniques are its comparatively low cost and simplicity of use. However, the automatic processing of US images has not developed to the same extent as other medical imaging modalities, probably due to the low signal to noise ratio of US images.

A. CONTEXT

In this chapter, we present a feasibility study of a tracking tool for brain deformations based on intraoperative 3D US image sequences. This work was performed within the framework of the European project ROBOSCOPE, a collaboration between the Fraunhofer Institute (Germany), Fokker Control System (Netherlands), Imperial College (UK), INRIA (France), ISM-Salzburg and Kretz Technik (Austria). The goal of the project is to assist neurosurgical operations using real time 3D US images and a robotic manipulator arm (Figure 4.1).

For the operation being planned on a preoperative MRI (MR1), the idea is to track in real time the deformation of anatomical structures using 3D US images acquired during surgery. To calibrate the system (i.e., to relate the MR

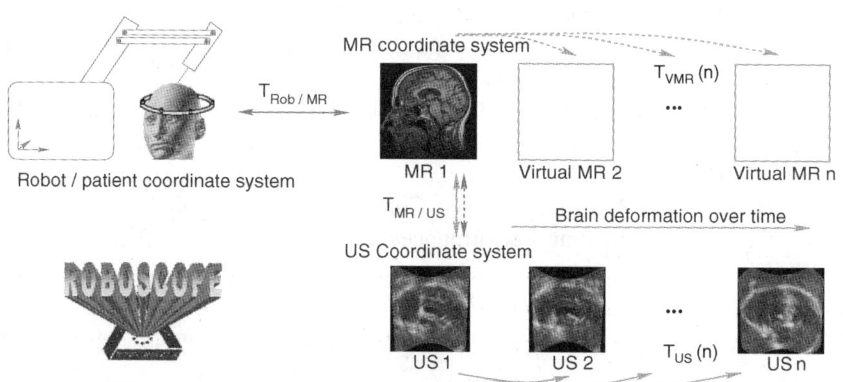

FIGURE 4.1 Overview of the image analysis part of the Roboscope project.

and the US co-ordinate systems) and possibly to correct for the distortions of the US acquisition device, a first US image (US1) has to be acquired with dura mater still closed in order to perform a rigid registration with the preoperative MR. Then, peroperative 3D US images are continuously acquired during surgery to track the brain deformations. From these deformations, one can update the preoperative plan and synthesize a virtual MR image that matches the current brain anatomy.

B. MR/US REGISTRATION

The idea of MR/US registration is already present in Ref. 1,2 where the US probe is calibrated (i.e., registered to the surgical space) and then tracked using an optical device. Using standard stereotactic neurosurgical procedures, they also register the MR image to the surgical space, thus relating the MR and US co-ordinate systems. This system has been improved[3] with the design of a real time, low cost US imaging system based on a PCI bus, while other[4] have used a DC magnetic position sensor to track the US probe. Some authors[5] proposed to interactively delineate corresponding surfaces in both MR and US images and perform the registration by visually fitting the surfaces using a 6D space-mouse. In addition, the outline of the 2D US image can be registered to the MR surface using a Chamfer matching technique.[6] All of these techniques only perform a rigid registration of the MR and the US image (i.e., the calibration of their co-ordinate systems) using external apparatus or interactive techniques.

For a nonrigid registration (i.e., a brain shift estimation),[7–9] the 2D US probe is still optically and rigidly tracked but the corresponding MR slice is displayed to the user who marks corresponding points on MR and US slices. Then, a thin plate spline warp is computed to determine the brain shift. This method was also enhanced with the possibility of using 3D US images and a deformation computed using a spring model instead of splines.[10] More recently, Ionescu et al.[11] registered US with CT data after automatically extracting contours from the US image using a watershed segmentation. In these studies, there is no processing of a full time sequence of US images. Rather, the brain shift estimation is limited to a few samples from time to time as the user interaction is required at least to define the landmarks.

To our knowledge, only one instance of automatic nonrigid MR/US registration has been recorded.[12] The idea is to register a surface extracted from the MR image to the 3D US image using a combination of the US intensity and the norm of its gradient in a Bayesian framework. The registration is quite fast (about 5 min), even if the compounding of the 3D US and the computation of its gradient takes much longer. However, experiments are presented only on phantom data and our experience (see Section IV) is that real US images may lead to quite different results.

Nevertheless, the first phase of all MR/US registrations is always a calibration step; that is, a rigid registration in order to relate the US probe position with the MR image co-ordinate system. Thus, we chose to split it into two subproblems. First, a rigid registration is performed between the MR image and a first US image acquired before the opening of the dura matter (there is no brain shift yet, so the transformation may safely be considered as rigid). Second, we look for the nonrigid motion within the US time-sequence.

C. TRACKING METHODS IN SEQUENCES OF US IMAGES

There are few articles on the registration of 3D US images. Strintzis and Kokkinidis[13] use a maximum likelihood approach to deduce a similarity measure for US images corrupted by a Rayleigh noise and a block matching strategy to recover the rigid motion. In Rohling et al.,[14] the correlation of the norm of the image gradient is used as the similarity measure to rigidly register two US images in replacement of the landmark-based RANSAC registration of Rohling et al.[15] However, these methods only deal with rigid motion and consider only two images, eluding the tracking problem.

One has to move to cardiac application to find some real tracking of nonrigid motion in US images. In Ref. 16, the endo- and epicardial surfaces are interactively segmented on each 2D image plane. Then, a shape-memory deformable model determines the correspondences between the points of the 3D surfaces of successive images. These correspondences are used to update an anisotropic linear elastic model (finite element mesh). The approach is appealing but relies once again on an interactive segmentation. In Sanchez-Ortiz et al.,[17] a combination of feature point extraction (phase based boundaries), and a multiscale fuzzy clustering algorithm (classifying the very low intensities of intraventricular pixels) is used to segment the surface of the left ventricular cavity. This process is done in 2D + time and then reconstructed in 3D. Thus, it exploits the whole sequence before tracking the motion itself, which is not possible for our application. These two methods are well suited for the shape of the cardiac ventricle using dedicated surface models. They could be adapted to the brain ventricles, but it seems difficult to extend them to the tracking of the volumetric deformations of the whole brain.

D. INTENSITY BASED NONRIGID REGISTRATION ALGORITHMS

Since feature or surface extraction is especially difficult in US images, we believe that an intensity based method can more easily yield an automatic algorithm. Over recent years, several nonrigid registration techniques have been proposed. Bajcsy and Kovačič[18] differentiated the linear correlation criterion and used a fixed fraction of its gradient as an external force to interact with a linear elasticity model.

Christensen[19] shows that the linear elasticity, valid for small displacements, cannot guarantee the conservation of the topology of the objects when the displacements become larger: the Jacobian of the transformation can become negative. Thus, he proposed a viscous fluid model of transformations as it can handle larger displacement. This model is also linearized in practice. Bro Nielsen started from the fluid model of Christensen and used the linearity of partial derivative equations to establish a regularization filter, several orders of magnitude faster than the previous finite element method.[20] He also justified his forces as the differential of the sum of square intensity differences (SSD) criterion, but he still used a fixed fraction of this gradient, and shows that Gaussian smoothing is an approximation of the linear elastic model.

Some authors[21] tried to apply to nonrigid registration some criteria developed for rigid or affine matching using block matching techniques. However, these criteria require a minimal window size, thus limiting the resolution of the result. Moreover, the regularization of the displacement field is usually implicit (i.e., only due to the integration of the criterion over the window), which means that it is difficult to explicitly control the regularity of the sought transformation.

Thirion[22] proposed to consider nonrigid registration as a diffusion process. He introduced in the images entities (demons) that push according to local characteristics of the images in a similar way Maxwell did for solving the Gibbs paradox in thermodynamics. The forces he proposed were inspired from the optical flow equations. This algorithm is increasingly used in several teams.[23-26] In Pennec et al.,[27] we investigated the nonrigid registration using gradient descent techniques. Differentiating the SSD criterion, we showed that the demons forces are an approximation of a second order gradient descent on this criterion. The same gradient descent techniques were applied to a more complex similarity measure in Cachier and Pennec[28] — the sum of Gaussian-windowed local correlation coefficients (LCC).

E. OVERVIEW OF THE ARTICLE'S ORGANIZATION

Section II of this chapter deals with the rigid registration (calibration) of the preoperative MR image with the first US image. The proposed method expands on the correlation ratio registration method[29] in order to deal with the specificities of the US acquisition. In essence, we have improved the method in following three distinct axes[30,31] — using the gradient information from the MR image, reducing the number of intensity parameters to be estimated, and using a robust intensity distance.

Section III develops an automatic intensity based nonrigid tracking algorithm suited for real time US image sequences, based on encouraging preliminary results.[27,32] We first present the registration method for two US images and how the method is turned into a tracking algorithm.[33]

In Section IV, we present some results of the rigid MR/US registration on clinical data (a baby and an actual surgical case), along with the results of an original evaluation of the registration accuracy. Then, we present qualitative results of the tracking algorithm on a sequence of 3D US animal images and a qualitative evaluation of the complete tracking system on a sequence of images of an MR and US compatible phantom.

II. RIGID MR/US REGISTRATION

A. CLASSICAL INTENSITY BASED SIMILARITY METRIC

Given two images I and J (considered as intensity functions of the position in space), the basic principle of intensity based registration methods is to search for a spatial transformation T such that the intensities of the transformed image $J \circ T$ are as similar as possible to the intensities of image I. The question is how to quantify the similarity between the two images intensities.

1. Sum of Squared Differences

If we assume that there are no intensity changes between the two images (except a Gaussian additive noise), then a good similarity metric (which may be justified by a maximum likelihood approach) is the SSD criterion:

$$\text{SSD}(T) = \|I - J \circ T\|^2 = \int [I(x) - J(T(x))]^2 \mathrm{d}x \qquad (4.1)$$

2. Correlation Coefficient (CC)

Now, let us assume that there is moreover a linear bias, so that the intensity $I(x)$ of image I at any point x may be modeled as $aJ(T(x)) + b$ plus some Gaussian noise, where a and b are some constant multiplicative and additive biases. A maximum likelihood approach leads to minimize the criterion

$$C(T, a, b) = \|I - aJ \circ T - b\|^2 = \int [I(x) - aJ(T(x)) - b]^2 \mathrm{d}x$$

Let us denote by $\bar{I}$ the mean intensity value of image I, $\text{Var}(I) = \|I - \bar{I}\|^2$ its variance, $\text{Cov}(I, J) = \int (I - \bar{I})(J - \bar{J}) \mathrm{d}x$ the covariance of image I and J, and, finally, $\rho(I, J) = \text{Cov}(I, J)/\sqrt{\text{Var}(I)\text{Var}(J)}$ the correlation coefficient between the two images. Computing the optimal values of the constants a and b and eliminating them, we obtain a criterion that only depends on the transformation:

$$C(T) = \text{Var}(I) - \frac{\text{Cov}(I, J \circ T)^2}{\text{Var}(J \circ T)} = \text{Var}(I)(1 - \rho(I, J \circ T)^2)$$

One problem is that we can compute this criterion only on the overlapping part of the transformed images. In order to avoid a minimum when the image overlap is small, we need to renormalize the criterion: a good choice, justified by Roche et al.,[34] is to look for a large variance of I in the overlapping region (we are trying to register informative parts), so that the criterion becomes:

$$C(T) = \frac{\|I - aJ \circ T - b\|^2}{\text{Var}(I)} = 1 - \rho(I, J \circ T)^2 \qquad (4.2)$$

where the integrals are computed over the intersection of the domains of I and $J{\circ}T$. Thus minimizing $C(T)$ yields the transformation that maximizes the squared or absolute value of the correlation coefficient $\rho(I, J{\circ}T)$ between the two images.

3. Correlation Ratio (CR)

Now, let us assume that the intensity mapping between image J and image I is an arbitrary functional relationship instead of being simply linear. Using the same arguments as above, the renormalized maximum likelihood criterion is now:

$$C(T,f) = \frac{\|I - f(J{\circ}T)\|^2}{\mathrm{Var}(I)} \tag{4.3}$$

This formulation is asymmetric in the sense that the cost function changes when permuting the roles of I and J. Since the positions and intensities of J actually serve to predict those of I, we will call J the "template image". In the context of US / MR image registration, we always choose the MR image as the template.

If no constraint is imposed to the intensity mapping f, then an important result is that the optimal f at fixed T enjoys an explicit form that is very fast to compute.[29] The minimization of Equation 4.3 may then be performed by traveling through the minima of $C(T,f)$ at fixed T. This yields the correlation ratio, $\eta^2_{I|J}(T) = 1 - \min_f C(T,f)$, a measure that reaches its maximum when $C(T,f)$ is minimal. In practice, the maximization of η^2 is performed using Powell's method.[35] Another important point is the discretization scheme used to compute the criterion, leading to the choice of an interpolation scheme.[34] In this paper, we use the partial volume interpolation,[36] which avoids many artifacts and local minima.[37,38]

B. BIVARIATE CORRELATION RATIO

US images are commonly said to be "gradient images" as they enhance the interfaces between anatomical structures. The physical reason is that the amplitudes of the US echoes are proportional to the squared difference of acoustical impedance caused by successive tissue layers. Ideally, the US signal should be high at the interfaces, and low within homogeneous tissues.

Thus, assuming that the MR intensities describe homogeneous classes of tissues amounts leads to considering the acoustic impedance Z as an unknown function of the MR intensities: $Z(x) = g(J(x))$. Now, when the US signal emitted from the probe encounters an interface (i.e., a high gradient of Z), the proportion of the reflected energy is $R = \|\nabla Z\|^2/Z^2$. Adding a very simple model of the log compression scheme used to visualize the US images, we obtain the following US image acquisition model: $I(x) = a \log(\|\nabla Z\|^2/Z^2) + b + \varepsilon(x)$.

Using $Z(x) = g(J(x))$ finally gives an unknown bivariate function: $I(x) = f(J(x), \|\nabla J(x)\|) + \varepsilon(x)$. Our new CR criterion is then:

$$C(T, f) = \frac{\|I - f(J \circ T, \|\nabla J \circ T\|)\|^2}{\mathrm{Var}(I)} \tag{4.4}$$

The MR gradient is practically computed by convolution with a Gaussian kernel.

In this rather simple modeling of the relationship between the two images, we ignored the incidence angle between the scan direction u of the US probe and the "surface normal" (the gradient). Ideally, we should use the cross product $\langle \nabla J \circ T, u \rangle$ instead of just the gradient norm $\|\nabla J \circ T\|$. However, recomputing the dot product image at each transformation trial entails a massive increase of the computation time in a straightforward implementation. Moreover, the received echo is actually much less anisotropic than would be the case with a perfectly specular reflection because of the diffraction of the US beam at the interfaces. Thus, we believe that ignoring the gradient direction is a good first order approximation on the theoretical side and a good trade-off from the computational point of view.

Another feature of the US images that has been occulted in our functional model is the presence of speckle. Reflection of the US waves do not only occur at acoustic impedance boundaries, but also within tissues due to small inhomogeneities (compared to the US wavelength) that are almost invisible in the MR. As a consequence, homogeneous tissue regions generally appear in the US image with a nonzero mean intensity and a strong texture. Also, since our model does not take attenuation into account, unrealistic intensity values are predicted in some areas, especially outside the skull. These limitations of our model strongly suggest to robustify our criterion, which will be done in Section II.D.

C. PARAMETRIC INTENSITY FIT

We are now looking for a bivariate intensity mapping f with floating values for the MR gradient component, instead of a simple function of already discretized intensity values in the standard correlation ratio registration method. The problem is that there is no natural way to discretize the MR gradient values and even if we found one that respects the dynamic of the gradient values, the number of parameters of the (discretized) function f grows as the square (e.g., 256^2 instead of 256 only for the classical CR). Thus, instead of the close-form solution of the discrete version, we chose to use a continuous but regularized description of the mapping f: we restrain our search to a polynomial function f of degree d. The number of parameters describing f then reduces to $(d + 1)(d + 2)/2$. In this paper, the degree was set to $d = 3$, implying that ten coefficients were estimated. Finding the coefficients of the polynomial minimizing 4.4 leads to a weighted least square linear regression problem, which is solved using the singular value decomposition (SVD).

However, this polynomial fitting procedure adds a significant extra computational cost and cannot be done for each transformation trial of a Brent/Powell minimization as before. Instead, we chose to perform an alternated optimization of the criterion along the transformation T and the intensity mapping f. In the first step, we find the best polynomial f (at a fixed T) and remap J and $\|\nabla J\|$ accordingly. In the second step, we minimize $C(T,f)$ with respect to T using Powell's method given the intensity corrected image $f(J, \|\nabla J\|)$. This iterative process is stopped when T and f do not evolve any more. This alternate minimization strategy saves us a lot of computation time (speed up factors are in the range of two to ten when setting the polynomial degree to $d = 3$). It is guaranteed to converge at least to a local maximum of the registration criterion. In practice, we did not observe any alteration of the performances with respect to the original technique.

D. ROBUST INTENSITY DISTANCE

Our method is based on the assumption that the intensities of the US may be well predicted from the information available in the MR. Due to several US artifacts, we do not expect this assumption to be perfectly true. As discussed above, shadowing, duplication or interference artifacts (speckle) may cause large variations of the US intensity from its predicted value, even when the images are perfectly registered. To reduce the sensitivity of the registration criterion to these outliers, we propose to use a robust estimation of the intensity differences using a one-step S-estimator.[39] The quadratic error function $\int [I(x) - f(J(T(x)))]^2 dx$ is then replaced with:

$$S^2(T,f) = \frac{S_0^2}{K} \int \Phi\left(\frac{[I(x) - f(J(T(x)))]}{S_0} \right) dx,$$

where K is a normalization constant that ensures consistency with the normal distribution, and S_0 is some initial guess of the scale. In our implementation, we have opted for the Geman-McClure redescending function $\Phi(x) = \frac{1}{2}x^2/(1 + x^2/c^2)$ for its computational efficiency and good robustness properties, to which we always set a cutoff distance $c = 3.648$ corresponding to 95% Gaussian efficiency.

The new registration criterion requires very few modifications in our alternate optimization scheme: the polynomial function f is now estimated using a simple iterative WLS procedure and the transformation is still found using Powell's method. Initially, the intensity mapping f is estimated in a nonrobust fashion. The starting value S_0 is then computed as the median of absolute intensity deviations. Due to the initial misalignment, it tends to be overestimated and may not efficiently reject outliers. For that reason, it is re-estimated after each alternated minimization step.

III. TRACKING DEFORMATIONS IN US IMAGES

Now that we have the method to register the MR image to the first US image, the goal is to track the brain deformations in 3D US time sequences. When analyzing the problem, we made the following observations. First, deformations are small between successive images in a real time sequence, but there are possibly large deformations around the surgical tools with respect to the preoperative image. Thus, the transformation space should allow large deformations, but only small deformations have to be retrieved between successive images. Second, there is a poor signal to noise ratio in US images and the absence of information in some areas. However, the speckle (inducing localized high intensities) is usually persistent in time and may produce reliable landmarks for successive images.[40] As a consequence, the transformation space should be able to interpolate in areas with little information while relying on high intensity voxels for registration of successive images. Last but not least, the algorithm is designed in view of a real time registration during surgery, which means that, at equal performances, one should prefer the fastest method.

Following the encouraging results obtained by Pennec et al.[27] for the intensity based free-form deformation (FFD) registration of two 3D US images, we have adapted the method in Pennec et al.[33] according to the previous observations. In the sequel of this section, we first detail the parameterization of our nonrigid transformations. Then we investigate the similarity criterion and the optimization strategy. Finally, we show how to turn this registration algorithm into a tracking tool suited for time sequences.[32]

A. PARAMETERIZATION OF THE TRANSFORMATION

Simple transformations, like rigid or affine ones, can be represented by a small number of parameters (6 and 12 in 3D). When it comes to FFD, we need to specify the co-ordinates $T(x)$ of each point x of the image after the transformation. Such a nonparametric transformation is usually represented by its displacement field $U(x) = T(x) - x$ (or $U = T - Id$), sampled at each voxel, and a regularization term has to be added to the criterion in order to reduce the effective number of degrees of freedom (DoF). This strategy proved to be successful in textured regions but induces some convergence problems in large uniform areas (as it is the case in the phantom sequence of Section IV.A.3).

We found that a reparameterization of the transformation was necessary to promote a better conditioning of the problem. In a standard FFD, we have a displacement t_i for each voxel position x_i. Let now t_i be a parameter of a smooth transformation defined by:

$$T(t_1, \ldots, t_n)(x) = \sum_i t_i G_\sigma(x - x_i) \qquad (4.5)$$

Note that when σ goes to 0, the parameterization tends toward the standard FFD description. Since the transformation is described as a sum of Gaussians, rather than a sum of Diracs, the gradient descent algorithm uses the derivatives of the similarity with respect to the displacement of an entire group of voxels, which is more robust to noise and may propagate the motion more rapidly in uniform areas. This new parameterization can be seen as a regularization of the gradient of the similarity energy (see Section III.D) and not as a regularization of the transformation as we can still interpolate any displacement value at each site x_i.

B. SIMILARITY ENERGY

Even if there is a poor signal to noise ratio in US images, the speckle is usually persistent in time and may produce reliable landmarks within the time sequence.[40] Hence, it is desirable to use a similarity measure which favors the correspondence of similar high intensities for the registration of successive images in the time-sequence. First experiments presented in Pennec et al.[27] indicated that the simplest one, the SSD, could be suited. Let I be the reference image and $J{\circ}T$ the transformed image to register; the criterion to minimize is:

$$\text{SSD}(T) = \int (I - J{\circ}T)^2 \mathrm{d}x,$$

In Cachier et al.,[28] we developed a more complex similarity measure — the sum of Gaussian windowed local correlation coefficients. Let $G{\star}f$ be the convolution of f by the Gaussian, $\bar{I} = (G{\star}I)$ the local mean, $\text{Var}(I) = G{\star}(I - \bar{I})^2$ the local variance, $\text{COV}(I, J{\circ}T) = G{\star}[(I - \bar{I})(J{\circ}T - \overline{J{\circ}T})]$ the local covariance between image I and image $J{\circ}T$, and finally $\rho(I, J{\circ}T) = \text{COV}(I, J{\circ}T)/\sqrt{\text{Var}(I)\text{Var}(J{\circ}T)}$ the LCC. Contrary to Section II.A, where the mean, variance, covariance, and correlation coefficient were global (i.e., constants computed for the whole image), these values are here functions of the position in space, just like the images, because of the locality of their computation. In particular, we cannot optimize the LCC directly: we have to integrate it to obtain the sum of the LCC:

$$\text{LCC}(T) = \int \rho(I, J{\circ}T)(x)\mathrm{d}x = \int \frac{\text{COV}(I, J{\circ}T)}{\sqrt{\text{Var}(I)\text{Var}(J{\circ}T)}}\,\mathrm{d}x$$

We have shown elsewhere[27,28] how the SSD and LCC criteria can be optimized using first and second order gradient descent techniques with a general FFD field by computing the gradient and the Hessian of the criteria. We summarize the main ideas in the following sections.

C. MINIMIZING THE SSD FOR A FREE-FORM DEFORMATION

Let us consider first the SSD criterion and the standard FFD description of the transformation. Let T be the current estimation of the transformation and $(\nabla_J{\circ}T)(x)[(H_J{\circ}T)(x)]$ be the transformed gradient (Hessian) of the image J.

The gradient and the Hessian of the SSD criterion are:

$$\nabla_{\text{SSD}}(T) = 2(J{\circ}T - I)(\nabla_J{\circ}T),$$

$$H_{\text{SSD}}(T) = 2(\nabla_J{\circ}T)(\nabla_J{\circ}T)^T + 2(J{\circ}T - I)(H_J{\circ}T)$$

Let us now approximate the criterion by its tangential quadratic form at the current transformation T, or equivalently consider a first order approximation of the criterion gradient in the first order for a small perturbation by a displacement field $u(x)$. We have: $\nabla_{\text{SSD}}(T + u) \simeq \nabla_{\text{SSD}}(T) + H_{\text{SSD}}(T)u$. Assuming that the Hessian matrix of the criterion is positive definite, the minimum is obtained for a null gradient; that is, for $u = -H_{\text{SSD}}^{(-1)}(T)\nabla_{\text{SSD}}(T)$. This formula requires one to invert the Hessian matrix $H_{SSD}(T)$ at each point x of the image. To speed up the process, we approximate this matrix by the closest scalar matrix. Using this approximation, we obtain the following adjustment vector field:

$$u \simeq \frac{-3(J{\circ}T - I)(\nabla_J{\circ}T)}{\|\nabla_J{\circ}T\|^2 + (J{\circ}T - I)(\Delta_J{\circ}T)}$$

In fact, when minimizing the reverse SSD criterion $\int (I{\circ}T^{(-1)} - J)^2 dx$, one finds that the optimal adjustment is given by:[27]

$$\hat{T} = T{\circ}(Id + u') \quad \text{with} \quad u' = \frac{3(I - J{\circ}T)\nabla_I}{\|\nabla_I\|^2 + (I - J{\circ}T)\Delta_I}$$

which justifies the empirical force used by Thirion's demons:

$$v = \frac{(I - J{\circ}T)\nabla_I}{\|\nabla_I\|^2 + \alpha(I - J{\circ}T)^2}$$

In practice, we have modified the Newton optimization scheme described above into a Levenberg–Marquardt method where the adjustment vector field is given at each step by $u = -(\lambda Id + H_{SSD})^{(-1)}\nabla_{SSD}$. Dropping the (possibly negative) second order terms in the Hessian, we are left with:

$$u = \frac{-3(J{\circ}T - I)(\nabla_J{\circ}T)}{\|\nabla_J{\circ}T\|^2 + \lambda^2} \tag{4.6}$$

The parameter λ performs a trade-off between a first order gradient descent ($\lambda \gg 1$ means that we do not trust the approximated Hessian matrix and we simply go along the gradient with a small time step) and a second order gradient descent ($\lambda \ll 1$ means that we use our simplified Hessian matrix). At each step, λ is divided by a fixed value α (typically 5) if the similarity criterion decreased, and the criterion is re-estimated with λ multiplied by α otherwise until the criterion decreases.

D. Minimizing the Similarity Energy for the New Parameterization

We now detail the differences induced by our new parameterization of the free-form transformation on the SSD criterion. Using the Gaussian parameterization of the transformation 4.5, t_i is now a parameter of the transformation. Deriving the SSD with regard to this parameter gives:

$$\nabla_{\text{SSD}}(T) = 2G_\sigma \star ((J \circ T - I)(\nabla J \circ T))$$

Thus, the Gaussian parameterization acts as a smoothing on the gradient of the energy. Therefore, it will be more robust and may escape from previous local minima. The minimization is performed as above with a Levenberg–Marquardt method using these regularized versions of the energy derivatives.

We now turn to the minimization of the LCC criterion. We have detailed elsewhere[28] how to compute efficiently this criterion and its gradient using convolutions. Our conclusion was that the following approximation of the gradient was performing an ideal trade-off with regard to computation time:

$$\nabla_{\text{LCC}} \simeq \left((I - \bar{I}) - (J \circ T - \overline{J \circ T}) \frac{(\overline{I J \circ T} - \bar{I}\,\overline{J \circ T})}{\sqrt{\text{Var}(J \circ T)}} \right) \frac{\nabla_J \circ T}{\sqrt{\text{Var}(I)}\sqrt{\text{Var}(J \circ T)}}$$

As above for the SSD, using our Gaussian parameterization of the displacement field amounts to convolving this gradient with G_σ.

However, as LCC is not a least square problem, it seems difficult to derive a Gauss–Newton minimization scheme as we have done for the SSD. Instead, we remarked that Equation 4.6 may be rewritten:

$$u(x) = -3 \frac{2E(x)\nabla E(x)}{\|\nabla E(x)\|^2 + 4\lambda^2 E(x)}$$

where $E(x) = (I(x) - (J \circ T)(x))^2$ is our local similarity energy and $\nabla E(x) = 2(I(x) - (J \circ T)(x))\nabla J \circ T(x)$ is its gradient with regard to the transformation $T(x)$. With this formulation, it is now easy to replace the local similarity energy by our LCC (and its gradient). In practice, we found that this adaptation performed very well despite its weak theoretical background.

E. Regularization Energy

As noted above, there are too many DoF with FFDs, and we have to regularize them. Thus, there is a trade-off to find between the similarity energy, reflected by the visual quality of the registration, and the smoothing energy, reflected by the regularity of the transformation.

In the regularization theory framework, one minimizes the weighted sum of the energies: $E_{\text{sim}} + \lambda E_{\text{reg}}$. This formulation has proven to be successful for data approximation, and has been used for various approaches of nonrigid registration algorithms.[41] However, there is an important difference between data approximation and image registration. In data approximation, both energies measure different properties of the same object (the similarity and the smoothness of the data), while the two energies relate to different objects in image registration (the intensities of the images for the matching energy and the transformation for the regularization energy). Thus, one has to find a nonlinear trade-off between the two energies.

Another widely employed method attempts to separate the image measure from the transformation measure, and could be compared with the approach of game theory. It consists of alternatively decreasing the similarity energy and the smoothing energy. This approach is chosen in many block-matching algorithms[42] and in some optical flow based techniques.[22] In view of a real time system, this is particularly well suited for the stretch energy (or membrane model) $E_{\mathrm{reg}} = \|\nabla T\|^2 = \int \mathrm{Tr}(\nabla T \cdot \nabla T^{\mathrm{T}})$ as the associated Euler–Lagrange evolution equation corresponds to the heat propagation in a homogeneous material. Thus, one step of gradient descent corresponds to convolution of the transformation by a Gaussian with a standard deviation linked to the time step of the gradient descent.[43] In this way, we obtain a simple regularization step that corresponds to the Gaussian smoothing of the transformation parameters t_i with a smoothing parameter (the σ_T of this Gaussian) that has a physical meaning. The final algorithm consists of alternatively performing one step of gradient descent on the similarity energy E_{sim} and one step of smoothing by Gaussian filtering of standard deviation σ_T.

F. FROM REGISTRATION TO TRACKING

In the previous sections, we studied how to register two US images together. We now have to estimate the deformation of the brain between the first image (which is assumed to correspond to the MR image of the preoperative brain) and the current image of the sequence. One could directly register US_1 (taken at time t_1) and US_n (at time t_n), but the deformations could be quite large and the intensity changes important. To constrain the problem, we need to exploit the temporal continuity of the deformation.

First, assuming that we already have the deformation $T_{\mathrm{US}}(n)$ from image US_1 to US_n, we register US_n with the current image US_{n+1}, obtaining the transformation $dT_{\mathrm{US}}(n)$. If the time step between two images is short with respect to the deformation rate (which should be the case in real time sequences at a rate ranging from one to five images per second), then this registration should be easy. Moreover, the intensity changes should be small. For this step, we believe that the SSD criterion is well adapted.

Then, composing with the previous deformation, we obtain a first estimation of $T_{\mathrm{US}}(n+1) \simeq dT_{\mathrm{US}}(n) \circ T_{\mathrm{US}}(n)$. However, the composition of deformation fields involves interpolations and just keeping this estimation would finally lead to a

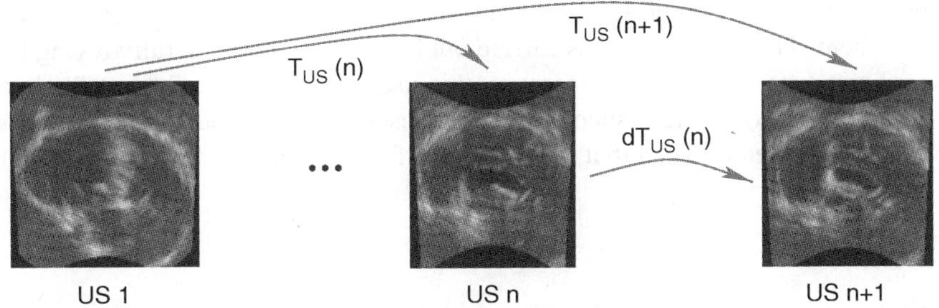

FIGURE 4.2 The deformations computed in the tracking algorithm.

disastrous cumulation of interpolation errors:

$$T_{US}(n + 1) = dT_{US}(n){\circ}dT_{US}(n - 1)...dT_{US}(2){\circ}dT_{US}(1)$$

Moreover, a small systematic error in the computation of $dT_{US}(n)$ leads to a huge drift in $T_{US}(n)$ as we go along the sequence (Figure 4.2).

Thus, we only use $dT_{US}(n){\circ}T_{US}(n)$ as an initialization for the registration of US_1 to US_n. Starting from this position, the residual deformation should be small (it corresponds to the correction of interpolation and systematic error effects) but the difference between homologous point intensities might remain important. In this case, the LCC criterion might be better than the SSD one, despite its worse computational efficiency. In practice, we run most of the experiments presented in Section I with the SSD and LCC criterion without finding significant differences. Since the LCC is still around 2 times slower than the SSD and since the computation time of the US–US nonrigid registration is a key issue for real time motion tracking, we preferred to keep the SSD criterion. We believe that this choice could be reconsidered if the sequence was to present some important intensity changes along time.

One of the main consequences of our tracking method is that the first US image has to be of very high quality since it will be the only reference for tracking deformations along the whole sequence. One possibility consists of acquiring several images of the still brain in order to compute a mean image of better quality. Another possibility consists of performing some anisotropic diffusion on US_1 to improve its quality.

IV. EXPERIMENTS

In this section, we present quantitative results of the rigid MR/US registration algorithm on real brain images, and qualitative results of the tracking algorithm and its combination with the MR/US registration on animal and phantom sequences. The location of the US probe being linked to the pathology and its orientation being arbitrary (the rotation may be superior to 90°), it was necessary to provide a rough initial estimate of the MR/US transformation. This was done using an interactive interface that allows one to draw lines in the images and match them. This procedure was carried out by a nonexpert, generally taking less than 2 min (see Figure 4.3). However, this user interaction could be alleviated using a calibration system such as the one described in Pagoulatos et al.[4] After initialization, we observed that the algorithm found residual displacements up to 10 mm and 10°.

A. DATA

All 3D-US images were acquired using a commercial 3D-US volume scanner Voluson 530 D from Kretz Technology (4 to 9 MHz, 90° aperture). It should be emphasized that all the US images provided to us in this project were already resampled in a Cartesian format. As a consequence, the images suffer from interpolation (blurring) artifacts in areas far from the probe, and the details close to the probe are averaged out. Moreover, border parts of the images were often clipped to fit into a cubic shape, which implied some important loss of information. In the future, it will be important to deal directly with the original geometry of the US images.

1. Baby Dataset

This clinical dataset was acquired to simulate the degradation of the US image quality with respect to the number of converters used in the probe. Here, we have one MR T1 image of a baby's head and five transfontanel US images with different percentages of converters used (40,60,70,90, and 100%). The MR image has $256 \times 256 \times 124$ voxels of size 0.9 mm^3. The Cartesian US images have $184 \times 184 \times 184$ voxels of size $0.29 \times 0.29 \times 0.29$ mm^3.

As we have no deformations within the images, we can only rigidly register all the US images onto our single MR. An example result is presented in Figure 4.4. The visual quality of the registration is very good. We did not notice a significant change of the registration result with respect to the number of converters used in the US probe. This emphasizes the robustness of the information used in the US image for the registration (basically the midsagittal plane and the ventricular surface). However, other experiments with neonates images showed that the quality of the MR was a crucial parameter for the registration to succeed, since it is difficult to keep a neonate motionless in a MRI scanner all along the acquisition time.

2. Patient Images During Tumor Resection

This dataset is an actual surgical case: two MR T1 images with and without a contrast agent were acquired before surgery. After craniotomy (dura mater still closed), a set of 3D US images was acquired to precisely locate the tumor to resect. The MR images have $256 \times 256 \times 124$ voxels of size $0.9 \times 0.9 \times 1.1$ mm^3, while the US images have various dimensions with a cubic voxel size ranging from 0.17 to 0.95 mm.

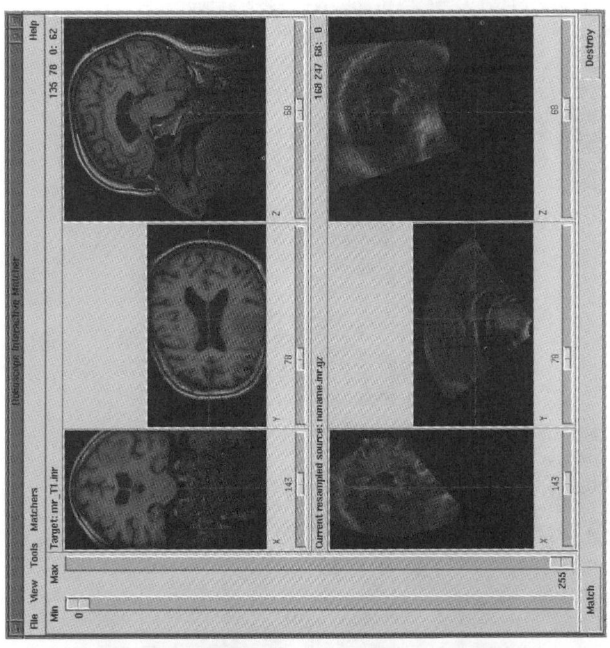

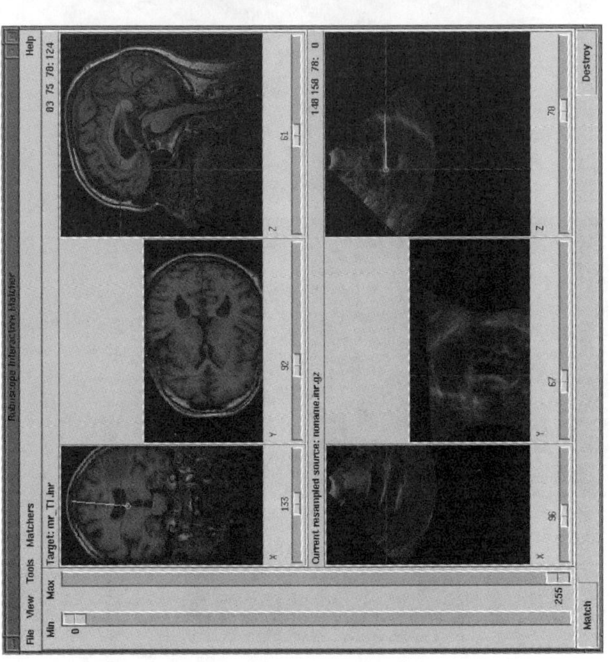

FIGURE 4.3 Interactive interface to perform the MR/US initial registration. Left: the users specifies a corresponding point and direction in one of the slice of each image. Right: the software computes the 3D rigid transformation that superimposes the two slices and the user features. The process is usually iterated twice with the remaining slices in order to better adjust the transformation. The whole process generally takes less than 2 min.

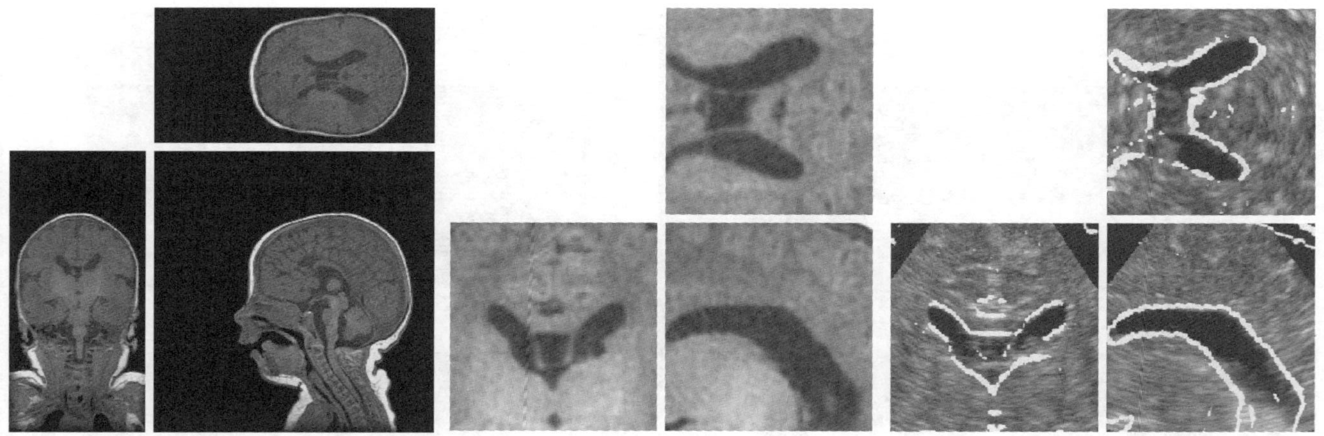

FIGURE 4.4 Example registration of MR and US images of the baby dataset. From left to right: original MR T1 image, closeup on the ventricle area, and registered US image with MR contours superimposed.

In this experiment, we use the three US images that are large enough to contain the ventricles. Unfortunately, we could only test for the rigid MR/US registration as we have no US images during surgery. An example of the registration results is presented in Figure 4.5. One can see that the ventricles and some of the sulci visible in the US image are very well matched. When looking at the image superimposition more closely (using a dynamic fusion visualization), it appears that a small residual deformation (about 1 to 2 mm maximum) is not corrected. This deformation could be due to the sound speed that varies depending on the brain tissues (it is assumed to be constant for the US image reconstruction), or to a real brain deformation during the opening of the skull. More experiments would be needed to determine the right hypothesis. Nevertheless, the validation scheme of Section IV.B.3 shows that a registration accuracy of 0.6 mm at the center and 1.6 mm in the whole brain area is achieved. This example is interesting as it demonstrate that the MR/US registration could be performed in clinical conditions with a sufficient accuracy.

3.   A Phantom Study

For the evaluation of the US images acquisitions, Professor Auer and his colleagues at ISM (Austria) developed an MR and US compatible phantom to simulate brain deformations. It is made of two balloons, one ellipsoid and one ellipsoid with a "nose", which can be inflated with known volumes. Each acquisition consists of one 3D MR and one 3D US image. Both balloons were initially filled with 40 ml of fluid. During a first phase, only the small ellipsoidal balloon was filled by steps of 10 ml (Acquisitions 1 to 4 in Figure 4.7). Then, this balloon was deflated, while the "nose" balloon was filled (still by steps of 10 ml), thus creating much larger deformations (acquisitions 5 and 6 in Figure 4.7). The MR images have $256 \times 256 \times 124$ voxels with a voxel size of $0.9 \times 0.9 \times 1 \text{ mm}^3$. The Cartesian US images are cubic with 184 voxels of size 0.41 mm on each side.

The first MR image was rigidly registered to the first US image with our bivariate CR method. An example result is provided in Figure 4.6. Then, deformations are estimated using the tracking algorithm on the US sequence, and the corresponding virtual MR image is computed. Since the US probe had to be removed from the phantom for each MR scans, we had to rigidly reregister all the US images before tracking in order to minimize the amount of global motion between US frames. Due to this resampling, and to the too focused conversion from polar to Cartesian co-ordinates by the US machine, one can observe some missing information in the borders (first row of Figure 4.7). Finally, the remaining MR images can be used to assess the quality of the tracking thanks to a rigid MR/US registration for each acquisition (third and fourth row of Figure 4.7).

Even if there are very few salient landmarks (all the information is located in the thick and smooth balloons boundaries, and thus the tracking problem is

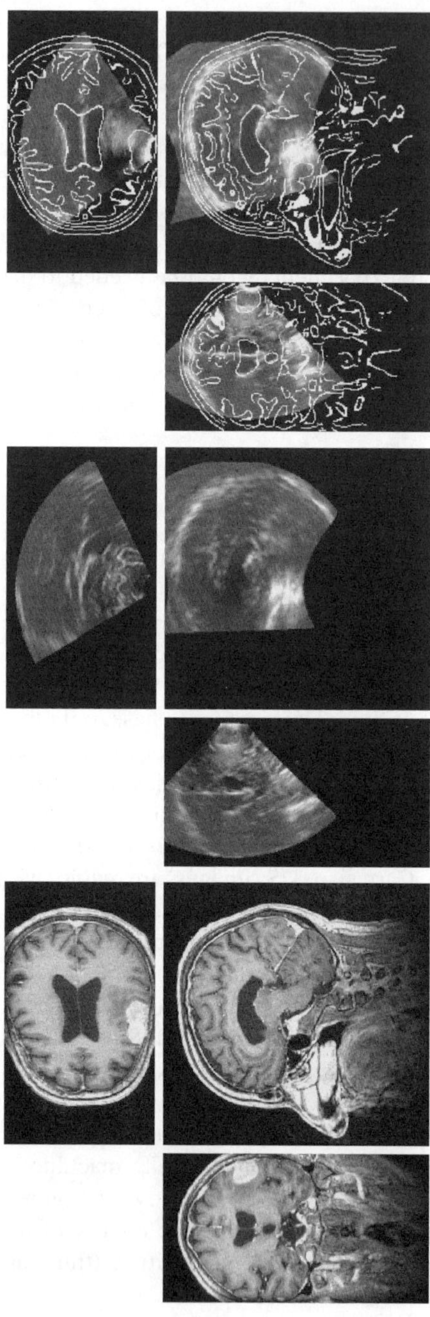

FIGURE 4.5 Example registration of MR and US images of the patient. From left to right: MR T1 image with a contrast agent, manual initialization of the US image registration, and result of the automatic registration of the US image with the MR contours superimposed.

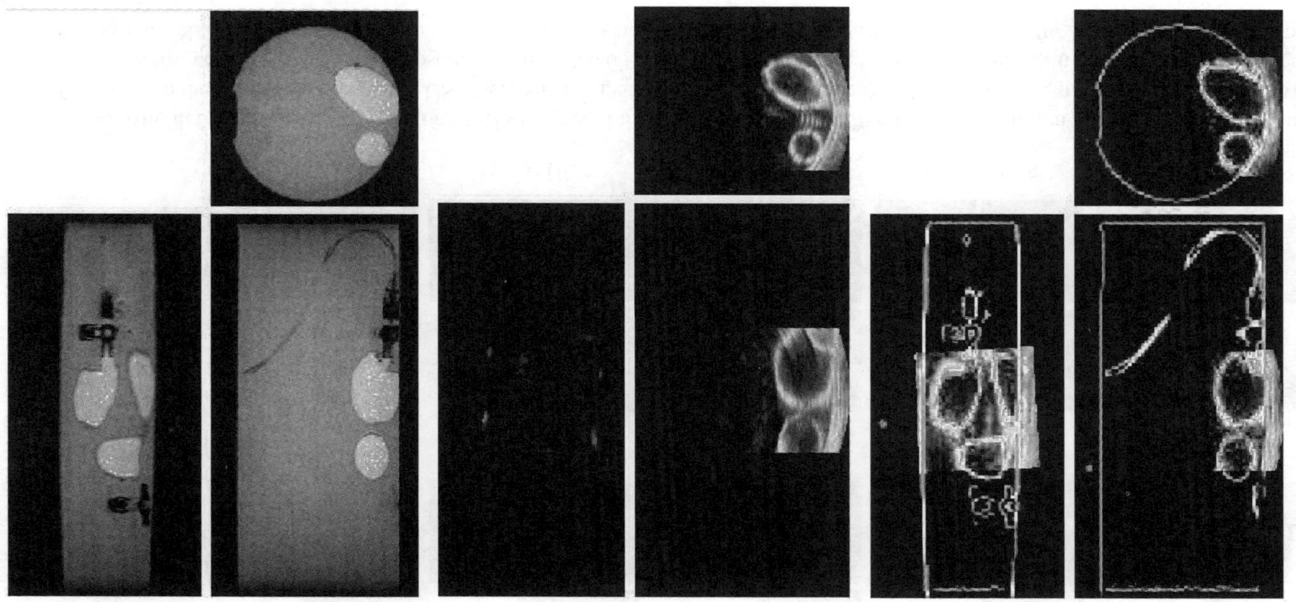

FIGURE 4.6 (See color insert following page 236) Example registration of MR and US images of the phantom. From left to right: MR image, manual initialization of the US image registration, and result of the automatic registration of the US image with the MR contours superimposed.

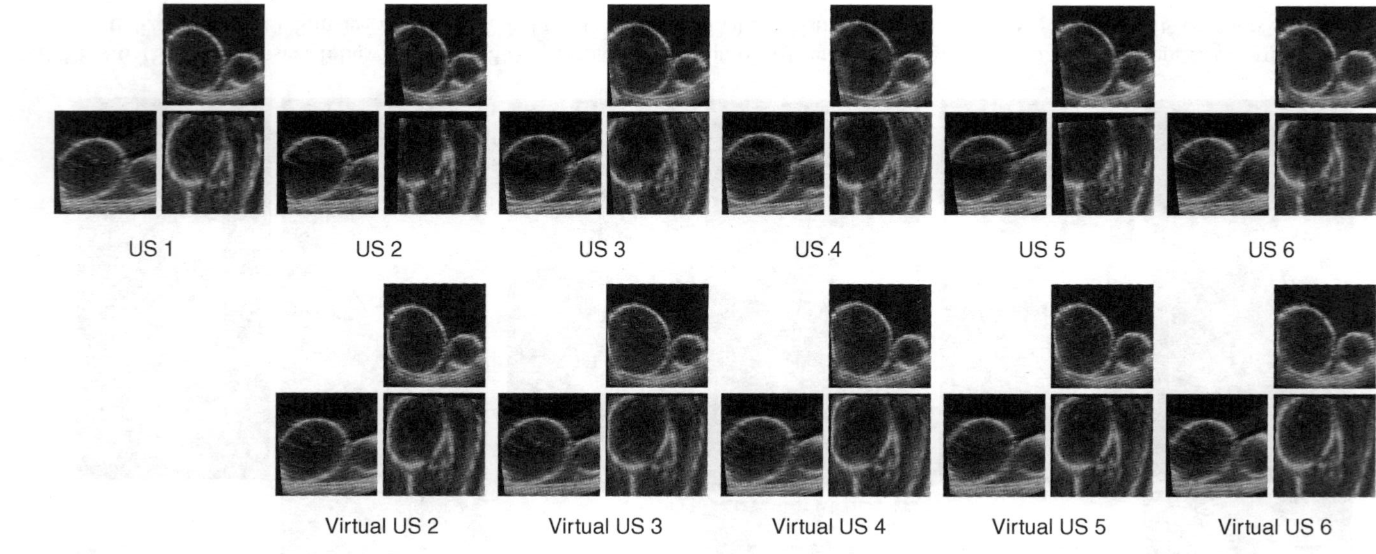

FIGURE 4.7 Tracking deformations on a phantom. 6 US images of the Phantom sequence after a rigid registration to compensate for the motion of the probe (on top) and the "virtual" US images (US 1 deformed to match the current US image) resulting from the tracking (bottom). The volume of the balloons ranges from 60 to 90 ml for the ellipsoid one and 40 to 60 ml for the more complex one. Reprinted, with permission, from Pennec, X., Cachier, P., and Ayache, N., Tracking brain deformations in time-sequences of 3D US images, *Pattern Recognit. Lett.*, 24(4–5), 801–803, Copyright Elsevier February 2003.

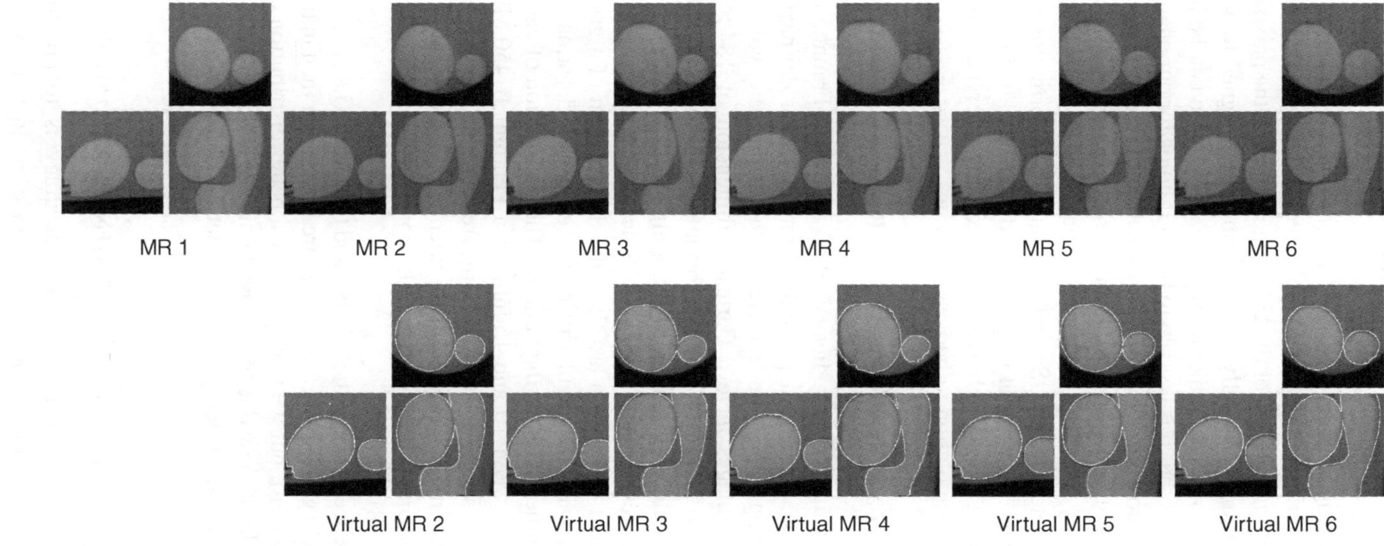

FIGURE 4.7 (Continued): On top, the "original" MR images (rigidly registered to the corresponding US images to correct for the probe motion and the phantom motion between MR acquisitions) and the virtual MR image synthesized using the deformation field computed on the US images on the bottom. To assess the quality of the tracking, we superimposed the contours of the "original" MR images.

loosely constrained), results are good all along the sequence. This shows that the SSD criterion correctly captures the information at edges and that our parameterized deformation interpolates reasonably well in uniform areas. However, when looking at the virtual MR in more detail, one can find some places where the motion is less accurately recovered; namely, the contact between the balloons and borders of the US images. Indeed, the parameterization of the transformation and especially its smoothing are designed to approximate the behavior of a uniform elastic like body. This assumption can be justified for the shift of brain tissues, but it is less obvious that it holds for our phantom where balloons are placed into a viscous fluid. In particular, the fluid motions between the two balloons cannot be recovered. On the borders of the US images, there is often a lack of intensity information (due to the inadequate conversion from polar to Cartesian co-ordinates by the US machine), and the deformation can only be extrapolated from the smoothing of neighboring displacements. Since we are not using a precise geometrical and physical model of the observed structures, one cannot expect this extrapolation to be very accurate.

4. US Images of a Balloon Inflated in a Pig Brain

This dataset was obtained by Dr. Ing. V. Paul at IBMT, Fraunhofer Institute (Germany) from a pig brain at a postlethal status. A cyst drainage has been simulated by deflating a balloon catheter with a complete volume scan at three steps. All US images have a Cartesian dimension of $184 \times 184 \times 184$ voxels of size $0.26 \times 0.26 \times 0.26$ mm^3. Unfortunately, we have no MR image in this dataset, so we could only run the tracking algorithm.

We present the results in Figure 4.3. Since we have no corresponding MR image, we present on the two last lines the deformation of a grid (a virtual MR image) to emphasize the regularity of the estimated deformation and the deformation of a segmentation of the balloon. The registration of each image of the sequence takes between 10 and 15 min on a Pentium II 450 Mhz running Linux. To visually assess the quality of the registration, we segmented the balloon on the first image. Then, this segmentation is deformed using the transformation found and superimposed to the corresponding original US image (last row of Figure 4.8).

The correspondence between the original and the virtual (i.e., deformed US 1) images is qualitatively good. In fact, the edges are less salient than in the phantom images (see next section), but we have globally a better distribution of intensity features over the field of view due to the speckle in these real brain images. One should also note on the deformed grid images that the deformation found is very smooth.

B. MR/US Rigid Registration Consistency Evaluation

The usual way of measuring a registration accuracy is to provide a target registration error (TRE). This is the mean RMS error in millimeters that the

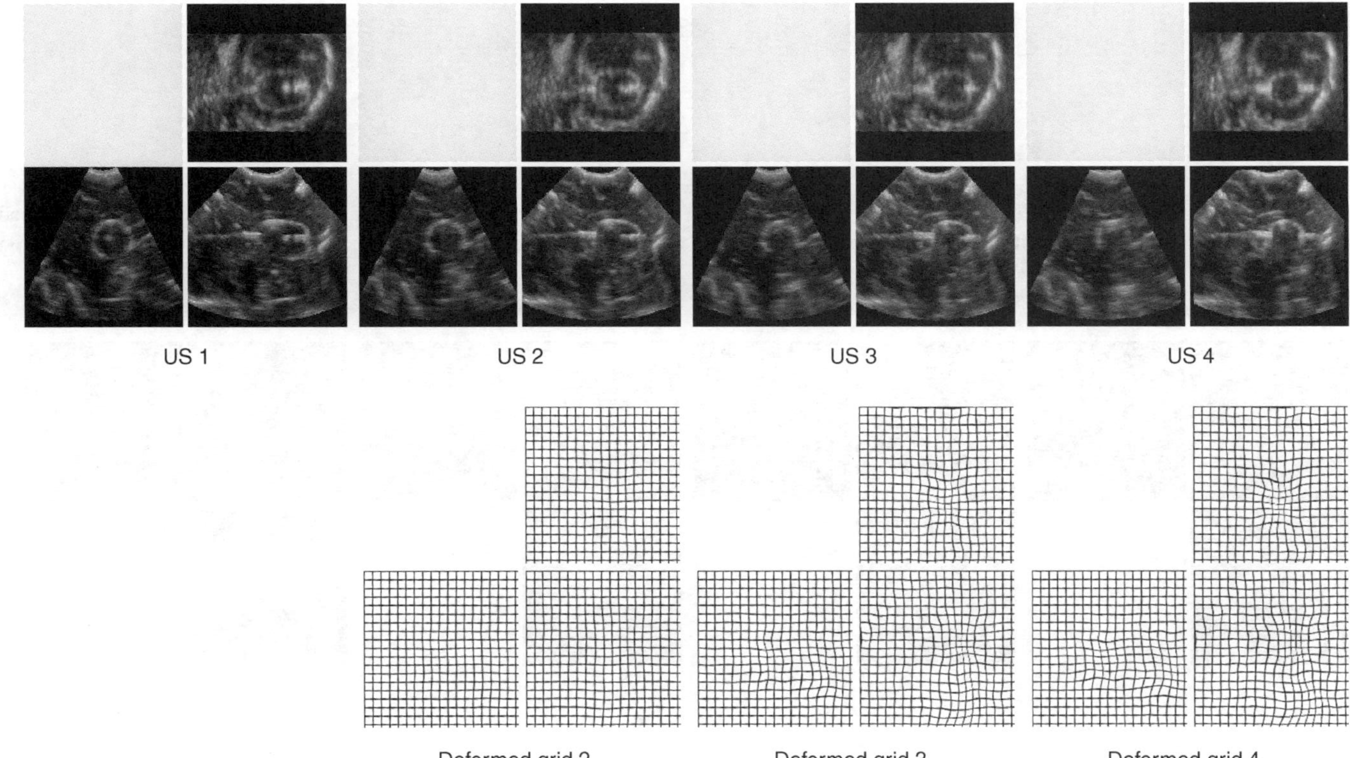

FIGURE 4.8 Tracking deformations on a pig brain. In this figure, each triplet of 2D images represents 3 orthogonal views resliced from the 3D image. **Top**: The 4 images of the pig brain with a deflating balloon simulating a cyst drainage. **Middle:** deformation of a grid to visualize more precisely the location of the deformations found. These images correspond to the deformation of an image of a 3D grid (a "virtual MR" image) with strips orthogonal to each 2D resliced plane: they allow to visualize the in-plane deformation for each 2D slice.

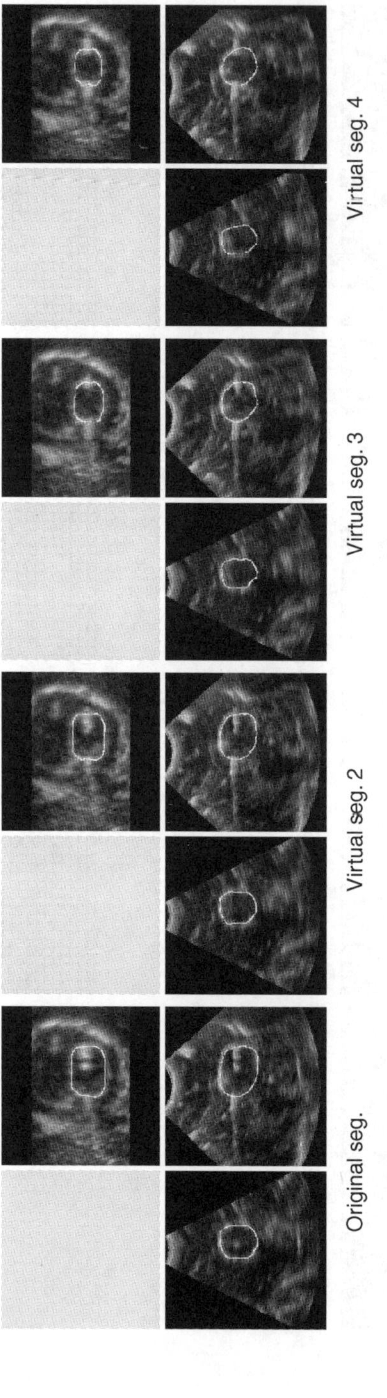

FIGURE 4.8 (Continued): We segmented the balloon on the first image. Then, this segmentation is deformed using the transformation found and superimposed to the corresponding original US image.

registration accomplished with regard to the ground truth on a specified set of target points. Using the methods developed in Pennec and Thirion,[44] we can also measure this error directly on the transformation in order to provide a full covariance matrix on the transformation parameters. As this matrix is rather difficult to interpret, we can simplify it and provide the mean rotation error σ_{rot} in degrees and the mean translation error σ_{trans} in mm. We may also propagate this covariance matrix onto a set of prespecified target or test points to obtain a TRE σ_{test}.

1. Registration Loops

It is very difficult to obtain a gold standard registration between MR and US images since an external marker cannot be seen in US images (US waves do not properly propagate in the air) and even internal markers detection is prone to inaccuracy. Moreover, if such a calibration setup was possible for phantom data in a controlled environment, it is not really feasible for clinical data. To circumvent the problem, the registration loop protocol introduced previously.[30,31,45–47]

The objective is to acquire several MR and US images of the same subject (or object), and to compose a series of registration transformations in a loop that lead ideally to the identity transformation. Typical loops in the case of the phantom data, sketched in Figure 4.9, are sequences such as $MR_i \rightarrow US_i \rightarrow US_j \rightarrow MR_j \rightarrow MR_i$. If we were given perfectly registered images within each modality, then applying the transformations of the loop in turn to a test point in MR_i would lead to a displacement of this test point that it due to the errors of the two MR/US registrations. Since the variances are additives, the observed TRE should be $\sigma_{loop}^2 = 2\sigma_{MR/US}^2$. In most cases, the intramodality transformations are not perfectly known and need to be estimated. Thus, we have to take into account their variability and the error on the loop becomes: $\sigma_{loop}^2 = 2\sigma_{MR/US}^2 + \sigma_{MR/MR}^2 + \sigma_{US/US}^2$. Finally, we can estimate the *expected accuracy* or consistency using:

$$\sigma_{MR/US} = \sqrt{\frac{1}{2}(\sigma_{loop}^2 - \sigma_{MR/MR}^2 - \sigma_{US/US}^2)}$$

Note that neglecting the intramodality errors and setting $\sigma_{MR/US} = \sigma_{loop}/\sqrt{2}$ yields a more conservative error estimation.

Using different MR and US images in the loop allows one to decorrelate the MR/US transformation (none of the data is common to the registrations), but this is not always possible. For instance, we have only one MR image in the baby dataset. Moreover, if the intramodality registration is also done using the images and not using external information (e.g., markers), then a special feature of one image may similarly affect the different registrations involved in the loop and, consequently, hides the bias due to the variability of this feature.

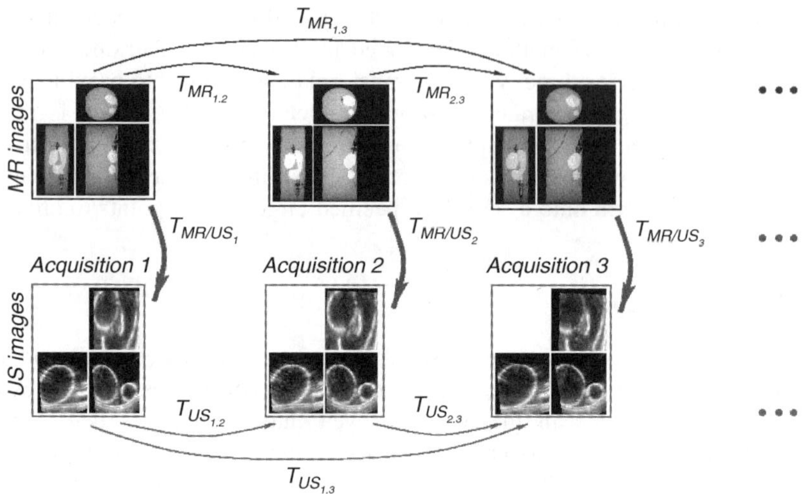

FIGURE 4.9 Registration loops used to estimate the registration consistency. Adapted, with permission, from Roche, A., Pennec, X., Malandain, G. and Ayache, N., Rigid registration of 3D ultrasound with MR images: a new approach combining intensity and gradient information, *IEEE Trans. Med. Imaging*, 20(10), 1038–1049, Copyright IEEE October 2001.

This is why the registration loop protocol only leads to a measure of consistency which is a lower bound on the accuracy.

2. Bronze Standard Registration

Our goal now is to compute the $n - 1$ most reliable transformations $\bar{T}_{i,i+1}$ that relate the n (successive) MR_i (or US_i) images. Estimations of these transformations are readily available by computing all the possible registrations $T_{i,j}$ between the MR images using m different methods. Then, the transformations $\bar{T}_{i,i+1}$ that best explain these measurements are computed by minimizing the sum of the squared distance between the observed transformations $T_{i,j}$ and the corresponding combination of the sought transformation $\bar{T}_{i,i+1} \circ \bar{T}_{i+1,i+2} \dots \bar{T}_{j-1,j}$. The distance between transformations is chosen as a robust variant of the left invariant distance on rigid transformation developed in Pennec et al.[45]

In this process, we do not only estimate the optimal transformation, but also the rotational and translational variance of the transformation measurements, which are propagated through the criterion to give an estimation of the variance of the optimal transformations.

The estimation $\bar{T}_{i,i+1}$ of the perfect registration $T_{i,i+1}$ is called *bronze standard* because the result converges toward the true transformation $T_{i,i+1}$, if there is a sufficiently high number of images (to decrease the noise level), up to the intrinsic bias (if there is any) introduced by the method. Now, using different registration procedures based on different methods, the intrinsic bias of each method also becomes a random variable, which we expect to be centered around zero and averaged out in the minimization procedure. The different bias of the methods is now integrated into the transformation variability. To fully reach this goal, it is important to use as many independent registration methods as possible.

In our setup, we used as many images as possible within each dataset, and up to three different methods with different geometric features or intensity measures. Two of these methods are intensity based and were used for both MR/MR and US/US registrations: the algorithm Aladin[42] has a block matching strategy where matches are determined using the coefficient of correlation, and the transformation is robustly estimated using a least trimmed squares; and the algorithm Yasmina uses the Powell algorithm to optimize the SSD or a robust variant of the CR metric between the images.[31] For the MR/MR registration, we added a feature based method, the crest lines registration described and thoroughly evaluated in Pennec et al.[48]

3. Consistency Results

We ran the MR/US registration on the phantom, baby, and patient datasets. Results are summarized in Table 4.1. The optimization of the bivariate CR was

TABLE 4.1

Estimated Registration Errors: Standard Deviations of the Residual Rotation (σ_{rot}), of the Translation (σ_{trans}), Which Corresponds to the Error of a Test Point at the Center of the Image, and on the Test Points that Enclose the Area of Interest (σ_{test})

		σ_{rot} (°)	σ_{trans} (mm)	σ_{test} (mm)
Phantom dataset	Multiple MR/MR	0.06	0.10	0.13
MR: $0.9 \times 0.9 \times 1$ mm^3	Multiple US/US	0.60	0.40	0.71
US: $0.41 \times 0.41 \times 0.41$ mm^3	Loop	1.62	1.43	2.07
	Expected MR/US	1.06	0.97	1.37
Baby dataset	Multiple US/US	0.10	0.06	0.12
MR: $0.9 \times 0.9 \times 0.9$ mm^3	Loop	1.71	0.51	1.27
US: $0.3 \times 0.3 \times 0.3$ mm^3	Expected MR/US	1.21	0.36	0.89
Patient data set	MR/MR	0.06	0.06	0.10
MR: $0.9 \times 0.9 \times 1.1$ mm^3	Loop	2.22	0.82	2.33
US: 0.63^3 and 0.95^3 mm^3	Expected MR/US	1.57	0.58	1.65

Reprinted with permission from Roche, A., Pennec, X., Malandain, G. and Ayache, N., Rigid registration of 3D ultrasound with MR images: a new approach combining intensity and gradient information, *IEEE Trans. Med. Imaging*, 20(10), 1038–1049, Copyright IEEE October 2001.

realized using Powell's method and took 5 to 10 min on a Pentium II 450 MHz running Linux.

For the phantom data, we used the 54 loops $US_i \rightarrow MR_i \rightarrow MR_j \rightarrow US_j \rightarrow US_i$. Eight test points were placed at the corners of the Cartesian US image in order to fully enclose the balloons area. One can see that the measured consistency is on the order of the MR voxel size. One could probably slightly reduce the uncertainty of the US / US registration by incorporating rigid markers in the phantom. This would strengthen our belief in this result, but we believe that it would not drastically change it.

In the case of the baby dataset, we used five loops $US_i \rightarrow MR \rightarrow US_j \rightarrow US_i$ since only one MR image was available. In this case, the loop may hide a MR induced bias as the two MR / US registrations share the same MR image. We indeed observe a consistency which is slightly less than the MR voxel size. This better result may also be due to US images that are much more focused on the ventricular area (we experimentally observed that the ventricles were the main intensity features practically used by the registration algorithm).

For the preoperative (patient) data, the three US images cannot be easily registered into the same co-ordinate system as they were acquired at different times of the operation. Thus, we were left with only three registration loops $US_i \rightarrow MR_1 \rightarrow MR_2 \rightarrow US_i$. Our validation scheme exhibits a registration accuracy of 0.6 mm at the center and 1.6 mm in the whole brain area (the test points are taken here at the corners of a $80 \times 80 \times 80$ mm^3 cube centered in the Cartesian US image so that they lie in the acquisition cone). However, when we look more carefully at the results, we find that the loop involving the smallest US image (real Cartesian size $150 \times 85 \times 100$ mm^3, voxel size 0.63^3 mm^3) is responsible for a test point error of 2.6 mm (0.85 mm at the center) while the loops involving the two larger US images (real size $170 \times 130 \times 180$, voxels size 0.95^3 mm^3) have a much smaller test point error of about 0.84 mm (0.4 mm at the center). We suspect that nonrigidity in the smallest US could account for the registration inaccuracy. Another explanation could be a misestimation of the sound speed for this small US acquisition leading to a false voxel size and once again the violation of the rigidity assumption.

C. 3D US Tracking Performances

When it comes to 3D deformations (in our case, 3D nonrigid tracking), it becomes much more difficult to assess the algorithm performances since they depend on the location in space.

Qualitatively, we expect to obtain good matches if we have some time and space consistent information (edges, speckle), but we need to regularize in other places. This means that the registration quality depends on the fit between the regularization model and the mechanical properties of the observed tissues. From a quantitative point of view, we would need to know the displacement of each point in the images in order to evaluate the accuracy of the estimated deformation. If it is sometimes possible to determine some fuzzy corresponding landmarks in our US images, then it seems difficult to generalize that approach to the whole image.

However, we can usually determine corresponding surfaces, and measure the distance between the surfaces or the difference in the volume they enclose. In the phantom case, we segmented the surface of the balloons in the MR images. The maximal distance between the deformed surface of the first MR and the surface of the current MR was 2 to 3 mm all along the sequence, while the mean distance was below 1 mm. This shows that the tracking is quite accurate at the places where we have information (here the balloon edges). With these data, we have no mean to assess the accuracy of the interpolated motion within and outside the balloons.

In the pig brain images, we have no MR and it is much more difficult to segment faithfully the edges of the inflated balloon in each US image. However, we know the theoretical volume of the balloon in each image. Thus, we (approximately) segmented the balloon in the first US image, and deformed its volume after deformation (bottom line of Figure 4.8). Since our segmentation tends to overestimate the balloon volume, we believe that it is more interesting to compare the ratio of the volumes than the volumes themselves.

Image number	1	2	3	4
Original balloon volume (cm^3)	1.25	1.00	0.75	0.5
Relative volume ratio		0.8	0.6	0.4
Measured balloon volume	1.28	1.10	0.80	0.67
Measured volume ratio		0.86	0.62	0.53

The measurements indicate that we are overestimating the volume (underestimating the deformation) by 7.5% for Image 2, by 3.3% for Image 3, and by 30% for Image 4. However, one should note that volume measurements are very sensitive as they relate to the cube of the balloon dimension. This corresponds to an error of less than 1 mm on the balloon diameter. This could be explained by an occlusion of the lower part of the balloon probably due to an air bubble trapped inside the balloon during the experience. On US 4, almost the entire lower half of the balloon is shadowed by the air bubble. In these conditions, one cannot expect a perfect retrieval. The estimated deformation at the occlusion being computed thanks to the regularization of the deformation field from neighboring structures, it is expected to be less than the real deformations (maximal at the balloon boundaries).

In both the pig brain and phantom experiments, reducing the smoothing of the transformation could allow the algorithm to find a closer fit. However, this could allow some unwanted high frequency deformations due to the noise in the US images. We believe that it is better to recover the most important deformations and miss some smaller parts than to try to match exactly the images and have the possibility to create some possibly large deformations.

V. DISCUSSION

We have presented a new automated method to rigidly register 3D US with MR images. It is based on a multivariate and robust generalization of the CR measure that allows one to better take into account the physics of the US images acquisition. The assumption is that the US signal may be approximated by a function of both the MR intensity and its gradient magnitude. This model does not take into account the speckle and attenuation effects, which are important aspects of the US physics. However, because we introduced a robust intensity distance measure in our bivariate CR criterion, the functional relationship assumption does not need to hold throughout the whole image.

Our implementation using Powell's optimization method was successful in registering more than 20 MR/US volume pairs from phantom and clinical data. To evaluate the registration performances, we designed an original approach to establish a bronze standard in the absence of ground truth. We found the worst registration errors (maximum error in the region of interest defined by the US cone) to be of the order of the MR voxel size (1 to 1.5 mm). A robustness study performed in Roche et al.[31] also showed that the bivariate CR significantly outperforms the conventional CR and MI similarity metrics. Incidentally, we believe that the generalized CR could be considered in other multimodal registration problems where conventional similarity measures exhibit a lack of robustness, such as CT or PET to MR registration.[49]

We believe that our algorithm may be improved in several ways to better take into account the specific nature of the US images. The first improvement would be using sampling techniques adapted to the polar geometry of the US image and the speckle size.[50] The second improvement would be to use the gradient orientation with regard to the US scan line in addition to its magnitude. Another interesting development would be to consider (and possibly estimate during registration) the spatial variation of the speed of the sound within tissues. This last development would allow for the correction of the main US image distortions without going for a full FFD.

In the second part of this chapter, we developed a tracking algorithm adapted to time sequences of US images and not only to the registration of two images. The algorithm is able to recover an important part of the deformations and issues a smooth deformation, despite the noisy nature of the US images. Experiments on animal and phantom data show that this allows one to simulate virtual MR images qualitatively close to the real ones. Quantitative measurement remains to be done, but it seems that an accuracy of 1 to 2 mm is achievable in the areas where there is an elastic deformation. This is encouraging since the accuracy of the clinicians without preoperative imaging is estimated to be around 3 to 5 mm. In our experiments, we observed that the SSD criterion was well adapted to the registration of successive US images of the time sequence, and also performed well for the update of the global transformation. However, the appearance of some tissues in US images is known to change with time, and we believe that the

LCC criterion will be more adapted to time sequences of brain US images in a clinical setup.

The computation times are still far from real time for a continuous tracking of deformations during surgery, but this implementation was focused on generic components in order to test different criteria and gradient descent approaches. A dedicated reimplementation of the method may gain a factor of 4 to 8, leading to a clinically useful tool for brain shift estimation (one estimation per minute or two). To be further accelerated and reach real time (video rate for instance), the algorithm needs to be parallelized. This has been partly realized in Stefanescu et al.,[51] with an acceleration of a factor of 11 on 15 processors. However, the limiting factor becomes then the latency of the network to transfer the input images ($184^3 = 6$ MB in the case of our US images) and more importantly the resulting deformation field (72 MB at full resolution). This last transfer could be avoided by only providing the current resampled MR image.

The type of transformation is a very sensitive choice for such a tracking algorithm. In this work, we made the assumption of a uniform elastic-like material. This may be adapted for the brain tissues (white and gray matter), but some improvements will be needed to cope with the nonelastic deformations that occur with the CSF leakage (particularly in the ventricles). Likewise, the introduction of surgical instruments will create shadows and artifacts that will hide important information in the US image, and may mislead the registration algorithm. Thus, we would need to work with dense transformations with a space varying regularization, depending on the underlying brain tissue type, and also a space varying trade-off between the similarity and regularization criteria, in order to use image information where it is reliable and regularization around and behind instruments. First steps toward these goals have been made recently with the introduction of an anatomically informed regularization of the deformation[52] with an adaptive similarity/regularization trade-off, implemented in a GRID compatible framework. Results are already very good on the registration of MR images and we believe that the adaptation to time series of US images will lead to excellent results.

ACKNOWLEDGMENTS

This work was partially supported by the EC funded ROBOSCOPE project HC 4018, a collaboration between the Fraunhofer Institute (Germany), Fokker Control System (Netherlands), Imperial College (UK), INRIA (France), ISM-Salzburg and Kretz Technik (Austria). The authors thank Dr. Ing. V. Paul at IBMT, Fraunhofer Institute for the acquisition of the pig brain images, and Prof. Auer and his colleagues at ISM for the acquisitions of all other images.

This chapter synthesizes and partly reprints material with permission from IEEE Transactions on Medical Imaging, 20(10), A. Roche, X. Pennec,

REFERENCES

1. Trobaugh, J., Richard, W., Smith, K., and Bucholz, R., Frameless stereotactic ultrasonography — method and applications, *Comput. Med. Imaging Graph.*, 18(4), 235–246, 1994.
2. Trobaugh, J., Trobaugh, D., and Richard, W., Three-dimensional imaging with stereotactic ultrasonography, *Comput. Med. Imaging Graph.*, 18(5), 315–323, 1994.
3. Richard, W., Zar, D., LaPresto, E., and Steiner, C., A low-cost PCI-bus-based ultrasound system for use in image-guided neurosurgery, *Comput. Med. Imaging Graph.*, 23(5), 267–276, 1999.
4. Pagoulatos, N., Edwards, W., Haynor, D., and Kim, Y., Interactive 3-D registration of ultrasound and magnetic resonance images based on a magnetic position sensor, *IEEE Trans. Inf. Technol. Biomed.*, 3(4), 278–288, 1999.
5. Erbe, H., Kriete, A., Jödicke, A., Deinsberger, W., and Böker, D.-K., 3D-ultrasonography and image matching for detection of brain shift during intracranial surgery, *Comput. Assist. Radiol.*, 225–230, 1996.
6. Hata, N., Suzuki, M., Dohi, T., Iseki, H., Takakura, K., and Hashimoto, D., Registration of ultrasound echography for intraoperative use: a newly developed multiproperty method, pp. 251–259. In *Proceedings of VBC'94*, SPIE 2359, Rochester, MN, USA, 1994.
7. Gobbi, D., Comeau, R., and Peters, T., Ultrasound probe tracking for real-time ultra-sound/MRI overlay and visualization of brain shift, pp. 920–927. In *Proceedings of MICCAI'99*, LNCS 1679, Cambridge, UK, 1999.
8. Gobbi, D., Comeau, R., and Peters, T., Ultrasound/MRI overlay with image warping for neurosurgery, pp. 106–114. In *Proceedings of MICCAI'00*, LNCS 1935. Pittsburgh, 2000.
9. Comeau, R., Sadikot, A., Fenster, A., and Peters, T., Intraoperative ultrasound for guidance and tissue shift correction in image-guided neurosurgery, *Med. Phys.*, 27(4), 787–800, 2000.
10. Bucholz, R. D., Yeh, D. D., Trobaugh, J., McDurmont, L. L., Sturm, C. D., Baumann, C., Henderson, J. M., Levy, A., and Kessman P., The correction of stereotactic inaccuracy caused by brain shift using an intraoperative ultrasound device, pp. 459–466. In *Proceedings of CVRMed-MRCAS'97*, LNCS 1205, 1997.
11. Ionescu, G., Lavallée, S., and Demongeot, J., Automated registration of ultrasound with ct images: application to computer assisted prostate radiotherapy and orthopedics, pp. 768–777. In *Proceedings MICCAI'99*, LNCS 1679, Cambridge (UK), 1999.

12. King, A., Blackall, J., Penney, G., Edwards, P., Hill, D., and Hawkes, D., Baysian estimation of intra-operative deformation for image-guided surgery using 3-D ultrasound, pp. 588–597. In *Proceedings of MICCAI'00*, LNCS 1935, 2000.

13. Strintzis, M. G., and Kokkinidis, I., Maximum likelihood motion estimation in ultrasound image sequences, *IEEE Signal Process. Lett.*, 4(6) 1997.

14. Rohling, R. N., Gee, A. H., and Berman, L., Automatic registration of 3-D ultrasound images, *Med. Biol.*, 24(6), 841–854, 1998.

15. Rohling, R. N., Gee, A. H., and Berman, L., Three-dimensional spatial compounding of ultrasound images, *Med. Image Anal.*, 1(3), 177–193, 1997.

16. Papademetris, X., Sinusas, A., Dione, D., and Duncan, J., 3D cardiac deformation from ultrasound images, pp. 421–429. In *Proceedings of MICCAI'99*, LNCS 1679, Cambridge, UK, 1999.

17. Sanchez-Ortiz, G., Declerck, J., Mulet-Parada, M., and Noble, J., Automatic 3D echocardiographic image analysis, pp. 687–696. In *Proceedings of MICCAI'00*, LNCS 1935, Pittsburgh, USA, 2000.

18. Bajcsy, R., and Kovačič, S., Multiresolution elastic matching, *Comput. Vis. Graph. Image Process.*, 46, 1–21, 1989.

19. Christensen, G. E., Joshi, S. C., and Miller, M. I., Volumetric transformation of brain anatomy, *IEEE Trans. Med. Imaging*, 16(6), 864–877, 1997.

20. Bro-Nielsen, M., 1996. Medical Image Registration and Surgery Simulation. Ph.D. thesis, Institut for Matematisk Modellering, Danmarks Tekniske Universitet, Lyngby, Denmark.

21. Maintz, J. B. A., Meijering, E. H. W., and Viergever, M. A., 1998. General multimodal elastic registration based on mutual information. In Hansom, K. M., editor, *Medical Imaging 1998 — Image Processing*, Volume 3338, pp. 144–154, SPIE, Bellingham, WA, 1998.

22. Thirion, J.-P., Image matching as a diffusion process: an analogy with Maxwell's demons, *Med. Image Anal.*, 2(3), 1998.

23. Dawant, B. M., Hartmann, S. L., and Gadamsetty, S., Brain atlas deformation in the presence of large space-occupying tumors, pp. 589–596. In *Proceedings of MICCAI'99*, LNCS 1679, Cambridge, UK, 1999.

24. Bricault, I., Ferretti, G., and Cinquin, P., Registration of real and CT-derived virtual bronchoscopic images to assist transbronchial biopsy, *Trans. Med. Imaging*, 17(5), 703–714, 1998.

25. Webb, J., Guimond, A., Roberts, N., Eldridge, P., Chadwick, P. E., Meunier, J., and Thirion, J.-P., Automatic detection of hippocampal atrophy on magnetic resonnance images, *Magn. Reson. Imaging*, 17(8), 1149–1161, 1999.

26. Prima, S., Thirion, J.-P., Subsol, G., and Roberts, N., Automatic analysis of normal brain dissymmetry of males and females in MR images, pp. 770–779. In *Proceedings of MICCAI'98*, LNCS 1496, 1998.

27. Pennec, X., Cachier, P., and Ayache, N., Understanding the "demon's algorithm": 3D non-rigid registration by gradient descent, pp. 597–605. In *Proceedings of MICCAI'99*, LNCS 1679, Cambridge, UK, 1999.

28. Cachier, P., and Pennec, X., 3D non-rigid registration by gradient descent on a Gaussian-windowed similarity measure using convolutions, pp. 182–189. In *Proceedings of MMBIA'00*. Hilton Head Island, USA: IEEE Comput. Society, 2000.

29. Roche, A., Malandain, G., Pennec, X., and Ayache, N., The correlation ratio as a new similarity measure for multimodal image registration, pp. 1115–1124. In *Proceedings of MICCAI'98*, LNCS 1496. Cambridge, USA, 1998.

30. Roche, A., Pennec, X., Rudolph, M., Auer, D. P., Malandain, G., Ourselin, S., Auer, L. M., and Ayache, N., Generalized correlation ratio for rigid registration of 3D ultrasound with MR images, pp. 567–577. In *Proceedings of MICCAI'00*, LNCS 1935, Submitted to IEEE TMI, 2000.

31. Roche, A., Pennec, X., Malandain, G., and Ayache, N., Rigid registration of 3D ultra-sound with MR images: a new approach combining intensity and gradient information, *IEEE Trans. Med. Imaging*, 20(10), 1038–1049, 2001.

32. Pennec, X., Cachier, P., and Ayache, P., Tracking brain deformations in time sequences of 3D US images, pp. 169–175. In *Proceedings of IPMI'01*, LNCS 2082, Insana, M., and Leahy, R., Eds., 2001.

33. Pennec, X., Cachier, P., and Ayache, N., Tracking brain deformations in time-sequences of 3D US images, *Pattern Recognit. Lett. — Spec. Issue Ultrason. Image Process. Anal.*, 24(4–5), 801–813, 2003.

34. Roche, A., Malandain, G., and Ayache, N., Unifying maximum likelihood approaches in medical image registration, *Int. J. Imaging Syst. Technol.: Spec. Issue 3D Imaging*, 11, 71–80, 2000.

35. Press, W., Flannery, B., Teukolsky, S., and Vetterling, W., *Numerical Recipices in C*, Cambridge University Press, Cambridge, 1991.

36. Maes, F., Collignon, A., Vandermeulen, D., Marchal, G., and Suetens, P., Multimodality image registration by maximization of mutual information, *IEEE Trans. Med. Imaging*, 16, 187–198, 1997.

37. Maes, F., Vandermeulen, D., and Suetens, P., Comparative evaluation of multiresolution optimization strategies for multimodality image registration by maximization of mutual information, *Med. Image Anal.*, 3(4), 373–386, 1999.

38. Pluim, J., Maintz, J., and Viergever, M., Mutual information matching and interpolation artefacts, pp. 56–65. In *Proceedings SPIE 3661*, 1999.

39. Rousseeuw, P. J., and Leroy, A. M., *Robust Regression and Outlier Detection*, Wiley, New York, 1987.

40. Meunier, J., and Bertrand, M., Ultrasonic texture motion analysis: theory and simulation, *IEEE Trans. Med. Imaging*, 14(2), 1995.

41. Ferrant, M., Warfield, S. K., Guttmann, C. R. G., Mulkern, R. V., Jolesz, F. A., and Kikinis, R., 3D Image matching using a finite element based elastic deformation model, pp. 202–209. In *Proceedings of MICCAI'99*, LNCS 1679, Cambridge, UK, 1999.

42. Ourselin, S., Roche, A., Prima, S., and Ayache, N., Block matching: a general framework to improve robustness of rigid registration of medical images, pp. 557–566. In *Proceedings of MICCAI'2000*, LNCS 1935, Springer, 2000.

43. Morel, J.-M., and Solimini, S., Variational methods in image segmentation, *Progress in Nonlinear Differential Equations and their Applications*, Birkhauser Boston Inc., Cambridge, MA, 1995.

44. Pennec, X., and Thirion, J.-P., A framework for uncertainty and validation of 3D registration methods based on points and frames, *Int. J. Comput. Vis.*, 25(3), 203–229, 1997.

45. Pennec, X., Guttmann, C., and Thirion, J-P., Feature-based registration of medical images: estimation and validation of the pose accuracy, pp. 1107–1114.

In *Proceedings of First International Conference on Medical Image Computing and Computer-Assisted Intervention (MICCAI'98)*, LNCS 1496, Springer Verlag, Cambridge, USA, 1998.

46. Holden, M., Hill, D., Denton, E., Jarosz, J., Cox, T., Rohlfing, T., Goodey, J., and Hawkes, D., Voxel similarity measures for 3D serial MR brain image registration, *IEEE Trams. Med. Imaging*, 19(2) 2000.

47. Penney, G., Blackall, J., Hayashi, D., Sabharwal, T., Adam, A., and Hawkes, D., Overview of an ultrasound to CT or MR registration system for use in thermal ablation of liver metastases. In *Proceedings of Medical Image Understanding and Analysis (MIUA'01)*, 2001.

48. Pennec, X., Ayache, N., and Thirion, J.-P., Landmark-based registration using features identified through differential geometry, In *Handbook of Medical Imaging*, Bankman, I., Ed., Academic Press, 2000, pp. 499–513, chap. 31.

49. Pluim, J., Maintz, J., and Viergever, M., Image registration by maximization of combined mutual information and gradient information, *IEEE Trans. Med. Imaging*, 19, 809–814, 2000.

50. Smith, W., and Fenster, A., Optimum scan spacing for three dimensionnal ultrasound by speckle statistics, *Ultrasound Med. Biol.*, 26(4), 551–562, 2000.

51. Stefanescu, R., Pennec, X., and Ayache, N., Parallel non-rigid registration on a cluster of workstations. In *Proceedings of HealthGrid'03*, Norager, S., Ed., European Commission: Lyon, DG Information Society, 2003.

52. Stefanescu, R., Pennec, X., and Ayache, N., Grid enabled non-rigid registration with a dense transformation and a priori information. In *Proceedings of MICCAI'03, Part II*, Ellis, R. E., and Peters, T. M., Eds., 2879 LNCS, Springer Verlag, Montreal, 2003, pp. 804–811.

5 Multisensor Data Inversion and Fusion Based on Shared Image Structure

Robert A. Weisenseel, William C. Karl, and Raymond C. Chan**

CONTENTS

---

* Co-first authors Weisenseel and Chan contributed equally to this work.

I. INTRODUCTION

Limited information quality from single modality observations often leads to the desire to combine data from multiple, complementary sensors. The hope is that information that is weakly present in each modality or sensor will reinforce each other when combined, thereby rising above the background and yielding more reliable estimates. Additional challenges occur when the observed data are not directly interpretable in their raw form (i.e., are not images), but must be inverted to obtain meaningful imagery. Examples include tomographic scans and magnetic resonance data, whose raw data consist of scene projections and Fourier samples, respectively. Finally, such multisensor observations are seldom aligned to the same grid or occur at the same resolution. As a consequence, the images corresponding to multiple sensors generally needs to be registered or aligned and some means of coping with the differing scales devised. Overall, then, the fusion of such multisensor data combines the significant individual challenges of image reconstruction, scene alignment, and resolution enhancement on the path to information fusion.

An example application for which information fusion plays an important role is the characterization of atherosclerotic plaques from multimodality vascular imagery. This is an area of active biomedical research interest since high-risk plaques are vulnerable to rupture, resulting in heart attacks or strokes which are the major causes of death and morbidity in the U.S. No single modality can currently provide unambiguous assessment of such vulnerable plaques. As a result, many research groups have turned to multimodality sensing for plaque imaging both *ex vivo* and *in vivo* as a means of interrogating different physicochemical properties of atherosclerotic lesions, and there is interest in evaluating the relative strengths of new vascular imaging techniques that are available. Information fusion allows for complementary measurements from each sensor to be combined, integrating the strengths of each modality to improve our ability to characterize the properties of vulnerable plaques and ultimately to improve clinical diagnosis and treatment.

Figure 5.1 illustrates an example of multimodality imaging of an excised atherosclerotic plaque with 16-slice multidetector computed tomography (MDCT), volume computed tomography (VCT), multispectral magnetic resonance imaging (MRI; the color channels correspond to T1, T2, and proton-density [PD] weighted sequences), and histology. Regions of interest corresponding to the plaque necrotic core and calcified nodules are indicated relative to the residual arterial lumen. MDCT imaging is extremely fast and

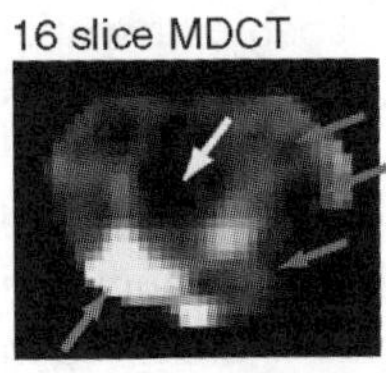

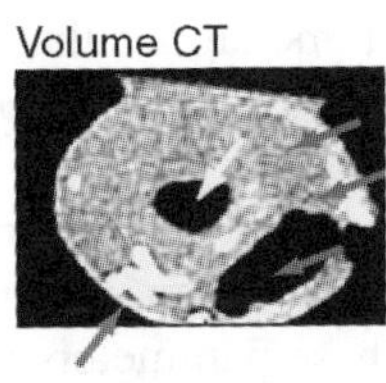

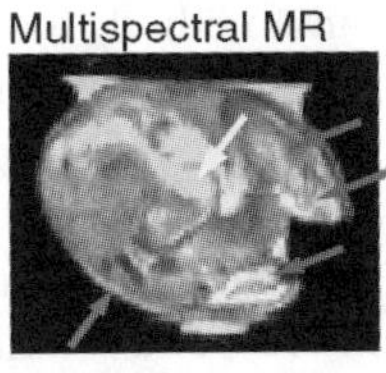

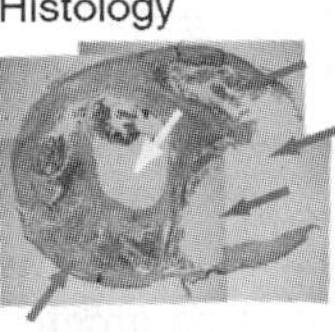

FIGURE 5.1 (See color insert following page 236) An example of multimodality imaging of an atherosclerotic plaque. Cross-sectional images of the plaque from 16-slice MDCT, VCT, multispectral MR, and histology are shown. The red arrows indicate the extent of a necrotic core, the green arrows indicate calcified tissue, and the white arrows indicate the residual lumen through the vessel.

has good dynamic range relative to VCT, a new imaging technology, with higher intrinsic spatial resolution than MDCT and MRI. MRI is slower relative to CT-based acquisition, yet it has higher intrinsic soft-tissue contrast. Methods for information fusion play an essential role in defining how these imaging technologies can be optimally combined and to improve our ability to characterize atherosclerotic lesion properties.

In this chapter we focus on a unified approach to the collection of problems arising in the fusion of data from imaging sensors by extracting, aligning, and fusing information from multiple heterogeneous imaging modalities simultaneously with the image reconstruction processes through a unified variational formulation. The approach is based on the belief that for many problems scene boundary structure is shared in different modalities, even though the different modalities may be based on different physical interaction mechanisms with the scene. We are primarily motivated by problems arising in medical imaging, as described above, though the techniques are applicable to multisensor fusion problems arising in other problem domains.

First, we give a statement of our problem and models and summarize related approaches in Section II. In Section III we present our unified multimodality fusion formulation. We discuss the optimization approach we take, including practical issues in making such a method feasible, in Section IV. In Section V we present examples of fusion applied to simulated vascular imaging and real vascular imaging with MDCT and VCT.

II. BACKGROUND AND PROBLEM STATEMENT

While combining the information from multiple, heterogeneous sensors is an evocative idea, there is no single, obvious approach to accomplish this aim or even an agreed-upon definition of sensor fusion. One common viewpoint focuses on the *level* at which the information from multiple sensors is fused, with a common set of levels being: measurement or signal-level, object-level, attribute or feature-level, and decision-level.[1-3] We can think of fusion as a form of data compression, and in this context these levels can be thought of as relating to the degree of independence in the processing of different pieces of data. Further, in performing fusion, many methods focus on modalities that are capable of spatial localization — which we term "imaging modalities". Our interest in this work is in such image modalities.

At one extreme we can attempt to combine the information from multiple sensors at the lowest signal level. In particular, we can imagine defining a common set of parameters of interest and relating these parameters to what is observed in each sensor modality through a set of physical, first-principle observation models. Consider the case of combining x-ray computed tomography (CT) and magnetic resonance (MR) data. Typically x-ray CT data are used to create an image of tissue x-ray attenuation, while MR data are used to create an image of a quantity such as proton density. Instead, we might imagine defining a single underlying field of tissue property parameters and then finding associated models that would predict both the CT projection data and the MR Fourier data corresponding to this single set of parameters. Given these associated sensor-observation models and the observed data, the problem of sensor fusion would be reduced to a conceptually straightforward data inversion problem, in which the combined set of observations is jointly processed to produce one set of parameters or image. Such an approach is analogous to a joint compression approach. It is conceptually attractive as a means of combining data from multiple sensors, as by using physical models to jointly process the entire ensemble of data we would expect to identify any weak, but complementary information from the different sensors.

Unfortunately, such an approach is intractable for most cases. Unified physical scene models simply do not exist for the majority of problems, or would need to be so fine grained as to render them useless for practical computation. Further, such models would be highly problem-specific and thus difficult to generalize to new situations.

To avoid such problems, one can go to the other extreme and process each sensor's output separately, and only later combine the resulting, individually obtained, outputs or images. This approach is analogous to compression through independent processing of each channel, and has the benefit of largely decoupling the different modalities. The price for this simplification is the possible loss of performance and information arising from this decoupled processing. Complementary information may not be exploited if it is masked by separate processing of individual modalities.

In practice, most approaches fall between these extremes. In fusing data from multiple imaging modalities, a common sequence of steps involves individual sensor image reconstruction or formation, followed by image enhancement or noise suppression, succeeded by image registration or alignment, and finally image fusion and feature extraction, as illustrated in Figure 5.2. Typically, each of the steps in such a chain is handled as a separate unrelated subproblem. In particular, the reconstructions of each modality are performed independently and before filtering to suppress noise, the alignment of the resulting images is done separately from the reconstruction, and so forth.

We will present a formulation in Section III which attempts to combine the reconstruction, enhancement, and alignment components (i.e., the first three levels)

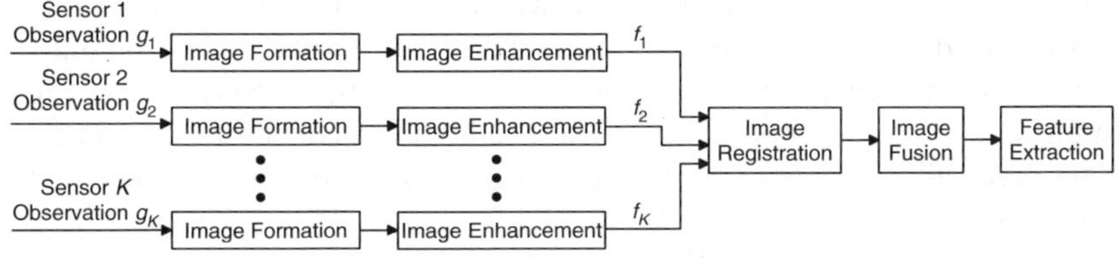

FIGURE 5.2 Common image fusion processing chain.

of this common processing structure into a single unified process. Such a unified approach moves these parts of the process closer to the signal or data level and thus should provide better information preservation. To proceed further, let us define some notation, provide a precise description of the problem elements, and discuss previous related work.

A. PROBLEM STATEMENT

We assume that we have observations g_k from K heterogeneous imaging sensors — that is, from sensors capable of localization or image formation. To each of these K sensors there is a natural corresponding property field, f_k, underlying sensor interaction for that modality. For example, in the case of computerized tomography, f_{CT} would correspond to the field of tissue x-ray attenuation and the resulting observations, g, would be integrals of this density. We further assume a linear relationship between each property field and its corresponding sensor data, so that, in the absence of noise:

$$g_k(x_k') = \int_{X_k} h_k(x_k', x_k) f_k(x_k) \mathrm{d}x_k \qquad (5.1)$$

for some kernel $h_k(x_k', x_k)$. In Equation 5.1 x_k is a potentially multi-dimensional spatial variable defined in the space of the property field, f_k, x_k' is the spatial variable for the corresponding data, and X_k is the spatial domain of definition of h_k. We will sometimes denote Equation 5.1 through operator notation as: $g_k = H_k f_k$. A separate linear relationship such as this exists for each sensor modality. For convenience, we let f represent the collection of all the property fields for all the modalities: $f = [f_1 ... f_K]^T$.

We have assumed here that g_k depends linearly on f_k, rather than allowing H_k to be a more general nonlinear mapping of the form $g_k = H_k(f_k)$. Such nonlinear mappings can arise, for example, in wave propagation problems, where we relate an observed set of reflected waves g_k to the material properties of the media through which the waves are transmitted (in general, such a dependence is nonlinear). In practice our linearity assumption is not as restrictive as it appears. Often, approaches for estimation in the presence of nonlinear observations involve sequentially linearizing the observation model around the current estimate. The resulting series of substeps thus involve a linearized observation model form comparable to Equation 5.1, with the exception that the linear kernel H_k, becomes dependent on the current estimate for f_k. Further, the techniques we describe are amenable to nonlinear observation models, at the expense of more complicated and costly optimization. Restricting ourselves to the case of linear observations allows us to avoid unnecessary complications while concentrating on our main points.

We further assume that the property fields f_k correspond to the spatial distribution of some physical quantity. As such, they exhibit a continuity or cohesiveness across their spatial variable x_k. In other words, points that are near each other spatially are likely to have similar values except at isolated discontinuities. Such an assumption is usually satisfied for physical property fields arising in medical imaging.

In practice, the observations for each modality are corrupted by noise or distortion. We capture these noise processes through specification of the corresponding log-likelihood of g_k given f_k: $\ln p(g_k|f_k)$. For example, in the common case of independent sampled observations with additive, white Gaussian noise we have:

$$\ln p(g_k|f_k) \propto \sum_{x_k'} [g_k(x_k') - H_k f_k(x_k')]^2, \tag{5.2}$$

where we have assumed that we make observations at a discrete set of points x_k'. Another common example arises if each sampled observation is independent and instead obeys a Poisson density. In this case we have:

$$\ln p(g_k|f_k) \propto \sum_{x_k'} \{g_k(x_k')\ln[H_k f_k(x_k')] - H_k f_k(x_k') - \ln[g_k(x_k')!]\} \tag{5.3}$$

Our overall goal is to process the observations g_k by performing image formation, noise suppression of the resulting images, and corresponding alignment given the multiple modality models H_k in Equation 5.1 and noise models such as those presented above. Because they are paradigms for many problems of interest, including our examples, we discuss the imaging models of two prominent types of sensors in more detail next.

1. Reconstruction from Projection Data

The first imaging model we discuss is the projection model associated with CT. The use of CT is a well-established diagnostic medical imaging tool.[4] There are a number of medical CT modalities (x-ray transmission, PET, SPECT, etc.) and geometries (parallel-beam, fan-beam, cone-beam, spiral).[5] We focus in this discussion on parallel beam transmission tomography.

The aim of x-ray transmission tomography is to reconstruct the spatial distribution of tissue attenuation from a set of projections through that volume, taken at many angles around the object being imaged. In x-ray transmission CT, the physical mechanism used to obtain a single projection element consists of illuminating the volume on one side with an x-ray source, and detecting the intensity of the x-rays on the other side of the volume. The amount of attenuation is measured relative to a calibration reference obtained along an unattenuated or minimally attenuated path. We can model the total attenuation as an integration over the tissue volume attenuation along the beam path. In the most common implementations of CT reconstruction, the reconstruction basis elements are everywhere large enough relative to the beam spacing that we may view these integrations approximately as line integrals through the space.

Assuming we have monochromatic radiation and neglecting scatter and beam width, the detected radiation intensity I_d corresponding to a source–detector pair is related to the source intensity, I_0, and the spatial distribution of the tissue attenuation coefficient $f(x)$ on the line connecting the source and detector through Beer's law:[5]

$$I_d(x') = I_0 \exp\left[-\int_L f(x)\mathrm{d}x \right] \tag{5.4}$$

where the integral is over the line L between the source and detector. Most modern machines process the raw measured intensity I_d and combine it with the known, calibrated value of I_0 through a log operation to yield the observed projection measurements:

$$g(x') = -\ln\left[\frac{I_d(x')}{I_0} \right] = \int_L f(x)\mathrm{d}x \tag{5.5}$$

which defines the standard tomography line integral observation kernel. This relationship is usually expressed as:

$$g(x') = \int \delta_{x'}(x)f(x)\mathrm{d}x \tag{5.6}$$

where $\delta_{x'}(x)$ denotes an impulse in the spatial variable x along the line connecting the source–detector pair corresponding to the variable x'. The associated imaging kernel $h(x',x)$ of Equation 5.1 can be seen to be linear, but shift variant.

Observations are taken for many source–detector pairs, resulting in many line integrals of f. The goal of tomography is to invert such a projection operation (integrations over volume elements at many angles along many paths through the space), to obtain the attenuation function $f(x)$. Many other types of tomography (e.g., PET, SPECT, SAR) can be modeled, at least approximately, through a similar line integral relationship. As a result, this basic line integral projection model also appears in many other application domains.

Since the sensor in x-ray transmission tomography is actually counting x-ray photons, the corresponding data likelihood follows a Poisson distribution. However, at relatively high dose levels (i.e., large numbers of x-ray photons), this is well approximated by a Gaussian distribution. In other cases (e.g., PET) count levels are lower and a Poisson model is more appropriate.

For x-ray transmission tomography there are several standard methods for inversion and image formation. In fact, in the limit of continuous parallel-beam detector geometry and a continuous set of projection viewing angles, the forward operator of x-ray transmission tomography reduces to the Radon transform, which has a known analytic inverse, the inverse Radon transform.[6] However, practical systems cannot collect a continuous set of projection data over continuous angles. In spite of this, the analytic inversion results, valid for continuous data, are applied in practical sampled-data situations, resulting in approximate inversion methods. They are used despite their nonoptimality because they can be computed relatively efficiently, and for many situations they work remarkably well.

Perhaps the most widely used of these is the filtered backprojection method (FBP).[7] As its name suggests, this approach consists of convolutionally filtering each projection (usually using Fourier methods) before backprojecting it. Backprojection is the adjoint of the forward projection operator, and is often described as smearing the filtered 1D projection function at each projection viewing angle back along the path of the projection line integrations in the 2D image space, followed by integrating these smeared results in the image space over all projection viewing angles.

FBPs filter noise through several approaches. One is to control the cutoff frequency in the Fourier-domain filtering process; another is to use low-order basis element parameterizations with relatively broad image-domain basis elements for the backprojection, so each image element is interpolated and averaged from multiple beam projections. Note there is no explicit inclusion of a noise model, simply low-pass filtering of the result.

The main downsides of filtered-backprojection methods are their inability to handle nonuniformly sampled geometries (e.g., limited-angle or sparse-angle tomography), non-Gaussian noise models that are critical to low-dose x-ray transmission and nuclear emission tomography models, and the inclusion of prior information.

Another class of approaches to tomographic image reconstruction is to directly model a sampled version of the forward model Equation 5.5 with a matrix relationship and then solve the resulting matrix equations. The solution of the resulting large collections of linear equations is done iteratively. Such an approach leads to methods such as the algebraic reconstruction technique (ART).[8,9] These iterative approaches for CT reconstruction are slower than approximations to analytic inverses (such as filtered backprojection), but apply to a wider range of cases and are generally more accurate. Such model-based approaches are also more readily adapted to allow the inclusion of both noise models and prior information into the reconstruction process.

2. Deblurring of Locally Sensed Data

Another common imaging model addresses partial volume effects which lead to blurring and resolution loss. For sensors in this class the kernel $h(x', x)$ captures the effects of local averaging and the sensor is assumed to be directly observing the quantity of interest (i.e., the spaces x' and x are similar). The spatial extent of $h(x', x)$ is generally limited to a small fraction of the entire domain (in contrast to the projection model of CT which extends throughout the field) so the corresponding operator is local. Often the operator is also shift invariant, so that $h(x', x) = h(x' - x)$, in which case the associated imaging operation is a convolution. When this is true, the sensor can be characterized by its response to a point source input, which is then termed the point spread function or PSF. In other cases, the kernel is not well approximated as shift invariant, though the locality of its effects is still satisfied. Perhaps the most famous example of this was the Hubble Space Telescope prior to the installation of the COSTAR corrective optics.

Many physical imaging sensors are directly modeled through such local, averaging behavior in the imaging space. For example, all optical imagers fall in this category. Further, the effects of many other physical systems are explained this way at an aggregate level. In such cases, an accepted front-end image formation process is lumped into the physical sensing process to create an overall end-to-end imaging system. The aggregate characteristics of the overall system of both sensor and image formation are captured through the local operator or PSF $h(x', x)$ of the system. In practice, a point source is used as the input into the imaging chain and the resulting response of the overall sensor/reconstruction system is recorded and used for the convolutional model $h(x' - x)$. Note that such aggregate blurring models can also include resolution loss effects, allowing them to form the basis for superresolution reconstructions.

For example, MRI is an extraordinarily powerful and versatile modality capable of producing high-resolution and high-contrast imagery of the soft tissues of the body. MR can focus in on a remarkable range of chemical properties and, in various operating modes under the right conditions, produce what are effectively vector images.[10] The physics of the MR sensor are such that the directly observed quantities are samples in the Fourier domain of the magnetic resonance properties, f_{MR}, of the underlying tissue. These Fourier samples can be efficiently inverted to obtain the original magnetic property field.[11,12] Proton-density-weighted MRI, for example, can be usefully modeled as directly observing the proton density field with an associated PSF characterizing the blurring and resolution effects of the overall process. This is the approach we take with our MR processing examples in Section V, though we could just as well have used a Fourier domain observation model.

B. Related Work in Image-Based Fusion

Here we discuss work in image reconstruction and image-based data fusion that is related to the approach we will describe in Section III. While there is a great body of literature on image formation as well as image fusion, we will only focus on those elements most similar to the method we present.

1. The Mumford–Shah Variational Approach to Image Processing

A common desire in image analysis is to locate objects in a scene. To this end, a universal approach is to use a gradient operator or high-pass filter to find scene edges, associated with object boundary information. Unfortunately, the presence of noise makes application of such high-pass operators directly on the raw data problematic, due to noise amplification. To mitigate this effect, raw data are typically smoothed first. But this smoothing blurs and degrades object boundary information. Thus, the two steps of smoothing and edge enhancement are interlinked and work in opposition, yet they are typically applied independently.

In Refs. 13,14 Mumford and Shah proposed performing these tasks in a unified manner. They modeled a scene as being composed of both pixel values, represented by f, as well as region boundaries, represented by S. They then proposed *jointly* estimating f and S from the noisy data as the solution of the following minimization problem:

$$[\hat{f}, \hat{S}] = \arg\min_{f,S} \underbrace{\int_X \gamma^2(g-f)^2 dx}_{\text{DataFidelity}} + \underbrace{\int_{X\backslash S} \lambda^2\|\nabla f\|^2 dx}_{\text{Inhomogeneous Smoothing}} + \underbrace{|S|}_{\text{Boundary Length}} \tag{5.7}$$

where X is the entire region of integration, S is a set of boundaries (typically intended to be a set of measure zero, e.g., line-like curves in two dimensions or plane-like surfaces in three dimensions), $X\backslash S$ indicates the set subtraction of S from X, and $|S|$ indicates the length of boundaries in two dimensions and the surface area of S in three. The first term (data fidelity) attempts to fit f to the noisy data image g. The second term smoothes f, except that this term suspends the smoothing at boundary locations, defining edges in the smoothed image f. The third term $|S|$, penalizes total boundary length so that the functional will not simply place boundaries everywhere, suspending smoothing throughout the image. Given a noisy image g, optimizing this functional for both f and S should produce an image f that is piecewise smooth while still preserving edges. This approach is often called segmentation because it can usually separate an image into disjoint homogeneous regions.

Ambrosio and Tortorelli proposed a modified version of this approach[15,16] which used a continuous function approximation $s(x)$ to the binary valued boundary set S of the original Mumford–Shah formulation:

$$[\hat{f}, \hat{s}] = \arg\min_{f,s} \int_X \underbrace{\gamma^2(g-f)^2}_{\text{DataFidelity}} + \underbrace{\lambda^2(1-s)^2\|\nabla f\|^2}_{\text{Inhomogeneous Smoothing}}$$

$$+ \underbrace{\rho\|\nabla s\|^2 + \frac{1}{\rho}s^2}_{\text{Boundary Length}} dx \tag{5.8}$$

where the field $s(x)$ is a continuously valued approximation to the set function represented by S. In the limit, as the parameter ρ approaches zero, this objective functional recovers the Mumford–Shah functional in Equation 5.7. Where $\|\nabla f\|^2$

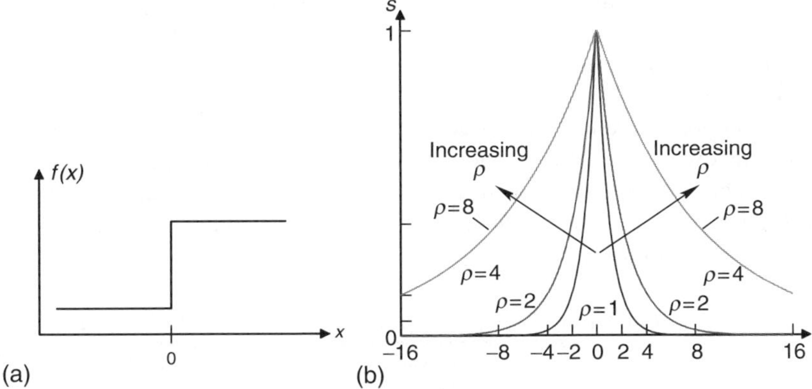

FIGURE 5.3 Illustration of Ambrosio–Tortorelli boundary function s, for a discontinuity in a 1D f, (a) a step function, (b) the corresponding boundary field s for various ρ.

is large, s will be driven to 1 by the inhomogeneous smoothing term and where $\|\nabla f\|^2$ is small, s will be driven to zero by the s^2 term. The term $\|\nabla s\|^2$ makes the boundary function s taper smoothly toward zero as we move away from discontinuities in f. The parameter ρ is sometimes called the boundary scale parameter because it controls how quickly the boundary field s tapers to zero away from discontinuities. To illustrate the idea, Figure 5.3(a) shows a discontinuity in one dimension; this could also be a 1D profile running perpendicularly through an infinite linear boundary in 2D or an infinite planar boundary in 3D. Figure 5.3(b) illustrates the corresponding s given this f for the Ambrosio–Tortorelli objective function in Equation 5.8 and boundary scale parameter $\rho \in \{1, 2, 4, 8\}$.

This image enhancement approach is capable of representing isolated boundary segments that are not part of closed curves. As presented, the approach does not perform image formation or inversion, or operates on a single modality, and thus does not perform fusion. As we will demonstrate, the boundary-scale parameter can be used to facilitate numerical optimization in the fusion process by reducing local minima, resulting in more robust estimates.

Other authors have developed single image variants of these methods. Recently, a popular method has been to represent the boundary set S in Equation 5.7 as a curve obtained as the level set of a surface. The optimal solution curve is then obtained through an evolution process in the gradient direction. Such curve evolution methods can be found in Refs. 17–19. One limitation of such curve evolution methods is that they cannot represent partial boundary segments, which are not part of a closed curve.

2. Single Parameter Image Fusion

Work in this area focuses on fusing data from multiple sensors, all sensitive to, or observing, the same underlying property field. As such all sensors are observing the same, single unknown parameter field. While there are many methods of combining data that could be viewed as fitting into this category (e.g., work in array processing),[20] we will concentrate only on the most relevant to our development.

The authors of Refs. 21,22 used an extension of Equation 5.8 to produce a fused piecewise smooth surface reconstruction from optical and range measurements using a shape-from-shading observation model. A common fused range or surface field f was found. The corresponding objective function was (in our notation):

$$[\hat{f}, \hat{s}] = \arg\min_{f,s} \int_{X'} \underbrace{\gamma_1^2(g_1 - H_1(f))^2}_{\substack{\text{Shape-from-Shading} \\ \text{Data Fidelitys}}} + \underbrace{\gamma_2^2(g_2 - H_2(f))^2}_{\text{Range Data Fidelity}} \, dx'$$

$$+ \int_X \underbrace{\lambda^2(1-s)^2\|\nabla f\|^2}_{\text{Inhomogenous Smoothing}} + \underbrace{\rho\|\nabla s\|^2 + \tfrac{1}{\rho}s^2 dx}_{\text{Boundary Length}} \tag{5.9}$$

where g_1 were intensity observations, g_2 were range observations, and H_1 and H_2 were corresponding observation models relating intensity and range to surface shape, respectively. In this approach no image alignment is considered.

3. Multiparameter Image Fusion

Work in this area focuses on fusing data from multiple sensors, each sensitive to, or observing, a different underlying property field. The observation models for such problems are most naturally described through the presence of multiple property fields, in contrast to the previous section.

Pien and Gauch, in Ref. 23, used a formulation derived from Equation 5.8 to fuse the reconstructed images of multichannel MR data. The images were T1 weighted, T2 weighted, and PD weighted MR images. The general formulation is given by:

$$[\hat{f}, \hat{s}] = \arg\min_{f_k,s} \int_X \sum_{k=1}^{K} \Big\{ \underbrace{\gamma_k^2(g_k - f_k)^2}_{\substack{\text{Independent} \\ \text{Data Fidelity}}}$$

$$+ \underbrace{\lambda_k^2(1-s)^2\|\nabla f_k\|^2}_{\substack{\text{Independent Inhomogenous} \\ \text{Smoothing, Given Boundary}}} \Big\} + \underbrace{\rho\|\nabla s\|^2 + \tfrac{1}{\rho}s^2}_{\substack{\text{Shared, Common} \\ \text{Boundary Model}}} dx \tag{5.10}$$

where $K = 3$ is the number of MR image channels, g_k is the noisy observed MR image from channel k, f_k is the corresponding smoothed physicochemical property image, and s was a shared, common boundary field, constructed with an

Ambrosio–Tortorelli boundary model. While all the images collected were from a single MRI system, there was no clear physical model relating the different physicochemical properties f_k imaged by each MR channel. The authors choose independent smoothing as well as independent data fidelity models for each channel, resulting in a restored image for each channel. A common tissue boundary term is used, however. No alignment or resolution matching is done. Kaufhold illustrated a similar model with an interpretation of Mumford–Shah edge estimation that is conducive to statistical fusion methods in Ref. 24.

Other multiparameter field fusion work includes Ref. 25, where the authors fused intravascular ultrasound imagery with angiograms in order to track the ultrasound catheter and correct the ultrasound imagery to account for bends as the catheter travels along the vessel. In Refs. 26,27, the author uses laser range information and ultrasound measurements to help constrain an ill-posed limited-angle x-ray tomography problem for nondestructive evaluation of an aircraft control surface constructed from sandwiched materials. More recently, Mohammad-Djafari also considered similar methods for fusing ultrasound boundary information with CT reconstructions using Bayesian methods.[28] Other researchers have combined audio and video data for speech recognition and speaker localization.[29–31] In these cases, the authors have exploited not only a shared spatial structure (i.e., the common location of the speech source for both audio and video data), but also a shared temporal structure. These authors demonstrate that there are significant temporal correlations between audio and video that we can use to improve speaker localization and speech recognition.

A number of works have also taken a curve evolution approach to this problem. In Ref. 17 a similar independent vector Mumford–Shah approach based on a curve evolution model for color imagery was proposed. In Ref. 32 a level-set curve evolution method is used to align and segment an MR image and CT image simultaneously, by first assuming that each image could be segmented with a single curve, and then assuming that these two curves are related by an unknown rigid-body co-ordinate transformation. In effect, the authors estimated a single, common boundary curve jointly from the MR and CT images. Their objective functional appeared as:

$$[f_{\text{in}}, f_{\text{out}}, f'_{\text{in}}, f'_{\text{out}}, C, T]$$

$$= \underset{f_{\text{in}}, f_{\text{out}}, f'_{\text{in}}, f'_{\text{out}} C, \phi}{\arg\min} \int_{C_{\text{in}}} (g - f_{\text{in}})^2 \, dx + \int_{C_{\text{out}}} (g - f_{\text{out}})^2 \, dx$$

$$+ \int_{C'_{\text{in}}} (g' - f'_{\text{in}})^2 \, dx + \int_{C'_{\text{out}}} (g' - f'_{\text{out}})^2 \, dx + \nu |C|$$

where C is the curve in the first modality, $C' = T(C)$ is the curve for the second modality, which is related to the first by the unknown co-ordinate transformation T, C_{in} and C_{out} are the regions inside and outside the first modality's curve, respectively, C'_{in} and C'_{out} are the regions inside and outside the second modality's curve, respectively. The data for first and second modalities are g and g', respectively. The values $f_{\text{in}}, f_{\text{out}}, f'_{\text{in}}$, and f'_{in} are the piecewise constant values of f inside and outside the curves for each modality. The term $\nu|C|$ is a curve length penalty. Thus, their approach simultaneously segmented two images with a single, common curve whose location in the images is related by an unknown co-ordinate transformation. To optimize this functional they model the curves as a zero-level set of a higher dimensional function, and evolve the curves by modifying this level-set function using the approach of Osher and Sethian.[33] The method did not address data inversion or resolution differences.

Other work which exploits boundary information includes Refs. 34,35 where the authors used boundary information extracted from MR imagery to help reconstruct and segment emission computed tomography (ECT) imagery. Their boundary curve model was a node-based polygon in a polar co-ordinate system with an origin whose location also required estimation. More recently, Hero, Piramuthu, Fessler, and Titus[36] used a B-spline boundary model estimated from MR imagery to help regularize ECT reconstruction estimates. They have also used other methods for exploiting MR to help regularize ECT reconstructions.[37,38] In a similar approach, Vemuri, Chen, and Wang segmented and smoothed an image using a variational framework similar to that of Mumford–Shah by aligning boundaries drawn from a segmentation atlas to an observed image.[39]

These works use strictly feed–forward information; that is, the boundaries from one modality (e.g., MR or an atlas) are used to enhance a second modality. Such a feed-forward approach effectively presumes that an exogenous modality provides significantly higher quality boundary data than that from the modality we are reconstructing, which may not be the case in general.

III. SHARED BOUNDARY FUSION FORMULATION

In this section we discuss our approach to combining information from multiple imaging modalities sensitive to possibly diverse underlying property fields. We do not directly model the relationships among the particular physical property fields. Rather, we focus on fusing the shared boundary structure of the different modalities in the process of forming images. These shared boundaries are the set of locations where material properties change rapidly. Our modeling assumption is that a discontinuity in one modality's reconstructed image will likely also correspond to a spatial discontinuity in another modality's reconstructed image of the same scene. By fusing estimates of these discontinuity locations and exploiting our improved knowledge of them simultaneously during the reconstruction process, we can obtain higher quality reconstructions than we could obtain by reconstructing each image independently. We perform this process of joint boundary estimation, multimodality image reconstruction, as well as alignment and resolution matching in a common, unified framework, thus combining elements from previous work that have remained isolated to date.

In terms of the problem specification of Section II.A, we additionally assume that the property fields f_k of the K modalities are independent, except that they share a common, normalized boundary or discontinuity field, which we represent as s. Large values of this boundary field indicate a discontinuity in at least one of the property fields, whereas small values indicate no observed discontinuity in any of the property fields. This shared boundary field captures the structure common to many fusion problems, wherein discontinuities are observed to exist across different modalities. For example, organ boundaries are observed in both tomographic and ultrasound images, even though the mechanisms of data acquisition underlying these two modalities are very different.

To account for alignment and scaling differences between different modalities, we assume that the kth property field f_k is related or aligned to the common boundary field s through an unknown, invertible, co-ordinate transformation T_k. This transformation maps the common co-ordinate space of our boundaries, x, to each sensor's co-ordinate space x_k. We refer to this set of transformations collectively as T. Without loss of generality we set T_1 to identity, so our boundary co-ordinates coincide with the co-ordinate frame-of-reference for the first sensor.

We will assume for simplicity that T_k is an affine transformation of the following form:

$$x_k = T_k(x) = A_k x + t_k \tag{5.11}$$

where A_k is an invertible matrix and the vector t_k captures co-ordinate translations or shifts. The matrix A_k can be further decomposed as the product of a rotation, scaling, and shearing matrices, each governed by a single scalar parameter. We will refer to the collection of the kth sensor's alignment parameters (which define A_k and t_k) as θ_k. We will refer to optimizations over the transformation T_k or its parameters θ_k interchangeably. Other parameterized transformation models are, of course, possible.

With these assumptions, our overall problem is to find the common boundary field s together with the k reconstructed property fields f_k (i.e., the formed images), and their associated alignment transformations T_k, based on noisy observations g_k from k individual modalities. The approach we take to combining this information is to jointly estimate these quantities in a unified framework by minimizing an energy function as follows:

$$\hat{f}, \hat{s}, \hat{T} = \arg\min_{f,s,T} E(f, s, T) \tag{5.12}$$

where the energy $E(f, s, T)$ is defined as:

$$E(f, s, T) = \sum_{k=1}^{K} [E_k^{\text{data}}(g_k, f_k) + E_k^{\text{smooth}}(f_k, s, T_k)$$

$$+ E_k^{\text{align}}(T_k)] + E^{\text{bndry}}(s) \tag{5.13}$$

This energy is composed of four components. The term E_k^{data} is a data fidelity term for modality k which takes into account our observation model for that modality. The second term E_k^{smooth} is a smoothing or noise-suppression term for the property field associated with modality k, which is dependent on the common boundary field s and the alignment of that modality T_k. This term links each field f_k and the global boundary structure captured in s. The third term E_k^{align} describes our prior knowledge of, or constraints on, the alignment co-ordinate transformations T_k. Finally, the last term E^{bndry} captures prior knowledge or constraints concerning the shared boundary field s.

In our formulation, the data from every modality directly contributes information to the estimate of the common boundary structuring field s. Thus, by using all of our sensors, we can obtain a better estimate of this shared boundary structure. In particular, boundaries not seen well by one sensor modality may be augmented by information from another modality. Conversely, the improved knowledge of s is used to extract novel information unique to each individual modality's estimated field f_k. Weak boundaries in a single modality are enhanced, and thus the corresponding field estimates are improved. This framework allows us to directly improve our fundamental estimates of the property value fields by unifying image formation, enhancement, and alignment. We now discuss each of the terms in our proposed functional.

A. Sensor Observation Model Term

For the data fidelity or observation term E_k^{data} we use the negative log-likelihood of the data for the kth modality, g_k:

$$E_k^{\text{fid}}(f_k) = -\ln p(g_k | H_k f_k) \tag{5.14}$$

For example, for sampled observations in independent Gaussian noise we would have c.f. Equation 5.2:

$$E_k^{\text{data}}(f_k) = \gamma_k^2 \sum_{x_k'} [g_k(x_k') - H_k f_k(x_k')]^2 \tag{5.15}$$

where $H_k f_k$ represents the linear forward mapping relating the field, f_k, to the data g_k, and γ_k is a real, nonzero weighting parameter. We have assumed in Equation 5.15 that we make observations at a discrete set of points x_k'. Other statistical observation models can be similarly incorporated. The case of Poisson data was presented in Equation 5.3. The term E_k^{data} ensures that the estimates are consistent with the data and the corresponding sensor observation model.

Our inclusion of the forward model H_k in Equation 5.14 means that we are building image formation or data inversion into our fusion framework. The presence of this inversion is important because our image estimates for heterogeneous modalities can now be reconstructed in a common spatial representation. This common representation enables alignment and fusion in a relatively straightforward way. The inversion into a common spatial representation decouples the data resolution for each modality from the modality's reconstructed image resolution. In particular, a common resolution can be chosen for all of the modalities' reconstructions, significantly simplifying alignment and fusion for modalities that have inherently different data resolutions.

B. NOISE SUPPRESSION TERM

We define the smoothing or noise-suppression term E_k^{smooth} for the kth modality, as:

$$
\begin{aligned}
E_k^{\text{smooth}}(f_k, s, T_k) &= \lambda_k^2 \int_{Xk} \|\nabla f_k\|^2 ([1 - \varepsilon_k][1 - s(T_k^{-1}(x_k))]^2 + \varepsilon_k)\|J_{T_k}\|^{-1} \mathrm{d}x_k \\
&= \lambda_k^2 \int_X \|\nabla f_k(T_k(x))\|^2 ([1 - \varepsilon_k][1 - s(x)]^2 + \varepsilon_k) \mathrm{d}x \qquad (5.16)
\end{aligned}
$$

where the co-ordinates x of the common boundary and the co-ordinates x_k of modality k are related by the transformation $x_k = T_k(x)$, X and X_k are the regions of support for the reconstruction in the common boundary space and the individual property field space, respectively, λ_k is a real, nonzero weighting parameter, and $\varepsilon_k \in [0, 1]$ with $\varepsilon_k \ll 1$. The term $\|\nabla f_k\|^2$ denotes the square of the 2-norm of the spatial gradient of f_k, i.e.,

$$
\|\nabla f_k\|^2 = \sum_{\ell} \left(\frac{\partial f_k}{\partial (x_k)_\ell} \right)^2 \qquad (5.17)
$$

in which $(x_k)_\ell$ denotes the lth spatial dimension of the x_k space. The term $|J_{T_k}|$ is the determinant of the Jacobian matrix for the transformation T_k. The first expression in Equation 5.16 is presented in terms of the spatial co-ordinates x_k of the property field for modality k while the second version is in terms of the spatial co-ordinates x of the common boundary term. Note that when the transformation is restricted to be affine, the only elements that impact the determinant of the Jacobian are those affecting scale. Changes impacting translation, rotation, and shear do not change this determinant.

The effect of the term E_k^{smooth} is to anisotropically smooth the corresponding estimated property fields f_k away from the common set of boundaries captured in s, while suspending this smoothing in the vicinity of these boundaries. Further, the edge terms $\|\nabla f_k\|^2$, and the common boundary field s, are intimately coupled by E_k^{smooth}. Where $\|\nabla f_k\|^2$ is large for one or more modalities (or is consistent across modalities) the shared boundary estimate s will be driven toward a value of one, because then $(1 - s)^2$ will be forced toward zero, reducing the smoothness penalty. Thus, this penalty term will either prefer to smooth the images f_k, reducing $\|\nabla f_k\|^2$ for every modality, or insert a boundary to suspend such smoothing for every modality. If any modality strongly favors placing a boundary, then every modality will gain the benefit of knowing the boundary location more accurately, even for modalities that do not observe the boundary clearly.

This common boundary model is what constitutes fusion in our model-based approach. The modalities combine edges synergistically to estimate boundaries. This combination will let us make the boundary estimation less sensitive to any particular modality's errors, giving us more stable and reliable detected boundaries, and fewer spuriously identified boundaries. Conversely, every modality gains the benefits of improved boundary knowledge in its estimation. This will lead to fewer noise driven artifacts where regions are homogeneous (i.e., no boundaries) in any particular modality.

Each individual data smoothing term in Equation 5.16 is similar to the corresponding term in the original Ambrosio and Tortorelli version of the Mumford–Shah segmentation framework[13,15] for single images. However, they are also different, due to the presence of the parameter ε_k. The addition of this new term is motivated by the ill-posed nature of the inversion or image formation component which arises in many problems. In particular, in finding the optimal solution to Equation 5.12 we generally solve a sequence of problems for the individual fields f_k and the common boundary s. During this process, artifacts occurring in our early reconstructions can produce large gradients, which then create persistent but artificial image structure leading to local optimization minima. The addition of the parameter ε_k prevents the formulation from getting overly aggressive in its placement of edges, by producing slightly leaky boundaries.

In practice, we exploit the presence of ε_k in the optimization process. In particular, we use ε_k as a kind of deterministic annealing parameter to help us avoid undesirable local minima by adaptively varying its value during numerical solution. This approach allows us to gradually progress from a more blurred but convex solution to a sharper but less convex solution. As a result, our final solutions will be less dependent on our initialization, and have a reduced sensitivity to the other weighting parameters of our criterion, a commonly known difficulty with Mumford–Shah-based techniques. These issues are discussed in Section IV.

The inclusion of parameter ε_k is also advantageous for the alignment process. Large values of ε_k tend to produce broader edges in our reconstructions f_k, so $\|\nabla f_k\|^2$ will take on smaller values overall but the large values of $\|\nabla f_k\|^2$ will occur over broader bands of locations in space. These broader edges help to reduce the alignment sensitivity to small location shifts around the boundaries; a small variation in a location parameter will not produce a large variation in the energy as $\|\nabla f_k\|^2$ becomes misaligned with s. This reduced sensitivity makes alignment optimizations easier by broadening local minima.

C. ALIGNMENT TERM

The next term is E_k^{align}, which captures constraints and prior knowledge concerning the alignment of the individual modalities to the co-ordinates of the common boundary field. For example, this term can reflect our prior knowledge of the geometrical relationships between the different modalities. For our current application, we have found it sufficient to treat each parameter separately, letting its contribution to this energy be identically zero over a relatively small fixed range of nominal values and very large outside this range. This choice corresponds to a box constraint on the set of possible parameter values. Such a constraint is given by:

$$E_k^{\text{align}}(T_k(x_k; \theta_k)) = \frac{1}{\Delta} u(|\theta_k - \bar{\theta}_k| - \beta_k) \tag{5.18}$$

where $u(\cdot)$ denotes a unit-step function, that is,

$$u(x) = \begin{cases} 0 & x_j < 0 \ \forall j \\ 1 & \text{otherwise} \end{cases} \tag{5.19}$$

$\bar{\theta}_k$ is the center of the parameter constraint range, β_k denotes the size of the range, and $\Delta \ll 1$ is a small positive number. Note that $\bar{\theta}_k$ and β_k are vectors, just as θ_k is, and the absolute-value applies element-by-element. In practice, we choose $\bar{\theta}_k$, the range center, as an exogenous alignment parameter estimate obtained as part of the algorithmic initialization process. This box bound limits alignment search around a nominal set of values, but does not bias the alignment within this constraint.

Other choices for this term are possible, which would capture other forms of prior knowledge about the values of the alignment transformations. For example, statistical models of the parameters could be easily incorporated by setting this term equal to the log-likelihood of the transformation prior.

D. BOUNDARY TERM

Lastly, we define the common boundary term, $E^{\text{bndry}}(s)$, which captures prior knowledge of the behavior of the boundary s. We have chosen to use the similar term arising in the Ambrosio–Tortorelli model[15] of Equation 5.8:

$$E^{\text{bndry}}(s) = \int_X \rho \|\nabla s(x)\|^2 + \frac{1}{\rho} s^2(x) \mathrm{d}x \tag{5.20}$$

where $\|\nabla s\|^2$ is defined as in Equation 5.1, ρ is a strictly positive weighting parameter, and X is the region of support in the boundary space. This boundary term captures the notion that we should minimize the total length of boundaries, while using a boundary model that is smooth. The parameter ρ has the effect of controlling the scale of boundaries. Larger values of ρ give relatively broad boundary estimates while smaller values narrow the boundaries, giving more precise estimates of their locations, as we discussed in relation to Figure 5.3.

Similar to our use of the edge leakage parameter, ε, we also exploit ρ to aid our numerical solution of the alignment problem. In general, larger values of ρ and broader boundaries will increase the capture radius of neighboring edges across modalities. This greater interaction among modalities' edges over larger distances will tend to reduce local minima, helping us to attain a better optimum than we might obtain with narrow boundaries. Thus, ρ tends to act as a controllable multiscale parameter for alignment. We discuss this use in detail in Section IV.

IV. OPTIMIZATION APPROACH

We have presented a formulation for heterogeneous multisensor fusion that requires minimization of the objective function in Equation 5.12. This formulation contains elements that address a number of significant challenges in heterogeneous multisensor fusion. To be of real value, however, we also need an optimization method that is both numerically efficient as well as robust to challenges such as local minima — a common issue with nonquadratic functions such as ours. We present such an approach here, matched to the challenges of imaging sensors and which exploits the structure of our fusion formulation. In particular, we exploit the multiscale and smoothing effect of certain parameters to both speed computation and avoid false minima, as well as to mitigate the difficulties that arise in solving ill-posed image formation problems.

At the highest level, we use a block-coordinate minimization scheme to optimize Equation 5.12, alternately estimating the multiple property fields f_k, the common boundary field s, and the alignments T_k until convergence. Such iterative block minimization methods are a common approach to complex minimization problems.[40] We briefly discuss the challenges of each piece next.

The block energy minimization with respect to f_k is essentially an image formation problem with an anisotropic regularization. It will possess the typical challenges that ill-posed inverse problems are subject to, such as problem size and solution sensitivity. In addition, because we are performing multiple such inversions and linking them together through shared boundary structure, additional problems arise. In particular, reconstruction artifacts in one modality can produce spurious structure in the shared boundary field, which can then affect other modalities. We need additional methods to mitigate such effects, which are specific to the inversion part of our problem. We use the parameter ε to achieve a type of annealing, thereby stabilizing our inversions to such cross talk and reducing sensitivity to inversion artifacts.

The energy minimization over the boundary field s is a quadratic optimization, and, thus, possesses a unique minimizer. This minimizer is straight forward to compute as the solution of an associated set of linear normal equations.

Finally, we have the optimization involved in estimating the alignment terms T_k. This is a challenging aspect due to the possibility of multiple local minima. However, we find in practice that we can avoid convergence problems by using a good initial alignment and by exploiting the multiscale effect of the parameter ρ, as we will detail below.

Overall, our fusion problem has the following elements. The edge leakage parameter ε, allows for a *soft* choice of boundary placement. The scale parameter ρ, controls the scale of reconstructed boundaries. In general, image formation to produce f_k from g_k is computationally expensive, while calculation of the shared boundary s and set of alignment transformations T_k are comparatively inexpensive. The shared boundary field s can greatly affect estimation of the alignment transformation. These elements have led us to develop a particular grouping of optimization steps that yields both efficiency and robustness. These optimization steps, outlined in Figure 5.4, are given by:

1. Initialize f_k, s, T_k, ε, ρ.
2. Jointly form images and find boundary structure by iterating until convergence
   a. Estimate property fields f_k.
   b. Estimate shared boundary field s.
3. Artificially increase ρ, then jointly find alignments and boundary structure by iterating until convergence
   a. Estimate transformations T_k and update s
   b. Gradually reduce ρ back to its nominal value.
4. Reduce ε and go to 1.

There are two major suboptimizations that are performed. The first is over both the property fields f_k and the shared boundary s. The second is over the alignments T_k and the shared boundary s. Both suboptimizations involve updating of the shared boundary, but in different ways matched to the structure of the problem. Similarly, there are two main loops in the process. An inner loop which performs image formation and an outer loop that alternates between updating reconstructed imagery and alignment. We discuss each of the optimization steps and our solution to the associated challenges which arise next.

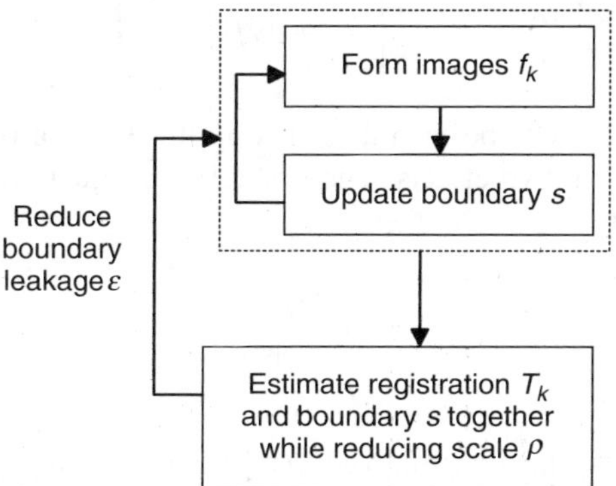

FIGURE 5.4 Block diagram of the optimization algorithm.

A. Shared Boundary Estimation

We first discuss optimization with respect to the shared boundary structure s, assuming the property fields f_k and alignment transformations T_k are given. The solution presented here occurs in both the inner image formation step as well as the alignment step, though it is used in different ways. To this end, note that minimization of Equation 5.13 over the shared boundary s for a fixed alignment T involves minimizing the following energy function:

$$
\begin{aligned}
E(s|f_k, T_k) &= \sum_{k=1}^{K} [E_k^{\text{smooth}}(f_k, s, T_k)] + E^{\text{bndry}}(s) \\
&= \int_X (1 - s(x))^2 \sum_{k=1}^{K} [\lambda_k^2 (1 - \varepsilon_k) \|\nabla f_k(T_k(x))\|^2] + \rho \|\nabla s(x)\|^2 \\
&\quad + \frac{1}{\rho} s^2(x) dx
\end{aligned}
\tag{5.21}
$$

To gain further insight into this optimization, let us define the following average of the registered gradient fields:

$$
\overline{\|\nabla f\|^2} \equiv \sum_{k=1}^{K} [\lambda_k^2 (1 - \varepsilon_k) \|\nabla f_k(T_k(x))\|^2]
\tag{5.22}
$$

This quantity can be seen to be an average of the squared-norms of the gradient fields of the reconstructed imagery obtained from all the modalities and thus represents a fusion of their boundary information. In terms of this quantity, one can show that minimization of Equation 5.21 is equivalent to minimization of the following energy (they only differ by a constant which is independent of s):

$$
E'(s|f_k, T_k) = \int_X \left(\frac{1}{\rho} + \overline{\|\nabla f\|^2} \right) \left[\frac{\overline{\|\nabla f\|^2}}{\frac{1}{\rho} + \overline{\|\nabla f\|^2}} - s \right]^2 + \rho \|\nabla s(x)\|^2 dx
\tag{5.23}
$$

From Equation 5.23 it can be seen that the optimization for the shared boundary structure s depends on two effects. The first term in Equation 5.23 forces s to be close to the quantity

$$
\frac{\overline{\|\nabla f\|^2}}{\frac{1}{\rho} + \overline{\|\nabla f\|^2}}
\tag{5.24}
$$

This quantity is a nonlinear normalization of the fused image gradients $\overline{\|\nabla f\|^2}$ into the range [0,1]. The breakpoint of this mapping is at the point $1/\rho$ — that is, values of fused image gradient that are small compared to $1/\rho$ are mapped to values near zero, while values of fused image gradient that are large compared to $1/\rho$ are mapped to values near one. The second term in Equation 5.23 provides smoothing of the estimated common boundary field s, thus suppressing noise and artifacts.

Overall, the optimization Equation 5.23 with respect to s is a quadratic minimization in s. It can be accomplish by solving the following set of linear normal equations for s:

$$\left(\rho\nabla^2 + \frac{1}{\rho} + \overline{\|\nabla f\|^2}\right)s = \overline{\|\nabla f\|^2} \tag{5.25}$$

or, equivalently in terms of the original problem elements:

$$\left(\rho\nabla^2 + \frac{1}{\rho} + \sum_{k=1}^{K}[\lambda_k^2(1 - \varepsilon_k)\|\nabla f_k(T_k(x))\|^2]\right)s$$

$$= \sum_{k=1}^{K}[\lambda_k^2(1 - \varepsilon_k)\|\nabla f_k(T_k(x))\|^2] \tag{5.26}$$

Note that the role of ρ is to increase the effect of the smoothing term, basically producing a scale dependent behavior to the resulting boundary field. Larger values of ρ correspond to the estimation of larger scale, diffuse boundary structure and smaller values of ρ produce smaller scale, more focused boundary structure. This multiscale effect of the parameter ρ can be seen in Figure 5.3 for an ideal 1D edge.

For a solution of Equation 5.26, we discretize the partial differential equations. The resulting linear set of equations exhibits a highly sparse and banded matrix structure. This structure leads to efficient matrix solution. We use an iterative method for their solution which combines sweeps of successive over relaxation (SOR) with multigrid for extremely fast and efficient solution.[41-44] Note that for any particular alignment T_k and set of fields f_k there is precisely one corresponding boundary field that minimizes the energy Equation 5.23.

B. BOUNDARY AWARE IMAGE FORMATION

In this step of the optimization, we jointly estimate each property field f_k and the shared boundary field s, assuming alignments T_k are given. We accomplish this joint suboptimization through a straight forward cyclical block co-ordinate descent, where we sequentially optimize with respect to each individual property field f_k holding s fixed and then optimize with respect to the shared boundary structure field s while holding all the f_k fixed. This latter optimization with respect to the shared boundary s has already been discussed in Section IV.1. Here we focus on the former task of performing boundary aware image formation.

Note that the individual fields f_k are only linked through the shared boundary structure s. If s is held fixed, joint optimization over all f_k is equivalent to independent optimization over each f_k individually. Thus, this suboptimization can equivalently be viewed as alternating between updating all the fields f_k jointly and then optimizing the boundary s. The minimization with respect to each f_k can take place separately with no loss of optimality. In particular, then, the key operation of this stage is to individually minimize each of the following energies with respect to f_k:

$$E(f_k|T_k, s) = E_k^{\mathrm{data}}(f_k) + E_k^{\mathrm{smooth}}(f_k, s, T_k) = -\ln p(g_k|H_k f_k)$$
$$+ \lambda_k^2 \int_{X_k} \|\nabla f_k(x_k)\|^2 ([1 - \varepsilon_k][1 - s(T_k^{-1}(x_k))]^2 + \varepsilon_k)|J_{\phi_k}|^{-1}\mathrm{d}x_k$$

$$(5.27)$$

The addition of the small term ε_k ensures that there is always a minimal amount of smoothing in the reconstruction process. This modification helps stabilize our reconstruction of f_k, even in regions where the boundary field is particularly dense and the image formation problem, as reflected by H_k, is poorly conditioned. During solution, we start the outer optimization loop with a large value of ε_k and then reduce its value during successive iterations of the outer loop. We have found that a geometric reduction schedule of ε_k works well in practice.

Let us define the following inhomogeneous weighting function:

$$w(x) = ([1 - \varepsilon][1 - s(T^{-1}(x))]^2 + \varepsilon)|J_{\phi_k}|^{-1} \qquad (5.28)$$

In terms of this function, our image formation problem can be phrased as minimization of the following energy:

$$E(f_k|T_k, s) = -\ln p(g_k|H_k f_k) + \lambda_k^2 \int_{X_k} w(x_k)\|\nabla f_k(x_k)\|^2 \mathrm{d}x_k \qquad (5.29)$$

This suboptimization can be recognized as a penalized likelihood image formation subproblem, for which a large literature exists and can be brought to bear.[37,35,45] While the weight $w(x)$ is spatially varying, the second term is quadratic in the unknown f_k. Thus, we do not have the concerns that often arise with nonquadratic penalty functions.

Further, if the data likelihood is also quadratic, as it is in the common case of sampled observations in additive Gaussian noise, then the overall problem is quadratic:

$$E(f_k|T_k, s) = \sum_{x_k'} [g_k(x_k') - H_k f_k(x_k')]^2$$
$$+ \lambda_k^2 \int_{X_k} w(x_k)\|\nabla f_k(x_k)\|^2 \mathrm{d}x_k \qquad (5.30)$$

where the sum is over the sample points in the data space. This is the case of most interest to us in fusing x-ray tomographic data with modalities such as MR or with other tomographic data. In this situation, a unique solution can be found by solving the associated set of linear normal equations:

$$(H_k^\dagger H_k + L_k^\dagger L_k)f_k = H_k^\dagger g_k \tag{5.31}$$

where H_k denotes the forward sensor kernel discussed c.f. (1), $\dagger$ denotes the adjoint operator, and L_k is a regularizing operator, given by:

$$L_k = \lambda_k \sqrt{w(x_k)}\nabla \tag{5.32}$$

This regularizing operator is a spatial gradient operator with an anisotropic weight dependent on the common boundary field s.

The large size of the fields f_k associated with many imaging modalities makes solution of these problems challenging. For example, one of our modalities of interest is high-resolution x-ray tomography with millions of pixels. When the normal equations (Equation 5.31) are discretized, both H_k and L_k exhibit a neighborhood structure that extends throughout the field, causing both storage and computational challenges. The spatially varying nature of the anisotropic operator L_k prevents the use of conventional tomographic inversion approaches, such as filtered back-projection.

We use the efficient preconditioned conjugate gradient (PCG) algorithm[43,44] to address such inversion challenges and to solve the corresponding discretized system of linear equations in Equation 5.31. We use a matrix preconditioner M_k constructed as follows:

$$M_k = L_k^T L_k + \text{diag}([H_k^T H_k]_{ii}) + \delta I \tag{5.33}$$

where $\text{diag}(M_{ii})$ denotes the diagonal matrix formed from the diagonal elements of the matrix M, and δ is a small value. We set δ to be 100 times smaller than the smallest nonzero diagonal element of $H_k^T H_k$.

This PCG approach significantly accelerates the convergence of the iterative algorithm for our tomographic reconstructions. Since our algorithm can involve a large number of such reconstructions, the computational savings enabled by this preconditioner are significant.

C. Multimodal Alignment

This step of the optimization is focused on jointly estimating the multimodality alignments T_k together with the shared boundary field s given the set of formed multimodality images f_k. This suboptimization is equivalent to jointly minimizing the following function with respect to s and T_k:

$$E(T_k,s|f_k) = \sum_{k=1}^{K} [E_k^{\text{smooth}}(f_k,s,T_k) + E_k^{\text{align}}(T_k)] + E^{\text{bndry}}(s) \tag{5.34}$$

where these terms were described in detail in Section III. Simultaneously optimizing alignment and the common boundary field is logical because the alignments obtained this way not only align the edges in the different modality reconstructions, but also produce a solution that minimizes the total boundary length — that is, good alignments result in fewer boundaries than poor ones.

As discussed in Section III. C, we have chosen E_k^{align} as a simple box bound on the alignment parameters, reflecting the fact that we are *a priori* neutral to different alignments in a range around our initialization. Such terms can be captured as limit constraints confining the parameters of T_k to a range $\theta_i \in [\bar{\theta}_i - \beta_i, \bar{\theta}_i + \beta_i]$ for each sensor k. The optimization problem implied by the energy function in Equation 5.34 is a limit-constrained optimization over s and T_k:

$$\hat{T},\hat{s} = \arg \min_{\substack{s,T(x;\theta)(2) \\ \theta \in [\bar{\theta}-\beta,\bar{\theta}+\beta]}} \left[\sum_{k=1}^{K} E_k^{\text{smooth}}(f_k,s,T_k) + E^{\text{bndry}}(s) \right] \tag{5.35}$$

where $\bar{\theta}$ and β collectively represent the K constraint range centers, $\bar{\theta}_k$ and constraint range widths, β_k, as described in Equation 5.18.

We could again accomplish the joint suboptimization over T_k and s in Equation 5.35 through cyclical block co-ordinate descent, alternating between updates of the T_k and s, as was done in Section IV.B. We have found it significantly faster, however, to use an iterative gradient descent-based approach for finding T_k with an embedded optimization of s occurring within each gradient update of T_k. This approach is feasible since the solution for the optimal s given T_k is relatively efficient in comparison with solution for the optimal T_k given s. We have already detailed this optimization over s for given T_k in Section IV.A.

Our gradient–descent-based approach to optimizing over T_k requires calculation of the gradient of the energy in with respect to T. The only terms remaining in Equation 5.35 that depend on T are the terms $E_k^{\text{smooth}}(f_k, s, T_k)$, which are defined in Equation 5.16. The descent direction d_k of the transformation T_k is the negative of the gradient of $E_k^{\text{smooth}}(f_k, s, T_k)$ with respect to its registration parameters θ_k:

$$d_k = -\frac{\partial E_k^{\text{smooth}}(f_k, s, T_k(x; \theta_k))}{\partial \theta_k} = -\int_X \left\langle q_k(x), \frac{\partial}{\partial \theta_k} T_k(x; \theta_k) \right\rangle dx \tag{5.36}$$

where $(\partial / \partial \theta_k)(T_k(x; \theta_k))$ is the spatially varying gradient of the transformation T_k with respect to its parameters θ_k and $\langle \cdot, \cdot \rangle$ indicates an inner product of the vectors. The vector quantity $q_k(x)$ is given by:

$$q_k(x) = \lambda_k^2 [1 - \varepsilon_k](1 - s(x))^2 \left[\frac{\partial}{\partial x} \|\nabla f_k(T_k(x; \theta_k))\|^2 \right] \tag{5.37}$$

and is the vector gradient of the norm-squared gradients of the field f_k transformed into the space of x using the current estimate of the co-ordinate transformation T_k, and weighted by the squared complement of the boundary field, that is, $(1 - s)^2$.

The negative of the vector q_k indicates which direction an edge in the reconstruction corresponding to modality k should move in order to merge with nearby boundaries. Further, the gradient of T_k with respect to the parameters θ_k, is a vector at each point in space indicating the direction in which that point would move for a positive change in transformation parameter.

To optimize the alignment parameters we descend the energy surface in the direction indicated by the gradient in Equation 5.36. Given a current set of alignment parameters $\theta^{(t)}$, we find a new set of alignment parameters, $\theta^{(t+1)}$, by moving along the descent direction, $d^{(t)}$ while simultaneously estimating the energy minimizing s for this new value of the alignment parameters:

$$\theta^{(t+1)} = \theta^{(t)} + \alpha^{(t)} d^{(t)} \tag{5.38}$$

where $d^{(t)}$ is the collection of all of the parameter descent directions $d_k^{(t)}$. We choose $\alpha^{(t)}$ to ensure that the new parameter values $\theta^{(t+1)}$, both stay within the constraints, $[\bar{\theta} - \beta, \bar{\theta} + \beta]$ and reduce the cost.

In summary, to jointly solve for the optimal set of T_k and s in Equation 5.34, we take a step in the direction of the cost gradient with respect to T_k according to Equation 5.38, we then solve for the unique s corresponding to the new value of T by solving Equation 5.25. We choose the size of the steps to ensure that the cost is reduced and that the box constraints are not violated. We iteratively repeat this process until convergence. Thus, while minimizing with respect to T_k we maintain the optimal solution for s throughout.

In performing this joint alignment optimization process for T_k and s, the main challenge is the presence of many local minima in the alignment parameters. Such local minima would trap a purely gradient-based technique away from the global optimum. To mitigate this problem a number of steps are taken. First, an initial alignment is performed using knowledge of the modalities involved and their coarse geometrical relationship. Box bounds on the set of possible alignment parameters are created from this information. These bounds globally constrain the search space through Equation 5.18. The energy minimization for alignment is thus started at a value near the correct ones, eliminating many false minima from consideration.

Another way in which we reduce the impact of local minima is through the multiscale parameters ε and ρ. A larger value of the boundary leakage parameter ε serves to smooth discontinuities in the reconstructions, reducing the optimization's sensitivity to large discontinuities $\|\nabla f\|^2$. As we have discussed, in our outer loop we start with a relatively large value of ε and reduce it as the iterations proceed.

In addition to global manipulation of ε, we also vary the boundary scale parameter ρ in the alignment suboptimization. In particular, at the start of each alignment suboptimization the value of ρ is increased from its nominal value. As the alignment iterations in the suboptimization proceed it is then returned to its nominal value. Large values of the ρ boundary scale parameter increase long-range interactions among boundaries, broadening the basin of attraction for the global optimum. We have found that a geometric reduction of ρ as the iterations proceed works well in practice.

Figure 5.5 illustrates how such variation of ε and ρ can help reduce the impact of local minima in the alignment optimization while maintaining

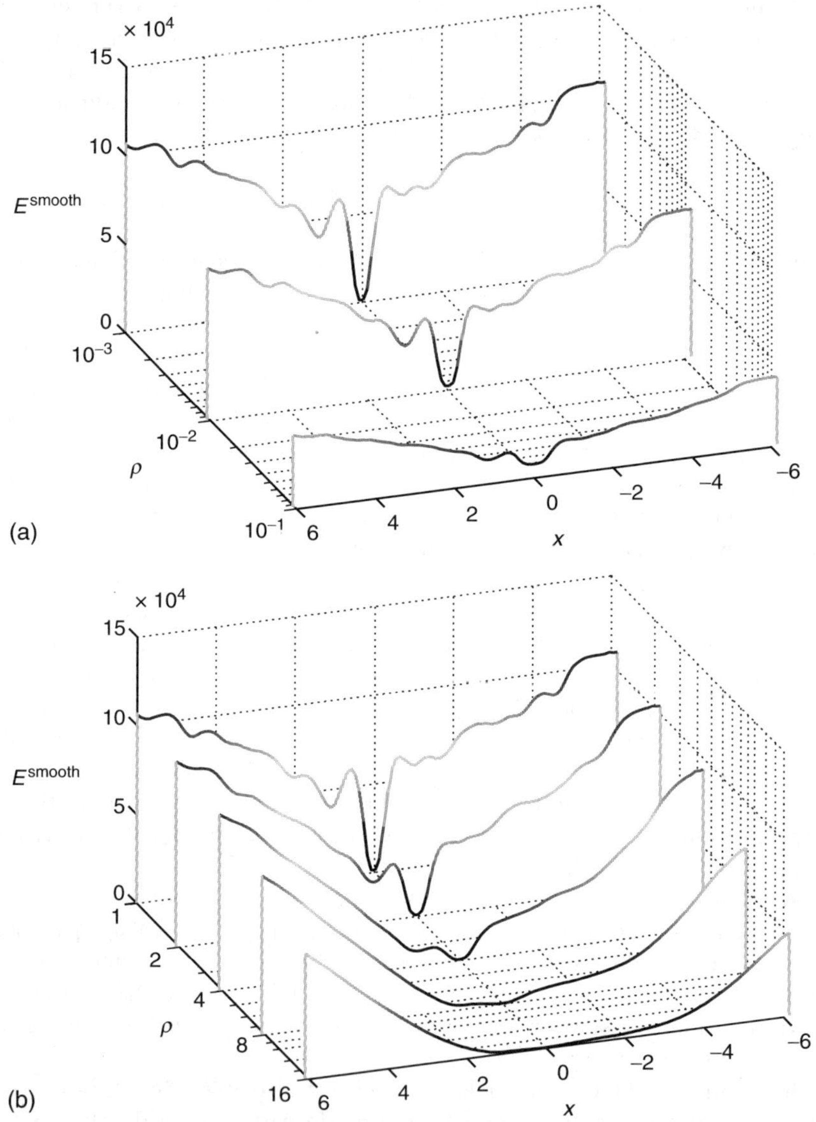

FIGURE 5.5 Slices of the alignment subenergy along the translation axis for various $\varepsilon(\varepsilon = \{0.001, 0.01, 0.1\})$ and $\rho(\rho = \{1, 2, 4, 8, 16\})$. Slices were taken through the true optimum. Larger values of ε and ρ reduce alignment local minima, improving the likelihood of locating the globally optimal alignment.

the global optimum. Each slice of the waterfall plots is a cross-section through the E^{smooth} energy surface along a translation direction through the true energy minimum for a different value of ε or ρ. The left graph in Figure 5.5 shows the effect on the alignment energy of varying ε, while the right graph in Figure 5.5 shows the effect of varying ρ on the alignment energy. Note, for example, that for small values of the boundary scale ρ there are significantly more local minima in the alignment energy surface than there are for larger values of ρ. This effect enables us to find an attainable global minimum for large values of ρ that is reasonably close to the global minimum for smaller values of ρ.

Our solution exploits the nature of the parameters in the variational formulation to avoid false minima and speed convergence. In our experience, the resulting algorithm has always performed efficiently and converged to good estimates of the alignment parameters for reasonable alignment initializations and starting points for ρ and ε. For a more complete discussion of parameter selection and multiscale optimization, see Ref. 41.

V. RESULTS

We demonstrate our multisensor fusion framework for joint reconstruction, alignment, and resolution matching in two examples. The first is based on simulated imaging of a 2D vessel cross-section, with observations of projection data for MDCT and formed image observations for proton-density-weighted MRI. This example illustrates simultaneous alignment with tomographic reconstruction for MDCT and superresolution reconstruction for MRI. In the second example, we perform fusion with real data from formed 2D imagery derived from MDCT and VCT imaging of an excised plaque specimen. This example draws on the measured point spread function from both imaging systems to illustrate joint reconstruction and alignment combining the high-resolution of VCT with the better dynamic range of MDCT.

A. FUSION OF SIMULATED MDCT–MR PLAQUE IMAGING

1. Observation and Inversion Model

We used the reference histology image of a diseased arterial cross-section in Figure 5.6 to simulate the property fields for CT and MR shown in Figure 5.6. The property fields for both modalities were piecewise constant; for CT, the property field reflects the distribution of x-ray attenuation coefficients whereas for MR, the property field reflects the proton density distribution. The reference histology was hand segmented to delineate the fibrous arterial wall, the vessel lumen, a lipid-rich necrotic core, perivascular fat, and a surrounding medium. For CT, an x-ray attenuation coefficient typical of biological tissue was assigned to the arterial wall, whereas a lower-density attenuation coefficient corresponding to oil was assigned to the lipid pool and perivascular fat. The lumen and

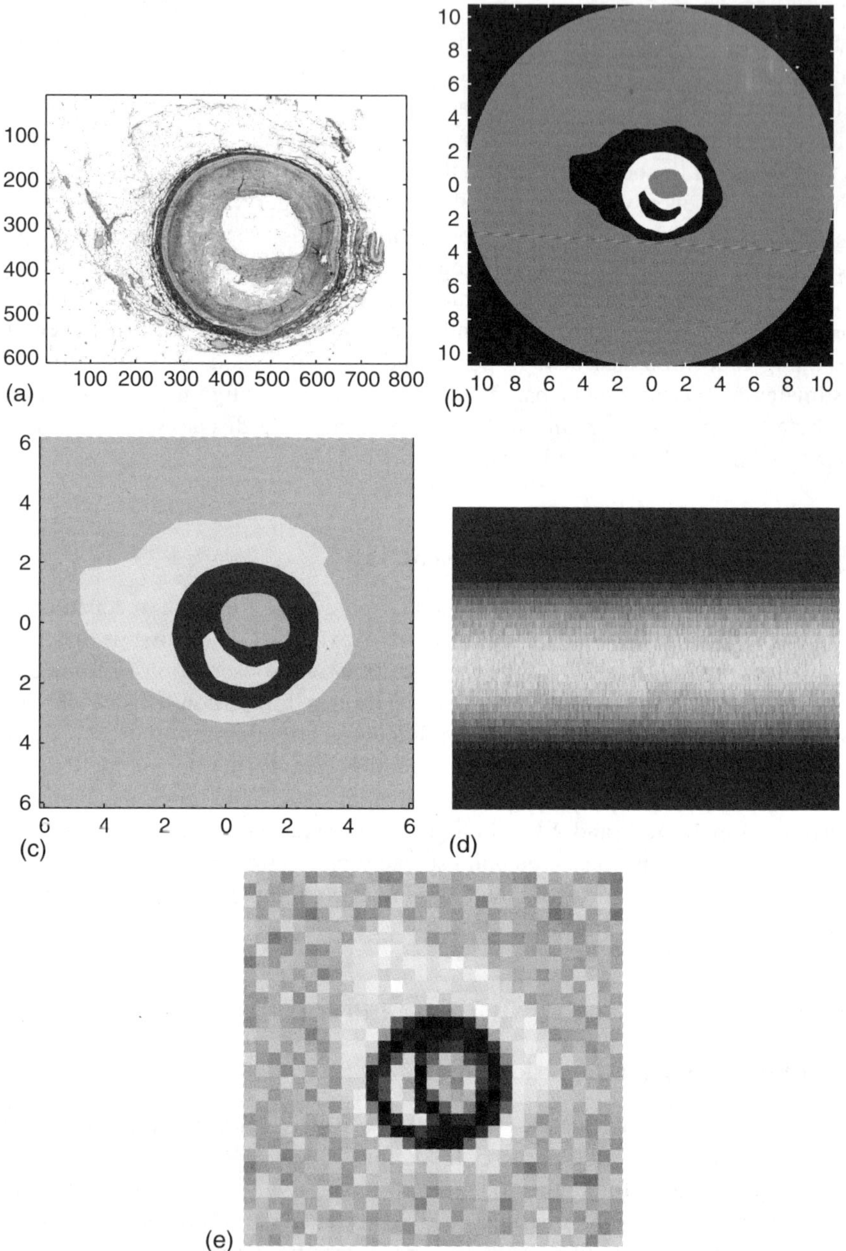

FIGURE 5.6 (See color insert) Histology used for generating blood vessel phantoms, corresponding simulated MDCT and MR blood vessel property fields, and misaligned data observations.

surrounding medium were assigned an attenuation coefficient corresponding to a hydrogel medium within which the vessel specimen was embedded. For MR, the property values for these regions were approximated by using empirical measurements averaged from MR images of fibrous tissue, lipid, and hydrogel.

The CT property field f_{CT} in Figure 5.6 is a 900×900 image, with a resolution of 24 μm per pixel. The field of view (FoV) shown is 21.6 mm in diameter, with its center coinciding with the CT isocenter. The source-to-isocenter distance R_{iso}, for the modeled MDCT geometry was 541 mm and the angular spacing of the detectors is 1.08 mrad to emulate a clinical scanner. We chose the angular range of the detectors, 43.2 mrad, to be wide enough to completely contain the FoV within the beams between the source and the detector and to have an even number of detectors (40 detectors). We collected 984 tomographic projections g_{CT}, uniformly spaced $\Delta\theta = 2\pi/984 \approx 6.4$ mrad apart in the range from $\theta \in [0, 2\pi - \Delta\theta]$. These data, Figure 5.6, were generated from the CT property field by using a fan-beam forward projection model, H_{CT}, and additive, independent, identically distributed, zero-mean Gaussian noise, n_{CT}, with a standard deviation of 2×10^{-7}:

$$g_{CT} = H_{CT}f_{CT} + n_{CT} \qquad (5.39)$$

The CT model used in our unified inversion process corresponds to a 212×212 reconstruction image of the attenuation property field with a spacing of 102 μm per pixel, chosen to match typical CT parameters.

For the MR imaging modality we directly used an aggregate point-spread function model matched to typical scanner resolution, as described in Section II.A. In particular, we generated observations, Figure 5.6, by first convolving a true, underlying MR property field f_{MR}, with a point spread function derived from a clinical scanner and then downsampling the result to the appropriate discrete sampling interval prior to the addition of independent and identically distributed zero-mean Gaussian noise, estimated empirically from the reference MR data:

$$g_{MR} = \downarrow (PSF_{MR} * f_{MR}) + n_{MR} \qquad (5.40)$$

The MR data sample spacing was 0.408 mm per pixel over a 12.6 mm $\times$ 12.6 mm field of view. Note that we could also have used a physically based Fourier imaging model.

For our MR reconstruction or inversion model, we use a computationally less expensive convolutional model, which assumes that each sample in our data is a simple average over a 4×4 block of elements of f_{MR} centered at our observation sample. This block-averaging model is also a convolution combined with subsampling, with the additional advantage that both it and its adjoint are easier to implement and more efficient to compute than the convolutional model used for generation of the observation. This model is sufficient for approximating the true property field f_{MR}, on a factor of four finer grid than the observed data g_{MR}, and reflects the stability of the approach to imperfections in our inversion model. The reconstructed MR field consisted of 212 mm × 212 mm samples to match the full CT image field of view. We modeled regions outside the MR observation 12.6 mm × 12.6 mm field of view simply as missing data, truncating these regions in our observation model to match the available data.

Finally, misalignment between the MDCT and MR datasets was introduced by rotating and translating the true, higher resolution MR property field f_{MR} prior to computing the MR observations g_{MR}. The misalignment in this experiment consisted of a 30° rotation and a translation of (2.04, 1.53) mm of the MR image relative to the CT image.

2. Fusion Results

We applied our shared structure approach to jointly estimate the underlying CT density field, the underlying high-resolution MR field, the modality alignment and the shared boundary structure. We use nominal values of $\rho = 2$, $\lambda_{CT} = 14.1$, $\lambda_{MR} = 70.7 \times 10^{-3}$, $\gamma_{CT} = 3.54$, and $\gamma_{MR} = 17.7 \times 10^{-3}$ for the weighting parameters in our fusion framework. We set $\varepsilon_{CT} = \varepsilon_{MR} = \varepsilon$ and reduced ε as the outer loop proceeded with $\varepsilon = 0.5^2, 0.5^3, 0.5^4, 0.5^5, 0.5^6$ for the five iterations of the outer loop needed for convergence. As we have described, early in the optimization this approach yields somewhat oversmoothed initial reconstructions which also aid image alignment (Figure 5.7).

To illustrate the effect of such parameter variation early in the iteration process Figure 5.8 shows the initial misaligned estimates, $\hat{f}_{CT}$ and $\hat{f}_{MR}$ after the first reconstruction step. Figure 5.8 shows the associated shared boundary field, $\hat{s}$,

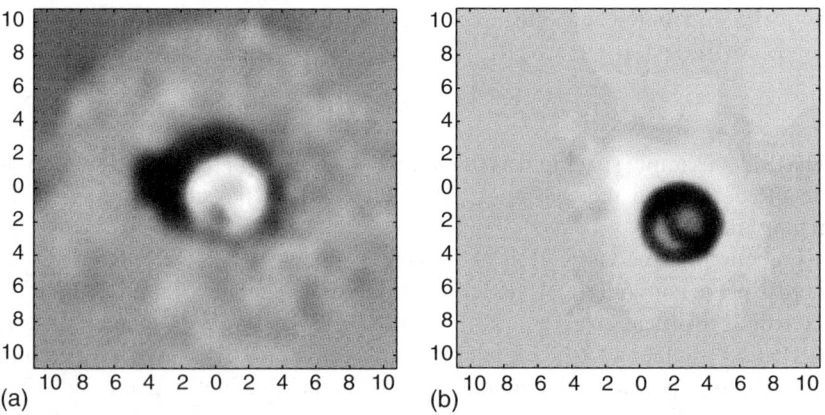

FIGURE 5.7 Initial estimates after first reconstruction step for misaligned MR and CT images, true misalignment: 30° rotation and (2.04, 1.53) mm translation.

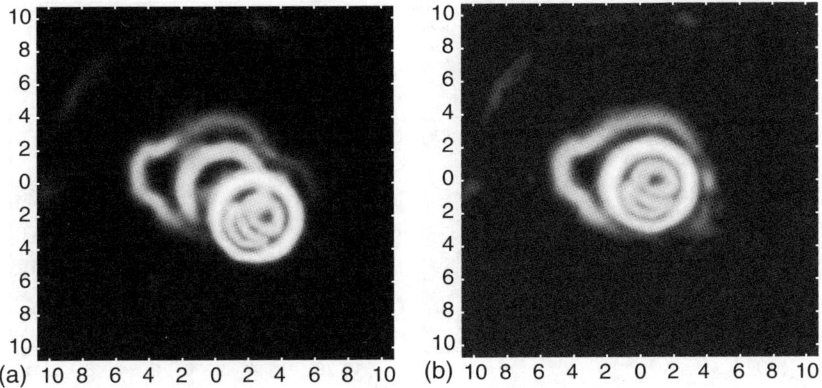

FIGURE 5.8 Boundary estimates $\hat{s}$, before and after initial rigid-body alignment.

that we estimate simultaneously with the $\hat{f}_{CT}$ and $\hat{f}_{MR}$ reconstructions shown in Figure 5.7, also prior to estimating alignment. As can be seen, the reconstructions are quite smooth and the associated shared boundary field prior to alignment exhibits misregistration.

The estimated shared boundary field after the end of the first alignment stage is shown in Figure 5.8. In each alignment suboptimization, we used a decreasing sequence of ρ values with $\rho = 2^4, 2^3, 2^2, 2$ for the four iterations needed for convergence. We have already done a good job aligning the two modalities at this first iteration. The corresponding estimates at this stage for rotation are 29.7° and for translation are (2.12, 1.46) mm.

Figure 5.9 shows the final results for this simulation example after five iterations of the outer loop with the parameter reduction schedule described above. Figure 5.9 shows the reconstructed CT field $\hat{f}_{CT}$, while Figure 5.9 shows the reconstructed MR field as co-ordinate-transformed and approximated on the boundary-CT grid, that is, $\hat{f}'_{MR} \approx \hat{f}_{MR}(\hat{\phi}(x_1))$. Figure 5.9 shows the final shared, common boundary field for the two modalities. For comparison, Figure 5.10 illustrates typical results from standard single modality methods. Figure 5.10 shows a Tikhonov-based reconstruction of the tomographic data, which produces images comparable to filtered back-projection for this geometry. Figure 5.10 shows the standard MR image, which is our observation. Note the improvement in noise suppression and structural detail that is present following fusion processing, as opposed to standard processing.

With unified processing, it is also possible to simplify interpretation of fusion results since reconstruction and alignment occur at a common resolution. For example, Figure 5.9 illustrates a false color composite that we have constructed by assigning the information in Figure 5.9 to each color channel, effectively visually presenting the fused information from all channels.

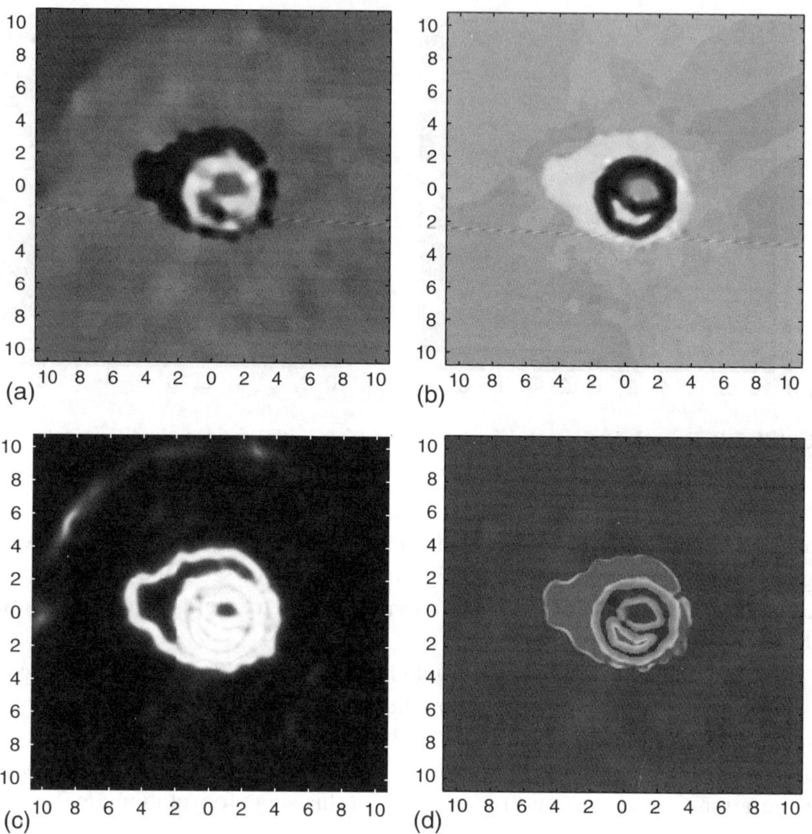

FIGURE 5.9 **(See color insert)** Final1 reconstructions and boundaries.

B. FUSION OF REAL MDCT–VCT PLAQUE IMAGING

1. Data Acquisition

We obtained a carotid atherosclerotic plaque from surgical endarterectomy and imaged the specimen in air using a Siemens Somatom Sensation 16 cardiac MDCT scanner (Siemens Medical Solutions, Forcheim, Germany). A standard cardiac imaging protocol was used (80 kVp, 500 mAs, 16 × 0.75 mm collimation, 0.75 s rotation time, 6 mm/s table feed, 22 cm field-of-view, and a 512 × 512 pixel matrix) in combination with a simulated ECG signal, which provided a 60 bpm trigger for image reconstruction. MDCT images were reconstructed with 0.75 mm thick slices, 0.3 mm increment, and a medium smooth image reconstruction filter (B35F). Following MDCT, we imaged the specimen using a prototype, high-resolution, digital-flat-panel-based, VCT system (Siemens Medical Solutions, Forcheim, Germany) which provides an isotropic spatial resolution of $0.2 \times 0.2 \times 0.2$ mm^3 (80 kVp, 30 mA, and a 512 × 512 pixel matrix). VCT images

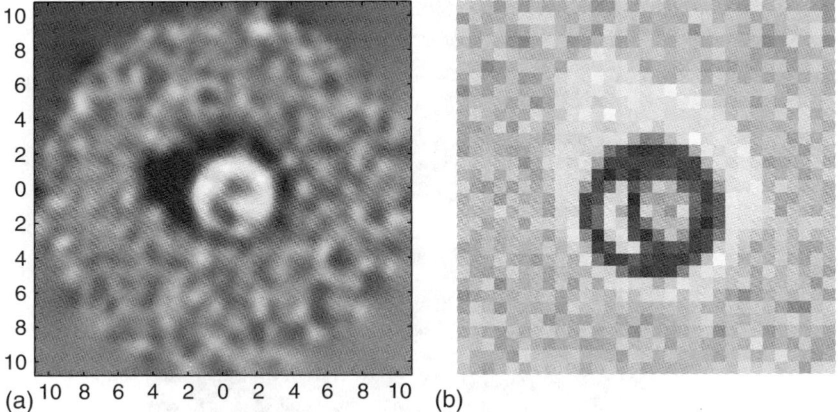

FIGURE 5.10 Standard reconstructions. The CT image is obtained by solving $[T^\dagger T + (\lambda/\gamma)^2\nabla^2]\hat{f} = T^\dagger g$ and is comparable to Filtered Back-Projection reconstruction. The MR image is our observation in this case.

were reconstructed with a 0.2 mm slice thickness using a sharp reconstruction kernel (H95A). The 2D point spread functions for MDCT and VCT image formation with the specified reconstruction kernels were characterized using a standard quality control resolution phantom. Since the raw projections from these acquisitions were not available, we focus here on the fusion of formed 2D imagery from MDCT and VCT. In this case, the inversion problem becomes one of using the measured MDCT and VCT point spread functions for superresolution reconstruction of the underlying property fields.

2. Fusion Results

Figure 5.11 illustrates the observed misaligned MDCT and VCT images. Note that in the original MDCT image, regions with varying CT density can be appreciated, however structural boundaries within the thickened eccentric plaque are extremely difficult to localize due to blurring in the image formation and reconstruction process. Furthermore, CT densities within the thin portion of the plaque appear washed out due to partial voluming. In the VCT observations, boundaries appear sharper due to its higher spatial resolution. However, imaging noise is higher and the dynamic range of CT density variation appears lower than in the corresponding MDCT image. We performed fusion using the processing scheme and weighting parameter assignment strategy described previously to visually optimize the appearance of the fused estimates.

For parameters, we selected $\lambda_{\mathrm{VCT}} = 0.05$ and $\gamma_{\mathrm{VCT}} = 0.01$, while we selected $\lambda_{\mathrm{MDCT}} = 0.02$ and $\gamma_{\mathrm{VCT}} = 0.01$. In the outer loop of the optimization we set $\varepsilon_{\mathrm{MDCT}} = \varepsilon_{\mathrm{VCT}} = \varepsilon$ and reduced ε as the outer loop proceeded with $\varepsilon = \{0.95, 0.2, 0.1, 0.05, 0.02\}$ for the five iterations of the outer loop needed for convergence suitable to the application. Similarly, in each alignment

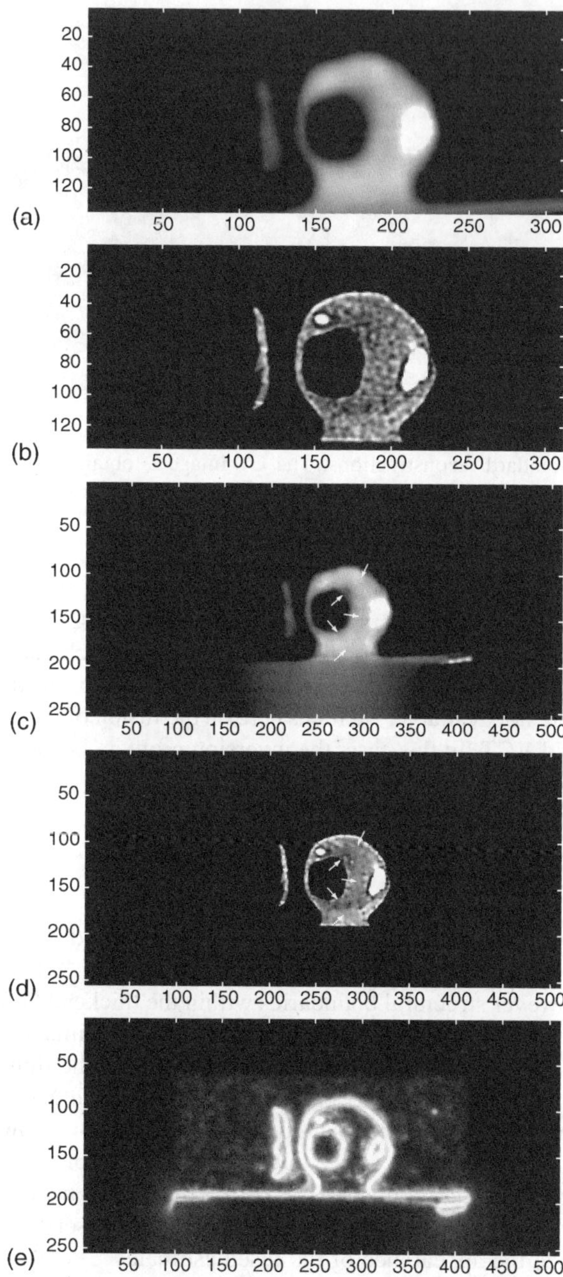

FIGURE 5.11 Observed plaque imagery from MDCT and VCT scanning, along with final estimates of property fields from MDCT and VCT, and the fused boundary field. Arrows in the fused images highlight example regions where enhanced structural detail can be better appreciated following fusion processing.

suboptimization we used a decreasing sequence of ρ values with $\rho = \{4, 2\}$ for the two iterations needed for convergence.

The superresolved and aligned MDCT property field shown in Figure 5.11 shows more distinct structure boundaries within the eccentric plaque. Not only are the regions of the bright calcified nodule more sharply defined, but boundaries between darker regions of lipid-rich tissue and moderately brighter bands of fibrous tissue can be better appreciated following multisensor fusion. The thin band of tissue that appears washed out in the original MDCT image appears more homogeneous and less blurred, with more distinct boundaries between tissue and the surrounding air.

The corresponding superresolved and aligned VCT property field also shows an improvement over the original VCT observations in the sense that the darker region of lipid-rich plaque is far less noisy than in the original imagery. Moderately brighter CT densities corresponding to bands of fibrous tissue in the plaque can also be better appreciated due to noise suppression following multisensor fusion. These early results are encouraging and work is proceeding to use this multisensor fusion framework in conjunction with a larger validation set of plaque imagery to help improve multimodality characterization of atherosclerotic plaques.

ACKNOWLEDGMENTS

The authors would like to thank Jennifer Lisauskas, M.S., for her contribution to the sample preparation and imaging experiment described. Furthermore, the authors would like to thank Denise P. Hinton, Ph.D, Rajiv Gupta, M.D., Ph.D, and Thomas J. Brady, M.D., for the support of the Cardiovascular MR and CT Program at the Massachusetts General Hospital.

REFERENCES

1. Wald, L., Some terms of reference in data fusion, *IEEE Trans. Geosci. Remote Sens.*, 37(3), 1190–1193, 1999, May.
2. Brooks, R. R., and Iyengar, S. S., *Multi-Sensor Fusion*, Prentice-Hall, Upper Saddle River, NJ, 1998.
3. Aggarwal, J. K., Ed., *Multisensor Fusion for Computer Vision, Number 99 in NATO ASI Series*, Springer, New York, 1993.
4. Kak, A.C., and Slaney, M., *Principles of Computerized Tomographic Imaging*, IEEE Press, Piscataway, NJ, 1987.
5. Webb, S., Ed., *The Physics of Medical Imaging, Medical Science Series*, Institute of Physics Publishing, Bristol, 1988.
6. Ludwig, D., The Radon transform on Euclidean space, *Commun. Pure Appl. Math.*, 19, 49–81, 1966.
7. Macovski, A., Medical imaging systems, *Prentice-Hall Information and System Sciences Series*, Prentice-Hall, Upper Saddle River, NJ, 1983.

8. Gordon, R., Bender, R., and Herman, G. T., Algebraic reconstruction techniques (art) for three-dimensional electron and x-ray photography, *J. Theor. Biol.*, 29, 471–481, 1970.

9. Mueller, K., and Yagel, R., Rapid 3-D cone-beam reconstruction with the simultaneous algebraic reconstruction technique (SART) using 2-D texture mapping hardware, *IEEE Trans. Med. Imaging*, 19(12), 1227–1237, 2000, December.

10. Liang, Z. P., and Lauterbur, P. C., Principles of magnetic resonance imaging: A signal processing perspective, *IEEE Press Series in Biomedical Imaging*, IEEE Press, Piscataway, NJ, 2000.

11. Fessler, J. A., and Sutton, B. P., Nonuniform fast Fourier transforms using min–max interpolation, *IEEE Trans. Signal Process.*, 51(2), 560–574, 2003, February.

12. Sutton, B. P., Noll, D. C., and Fessler, J. A., Fast, iterative, field-corrected image reconstruction for MRI, *IEEE Trans. Med. Imaging*, 22(2), 178–188, 2003.

13. Mumford, D., and Shah, J., Boundary detection by minimizing functionals, I, pp. 22–26. In *Proceedings of the IEEE Conference on Computer Vision and Pattern Recognition*. San Francisco, CA. IEEE, New York, 1985.

14. Mumford, D., and Shah, J., Optimal approximations by piecewise smooth functions and associated variational problems, *Commun. Pure Appl. Math.*, 42, 577–684, 1989.

15. Ambrosio, L., and Tortorelli, V. M., Approximation of functionals depending on jumps by elliptic functionals via Γ-convergence, *Commun. Pure Appl. Math.*, 43(8), 999–1036, 1990, December.

16. Ambrosio, L., and Tortorelli, V. M., On the approximation of functionals depending on jumps by quadratic, elliptic functionals, *Boll. Un. Mat.Ital.*, 1992.

17. Tsai, A., Yezzi, A. Jr., and Willsky, A. S., Curve evolution implementation of the Mumford–Shah functional for image segmentation, denoising, interpolation, and magnification, *IEEE Trans. Image Process.*, 10(8), 1169–1186, 2001, August.

18. Chan, T. F., and Vese, L. A., Active contour and segmentation models using geometric PDE's for medical imaging, In *Geometric Methods in Bio-Medical Image Processing, Mathematics and Visualization*, Malladi, R., Ed., Springer, Berlin, pp. 63–75, 2002.

19. Vese, L. A., and Chan, T. F., A multiphase level set framework for image segmentation using the Mumford and Shah model, *Int. J. Comput. Vis.*, 50(3), 271–293, 2002.

20. Haykin, S., Ed., *Advances in Spectral Analysis and Array Processing*, Prentice-Hall, Englewood Cliffs, NJ, 1991.

21. Pien, H. H., and Gauch, J. M., A variational approach to multi-sensor fusion of images, *Appl. Intell.*, 5, 217–235, 1995.

22. Shah, J., Pien, H. H., and Gauch, J. M., Recovery of surfaces with discontinuities by fusing shading and range data within a variational framework, *IEEE Trans. Image Process.*, 5(8), 1243–1251, 1996, August.

23. Pien, H. H., and Gauch, J. M., Variational segmentation of multi-channel MRI images, pp. 508–512. In *Proceedings of the IEEE International Conference on Image Processing*. Austin, TX. IEEE, New York. November, 1994.

24. Kaufhold, J. P., Energy Formulations of Medical Image Segmentations. Ph.D. thesis, Boston University, Boston, MA, 2001.

25. Wahle, A., Prause, G. P. M., DeJong, S. C., and Sonka, M., Geometrically correct 3-D reconstruction of intravascular ultrasound images by fusion with biplane

angiography — methods and validation, *IEEE Trans. Med. Imaging*, 18(8), 686–689, 1999, August.

26. Boyd, J. E., Limited-angle computed tomography for sandwich structure using data fusion, *J. Nondestr. Eval.*, 14(2), 61–76, 1995, June.

27. Boyd, J. E., and Little, J. J., Complementary data fusion for limited-angle tomography, pp. 288–294. In *Proceedings of the IEEE Conference on Computer Vision and Pattern Recognition*. Seattle, WA. IEEE, New York, 1994, June.

28. Mohammad-Djafari, A., Fusion of X-ray radiographic data and anatomical data in computed tomography, pp. 461–464. In *Proceedings of the IEEE International Conference on Image Processing*, Vol. 2, Rochester, NY. IEEE, New York, 2002, September.

29. Chen, T., Audiovisual speech processing: lip reading and synchronization, *IEEE Signal Process. Mag.*, 18(1), 9–21, 2001, January.

30. Strobel, N., Spors, S., and Rabenstein, R., Joint audio–video object localization and tracking, *IEEE Signal Process. Mag.*, 18(1), 22–31, 2001, January.

31. Fisher, J. W., III, Darrell, T., Freeman, W. T., and Viola, P., Learning joint statistical models for audio-visual fusion and segregation, *Advances in Neural Information Processing Systems*, Denver, CO. pp. 772–778, 2000, November.

32. Yezzi, A., Zöllei, L., and Kapur, T., A variational framework for joint segmentation and registration, pp. 44–51. In *Workshop on Mathematical Methods in Biomedical Image Analysis*. IEEE, New York, 2001, December.

33. Osher, S., and Sethian, J., Fronts propagating with curvature dependent speed: algorithms based on the Hamilton–Jacobi formulation, *J. Comput. Phys.*, 79(1), 12–49, 1988.

34. Chiao, P. C., Rogers, W. L., Clinthorne, N. H., Fessler, J. A., and Hero, A. O., Model-based estimation for dynamic cardiac studies using ECT, *IEEE Trans. Med. Imaging*, 13(2), 217–226, 1994, June.

35. Chiao, P. C., Rogers, W. L., Fessler, J. A., Clinthorne, N. H., and Hero, A. O., Model-based estimation with boundary side information or boundary regulariz-ation, *IEEE Trans. Med. Imaging*, 13(2), 227–234, 1994, June.

36. Hero, A. O., Piramuthu, R., Fessler, J. A., and Titus, S. R., Minimax emission computed tomography using high-resolution anatomical side information and B-spline models, *IEEE Trans. Inf. Theory*, 45(3), 920–938, 1999, April.

37. Titus, S., Hero, A. O., and Fessler, J. A., Penalized likelihood emission image reconstruction with uncertain boundary information, pp. 2813–2816. In *Proceedings of the IEEE International Conference on Acoustics, Speech, and Signal Processing*, Vol. 4, Munich. IEEE, New York, 1997, April.

38. Hero, A. O., and Piramuthu, R., A method for ECT image reconstruction with uncertain MRI side information using asymptotic marginalization. In *Proceedings of IEEE/EURASIP Workshop on Nonlinear Signal and Image Processing*. Mackinac Island, MI, 1997, September.

39. Vemuri, B. C., Chen, Y., and Wang, Z., Registration assisted image smoothing and segmentation, in *European Conference on Computer Vision, Number 2353 in Lecture Notes in Computer Science*, Heyden, A., Sparr, G., Nielsen, M., and Johansen, P., Eds., Springer, Berlin, pp. 546–559, 2002, May.

40. Bertsekas, D. P., *Nonlinear Programming*, Athena Scientific, Belmont, MA, 1995.

41. Weisenseel, R. A., Exploiting Shared Image Structure Fusion in Multi-modality Data Inversion for Atherosclerotic Plaque Characterization, Ph.D. thesis, Boston University, Boston, MA, 2004, January.

42. Briggs, W., A Multigrid Tutorial, *Society for Industrial and Applied Mathematics*, Philadelphia, PA, 1987.
43. Barrett, R., Berry, M., Chan, T. F., Demmel, J., Donato, J., Dongarra, J., Eijkhout, V., Pozo, R., Romine, C., and Van der Vorst, H., *Templates for the Solution of Linear Systems: Building Blocks for Iterative Methods*, 2nd ed., SIAM, Philadelphia, PA, 1994; http://www.netlib.org/linalg/html_templates/report.html.
44. Saad, Y., *Iterative Methods for Sparse Linear Systems*, *PWS Series in Computer Science*, PWS Publishing Co., Boston, MA, 1996, PWS — Division of International Thomson Publishing, Inc.
45. Yu, D. F., and Fessler, J. A., Edge-preserving tomographic reconstruction with nonlocal regularization, *IEEE Trans. Med. Imaging*, 21(2), 159–179, 2002, February.

6 Entropic Graphs for Registration

Huzefa Neemuchwala and Alfred Hero

CONTENTS

I. INTRODUCTION

Given 2D or 3D images gathered via multiple sensors located at different positions, the multisensor image registration problem is to align the images so that they have an identical pose in a common co-ordinate system (Figure 6.1). Image registration is becoming a challenging multisensor fusion problem due to the increased diversity of sensors capable of imaging objects and their intrinsic properties. In medical imaging, cross-sectional anatomic images are routinely acquired by magnetic induction (magnetic resonance imaging; MRI), absorption of accelerated energized photons (x-ray computed tomography; CT) and ultra high frequency sound (ultrasound) waves. Artifacts such as motion, occlusion, specular refraction, noise, and inhomogeneities in the object and imperfections in the transducer compound the difficulty of image registration. Cost and other physical considerations can constrain the spatial or spectral resolution and the signal-to-noise ratio (SNR). Despite these hindrances, image registration is now commonplace in medical imaging, satellite imaging, and stereo vision. Image registration also finds widespread usage in other pattern recognition and computer vision applications such as image segmentation, tracking, and motion compensation. A comprehensive survey of the image registration problem, its applications, and implementable algorithms can be found in Refs. 51,52. Image fusion is defined as the task of extracting cooccurring information from multisensor images. Image registration is hence a precursor to fusion. Image fusion finds several applications in medical imaging where it is used to fuse anatomic and metabolic information,[24,53,72] and build global anatomical atlases.[80]

The three chief components of an effective image registration system (Figure 6.2) are (1) definition of features that discriminate between different

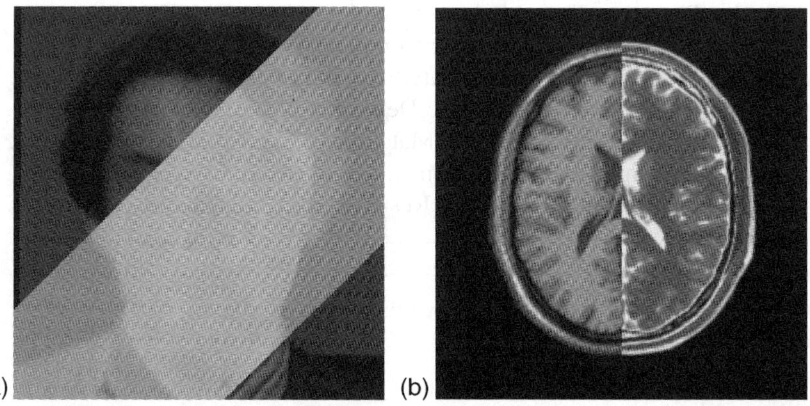

FIGURE 6.1 Image fusion: (a) co-registered images of the face acquired via visible light and longwave sensors. (b) Registered brain images acquired by time-weighted responses. Face and brain images courtesy Refs. 23 and 16, respectively.

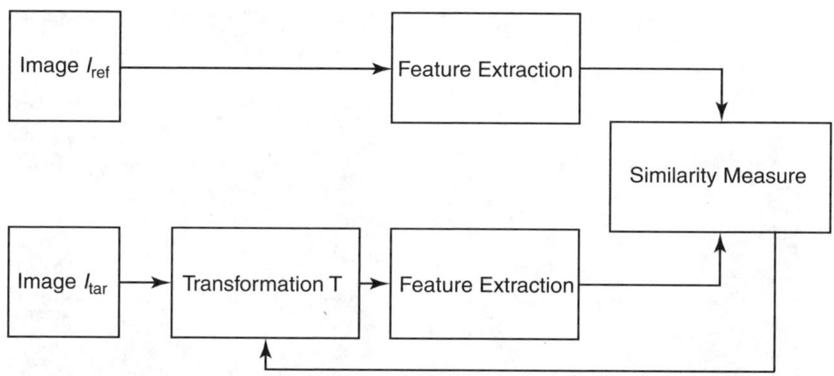

FIGURE 6.2 Block diagram of an image registration system.

image poses, (2) adaptation of a matching criterion that quantifies feature similarity, is capable of resolving important differences between images, yet is robust to image artifacts, and (3) implementation of optimization techniques which allow fast search over possible transformations. In this chapter, we shall be principally concerned with the first two components of the system. In a departure from conventional pixel-intensity features, we present techniques that use higher dimensional features extracted from images. We adapt traditional pixel matching methods that rely on entropy estimates to include higher dimensional features. We propose a general class of information theoretic feature similarity measures that are based on entropy and divergence and can be empirically estimated using entropic graphs, such as the minimal spanning tree (MST) or k-nearest neighbor graph (kNNG), and do not require density estimation or histograms.

Traditional approaches to image registration have included single pixel gray level features and correlation type matching functions. The correlation coefficient is a poor choice for the matching function in multisensor fusion problems. Multisensor images typically have intensity maps that are unique to the sensors used to acquire them and a direct linear correlation between intensity maps may not exist (Figure 6.3). Several other matching functions have been suggested in the literature.[37,42,66] Some of the most widespread techniques are: histogram matching;[39] texture matching;[2] intensity cross correlation;[52] optical flow matching;[47] kernel-based classification methods;[17] boosting classification methods;[19,44] information divergence minimization;[29,77,81] and mutual information (MI) maximization.[11,28,53,84] The last two methods can be called "entropic methods" since both use a matching criterion defined as a relative entropy between the feature distributions. The main advantage of entropic methods is that they can capture nonlinear relations between features in order to improve discrimination between poor and good image matches. When combined with a highly discriminatory feature set, and reliable prior information, entropic

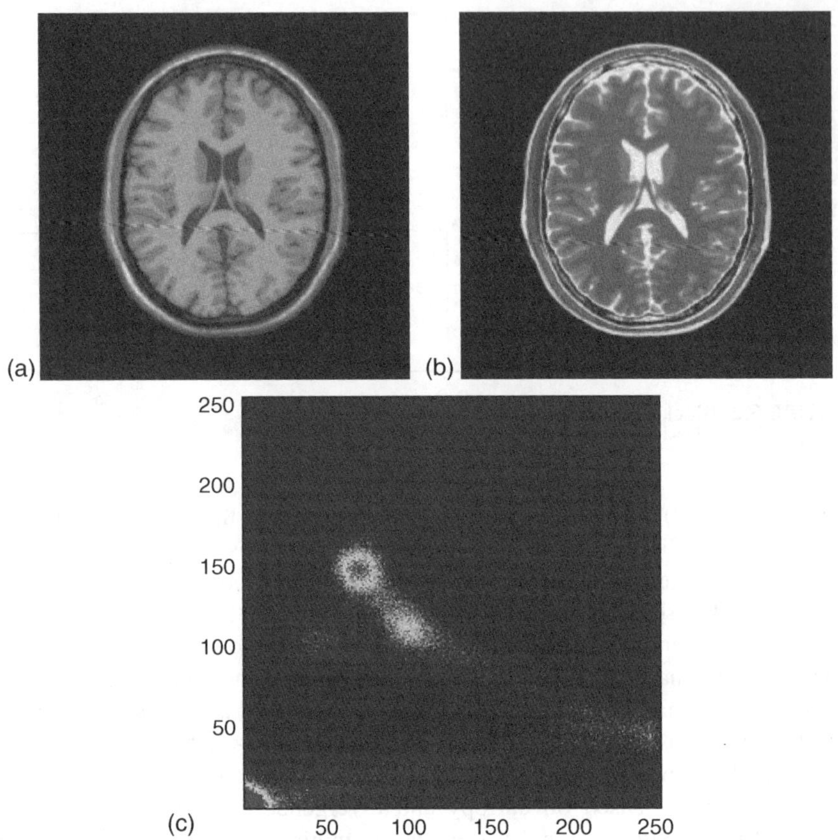

FIGURE 6.3 MRI images of the brain, with additive noise. (a) T1 weighted I_1, (b) T2 weighted I_2. Images courtesy Ref. 16. Although acquired by a single sensor, the time weighting renders different intensity maps to identical structures in the brain. (c) Joint gray-level pixel coincidence histogram is clustered and does not exhibit a linear correlation between intensities.

methods are very compelling and have been shown to be virtually unbeatable for some multimodality image registration applications.[37,48,53] However, due to the difficulty in estimating the relative entropy over high dimensional feature spaces, the application of entropic methods have been limited to one or two feature dimensions. The independent successes of relative entropy methods, for example, MI image registration, and the use of high dimensional features, for example, support vector machines (SVM's) for handwriting recognition, suggest that an extension of entropic methods to high dimensions would be worthwhile. Encouraging initial studies on these methods have been conducted by these authors and can be found in Refs. 58,60.

Here we describe several new techniques to extend methods of image registration to high dimensional feature spaces. Chief among the techniques is the introduction of entropic graphs to estimate a generalized α-entropy: Rényi's α-entropy. These entropic graph estimates can be computed via a host of combinatorial optimization methods including the MST and the kNNG. The computation and storage complexity of the MST and kNNG-based estimates increase linearly in feature dimension as opposed to the exponential rates of histogram-based estimates of entropy. Furthermore, as will be shown, entropic graphs can also be used to estimate more general similarity measures. Specific examples include the α-mutual information (α-MI), α-Jensen difference divergence, the Henze–Penrose (HP) affinity, which is a multi-dimensional approximation to the Wald–Wolfowitz test,[85] and the α-geometric-arithmetic (α-GA) mean divergence.[79] To our knowledge, the last two divergence measures have never been utilized in the context of image registration problems. We also explore variants of entropic graph methods that allow estimation with faster asymptotic convergence properties and reduced computational complexity.

The α-entropy of a multivariate distribution is a generalization of the better known Shannon entropy. Rényi introduced the α-entropy in a 1961 paper[71] and since then many important properties of α-entropy have been established.[4] From Rényi's α-entropy, the Rényi α-divergence and the Rényi α-mutual information (α-MI) can be defined in a straightforward manner. For $\alpha = 1$ these quantities reduce to the standard (Shannon) entropy, (Kullback–Liebler) divergence, and (Shannon) MI, respectively. Another useful quantity that can be derived from the α-entropy is the α-Jensen difference, which is a generalization of the standard Jensen difference and has been used here in our extension of entropic pattern matching methods to high feature dimension. As we will show, this generalization allows us to define an image matching algorithm that benefits from a simple estimation procedure and an extra degree of freedom (α).

Some additional comments on relevant prior work by us and others are in order. Various forms of α-entropy have been exploited by others for applications, including: reconstruction and registration of interferometric synthetic aperture radar (ISAR) images;[26,29] blind deconvolution;[25] and time-frequency analysis.[3,86] Again, our innovation with respect to these works is the extension to high dimensional features via entropic graph estimation methods. On the other hand, the α-entropy approaches described here should not be confused with entropy-α classification in SAR processing[15] which has no relation, whatsoever, to our work. A tutorial introduction to the use of entropic graphs to estimate multivariate α-entropy and other entropy quantities was published by us in a recent survey article.[35] As introduced in Ref. 36 and studied in Refs. 34,35, an entropic graph is any graph whose normalized total weight (sum of the edge lengths) is a consistent estimator of α-entropy. An example of an entropic graph is the MST, and due to its low computational complexity it is an attractive entropic graph algorithm. This graph estimator can be viewed as a multi-dimensional generalization of the Vasicek–Shannon entropy estimator for one-dimensional features.[7,83]

We have developed experiments that allow the user to examine and compare our methods with other methods currently used for image fusion tasks. The applications presented in this chapter are primarily selected to illustrate the flexibility of our method, in terms of selecting high dimensional features. However, they help us compare and contrast multi-dimensional entropy estimation methods. In the first example we perform registration on images obtained via multiband satellite sensors. Images acquired via these geostationary satellites are used in research related to heat dissipation from urban centers, climactic changes, and other ecological projects. Thermal and visible light images captured for the Urban Heat Island[68] project form a part of the database used here. NASA's visible earth project[57] also provides images captured via different satellite sensors, and such multiband images have been used here to provide a rich representative database of satellite images. Thermal and visible-light sensors image different bands in the electromagnetic spectrum, and thus have different intensity maps, removing any possibility of using correlation-based registration methods.

As a second example we apply our methods to registering medical images of the human brain acquired under dual modality (T1, T2 weighted) MRI. Simulated images of the brain under different time-echo responses to magnetic excitation are used. Different areas in the brain (neural tissue, fat, and water) have distinct magnetic excitation properties. Hence, they express different levels of excitation when appropriately time-weighted. This example qualifies as a multisensor fusion example due to the disparate intensity maps generated by the imaging sequence, commonly referred to as the T1 and T2 time-weighted MRI sequences. We demonstrate an image matching technique for MRI images sensitive to local perturbations in the image.

Higher dimensional features used for this work include those based on independent component analysis (ICA) and multi-dimensional wavelet image analysis. Local basis projection coefficients are implemented by projecting local 8×8 subimages of the image onto the ICA basis for the local image matching example from medical imaging. Multiresolution wavelet features are used for registration of satellite imagery. Local feature extraction via basis projection is a commonly used technique for image representation.[74,82] Wavelet bases are commonly used for image registration as is evidenced in Refs. 43,78,87. ICA features are somewhat less common but have been similarly applied by Olshausen, Hyvärinen, and others.[41,49,64] The high dimensionality ($=64$ for local basis projections) of these feature spaces precludes the application of standard entropy-based pattern matching methods and provides a good illustration of the power of our approach. The ability of the wavelet basis to capture spatial-frequency information in a hierarchical setting makes them an attractive choice for use in registration.

The paper is organized as follows: Section II introduces various entropy and α-entropy based similarity measures such as Rényi entropy and divergence, mutual information and α-Jensen difference divergence. Section III describes continuous Euclidean functionals such as the MST and the kNNG that asymptotic converge to the Rényi entropy. Section IV presents the entropic graph estimate of the Henze–Penrose affinity whereas Section V presents the entropic graph estimates of α-GA divergence α-MI. Next, Section VI describes, in detail, the feature based matching techniques used in this work, different types of features used and the advantages of using such methods. Computational considerations involved in constructing graphs are discussed in Section VII. Finally, Sections VIII and IX present the experiments we conducted to compare and contrast our methods with other registration algorithms.

II. ENTROPIC FEATURE SIMILARITY/DISSIMILARITY MEASURES

In this section we review entropy, relative entropy, and divergence as measures of dissimilarity between probability distributions. Let Y be a q-dimensional random vector and let $f(y)$ and $g(y)$ denote two possible densities for Y. Here Y will be a feature vector constructed from the reference image and the target image to be registered, and f and g will be multi-dimensional feature densities. For example, information divergence methods of image retrieval[21,76,82] specify f as the estimated density of the reference image features and g as the estimated density of the target image features. When the features are discrete valued the densities f and g are interpreted as probability mass functions.

A. Rényi Entropy and Divergence

The basis for entropic methods of image fusion is a measure of dissimilarity between densities f and g. A very general entropic dissimilarity measure is the Rényi α-divergence, also-called the Rényi α-relative entropy, between f and g of fractional order $\alpha \in (0, 1)$:[4,18,71]

$$D_\alpha(f\|g) = \frac{1}{\alpha - 1}\log \int g(z)\left(\frac{f(z)}{g(z)}\right)^\alpha dz = \frac{1}{\alpha - 1}\log \int f^\alpha(z)g^{1-\alpha}(z)dz \qquad (6.1)$$

When the density f is supported on a compact domain and g is uniform over this domain the α-divergence reduces to the Rényi α-entropy of f:

$$H_\alpha(f) = \frac{1}{1 - \alpha}\log \int f^\alpha(z)dz \qquad (6.2)$$

When specialized to various values of α the α-divergence can be related to other well known divergence and affinity measures. Two of the most important examples are the Hellinger dissimilarity $-2\log \int \sqrt{f(z)g(z)}\,dz$ obtained when

$\alpha = 1/2$, which is related to the Hellinger–Battacharya distance squared,

$$
\begin{aligned}
D_{\text{Hellinger}}(f\|g) &= \int \left(\sqrt{f(z)} - \sqrt{g(z)} \right)^2 dz \\
&= 2\left(1 - \exp\left(\tfrac{1}{2} D_{1/2}(f\|g) \right) \right)
\end{aligned}
\tag{6.3}
$$

and the Kullback–Liebler (KL) divergence,[46] obtained in the limit as $\alpha \to 1$,

$$
\lim_{\alpha \to 1} D_\alpha(f\|g) = \int g(z)\log \frac{g(z)}{f(z)} dz
\tag{6.4}
$$

B. Mutual Information and α-Mutual Information

The mutual information (MI) can be interpreted as a similarity measure between the reference and target pixel intensities or as a dissimilarity measure between the joint density and the product of the marginals of these intensities. The MI was introduced for gray scale image registration[84] and has since been applied to a variety of image matching problems.[28,48,53,69] Let X_0 be a reference image and consider a transformation of the target image (X_1), defined as $X_T = T(X_1)$. We assume that the images are sampled on a grid of $M \times N$ pixels. Let (z_{0k}, z_{Tk}) be the pair of (scalar) gray levels extracted from the kth pixel location in the reference and target images, respectively. The basic assumption underlying MI image matching is that $\{(z_{0k}, z_{Tk})\}_{k=1}^{MN}$ are independent identically distributed (i.i.d.) realizations of a pair $(Z_0, Z_T), (Z_T = T(Z_1))$ of random variables having joint density $f_{0,1}(z_0, z_T)$. If the reference and the target images were perfectly correlated (e.g., identical images), then Z_0 and Z_T would be dependent random variables. On the other hand, if the two images were statistically independent, the joint density of Z_0 and Z_T would factor into the product of the marginals $f_{0,1}(z_0, z_T) = f_0(z_0)f_1(z_T)$. This suggests using the α-divergence $D_\alpha(f_{0,1}(z_0, z_T)\|f_0(z_0)f_1(z_T))$ between $f_{0,1}(z_0, z_T)$ and $f_0(z_0)f_1(z_T)$ as a similarity measure. For $\alpha \in (0, 1)$ we call this the α-mutual information (or α-MI) between Z_0 and Z_T and it has the form

$$
\begin{aligned}
\alpha\text{MI} &= D_\alpha(f_{0,1}(Z_0, Z_T)\|f_0(Z_0)f_1(Z_T)) \\[2mm]
&= \frac{1}{\alpha - 1}\log \int f_{0,1}^\alpha(z_0, z_T)f_0^{1-\alpha}(z_0)f_i^{1-\alpha}(z_T)\mathrm{d}z_0\,\mathrm{d}z_T
\end{aligned}
\tag{6.5}
$$

When $\alpha \to 1$ the α-MI converges to the standard (Shannon) MI

$$
\text{MI} = \int f_{0,1}(z_0, z_T)\log\left(\frac{f_{0,1}(z_0, z_T)}{f_0(z_0)f_1(z_T)} \right)\mathrm{d}z_0\,\mathrm{d}z_T
\tag{6.6}
$$

For registering two discrete $M \times N$ images, one searches over a set of transformations of the target image to find the one that maximizes the MI Equation 6.6 between the reference and the transformed target. The MI is defined using features $(Z_0, Z_T) \in \{z_{0k}, z_{Tk}\}_{k=1}^{MN}$ equal to the discrete-valued intensity levels at common pixel locations (k, k) in the reference image and the rotated target image. We call this the "single pixel MI". In Ref. 84, the authors empirically approximated the single pixel MI Equation 6.6 by histogram plug-in estimates, which when extended to the α-MI gives the estimate

$$\widehat{\alpha\text{MI}} \overset{def}{=} \frac{1}{\alpha - 1} \log \sum_{z_0, z_T = 0}^{255} \hat{f}_{0,1}^{\alpha}(z_0, z_T)(\hat{f}_0(z_0)\hat{f}_1(z_T))^{1-\alpha} \tag{6.7}$$

In Equation 6.7 we assume 8 bit gray level $\hat{f}_{0,1}$ denotes the joint intensity level coincidence histogram

$$\hat{f}_{0,1}(z_0, z_T) = \frac{1}{MN} \sum_{k=1}^{MN} I_{z_{0k}, z_{Tk}}(z_0, z_T) \tag{6.8}$$

and $I_{z_{0k}, z_{Tk}}(z_0, z_T)$ is the indicator function equal to one when $(z_{0k}, z_{Tk}) = (z_0, z_T)$ and equal to zero otherwise. Other feature definitions have been proposed including gray level differences[11] and pixel pairs.[73]

Figure 6.4 illustrates the MI alignment procedure through a multisensor remote sensing example. Aligned images acquired by visible and thermally sensitive satellite sensors, generate a joint gray level pixel coincidence histogram $f_{0,1}(z_0, z_1)$. Note, that the joint gray-level pixel coincidence histogram is not concentrated along the diagonal due to the multisensor acquisition of the images. When the thermal image is rotationally transformed, the corresponding joint gray-level pixel coincidence histogram $f_{0,1}(z_0, z_T)$ is dispersed, thus yielding lower mutual information than before.

(1) *Relation of α-MI to Chernoff Bound*: The α-MI Equation 6.5 can be motivated as an appropriate registration function by large deviations theory through the Chernoff bound. Define the average probability of error $P_e(n)$ associated with a decision rule for deciding whether Z_T and Z_0 are independent (hypothesis H_0) or dependent (hypothesis H_1) random variables based on a set of i.i.d. samples $\{z_{0k}, z_{Tk}\}_{k=1}^{n}$, where $n = MN$. For any decision rule, this error probability has the representation:

$$P_e(n) = \beta(n)P(H_1) + \alpha(n)P(H_0) \tag{6.9}$$

where $\beta(n)$ and $\alpha(n)$ are the probabilities of Type II (say H_0 when H_1 true) and Type I (say H_1 when H_0 true) errors, respectively, of the decision rule and $P(H_1) = 1 - P(H_0)$ is the prior probability of H_1. When the decision rule is the optimal minimum probability of error test the Chernoff bound implies that:[20]

$$\lim_{n \to \infty} \frac{1}{n} \log P_e(n) = - \sup_{\alpha \in [0,1]} \{(1 - \alpha)D_\alpha(f_{0,1}(z_0, z_T) \| f_0(z_0)f_1(z_T)\} \tag{6.10}$$

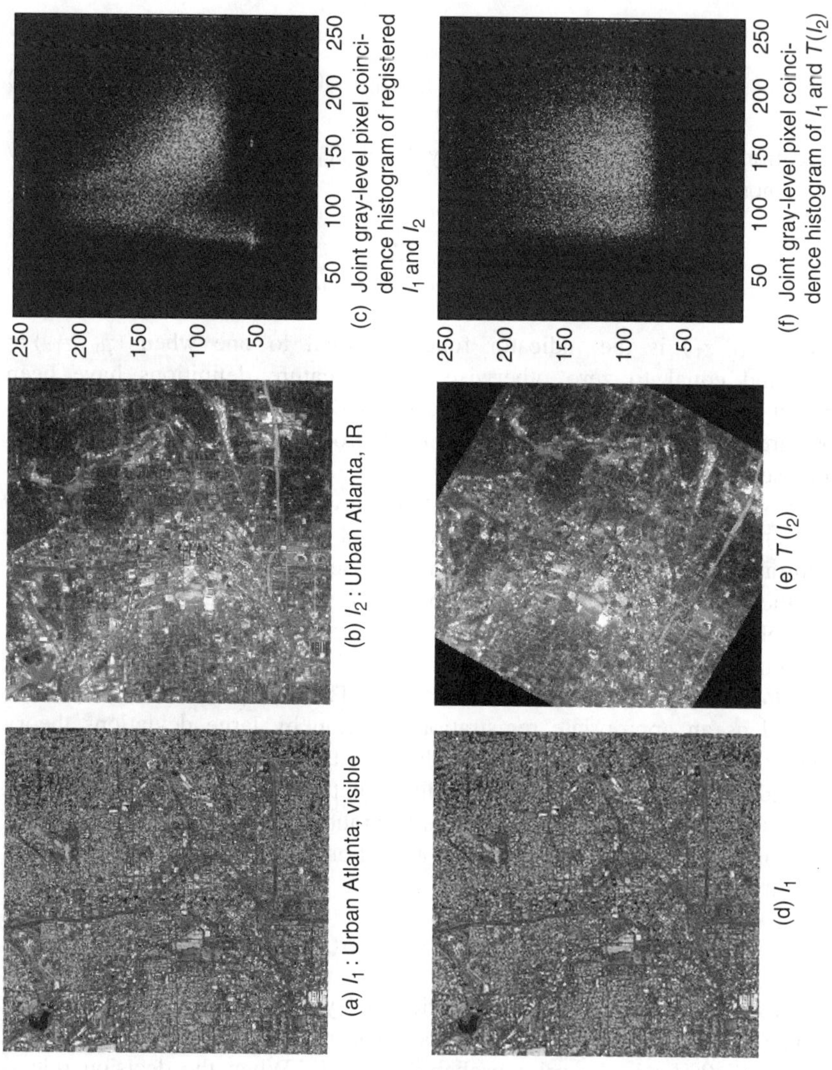

(a) I_1: Urban Atlanta, visible

(b) I_2: Urban Atlanta, IR

(c) Joint gray-level pixel coincidence histogram of registered I_1 and I_2

(d) I_1

(e) $T(I_2)$

(f) Joint gray-level pixel coincidence histogram of I_1 and $T(I_2)$

Thus, the mutual α-information gives the asymptotically optimal rate of exponential decay of the error probability for testing H_0 vs. H_1 as a function of the number $n = MN$ of samples. In particular, this implies that the α-MI can be used to select optimal transformation T that maximizes the right side of Equation 6.10. The appearance of the maximization over α implies the existence of an optimal parameter α ensuring the lowest possible registration error. When the optimal value α is not equal to 1 the MI criterion will be suboptimal in the sense of minimizing the asymptotic probability of error. For more discussion of the issue of optimal selection of α we refer the reader to Ref. 33.

C. α-JENSEN DISSIMILARITY MEASURE

An alternative entropic dissimilarity measure between two distributions is the α-Jensen difference. This function was independently proposed by Ma[32] and He et al.[29] for image registration problems. It was also used by Michel et al. in Ref. 54 for characterizing complexity of time-frequency images. For two densities f and g the α-Jensen difference is defined as[4]

$$\Delta H_\alpha(p, f, g) = H_\alpha(pf + qg) - [pH_\alpha(f) + qH_\alpha(g)] \qquad (6.11)$$

where $\alpha \in (0, 1)$, $p \in [0, 1]$, and $q = 1 - p$. As the α-entropy $H_\alpha(f)$ is strictly concave in f, Jensen's inequality implies that $\Delta H_\alpha(p, f, g) > 0$ when $f \neq g$ and $\Delta H_\alpha(p, f, g) = 0$ when $f = g$ (almost everywhere). Thus the α-Jensen difference is a bone fide measure of dissimilarity between f and g.

The α-Jensen difference can be applied as a surrogate optimization criterion in place of the α-divergence. One identifies $f = f_1(z_T)$ and $g = f_0(z_0)$ in Equation 6.11. In this case an image match occurs when the α-Jensen difference is minimized over i. This is the approach taken by He et al.[29] Hero et al.[32] for image registration applications and discussed in more detail below.

D. α-GEOMETRIC-ARITHMETIC MEAN DIVERGENCE

The α-geometric-arithmetic (α-GA) mean divergence[79] is another measure of dissimilarity between probability distributions. Given continuous distributions

---

FIGURE 6.4 Mutual information based registration of multisensor, visible and thermal infrared, images of Atlanta acquired via satellite.[68] Top row (in-registration): (a) visible light image I_1, (b) thermal image I_2, (c) joint gray-level pixel coincidence histogram $\hat{f}_{0,1}(z_0, z_1)$. Bottom row (out-of-registration): (d) visible light image, unaltered I_1, (e) rotationally transformed thermal image $T(I_2)$, and (f) joint gray-level pixel coincidence histogram shows wider dispersion $\hat{f}_{0,1}(z_0, z_T)$.

f and g, the α-GA:

$$\alpha D_{\mathrm{GA}}(f,g) = D_\alpha(pf + qg \| f^p g^q)$$

$$= \frac{1}{\alpha - 1} \log \int (pf(z) + qg(z))^\alpha (f^p(z) g^q(z))^{1-\alpha} \, dz \qquad (6.12)$$

The α-GA divergence is a measure of the discrepancy between the arithmetic mean and the geometric mean of f and g, respectively, with respect to weights p and $q = 1 - p$, $p \in [0,1]$. The α-GA divergence can thus be interpreted as the dissimilarity between the weighted arithmetic mean $pf(x) + qg(x)$ and the weighted geometric mean $f^p(x) g^q(x)$. Similarly to the α-Jensen difference Equation 6.11, the α-GA divergence is equal to zero if and only if $f = g$ (a.e.) and is otherwise greater than zero.

E. HENZE–PENROSE AFFINITY

While divergence measures dissimilarity between distributions, similarity between distributions can be measured by affinity measures. One measure of affinity between probability distributions f and g is

$$A_{HP}(f,g) = 2pq \int \frac{f(z)g(z)}{pf(z) + qg(z)} \, dz \qquad (6.13)$$

with respect to weights p and $q = 1 - p$, $p \in [0,1]$. This affinity measure was introduced by Henze and Penrose[30] as the limit of the Friedman–Rafsky statistic[27] and we shall call it the Henze–Penrose (HP) affinity. The HP affinity can be related to the divergence measure:

$$D_{HP}(f\|g) = 1 - A_{FR}(f,g) = \int \frac{p^2 f^2(z) + q^2 g^2(z)}{pf(z) + qg(z)} \, dz \qquad (6.14)$$

All of the above divergence measures can be obtained as special cases of the general class of f-divergences, for example, as defined in Refs. 4,18. In this article we focus on the cases for which we know how to implement entropic graph methods to estimate the divergence. For motivation consider the α-entropy Equation 6.2 which could be estimated by plugging in feature histogram estimates of the multivariate density f. A deterrent to this approach is the curse of dimensionality, which imposes prohibitive computational burden when attempting to construct histograms in large feature dimensions. For a fixed resolution per co-ordinate dimension the number of histogram bins increases geometrically in feature vector dimension. For example, for a 32 dimensional feature space even a coarse 10 cells per dimension would require keeping track of 10^{32} bins in the histogram, an unworkable and impractically large burden for any envisionable digital computer. As high dimensional feature spaces can be more discriminatory, this creates a barrier to performing robust high resolution histogram-based entropic registration. We circumvent this barrier by estimating the α-entropy via an entropic graph whose vertices are the locations of the feature vectors in feature space.

III. CONTINUOUS QUASIADDITIVE EUCLIDEAN FUNCTIONALS

A principal focus of this article is the use of minimal graphs over the feature vectors $Z_n = \{z_1, ..., z_n\}$, and their associated minimal edge lengths, for estimation of entropy of the underlying feature density $f(z)$. For consistent estimates we require convergence of minimal graph length to an entropy related quantity. Such convergence issues have been studied for many years, beginning with Beardwood et al.[6] The monographs of Steele[75] and Yukich[88] cover the interesting developments in this area. In the general unified framework of Redmond and Yukich[70] a widely applicable convergence result can be invoked for graphs whose length functionals can be shown to Euclidean, continuous, and quasiadditive. This result can often be applied to minimal graphs constructed by minimizing a graph length function L_γ of the form:

$$L_\gamma(Z_n) = \min_{E \in \Omega} \sum_{e \in E} \|e(Z_n)\|^\gamma$$

where Ω is a set of graphs with specified properties, for example, the class of acyclic or spanning graphs, e is an edge in Ω, $\|e\|$ is the Euclidean length of e, γ is called the edge exponent or the power weighting constant, and $0 < \gamma < d$. The determination of L_γ requires a combinatorial optimization over the set Ω.

If $Z_n = \{z_1, ..., z_n\}$ is a random i.i.d. sample of d-dimensional vectors drawn from a Lebesgue multivariate density f and the length functional L_γ is continuous quasiadditive, then the following limit holds[70]

$$\lim_{n \to \infty} L_\gamma(Z_n)/n^\alpha = \beta_{d,\gamma} \int f^\alpha(z)dz, \quad (a.s.) \tag{6.15}$$

where $\alpha = (d - \gamma)/d$ and $\beta_{d,\gamma}$ is a constant independent of f. Comparing this to the expression Equation 6.2 for the Rényi entropy it is obvious that an entropy estimator can be constructed as $(1 - \alpha)^{-1}\log(L_\gamma(Z_n)/n^\alpha) = H_\alpha(f) + c$, where $c = (1 - \alpha)^{-1} \log \beta_{d,\gamma}$ is a removable bias. Furthermore, it is seen that one can estimate entropy for different values of $\alpha \in [0, 1]$ by adjusting γ. In many cases the topology of the minimal graph is independent of γ and only a single combinatorial optimization is required to estimate H_α for all α.

A few words are in order concerning the sufficient conditions for the limit Equation 6.15. Roughly speaking, continuous quasiadditive functionals can be approximated closely by the sum of the weight functionals of minimal graphs constructed on a uniform partition of $[0, 1]^d$. Examples of graphs with continuous quasiadditive length functionals are the Euclidean MST, the traveling salesman tour solving the traveling salesman problem (TSP), the Steiner tree, the Delaunay triangulation, and the kNNG. An example of a graph that does not have a continuous quasiadditive length functional is the k-point MST (kMST) discussed in Ref. 36.

Even though any continuous quasiadditive functional could in principle be used to estimate entropy via relation Equation 6.15, only those that can be simply computed will be of interest to us here. An uninteresting example is the TSP length functional $L_\gamma^{\mathrm{TSP}}(Z_n) = \min_{C \in c} \sum_{e \in C} \|e\|^\gamma$, where C is a cyclic graph that spans the points Z_n and visits each point exactly once. Construction of the TSP is NP hard and hence is not attractive for practical image fusion applications. The following sections describe, in detail, the MST and kNN graph functionals.

A. A MINIMAL SPANNING TREE FOR ENTROPY ESTIMATION

A spanning tree is a connected acyclic graph which passes through all n feature vectors in Z_n. The MST connect these points with $n - 1$ edges, denoted $\{e_i\}$, in such a way as to minimize the total length:

$$L_\gamma(Z_n) = \min_{e \in T} \sum_e \|e\|^\gamma \qquad (6.16)$$

where T denotes the class of acyclic graphs (trees) that span Z_n. See Figure 6.5 and Figure 6.6 for an illustration when Z_n are points in the unit square. We adopt $\gamma = 1$ for the following experiments.

The MST length $L_n = L(Z_n)$ is plotted as a function of n in Figure 6.7 for the case of an i.i.d. uniform sample (panel b) and nonuniform sample (panel a) of $n = 100$ points in the plane. It is intuitive that the length of the MST spanning the more concentrated nonuniform set of points increases at a slower rate in n than does the MST spanning the uniformly distributed points. This observation has motivated the MST as a way to test for randomness in the plane[38]. As shown in Ref. 88, the MST length is a continuous quasiadditive functional and satisfies the limit Equation 6.15. More precisely, with $\alpha \overset{\mathrm{def}}{=} (d - \gamma)/d$ the log of the length function normalized by n^α converges (a.s.) within a constant factor to the α-entropy.

$$\lim_{n \to \infty} \log\left(\frac{L_\gamma(Z_n)}{n^\alpha} \right) = H_\alpha(f) + c_{MST}, \quad \text{(a.s.)} \qquad (6.17)$$

Therefore, we can identify the difference between the asymptotes shown on Figure 6.7a as the difference between the α-entropies of the uniform and nonuniform densities ($\alpha = 1/2$). Thus, if f is the underlying density of Z_n, the

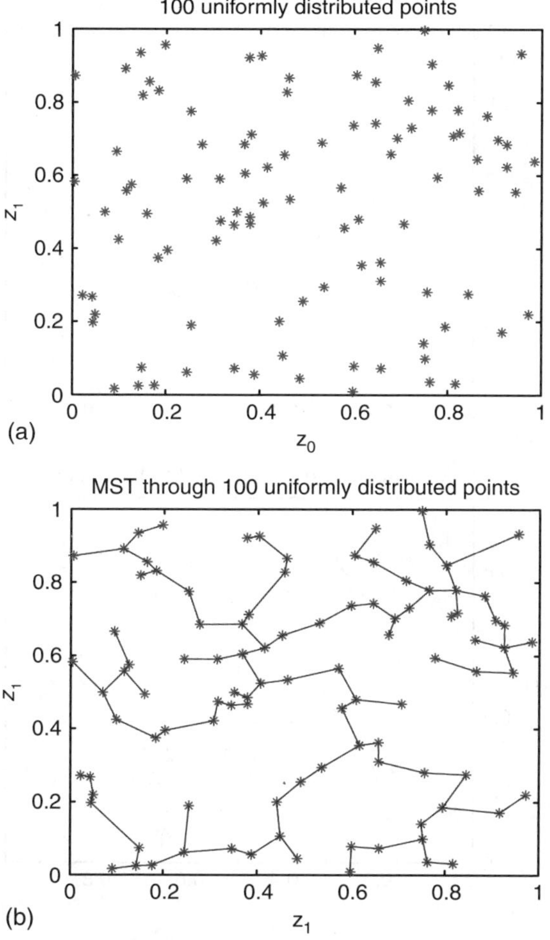

FIGURE 6.5 A set of $n = 100$ uniformly distributed points $\{Z_i\}$ in the unit square in $\mathbb{R}^2$ (a) and the corresponding minimal spanning tree (MST) (b).

α-entropy estimator

$$\hat{H}_\alpha(Z_n) = 1/(1 - \alpha)[\log L_\gamma(Z_n)/n^\alpha - \log \beta_{d,\gamma}] \qquad (6.18)$$

is an asymptotically unbiased and almost surely consistent estimator of the α-entropy of f where $\beta_{d,\gamma}$ is a constant which does not depend on the density f.

The constant $(c_{\mathrm{MST}} = (1 - \alpha)^{-1}\log \beta_{d,\gamma})$ in Equation 6.17 is a bias term that can be estimated offline. The constant $\beta_{d,\gamma}$ is the limit of $L_\gamma(Z_n)/n^\alpha$ as $n \to \infty$ for a uniform distribution $f(z) = 1$ on the unit cube $[0, 1]^d$. This constant can be

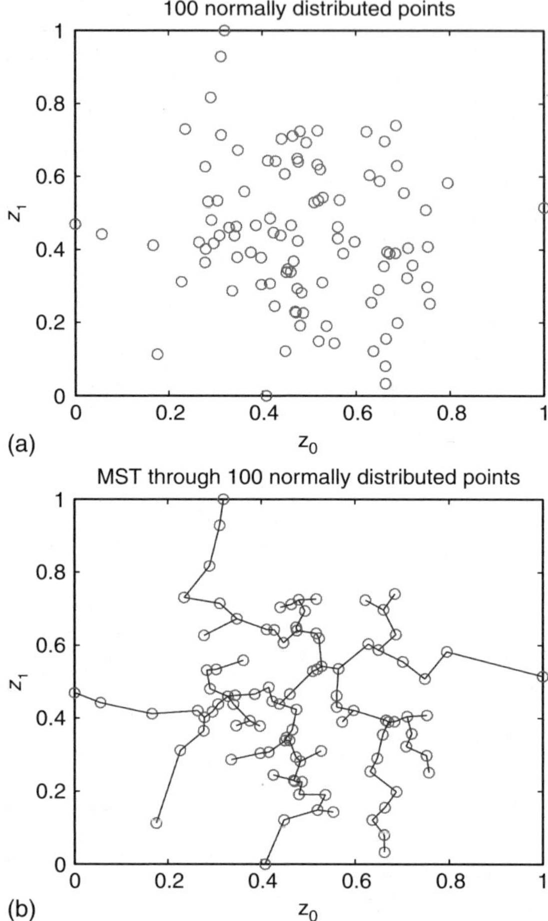

FIGURE 6.6 A set of $n = 100$ normally distributed points $\{Z_i\}$ in the unit square in $\mathbb{R}^2$ (a) and the corresponding minimal spanning tree (MST) (b).

approximated by Monte Carlo simulation of mean MST length for a large number of uniform d-dimensional random samples.

The MST approach to estimating the α-Jensen difference between the feature densities of two images can be implemented as follows. Assume two sets of feature vectors $Z_0 = \{z_0^{(i)}\}_{i=1}^{n_0}$ and $Z_1 = \{z_1^{(i)}\}_{i=1}^{n_1}$ are extracted from images X_0 and X_1 and are i.i.d. realizations from multivariate densities f_0 and f_1, respectively. In the applications explored in this paper $n_0 = n_1$, but it is worthwhile to maintain this level of generality. Define the set union $Z = Z_0 \cup Z_1$ containing $n = n_0 + n_1$ unordered feature vectors. If n_0, n_1 increase at constant rate as a function of n then any consistent entropy estimator constructed from the vectors $\{Z^{(i)}\}_{i=1}^{n_0+n_1}$ will converge to $H_\alpha(pf_0 + qf_1)$ as $n \to \infty$,

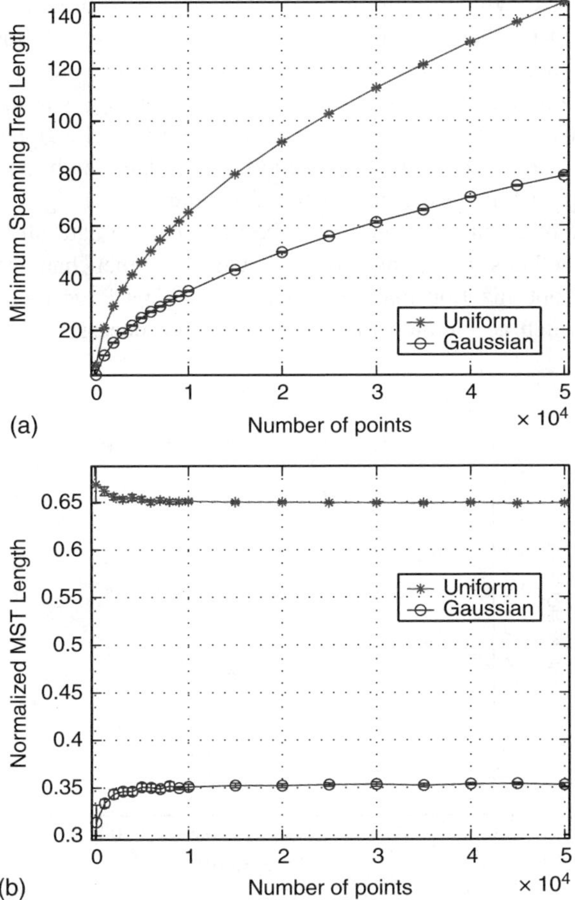

FIGURE 6.7 Mean length functions L_n of MST implemented with $\gamma = 1$ (a) and $L_n/\sqrt{n}$ (b) as a function of n for uniform and normal distributed points.

where $p = \lim_{n \to \infty} n_0/n$. This motivates the following finite sample entropic graph estimator of α-Jensen difference

$$\Delta \hat{H}_\alpha(p, f_0, f_1) = \hat{H}_\alpha(Z_0 \cup Z_1) - [p\hat{H}_\alpha(Z_0) + q\hat{H}_\alpha(Z_1)] \qquad (6.19)$$

where $p = n_0/n$, $\hat{H}_\alpha(Z_0 \cup Z_1)$ is the MST entropy estimator constructed on the n point union of both sets of feature vectors and the marginal entropies $\hat{H}_\alpha(Z_0), \hat{H}_\alpha(Z_1)$ are constructed on the individual sets of n_0 and n_1 feature vectors, respectively. We can similarly define a density-based estimator of α-Jensen difference. Observe that for affine image registration problems the marginal entropies $\{H_\alpha(f_i)\}_{i=1}^K$ over the set of image transformations will be identical, obviating the need to compute estimates of the marginal α-entropies.

As contrasted with histogram or density plug-in estimator of entropy or Jensen difference, the MST-based estimator enjoys the following properties:[31,33,36] it can easily be implemented in high dimensions; it completely bypasses the complication of choosing and fine tuning parameters such as histogram bin size, density kernel width, complexity, and adaptation speed; as the topology of the MST does not depend on the edge weight parameter γ, the MST α-entropy estimator can be generated for the entire range $\alpha \in (0, 1)$ once the MST for any given α is computed; the MST can be naturally robustified to outliers by methods of graph pruning. On the other hand, the need for combinatorial optimization may be a bottleneck for a large number of feature samples for which accelerated MST algorithms are necessary.

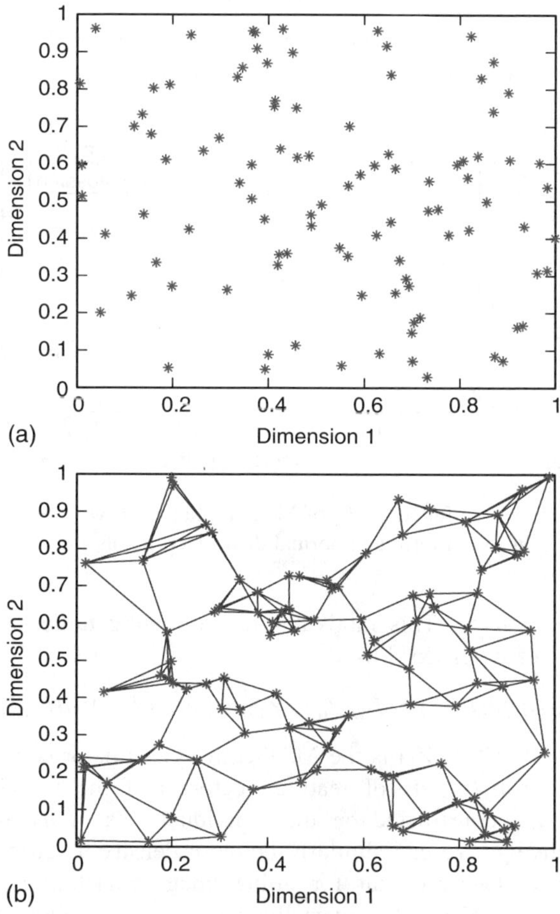

FIGURE 6.8 A set of $n = 100$ uniformly distributed points $\{Z_i\}$ in the unit square in $\mathbb{R}^2$ (a) and the corresponding kNNG ($k = 4$) (b).

B. NEAREST NEIGHBOR GRAPH ENTROPY ESTIMATOR

The kNNG is a continuous quasiadditive power weighted graph is a computationally attractive alternative to the MST. Given i.i.d vectors Z_n in $\mathbb{R}^d$, the 1 nearest neighbor of z_i in Z_n is given by

$$\arg \min_{z \in Z_n \setminus \{z_i\}} \|z - z_i\| \qquad (6.20)$$

where $\|z - z_i\|$ is the usual Euclidean (L_2) distance in $\mathbb{R}^d$. For general integer $k \geq 1$, the kNN of a point is defined in a similar way.[8,12,62] The kNNG puts a single edge between each point in Z_n and its kNN (Figure 6.8 and Figure 6.9). Let $N_{k,i} = N_{k,i}(Z_n)$ be the set of kNN of z_i in Z_n. The kNN problem consists of finding the set $N_{k,i}$ for each point z_i in the set $Z_n - \{z\}$.

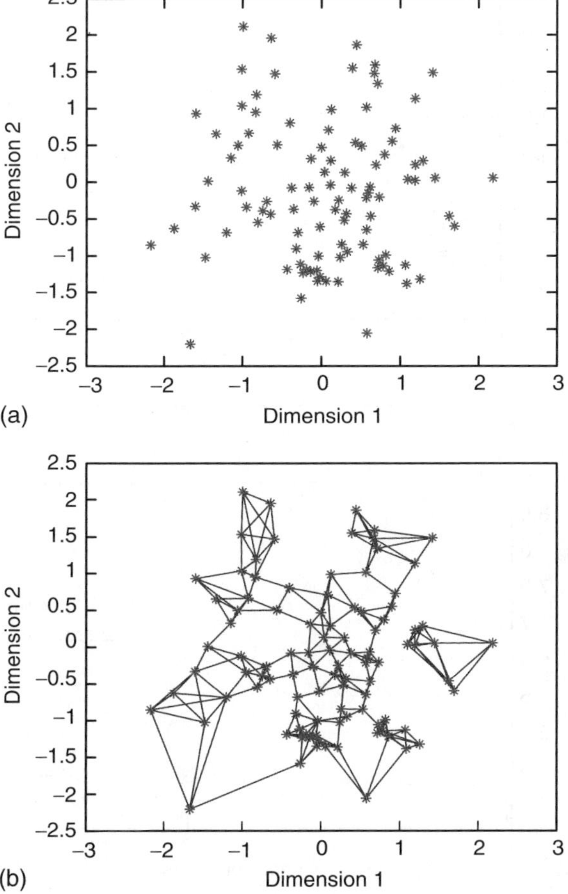

FIGURE 6.9 A set of $n = 100$ normally distributed points $\{Z_i\}$ in the unit square in $\mathbb{R}^2$ (a) and the corresponding kNNG ($k = 4$) (b).

This problem has exact solutions which run in linear-log-linear time and the total graph length is:

$$L_{\gamma,k}(Z_n) = \sum_{i=1}^{N} \sum_{e \in N_{k,i}} \|e\|^{\gamma} \tag{6.21}$$

In general, the kNNG will count edges at least once, but sometimes count edges more than once. If two points X_1 and X_2 are mutual kNN, then the same edge between X_1 and X_2 will be doubly counted.

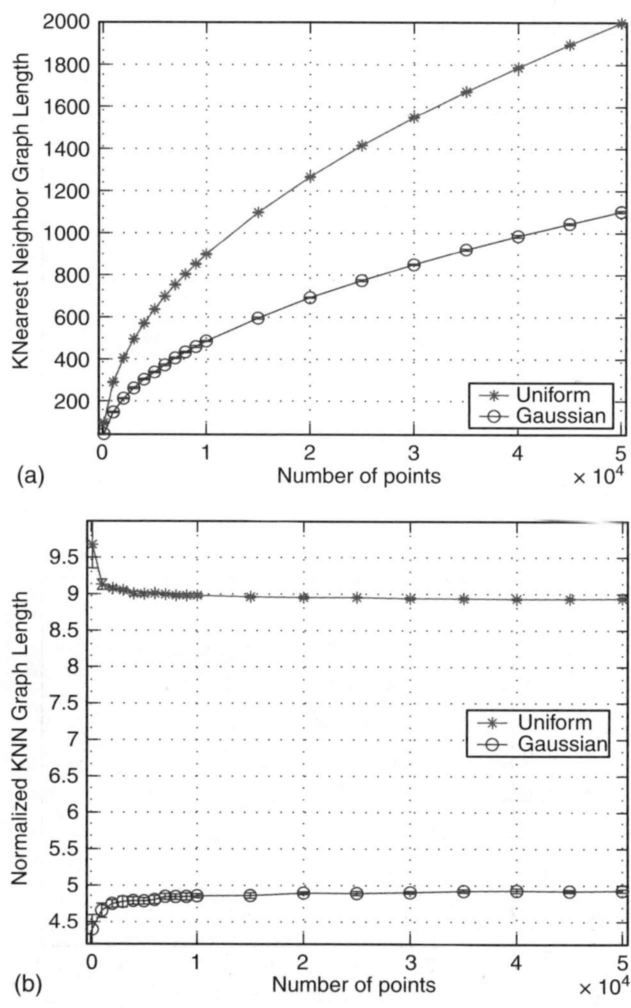

FIGURE 6.10 Mean Length functions L_n of kNNG implemented with $\gamma = 1$ (a) and $L_n/\sqrt{n}$ (b) as a function of n for uniform and Gaussian distributed points.

Analogously to the MST, the log length of the kNNG has a limit (Figure 6.10):

$$\lim_{n\to\infty} \log\left(\frac{L_{\gamma,k}(X_n)}{n^\alpha}\right) = H_\alpha(f) + c_{\mathrm{kNNG}}, \quad (\text{a.s.}) \qquad (6.22)$$

Once again this suggests an estimator of the Rényi α-entropy

$$\hat{H}_\alpha(Z_n) = 1/(1-\alpha)[\log L_{\gamma,k}(Z_n)/n^\alpha - \log \beta_{d,\gamma,k}] \qquad (6.23)$$

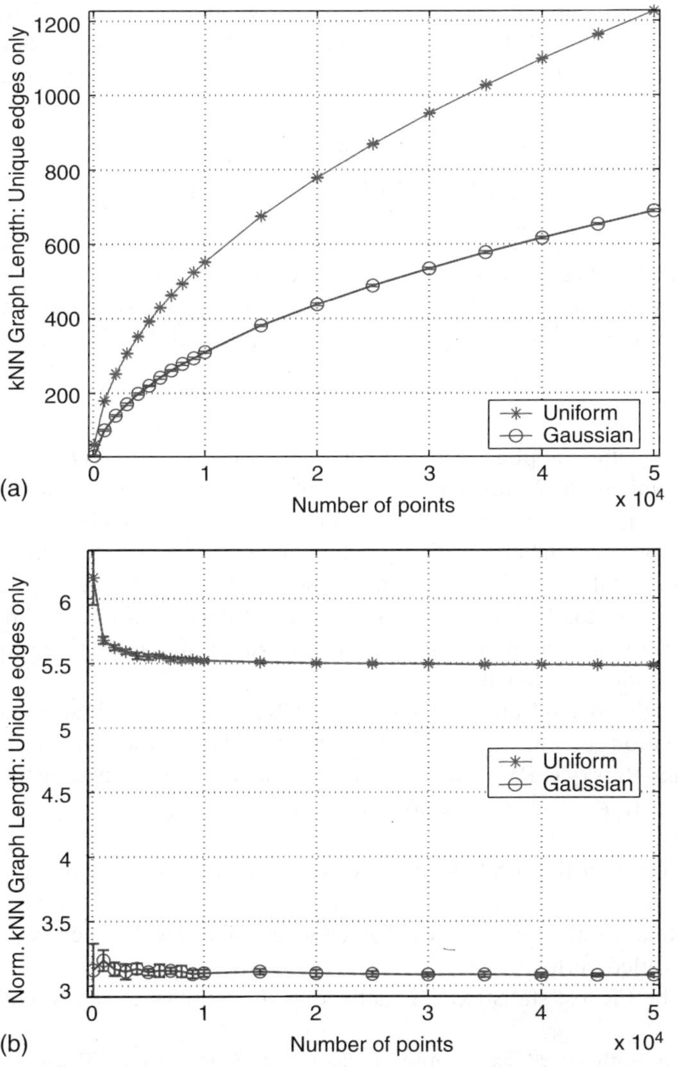

FIGURE 6.11 Mean length functions L_n of single-count kNNG implemented with $\gamma = 1$ (a) and $L_n/\sqrt{n}$ (b) as a function of n for uniform and normal distributed points.

As in the MST estimate of entropy, the constant $c_{kNNG} = (1 - \alpha)^{-1} \log \beta_{d,\gamma,k}$ can be estimated off-line by Monte Carlo simulation of the kNNG on random samples drawn from the unit cube. The complexity of the kNNG algorithm is dominated by the nearest neighbor search, which can be done in $O(n \log n)$ time for n sample points. This contrasts with the MST that requires a $O(n^2 \log n)$ implementation.

A related kNNG is the graph where edges connecting two points are counted only once. Such a graph eliminates one of the edges from each point pair that are mutual kNN. A kNNG can be built by pruning such that every unique edge contributes only once to the total length. The resultant graph has the identical appearance to the initial unpruned kNNG when plotted on the page. However, the cumulative lengths of the edges in the graphs differ, and so does their β factor (see Figure 6.11). We call this special pruned kNNG, the single-count kNNG.

IV. ENTROPIC GRAPH ESTIMATE OF HENZE–PENROSE AFFINITY

Friedman and Rafsky[27] presented a multivariate generalization of the Wald–Wolfowitz[85] runs statistic for the two sample problem. The Wald–Wolfowitz test statistic is used to decide between the following hypothesis based on a pair of samples $X, O \in \mathbb{R}^d$ with densities f_x and f_o respectively:

$$H_0 : f_x = f_o, \quad H_1 : J_x \neq J_o \tag{6.24}$$

The test statistic is applied to an i.i.d. random sample $\{X_i\}_{i=1}^{m}$, $\{O_i\}_{i=1}^{n}$ from f_x and f_o. In the univariate Wald–Wolfowitz test ($p = 1$), the $n + m$ scalar observations $\{Z_i\}_i = \{X_i\}_i, \{O_i\}_i$ are ranked in ascending order. Each observation is then replaced by a class label X or O depending upon the sample to which it originally belonged, resulting in a rank ordered sequence. The Wald–Wolfowitz test statistic is the total number of runs (runlength) R_ℓ of Xs or Os in the label sequence. As in runlength coding, R_ℓ is the length of consecutive sequences of length l of identical labels.

In Friedman and Rafsky's paper,[27] the MST was used to obtain a multivariate generalization of the Wald–Wolfowitz test. This procedure is called the Friedman–Rafsky (FR) test and is similar to the MST for estimating the α-Jensen difference. It is constructed as follows:

1. Construct the MST on the pooled multivariate sample points $\{X_i\} \cup \{O_i\}$.
2. Retain only those edges that connect an X labeled vertex to an O labeled vertex.
3. The FR test statistic, N, is defined as the number of edges retained.

The hypothesis H_1 is accepted for smaller values of the FR test statistic. As shown in Ref. 30, the FR test statistic N converges to the Henze–Penrose affinity (6.13) between the distributions f_x and f_o. The limit can be converted to the HP divergence by replacing N by the multivariate run length statistic, $R_\ell^{\mathrm{FR}} = n + m - 1 - N$.

For illustration of these graph constructions we consider two bivariate normal distributions with density functions f_1 and f_2 parameterized by their mean and covariance (μ_1, Σ_1), (μ_2, Σ_2). Graphs of the α-Jensen divergence calculated using MST (Figure 6.12), kNNG (Figure 6.13), and the Henze–Penrose affinity (Figure 6.14) are shown for the case where $\mu_1 = \mu_2$, $\Sigma_1 = \Sigma_2$. The "x" labeled points are samples from $f_1(x) = N(\mu_1, \Sigma_1)$, whereas the "o" labeled points are samples from $f_2(o) = N(\mu_2, \Sigma_2)$. μ_1 is then decreased so that $\mu_1 = \mu_2 - 3$.

V. ENTROPIC GRAPH ESTIMATORS OF α-GA AND α-MI

Assume for simplicity that the target and reference feature sets $O = \{o_i\}_i$ and $X = \{x_i\}_i$ have the same cardinality $m = n$. Here i denotes the ith pixel location in target and reference images. An entropic graph approximation to α-GA mean divergence (12) between target and reference is:

$$\widehat{\alpha D}_{\mathrm{GA}} = \frac{1}{\alpha - 1} \log \frac{1}{2n} \sum_{i=1}^{2n} \min\left\{ \left(\frac{e_i(o)}{e_i(x)} \right)^{\gamma/2}, \left(\frac{e_i(x)}{e_i(o)} \right)^{\gamma/2} \right\} \qquad (6.25)$$

where $e_i(o)$ and $e_i(x)$ are the distances from a point $z_i \in \{\{o_i\}^i, \{x_i\}^i\} \in \mathbb{R}^d$ to its nearest neighbor in $\{O_i\}_i$ and $\{X_i\}_i$, respectively. Here, as above, $\alpha = (d - \gamma)/d$.

Likewise, an entropic graph approximation to the α-MI (Equation 6.5) between the target and the reference is:

$$\widehat{\alpha \mathrm{MI}} = \frac{1}{\alpha - 1} \log \frac{1}{n^\alpha} \sum_{i=1}^{n} \left(\frac{e_i(o \times x)}{\sqrt{e_i(o)e_i(x)}} \right)^{2\gamma} \qquad (6.26)$$

where $e_i(o \times x)$ is the distance from the point $z_i = [o_i, x_i] \in \mathbb{R}^{2d}$ to its nearest neighbor in $\{Z_j\}_{j\neq i}$ and $e_i(o)(e_i(x))$ is the distance from the point $o_i \in \mathbb{R}^d$, $(x_i \in \mathbb{R}^d)$ to its nearest neighbor in $\{O_j\}_{j\neq i}(\{X_j\}_{j\neq i})$.

The estimators Equation 6.25 and Equation 6.26 can be derived from making a nearest neighbor approximation to the volume of the Voronoi cells constituting the kNN density estimator after plug-in to Equation 6.12 and Equation 6.5, respectively. The details are given in the Appendix. The theoretical convergence properties of these estimators are at present unknown.

Natural generalizations of Equation 6.25 and Equation 6.26 to multiple (> 2) images exist. The computational complexity of the α-MI estimator Equation 6.26 grows only linearly in the number of images to be registered while that of the α-GA estimator Equation 6.25 grows as linear log linear. Therefore, there is a significant complexity advantage to implementing α-MI via Equation 6.26 for simultaneous registration of a large number of images.

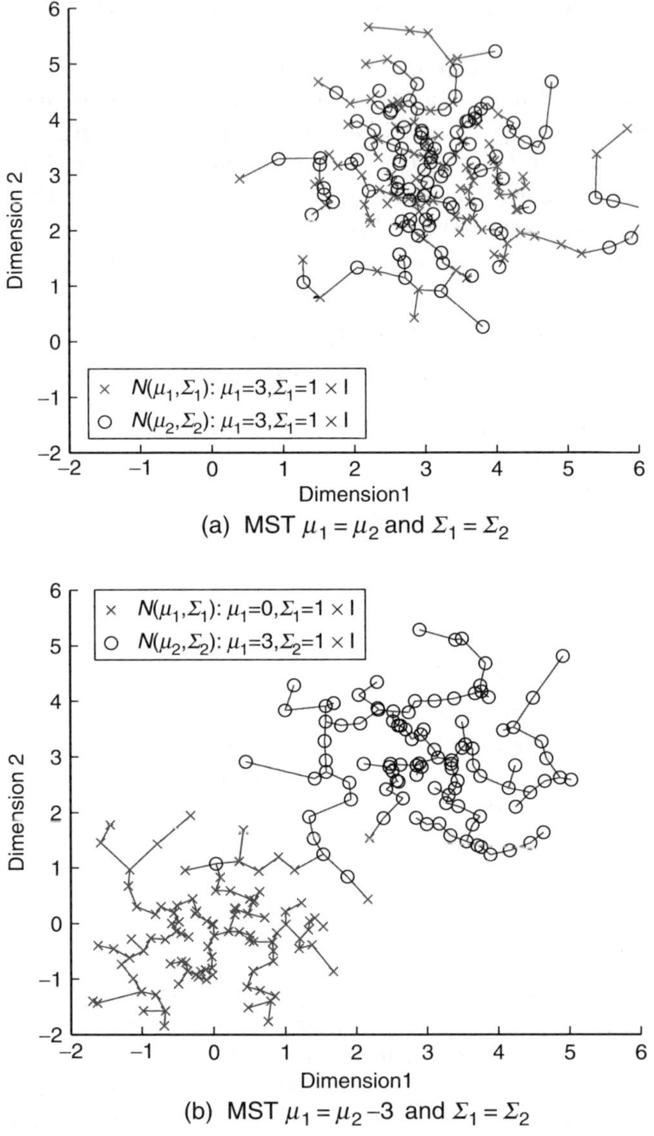

(a) MST $\mu_1=\mu_2$ and $\Sigma_1=\Sigma_2$

(b) MST $\mu_1=\mu_2-3$ and $\Sigma_1=\Sigma_2$

FIGURE 6.12 Illustration of MST for Gaussian case. Two bivariate normal distributions $N(\mu_1,\Sigma_1)$ and $N(\mu_1,\Sigma_1)$ are used. The "x" labeled points are samples from $f_1(x)=N(\mu_1,\Sigma_1)$, whereas the "o" labeled points are samples from $f_2(o)=N(\mu_2,\Sigma_2)$. (a) $\mu_1=\mu_2$ and $\Sigma_1=\Sigma_2$ and (b) $\mu_1=\mu_2-3$ while $\Sigma_1=\Sigma_2$.

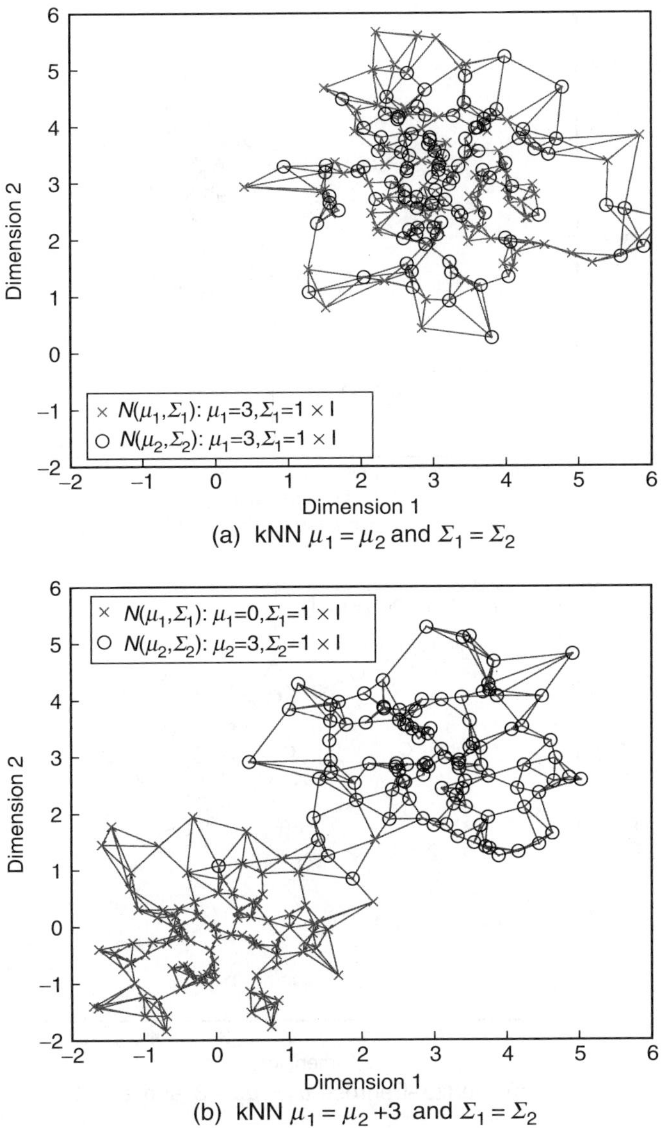

(a) kNN $\mu_1 = \mu_2$ and $\Sigma_1 = \Sigma_2$

(b) kNN $\mu_1 = \mu_2 + 3$ and $\Sigma_1 = \Sigma_2$

FIGURE 6.13 Illustration of kNN for Gaussian case. Two bivariate normal distributions $N(\mu_1, \Sigma_1)$ and $N(\mu_1, \Sigma_1)$ are used. The "x" labeled points are samples from $f_1(x) = N(\mu_1, \Sigma_1)$, whereas the "o" labeled points are samples from $f_2(o) = N(\mu_2, \Sigma_2)$. (a) $\mu_1 = \mu_2$ and $\Sigma_1 = \Sigma_2$ and (b) $\mu_1 = \mu_2 - 3$ while $\Sigma_1 = \Sigma_2$.

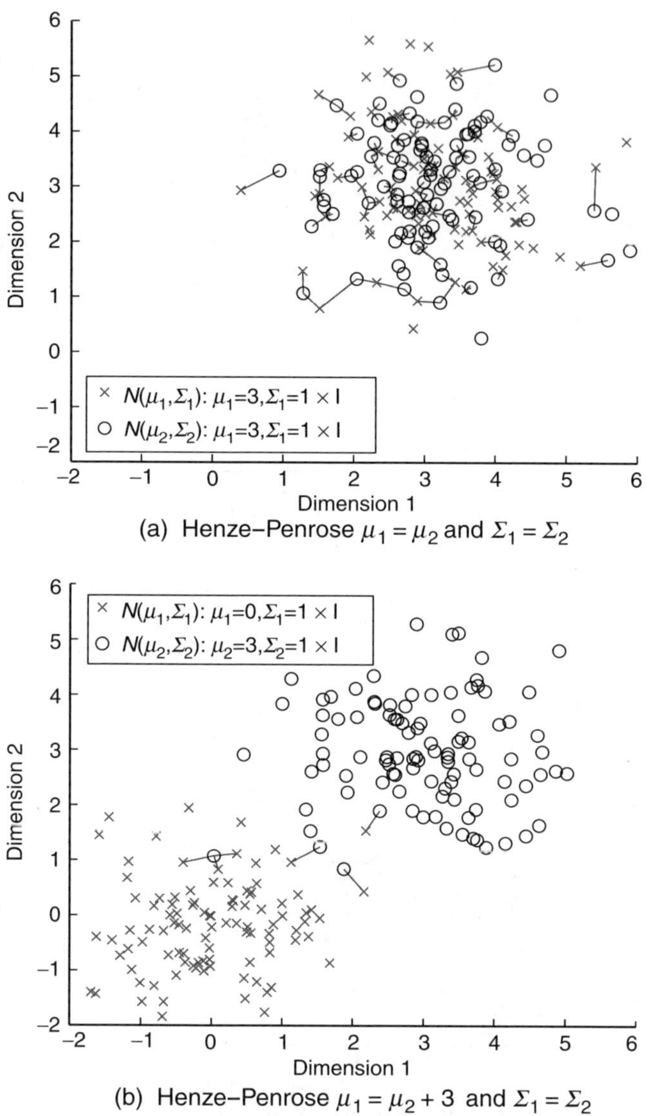

FIGURE 6.14 Illustration of Henze–Penrose affinity for Gaussian case. Two bivariate normal distributions $N(\mu_1, \Sigma_1)$ and $N(\mu_1, \Sigma_1)$ are used. The "x" labeled points are samples from $f_1(x) = N(\mu_1, \Sigma_1)$, whereas the "o" labeled points are samples from $f_2(o) = N(\mu_2, \Sigma_2)$. (a) $\mu_1 = \mu_2$ and $\Sigma_1 = \Sigma_2$ and (b) $\mu_1 = \mu_2 - 3$ while $\Sigma_1 = \Sigma_2$.

VI. FEATURE-BASED MATCHING

While scalar single pixel intensity level is the most popular feature for MI registration, it is not the only possible feature. As pointed out by Leventon and Grimson,[48] single pixel MI does not take into account joint spatial behavior of the coincidences and this can cause poor registration, especially in multi-modality situations. Alternative scalar valued features[11] and vector valued features[61,73] have been investigated for mutual information based image registration. We will focus on local basis projection feature vectors which generalize pixel intensity levels.

Basis projection features are extracted from an image by projecting local subimages onto a basis of linearly independent subimages of the same size. Such an approach is widely adopted in image matching applications, in particular with DCT or more general 2D wavelet bases.[21,22,50,74,82] Others have extracted a basis set adapted to image database using principal components (PCA) or independent components analysis (ICA).[41,49]

A. ICA Basis Projection Features

The ICA basis is especially well suited for our purposes since it aims to obtain vector features which have statistically independent elements that can facilitate estimation of α-MI and other entropic measures. Specifically, in ICA an optimal basis is found which decomposes the image X_i into a small number of approximately statistically independent components (subimages) $\{S_j\}$:

$$X_i = \sum_{j=1}^{p} a_{ij} S_j \qquad (6.27)$$

We select basis elements $\{S_j\}$ from an overcomplete linearly dependent basis using randomized selection over the database. For image i the feature vectors Z_i are defined as the coefficients $\{a_{ij}\}$ in Equation 6.27 obtained by projecting the image onto the basis.

In Figure 6.15 we illustrate the ICA basis selected for the MRI image database. ICA was implemented using Hyvarinen and Oja's[41] FastICA code (available from Ref. 40) which uses a fixed-point algorithm to perform maximum likelihood estimation of the basis elements in the ICA data model Equation 6.27. Figure 6.15 shows a set of 64 16×16 basis vectors which were estimated from over 100,000 16×16 training subimages randomly selected from five consecutive image slices, each from two MRI volumes scan of the brain, one of the scans was T1 weighted whereas the other is T2 weighted. Given this ICA basis and a pair of to-be-registered $M \times N$ images, coefficient vectors are extracted by projecting each 16×16 neighborhood in the images onto the basis set. For the 64 dimensional ICA bases shown in Figure 6.15, this yields a set of MN vectors in a 64 dimensional vector space which will be used to define features.

FIGURE 6.15 16×16 ICA basis set obtained from training on randomly selected 16×16 blocks in ten T1 and T2 time weighted MRI images. Features extracted from an image are the 64 dimensional vectors obtained by projecting 16×16 subimages of the image on the ICA basis.

B. MULTIRESOLUTION WAVELET BASIS FEATURES

Coarse-to-fine hierarchical wavelet basis functions describe a linear synthesis model for the image. The coarser basis functions have larger support than the finer basis; together they incorporate global and local spatial frequency information in the image. The multiresolution properties of the wavelet basis offer an alternative to the ICA basis, which is restricted to a single window size. Wavelet basis are commonly used for image registration,[43,78,87] and we briefly review them here.

A multiresolution analysis of the space of Lebesgue measurable functions, $L^2(\mathbb{R})$, is a set of closed, nested subspaces $V_j, j \in Z$. A wavelet expansion uses translations and dilations of one fixed function, the wavelet $\psi \in L^2(R)$. ψ is a wavelet if the collection of functions $\{\psi(x - l)|l \in Z\}$ is a Riesz basis of V_0 and its orthogonal complement W_0. The continuous wavelet transform of a function

$f(x) \in L^2(R)$ is given by:

$$Wf(a,b) = \langle f, \psi_{a,b} \rangle; \quad \psi_{a,b} = \frac{1}{\sqrt{|a|}} \psi\left(\frac{x-b}{a}\right) \qquad (6.28)$$

where $a, b \in \mathbb{R}$, $a \neq 0$.

For discrete wavelets, the dilation and translation parameters, b and a, are restricted to a discrete set, $a = 2^j$, $b = k$ where j and k are integers. The dyadic discrete wavelet transform is then given as:

$$Wf(j,k) = \langle f, \psi_{j,k} \rangle \psi_{j,k} = 2^{-j/2} \psi(2^{-j}x - k) \qquad (6.29)$$

where $j, k \in \mathbb{Z}$. Thus, the wavelet coefficient of f at scale j and translation k is the inner product of f with the appropriate basis vector at scale j and translation k. The 2D discrete wavelet analysis is obtained by a tensor product of two multiresolution analysis of $L^2(\mathbb{R})$. At each scale, j, we have one scaling function subspace and three wavelet subspaces. The discrete wavelet transform of an image is the projection of the image onto the scaling function V_0 subspaces and the wavelet subspaces W_0. The corresponding coefficients are called the approximate and detail coefficients, implying the low- and high-pass character-istics of the basis filters. The process of projecting the image onto the successively coarser spaces continues to achieve the approximation desired. The difference information sensitive to vertical, horizontal and diagonal edges are treated as the three dimensions of each feature vector. Several members of the discrete Meyer basis used in this work are plotted below in Figure 6.16.

VII. COMPUTATIONAL CONSIDERATIONS

A popular sentiment about graph methods, such as the MST and the kNNG, is that they could be computationally taxing. However, since the early days, graph theory algorithms have evolved and several variants with low time-memory complexity have been found. Henze-Penrose and the α-GA mean divergence metrics are based directly on the MST and kNNG, and first require the solution of these combinatorial optimization problems. This section is devoted to providing insight into the formulation of these algorithms and the assumptions that lead to faster, lower complexity variants of these algorithms.

A. REDUCING TIME-MEMORY COMPLEXITY OF THE MST

The MST problem has been studied since the early part of the 20th century. Owing to its widespread applicability in other computer science, optimization theory, and pattern recognition related problems, there have been, and continue to be, sporadic improvements in the time-memory complexity of the MST problem. Two principal algorithms exist for computing the MST, the Prim algorithm[67] and the Kruskal algorithm.[45] For sparse graphs the Kruskal algorithm is the fastest general purpose MST computation algorithm. Kruskal's algorithm maintains a

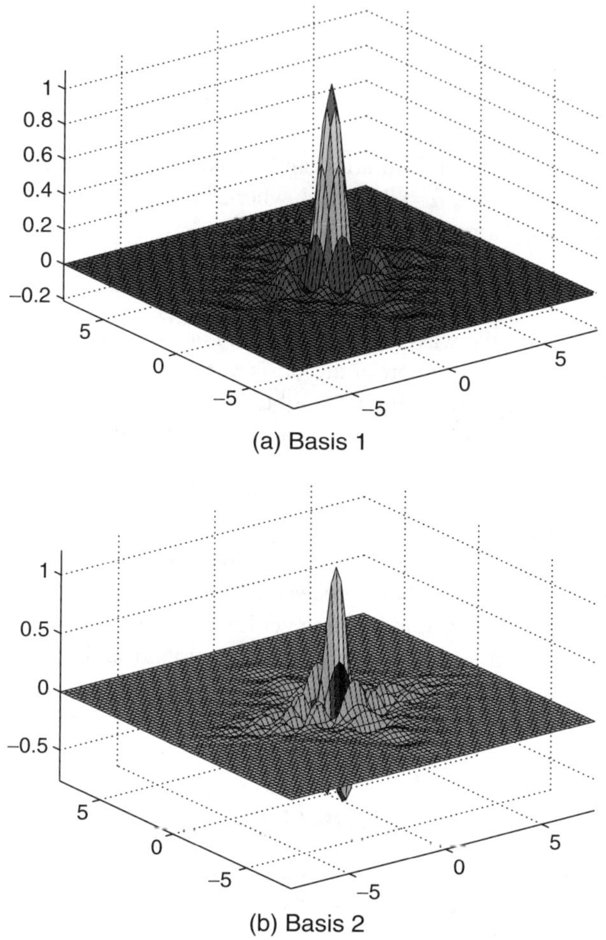

(a) Basis 1

(b) Basis 2

FIGURE 6.16 Wavelet decomposition: 2D discrete Meyer wavelet basis. Scale subspace (a) and three wavelet subspaces (b–d) at level 1 decomposition.

list of edges sorted by their weights and grows the tree one edge at a time. Cycles are avoided within the tree by discarding edges that connect two subtrees already joined through a prior established path. The time complexity of the Kruskal algorithm is of order $O(E \log E)$ and the memory requirement is $O(E)$, where E is the initial number of edges in the graph. Recent algorithms have been proposed that offer advantages over these older algorithms at the expense of increased complexity. A review can be found in Ref. 5.

An initial approach may be to construct the MST by including all the possible edges within the feature set. This results in N^2 edges for N points; a time requirement of $O(N^2)$ and a memory requirement of $O(N^2 \log N)$. The number of points in the graph is the total number of d-dimensional features

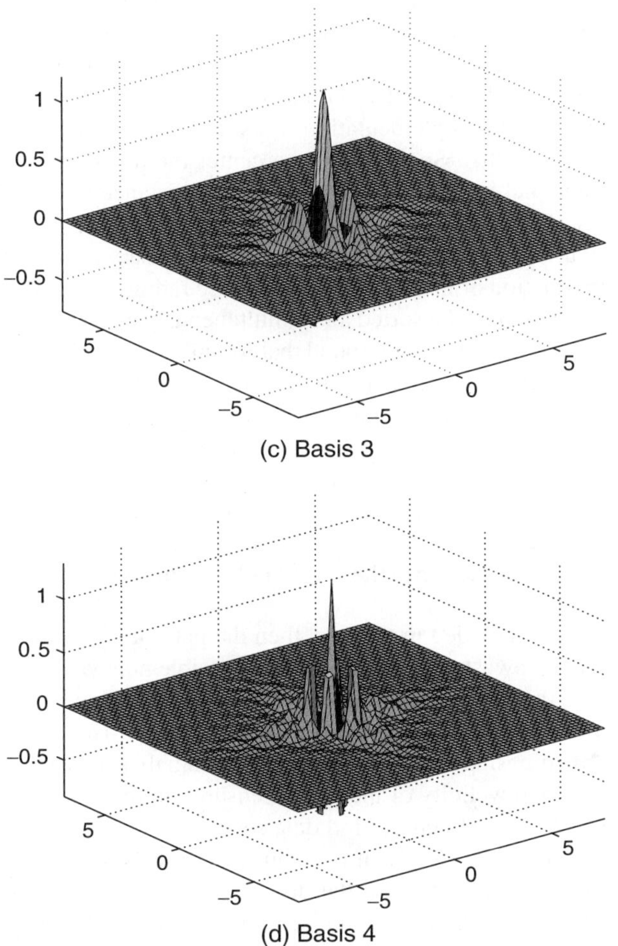

(c) Basis 3

(d) Basis 4

FIGURE 6.16 (Continued).

participating in the registration from the two images. If each image has $M \times N$ features (for, e.g., pixels), the total number of points in the graph is $2 \times M \times N \approx 150,000$ for images of size 256×256 pixels. The time and memory requirements of the MST are beyond the capabilities of even the fastest available desktop processors.

The earliest solution can be attributed to Bentley and Friedman.[10] Using a method to quickly find nearest neighbors in high dimensions they proposed building a minimum spanning tree using the assumption that local neighbors are more likely to be included in the MST than distant neighbors. Several improvements have been made on this technique, and have been proposed in Refs. 14,56. For our experiments we have been motivated by the adapted original

Bentley method, as explained below. This method achieves significant acceleration by sparsification of the initial graph before tree construction.

We have implemented a method for sparsification that allows MST to be constructed for several hundred thousand points in a few minutes of desktop computing time. This implementation uses a disc windowing method for constructing the edge list. Specifically, we center the discs at each point under consideration and pick only those neighbors whose distance from the point is less than the radius of the disc (See Figure 6.17 for illustration). A list intersection approach similar to Ref. 62 is adopted to prune unnecessary edges within the disc. Through a combination of list intersection and disc radius criterion we reduce the number of edges that must be sorted and simultaneously ensure that the MST thus built is valid. We have empirically found that for uniform distributions, a constant disc radius is best. For nonuniform distributions, the disc radius is better selected as the distance to the kNN. Figure 6.18 shows the bias of modified MST algorithm as a function of the radius parameter and the number of nearest neighbors for a uniform density on the plane.

It is straightforward to prove that, if the radius is suitably specified, our MST construction yields a valid minimum spanning tree. Recall that the Kruskal algorithm ensures construction of the exact MST.[45] Consider a point p_i in the graph.

1. If point p_i is included in the tree, then the path of its connection to the tree has the lowest weight amongst all possible noncyclic connections. To prove this is trivial: the disc criterion includes lower weight edge before considering an edge with a higher weight. Hence, if a path is found by imposing the disc, that path is the smallest possible noncyclic path. The noncyclicity of the path is ensured in the Kruskal algorithm through a standard Union-Find data set.

2. If point p_i is not in the tree, it is because all the edges between p_i and its neighbors considered using the disc criterion of edge inclusion have total edge weight greater than disc radius or have led to a cyclic path. Expanding the disc radius would then provide the path which is lowest in weight and noncyclic.

B. Reducing Time-Memory Complexity of the kNNG

Time-memory considerations in the nearest neighbor graph have prompted researchers to come up with various exact and approximate graph algorithms. With its wide-spread usage, it is not surprising that several fast methods exist for nearest neighbor graph constructions. Most of them are expandable to construct kNNGs. One of the first fast algorithms for constructing NNG was proposed by Bentley.[8,9] A comprehensive survey of the latest methods for nearest neighbor searches in vector spaces is presented in Ref. 12. A simple and intuitive method for nearest neighbor search in high dimensions is presented in Ref. 62.

Though compelling, the methods presented above focus on retrieving the exact nearest neighbors. One could hypothesize that for applications where the

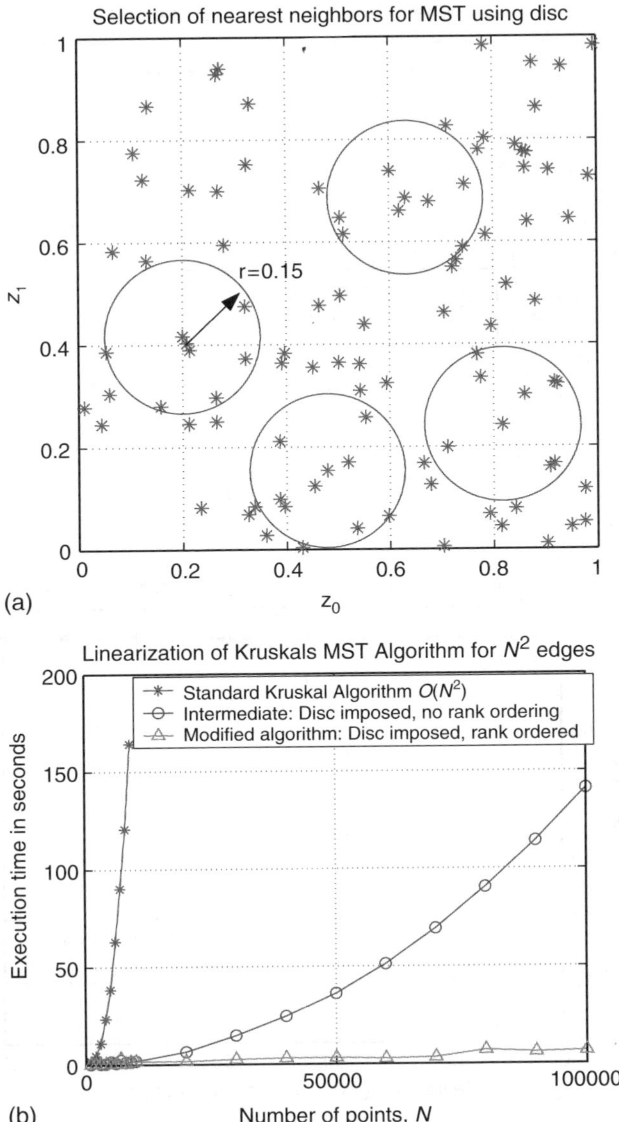

FIGURE 6.17 Disc-based acceleration of Kruskal's MST algorithm from $n^2 \log n$ to $n \log n$ (a), and comparison of computation time for Kruskal's standard MST algorithm with respect to our accelerated algorithm (b).

accuracy of the nearest neighbors is not critical, we could achieve significant speed-up by accepting a small bias in the nearest neighbors retrieved. This is the principal argument presented in Ref. 1. We conducted our own experiments on the approximate nearest neighbor method using the code provided

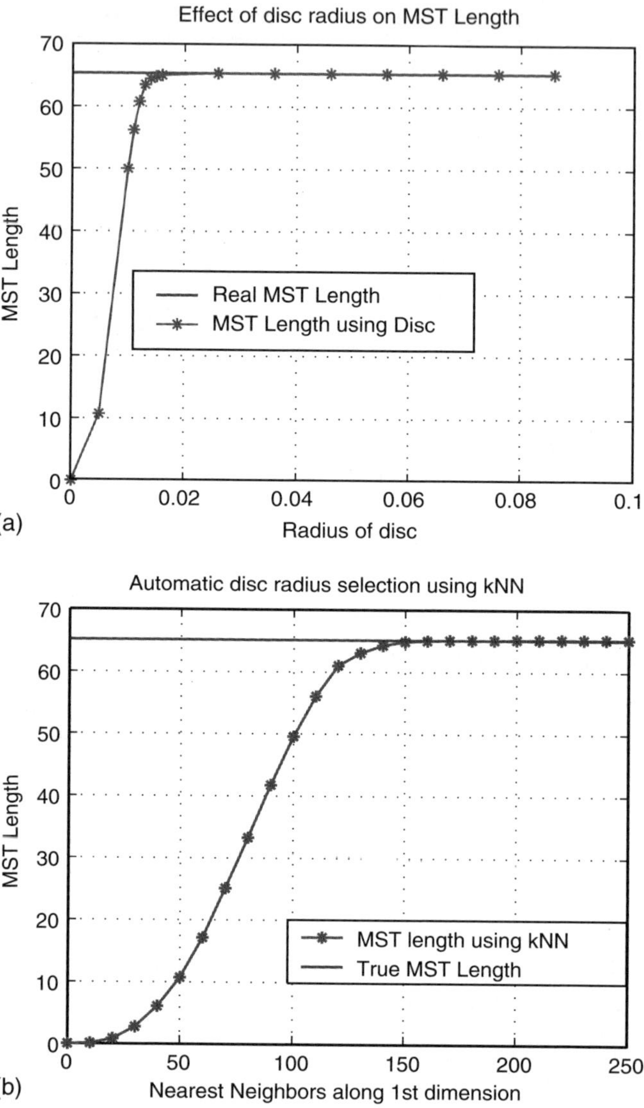

FIGURE 6.18 Bias of the $n \log n$ MST algorithm as a function of radius parameter (a), and as a function of the number of nearest neighbors (b) for uniform points in the unit square.

in Ref. 55 (Figure 6.19). We conducted benchmarks on uniformly distributed points in eight dimensional spaces. If the error incurred in picking the incorrect kth nearest neighbor $\leq \varepsilon$, the cumulative error in the length of the kNNG is plotted in Figure 6.19. Compared to an exact kNN search using k-d trees a significant reduction ($> 85\%$) in time can be obtained, through approximate nearest neighbor methods, incurring a 15% cumulative graph length error.

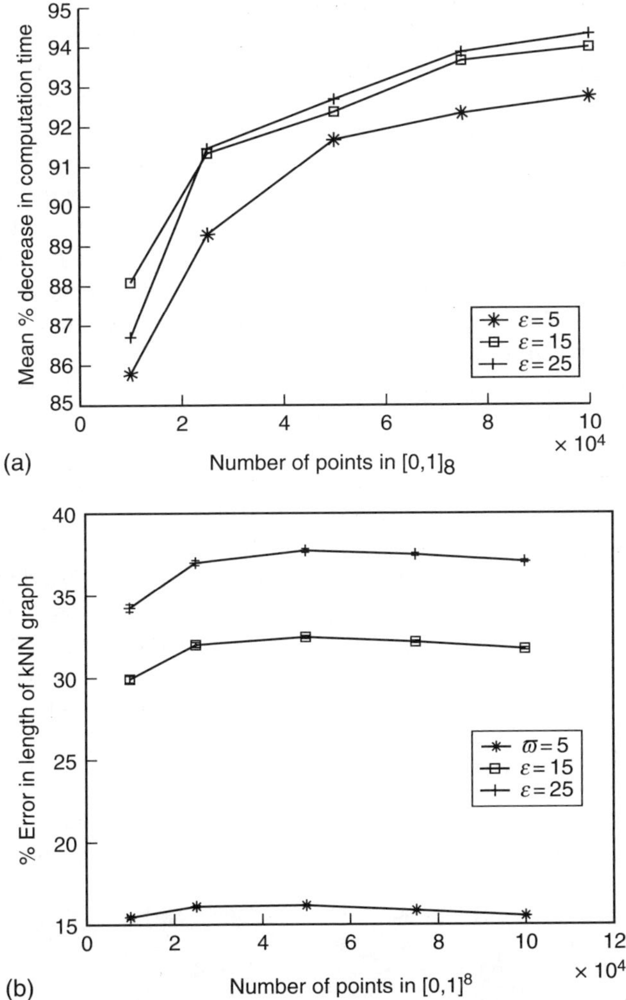

FIGURE 6.19 Approximate kNNG: (a) Decrease in computation time to build approximate kNNG for different ε, expressed as a percentage of time spent computing the exact kNNG over a uniformly distributed points in $[0, 1]^8$. An 85% reduction in computation time can be obtained by incurring a 15% error in cumulative graph length. (b) Corresponding error incurred in cumulative graph length.

VIII. APPLICATIONS: MULTISENSOR SATELLITE IMAGE FUSION

In this section, we shall illustrate entropic graph based image registration for a remote sensing example. Images of sites on the earth are gathered by a variety of geostationary satellites. Numerous sensors gather information in distinct frequency bands in the electromagnetic spectrum. These images help predict daily weather patterns, environmental parameters influencing crop cycles such as soil composition, water and mineral levels deeper in the Earth's crust, and may also serve as surveillance sensors meant to monitor activity over hostile regions. A satellite may carry more than one sensor and may acquire images throughout a period of time. Changing weather conditions may interfere with the signal. Images captured in a multisensor satellite imaging environment show linear deformations due to the position of the sensors relative to the object. This transformation is often linear in nature and may manifest itself as relative translational, rotational, or scaling between images. This provides a good setting to observe different divergence measures as a function of the relative deformation between images. We simulated linear rotational deformation in order to reliably test the image registration algorithms presented above.

Figure 6.20 shows two images of downtown Atlanta, captured with visible and thermal sensors, as a part of the Urban Heat Island project[68] that studies the creation of high heat spots in metropolitan areas across the U.S.A. Pairs of visible light and thermal satellite images were also obtained from NASA's Visible Earth website.[57] The variability in imagery arises due to the different specialized satellites used for imaging. These include weather satellites wherein the imagery shows heavy occlusion due to clouds and other atmospheric disturbances. Other satellites focus on urban areas with roads, bridges, and high rise buildings. Still other images show entire countries or continents, oceans and large geographic landmarks such as volcanoes and active geologic features. Lastly, images contain different landscapes such as deserts, mountains and valleys with dense foliage.

A. DEFORMATION AND FEATURE DEFINITION

Images are rotated through 0 to 32, with a step size adjusted to allow a finer sampling of the objective function near 0°. The images are projected onto a

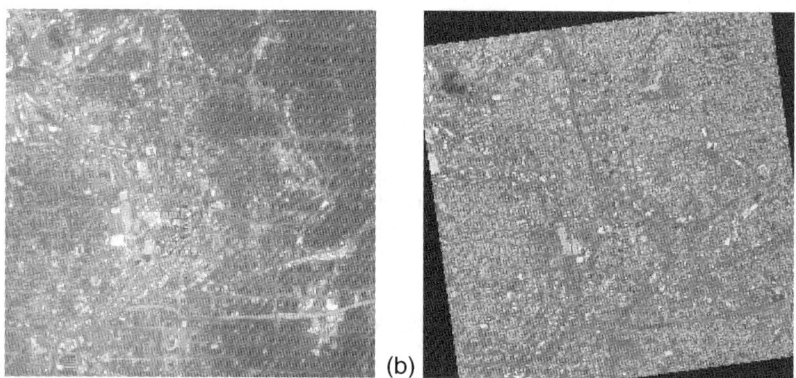

FIGURE 6.20 Images of downtown Atlanta obtained from Urban Heat Island project.[68] (a) Thermal image, (b) visible-light image under artificial rotational transformation. (From Project ATLANTA, NASA Marshall Space Flight Center, Huntsville, Alabama. With permission.)

Meyer wavelet basis, and the coefficients are used as features for registration. A feature sample from an image I in the database is represented as a tuple consisting of the coefficient vector, and a two-dimensional vector identifying the spatial co-ordinates of the origin of the image region it represents. For example $\{W_{(i,j)}, x_{(i,j)}, y_{(i,j)}\}$ represents the tuple from position $\{i,j\}$ in the image. Now, $\underline{W_{(i,j)}} \equiv \left\{w_{(i,j)}^{\text{Low}-\text{Low}}, w_{(i,j)}^{\text{Low}-\text{High}}, w_{(i,j)}^{\text{High}-\text{Low}}, w_{(i,j)}^{\text{High}-\text{High}}\right\}$, where the superscript identifies the frequency band in the wavelet spectrum. Features from both the images $\{Z_1, Z_2\}$ are pooled together to form a joint sample pool $\{Z_1 \bigcup Z_2\}$. The MST and kNNG are individually constructed on this sample pool.

Figure 6.21 shows the rotational mean-squared registration error for the images in our database, in the presence of additive noise. Best performance under the presence of noise can be seen through the use of the α-MI estimated using wavelet features and kNNG. Comparable performances are seen through the use of Henze–Penrose and α-GA mean divergences, both estimated using wavelet features. Interestingly, the single pixel Shannon MI has the poorest performance which may be attributed to its use of poorly discriminating scalar intensity features. Notice that the α-GA, Henze–Penrose affinity, and α-MI (wavelet-kNN estimate), all implemented with wavelet features, have significantly lower MSE compared to the other methods.

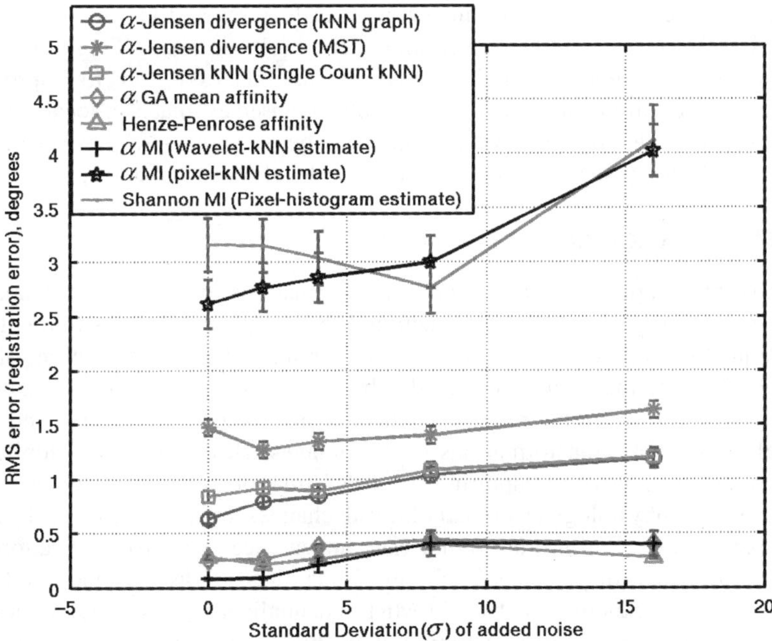

FIGURE 6.21 Rotational root mean squared error obtained from rotational registration of multisensor satellite imagery using six different image similarity/dissimilarity criteria. Standard error bars are as indicated. These plots were obtained from Monte Carlo trials consisting of adding i.i.d. Gaussian distributed noise to the images prior to registration.

Further insight into the performance of these wavelet-based divergence measures may be gained by considering the mean objective function over 750 independent trials. Figure 6.22a shows the α-MI, HP affinity, and the α-GA affinity and Figure 6.22b shows the α-Jensen difference divergence calculated using the kNNG and the MST. The sensitivity and robustness of the dissimilarity measures can be evaluated by observing the divergence function near zero rotational deformation (Figure 6.22).

IX. APPLICATIONS: LOCAL FEATURE MATCHING

The ability to discriminate differences between images with sensitivity to local differences is pivotal to any image matching algorithm. Previous work in these techniques has been limited to simple pixel based mutual information (MI) and pixel correlation techniques. In Ref. 65, local measures of MI outperform global MI in the context of adaptive grid refinement for automatic control point placement. However, the sensitivity of local MI deteriorates rapidly as the size of the image window decreases below 40×40 pixels in 2D.

The main constraints on these algorithms, when localizing differences, are (1) limited feature resolution of single pixel intensity based features, and (2) histogram estimators $h(X, Y)$ of joint probability density $f(X, Y)$ are noisy when computed with a small number of pixel features and are thus poor estimators of $f(X, Y)$ used by the algorithm to derive joint entropy $H(X, Y)$. Reliable identification of subtle local differences within images is key to improving registration sensitivity and accuracy.[59] Stable unbiased estimates of local entropy are required to identify sites of local mismatch between images. These estimates play a vital role in successfully implementing local transformations.

A. DEFORMATION LOCALIZATION

Iterative registration algorithms apply transformations to a sequence of images while minimizing some objective function. We demonstrate the sensitivity of our technique by tracking deformations that correspond to small perturbations of the image. These perturbations are recorded by the change in the mismatch metric.

Global deformations reflect a change in imaging geometry and are modeled as global transformations on the images. However, global similarity metrics are ineffective in capturing local deformations in medical images that occur due to physiological or pathological changes in the specimen. Typical examples are: change in brain tumor size, appearance of microcalcifications in breast, nonlinear displacement of soft tissue due to disease and modality induced inhomogeneities such as in MRI and nonlinear breast compression in x-ray mammograms. Most registration algorithms will not be reliable when the size of the mismatch site is insufficiently small, typically $(m \times n) \leq 40 \times 40$.[65] With a combination of ICA and α-entropy we match sites having as few as 8×8 pixels. Due to the limited number samples in the feature space, the faster convergence properties of the MST are better suited to this problem.

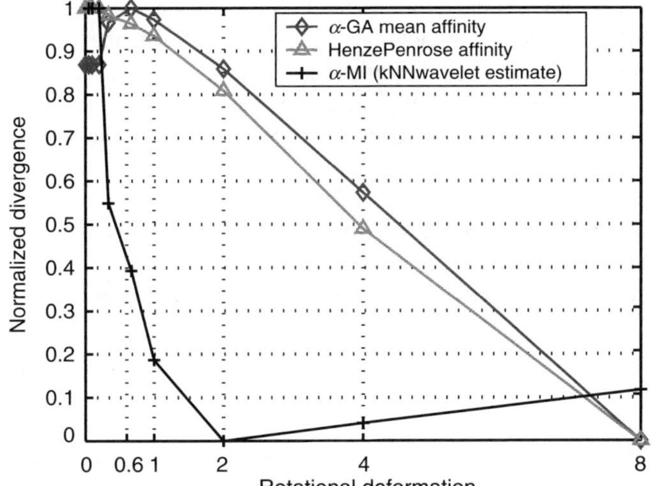

(a) Average α-GA affinity, HP affinity and α-MI (kNN-wavelet estimate. Rotation angle estimated by maximizing noisy versions of these objective functions.

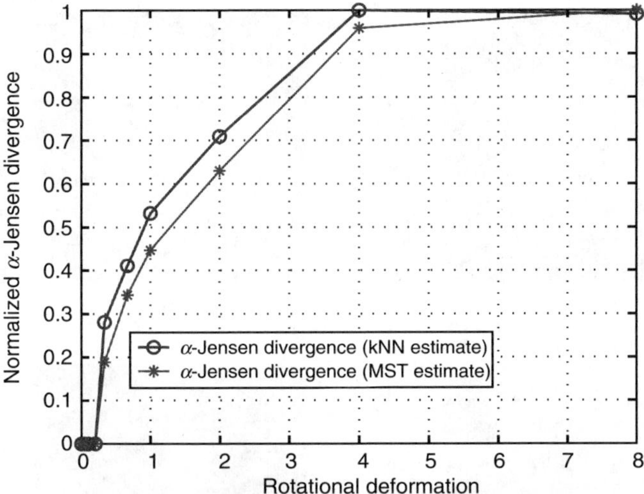

(b) Average α-Jensen divergence (kNN and MST estimate on wavelet features). Rotation angle estimated by minimizing noisy versions of these objective functions.

FIGURE 6.22 Average affinity and divergence, over all images, in the vicinity of zero rotation error: (a) α-Jensen (kNN) and α-Jensen (MST), (b) α-GA mean affinity, HP affinity and α-MI estimated using wavelet features and kNNG.

Although we do not estimate other divergence measures, α-Jensen calculated using the MST provides a benchmark for their performance.

In Figure 6.23, multimodal synthesized scan of T1 and T2 weighted brain MRI each of size 256×256 pixels[16] are seen. The original target images shall be deformed locally (see below) to generate a deformed target image.

1. *Locally deforming original image using B-splines*: B-spline deformations are cubic mapping functions that have local control and injective properties.[13] The 2D uniform tensor B-spline function F, is defined with a 4×4 control lattice ϕ in $\mathbb{R}^2$ as:

$$F(u, v) = \sum_{i=0}^{3} \sum_{j=0}^{3} B_i(u)B_j(v)\phi_{ij} \qquad (6.30)$$

where $0 \leq u, v \leq 1$, ϕ_{ij} is the spatial co-ordinates of the lattice, and B_i are the standard B-spline basis functions. The uniform B-spline basis functions used here are quite common in computer graphics literature

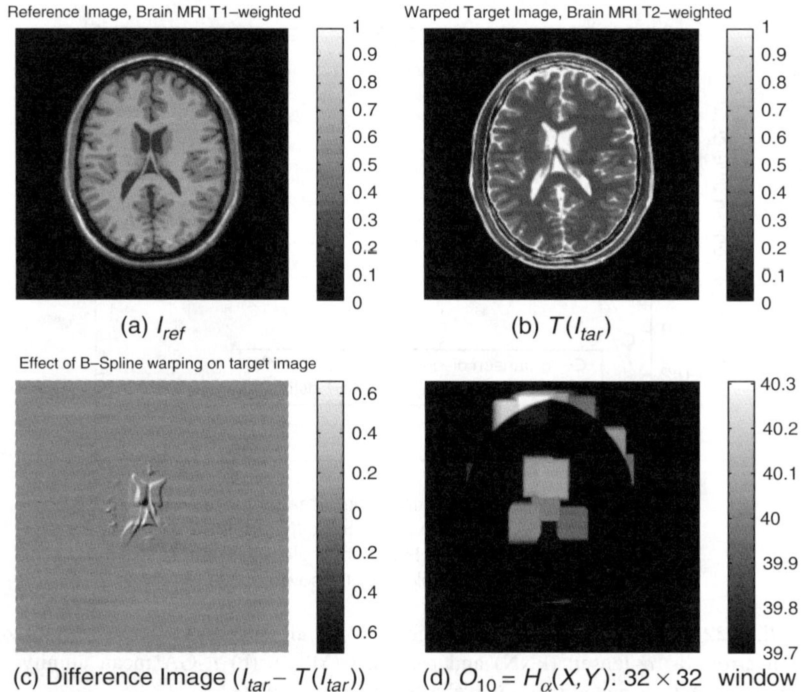

FIGURE 6.23 B-spline deformation on MRI images of the brain. (a) Reference image, (b) warped target, (c) true deformation, (d) $O_{10} = H_\alpha$ as seen with a 32×32 window, (e) 16×16 window and (f) 8×8 window, (g) $\nabla(O) = \nabla(H_\alpha) = O_{10} - O_0$ as seen with a 32×32, (h) 16×16, and (i) 8×8 window.

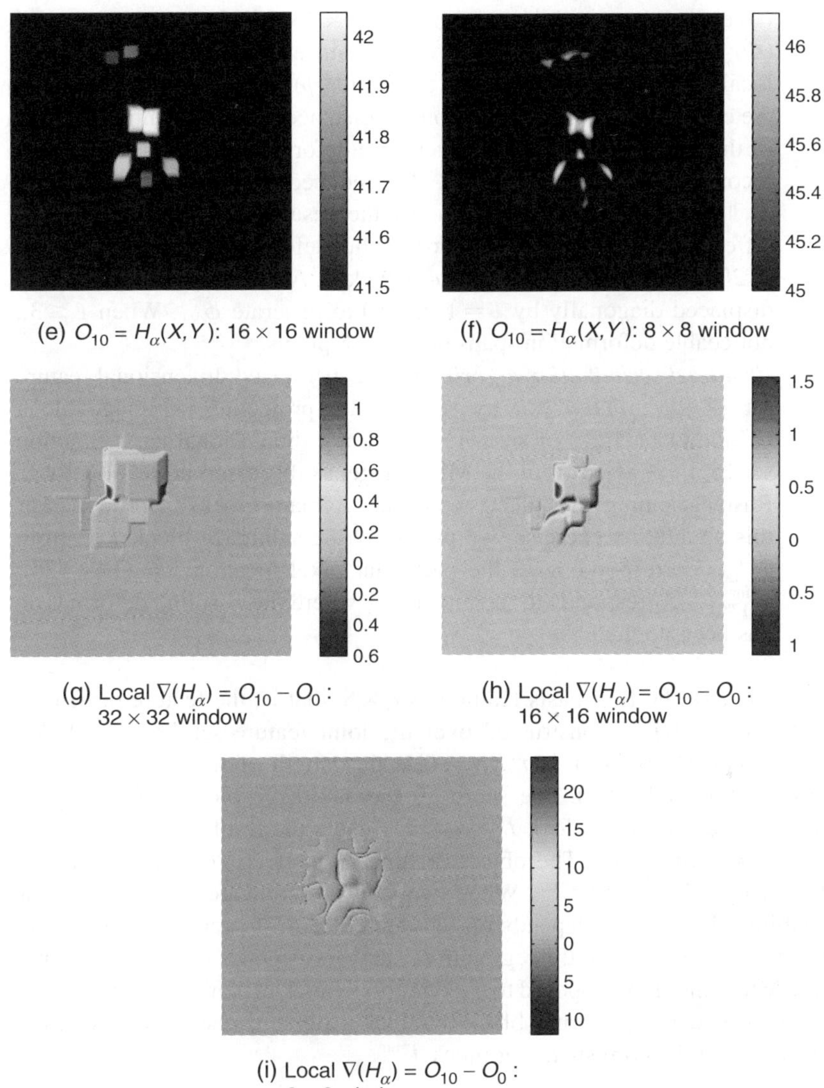

(e) $O_{10} = H_\alpha(X,Y)$: 16×16 window

(f) $O_{10} = H_\alpha(X,Y)$: 8×8 window

(g) Local $\nabla(H_\alpha) = O_{10} - O_0$: 32×32 window

(h) Local $\nabla(H_\alpha) = O_{10} - O_0$: 16×16 window

(i) Local $\nabla(H_\alpha) = O_{10} - O_0$: 8×8 window

FIGURE 6.23 (Continued).

and may be found in Ref. 13 are defined as:

$$B_0(u) = \frac{(1-u)^3}{6}, \quad B_1(u) = \frac{3u^3 - 6u^2 + 4}{6},$$

$$B_2(u) = \frac{-3u^3 + 3u^2 + 3u + 1}{6}, \quad B_3(u) = \frac{u^3}{6} \tag{6.31}$$

Given that the original images have 256×256 pixels, we impose a grid (Φ) of 10×10 control points on I_{tar}. Since the aim is to deform I_{tar} locally, not globally, we select a subgrid (ϕ) of 4×4 control points in the center of I_{tar}. We then diagonally displace, by $\ell = 10$ mm, only one of the control points in ϕ, to generate deformed grid ϕ_{def}. I_{tar} is then reconstructed according to ϕ_{def}. The induced deformation is measured as $\|\phi_{\text{def}} - \phi\|$. Figure 6.23 shows the resultant warped image and difference image $I_{\text{tar}} - T(I_{\text{tar}})$. For smaller deformations, Φ is a finer grid of 20×20 points, from which ϕ is picked. A control point in ϕ is then displaced diagonally by $\ell = 1, 2, \ldots 10$ to generate ϕ_{def}. When $\ell \leq 3$,, noticeable deformation spans only 8×8 pixels.

2. *Feature discrimination algorithm*: we generate a d-dimensional feature set $\{Z_i\}_{i=1}^{m \times n}$, $m \times n \geq d$ by sequentially projecting subimage block (window) $\{\Gamma_j\}_{j=1}^{M \times N}$ of size $m \times n$ onto a d-dimensional basis function set $\{S_k\}$ extracted from the MRI image, as discussed in Section VI.A. Raster scanning through I_{ref} we select subimage blocks $\{\Gamma_i^{\,ref}\}_{i=1}^{M \times N}$. For this simulation exercise, we pick only the subimage block $\Gamma^{\,tar}$ from $T(I_{\text{tar}})$ corresponding to the particular pixel location $k = (128, 128)$. $\Gamma_{128,128}^{\,tar}$ corresponds to the area in I_{tar} where the B-spline deformation has been applied.

The size of the ICA basis features is 8×8, that is, the feature dimension is $d = 64$. The MST is constructed over the joint feature set $\{Z_i^{\text{ref}}, Z_j^{\text{tar}}\}$. When suitably normalized with $1/n^{\alpha}$, $\alpha = 0.5$, the length of the MST becomes an estimate of $H_{\alpha}(Z_i^{\text{ref}}, Z_j^{\text{tar}})$. We score all the subimage blocks $\{\Gamma_i^{\text{ref}}\}_{i=1}^{M \times N}$ with respect to the subimage block $\Gamma_{128,128}^{\text{tar}}$. Let O_1 be the resultant $M \times N$ matrix of scores at deformation l. The objective function surface O_1 is a similarity map between $\{\Gamma^{\text{ref}}\}_{i=1}^{M \times N}$ and Γ^{tar}. When two sites are compared, the resulting joint probability distribution depends on the degree of mismatch. The best match is detected by searching for the region in I_{ref} that corresponds to Γ^{tar} as determined by the MST length. As opposed to the one-to-all block matching approach adopted here, one could also perform a block-by-block matching, where each block Γ_i^{ref} is compared with its corresponding block Γ_i^{tar}.

B. LOCAL FEATURE MATCHING RESULTS

Figure 6.23 shows O_{10} for $m \times n = 8 \times 8, 16 \times 16$ and 32×32. Similar maps can be generated for $\ell = \ell_1, \ell_2, \ldots \ell_p$. The gradient $\nabla(O) = O_{\ell_1} - O_{\ell_2}$ reflects the change in H_{α}, the objective function, when I_{tar} experiences an incremental change in deformation, from $\ell = \ell_1 \rightarrow \ell_2$. This gradient, at various subimage block size is seen in Figure 6.23, where $\ell_1 = 0$ and $\ell_2 = 10$. For demonstration purposes in Figure 6.23, we imposed a large deformation to I_{tar}. Smaller deformations generated using a control grid spanning only 40×40 pixels are used to generate Figure 6.24. It shows the ratio of the gradient of

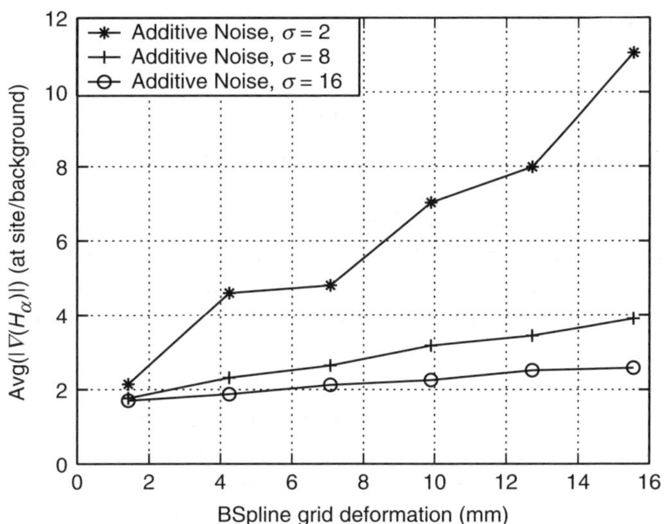

FIGURE 6.24 Ratio of $\nabla(H_\alpha) = \nabla O$ calculated over deformation site v/s background image for smaller deformation spanning $m \times n \geq 8 \times 8$.

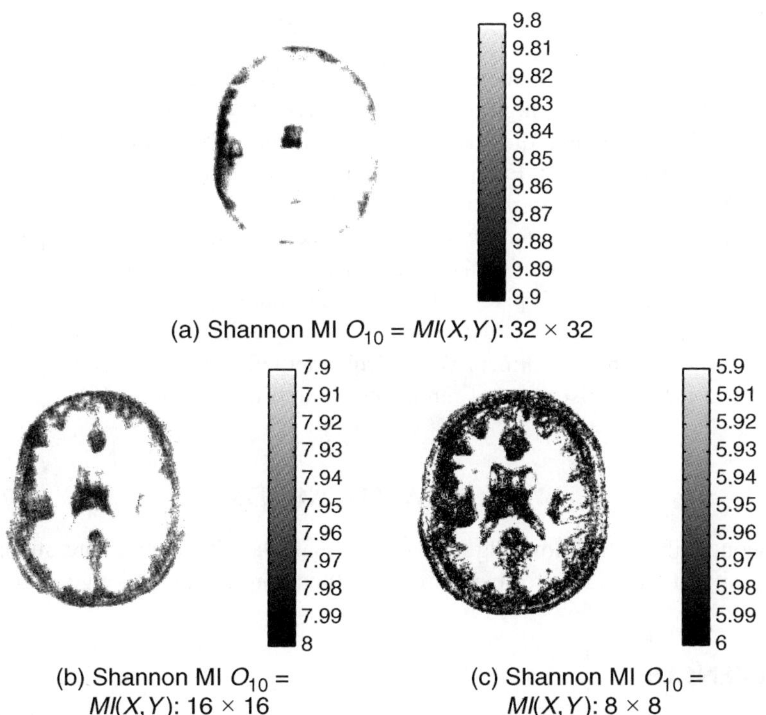

(a) Shannon MI $O_{10} = MI(X,Y)$: 32×32

(b) Shannon MI $O_{10} = MI(X,Y)$: 16×16

(c) Shannon MI $O_{10} = MI(X,Y)$: 8×8

FIGURE 6.25 Performance of Shannon MI, computed using pixel intensity histograms, on deformed MRI images: (a) 32×32 window, (b) 16×16 window, and (c) 8×8 window.

the objective function:

$$R = \frac{\frac{1}{(m \times n)} \sum_{i=1}^{m \times n} |\nabla(O(i))|}{\frac{1}{(M \times N - m \times n)} \sum_{i=1}^{M \times N - m \times n} |\nabla(O(i))|} \tag{6.32}$$

over the deformation site vs. background in the presence of additive Gaussian noise.

Figure 6.25 shows the similarity map O_ℓ when constructed using a histogram estimate of joint entropy calculated over subimage size $m \times n$ in Equation 6.6. At lower subimage sizes, the estimate displays bias and several local minima even under noise free conditions. It is thus unsuitable for detection of local deformation of I_{tar}.

The framework presented here could be extended to (1) enhance registration performance by sensitizing it to local mismatch, (2) automatically track features of interest, such as tumors in brain or microcalcifications in breast across temporal image sequences, (3) reliably match or register small images or image regions so as to improve disease diagnosis by locating and identifying small pathological changes in medical image volumes, and (4) automate control point placement to initiate registration.

X. CONCLUSION

In this paper we have presented several techniques to extend the multisensor image fusion problem to high dimensional feature spaces. Rényi's α-entropy is estimated directly in high dimensions through the use of entropic graph methods. These include the use of Euclidean functionals such as the α-Jensen, the HP divergence, and GA mean divergence. Graph theory methods such as the MST and the kNNG are central to our approach due to their quasiadditive properties in estimating Euclidean functionals. These methods provide a robust and viable alternative to traditional pixel intensity histograms used for estimating MI. Higher-dimensional features used for this work are the wavelet basis and ICA, where features are 64 dimensional. Our methods are validated through a demonstration of registration of multisensor satellite and medical imagery.

ACKNOWLEDGMENTS

We would like to acknowledge Sakina Zabuawala, EECS graduate student at University of Michigan, for benchmarking the approximate NN algorithm.

REFERENCES

1. Arya, S., Mount, D. M., Netanyahu, N. S., Silverman, R., and Wu, A. Y., An optimal algorithm for approximate nearest neighbor searching fixed dimensions, *J. ACM*, 45(6), 891–923, 1998.

2. Ashley, J., Barber, R., Flickner, M., Hafner, J. L., Lee, D., Niblack, W., and Petkovic, D., Automatic and semiautomatic methods for image annotation and retrieval in Query by Image Content (QBIC), In *Storage and Retrieval for Image and Video Databases III*, Niblack, W. and Jain, R., eds., February 5–10, 1995, San Diego/La Jolla, CA, USA. SPIE Proceedings Vol. 2420, pp. 24–35, 1995.

3. Baraniuk, R., Flandrin, P., Jensen, A. J. E. M., and Michel, O., Measuring time frequency information content using the Rényi entropies, *IEEE Trans. Inform. Theory*, IT-47(4) 2001, April.

4. Basseville, M., Distance measures for signal processing and pattern recognition, *Signal Processing*, 18, 349–369, 1989.

5. Bazlamaçci, C. F., and Hindi, K., Minimum-weight spanning tree algorithms: A survey and empirical study, *Comput. Operations Res.*, 28, 767–785, 2001.

6. Beardwood, J., Halton, J. H., and Hammersley, J. M., The shortest path through many points, *Proc. Cambridge Philos. Soc.*, 55, 299–327, 1959.

7. Beirlant, J., Dudewicz, E. J., Györfi, L., and van der Meulen, E. C., Nonparametric entropy estimation: an overview, *Intern. J. Math. Stat. Sci.*, 6(1), 17–39, 1997, June.

8. Bentley, J. L., Multidimensional binary search trees in database applications, *IEEE Trans. Software Eng.*, SE-5(4), 333–340, 1979.

9. Bentley, J. L., Multidimensional binary search trees used for associative searching, *Commun. ACM*, 18(9), 509–517, 1975, September.

10. Bentley, J. L., and Friedman, J. H., Fast algorithms for constructing minimal spanning trees in coordinate spaces, *IEEE Trans. Comput.*, C-27(2), 97–105, 1978.

11. Butz, T., and Thiran, J., Affine registration with feature space mutual information, *Lecture Notes in Computer Science 2208: MICCAI*, 2001, pp. 549–556.

12. Ch'avez, E., Navarro, G., Baeza-Yates, R., and Marroqu'in, J.L., Searching in metric spaces, *ACM Comput. Surv.*, 33(3), 273–321, 2001, September.

13. Choi, Y., and Lee, S., Injectivity conditions of 2D and 3D uniform cubic B-spline functions, *Graph. Models*, 62, 411–427, 2000.

14. Clarkson, K. L., An algorithm for geometric minimum spanning trees requiring nearly linear expected time, *Algorithmica*, 4, 461–469, 1989.

15. Cloude, S. R., and Pottier, E., An entropy based classification scheme for land applications of polarimetric SAR, *IEEE Trans. Geosci. Remot. Sens.*, 75, 68–78, 1997.

16. Cocosco, C. A., Kollokian, V., Kwan, R. K. S., and Evans, A. C., Brainweb: online interface to a 3D MRI simulated brain database, *NeuroImage*, 5(4) 1997.

17. Cristiani, N., and Shaw-Taylor, J., *Support Vector Machines and other Kernel-based Learning Methods*, Cambridge University Press, Cambridge, 2000.

18. Csisz'ar, I., Information-type measures of divergence of probability distributions and indirect observations, *Studia Sci. Math. Hung.*, 2, 299–318, 1967.

19. de Bonet, J. S., and Viola, P., Structure driven image database retrieval, Advances in Neural Information Processing, Vol. 10, 1997.

20. Dembo, A., and Zeitouni, O., *Large Deviations Techniques and Applications*, Springer-Verlag, New York, 1998.

21. Do, M. N., and Vetterli, M., Texture similarity measurement using Kullback–Liebler distance on wavelet subbands, pp. 367–370. In *IEEE International Conference on Image Processing*, Vancouver, BC, 2000.

22. Dunn, D., Higgins, W., and Wakeley, J., Texture segmentation using 2d gabor elementary functions, *IEEE Trans. Pattern Anal. Mach. Intelligence*, 16(2), 130–149, 1994.

23. Equinox Corporation. Human identification at distance project.

24. Erdi, Y., Rosenzweig, K., Erdi, A., Macapinlac, H., Hu, Y., Braban, L., Humm, J., Squire, O., Chui, C., Larson, S., and Yorke, E., Radiotherapy treatment planning for patients with non-small cell lung cancer using pet, *Radiother. Oncol.*, 62(1), 51–60, 2002.

25. Erdogmus, V., Prncipe, J., and Vielva, L., Blind deconvolution with minimum Rényi entropy, *EUSIPCO*, Toulouse, France, 2002.

26. Frieden, B. R., and Bajkova, A. T., Reconstruction of complex signals using minimum Rényi information. In *Proceedings of Meeting of International Society for Optical Engineering (SPIE)*, Vol. 2298, 1994.

27. Friedman, J. H., and Rafsky, L. C., Multivariate generalizations of the Wald-Wolfowitz and Smirnov two-sample tests, *Ann. Stat.*, 7(4), 697–717, 1979.

28. Gilles, S., Description and Experimentation of Image Matching Using Mutual Information, Technical Report, Oxford University, New York, 1996; www-rocq.inria.fr/~gilles/IMMMI/mutual_info.ps.gz.

29. He, Y., Hamza, A. B., and Krim, H., An information divergence measure for ISAR image registration, pp. 130–131, In *Workshop on Statistical Signal Processing*, Singapore, 2001, August.

30. Henze, N., and Penrose, M., On the multivariate runs test, *Ann. Stat.*, 27, 290–298, 1999.

31. Hero, A. O., Costa, J., and Ma, B., Asymptotic Relations Between Minimal Graphs and Alpha Entropy, Technical Report 334, March, Comm. and Sig. Proc. Lab. (CSPL), Dept. of EECS, University of Michigan, Ann Arbor, 2003; www.eecs.umich.edu/~hero/det_est.html.

32. Hero, O., Ma, B., and Michel. O., Imaging applications of stochastic minimal graphs. In *IEEE International Conference on Image Processing*, Thessaloniki, Greece, 2001, October.

33. Hero, O., Ma, B., Michel, O., and Gorman, J. D., Alpha-divergence for Classification, Indexing and Retrieval, Technical Report 328, Comm. and Sig. Proc. Lab. (CSPL), Dept. EECS, University of Michigan, Ann Arbor, 2001, July, www.eecs.umich.edu/~hero/det_est.html.

34. Hero, A. O., Costa, J., and Ma, B., Convergence rates of minimal graphs with random vertices, *IEEE Trans. Inform. Theory*, 2002, submitted for publication.

35. Hero, A. O., Ma, B., Michel, O., and Gorman, J., Applications of entropic spanning graphs, *IEEE Signal Proc. Mag.*, 19(5), 85–95, 2002, September.

36. Hero, A. O., and Michel, O., Asymptotic theory of greedy approximations to minimal k-point random graphs, *IEEE Trans. Inform. Theory*, IT-45(6), 1921–1939, 1999, September.

37. Hill, D., Batchelor, P., Holden, M., and Hawkes, D., Medical image registration, *Phys. Med. Biol.*, 26, R1–R45, 2001.

38. Hoffman, R., and Jain, A. K., A test of randomness based on the minimal spanning tree, *Pattern Recognition Letters*, 1, 175–180, 1983.

39. Huang, J., Kumar, S., Mitra, M., and Zhu., W., Spatial color indexing and applications, pp. 602–608. In *Proceedings of IEEE International Conference on Computer Vision ICCV'98*.

40. Hyvärinen, A., Fast ICA Code. www.cis.hut.fi/projects/ica/fastica/.

41. Hyvärinen, A., and Oja, E., Independent component analysis: algorithms and applications, *Neural Networks*, 13(4–5), 411–430, 1999.

42. Jenkinson, M., Bannister, P., Brady, M., and Smith, S., Improved Methods for the Registration and Motion Correction of Brain Images, Technical Report, Oxford University, New York, 2002.

43. Johnson, K., Cole-Rhodes, A., Zavorin, I., and Le Moigne, J., Multi-resolution image registration of remotely sensed imagery using mutual information. In *Proceedings of SPIE OE/Aerospace Sensing, Wavelet Applications VIII*, Orlando, FL, 2001.

44. Kieu, T., and Viola, P., Boosting image retrieval. In *IEEE Conference on Computer Vision and Pattern Recognition*, 2000.

45. Kruskal, J. B., On the shortest spanning subtree of a graph and the traveling salesman problem, *Proc. Amer. Math. Soc.*, 7, 48–50, 1956.

46. Kullback, S., and Leibler, R. A., On information and sufficiency, *Ann. Math. Statist.*, 22, 79–86, 1951.

47. Lef'ebure, M., and Cohen, L., Image registration, optical flow and local rigidity, *J. Math. Imaging Vis.*, 14(2), 131–147, 2001.

48. Leventon, M. E., and Grimson, W. E. L., Multi-modal Volume Registration using Joint Intensity Distributions, Technical Report, MIT AI Laboratory, 1998; www.ai.mit.edu/projects/vision-surgery.

49. Lewicki, M., and Olshausen, B., Probabilistic framework for the adaptation and comparison of image codes, *J. Opt. Soc. Am.*, 16(7), 1587–1601, 1999.

50. Ma, W., and Manjunath, B., Netra: A toolbox for navigating large image databased, pp. 568–571. In *Proceedings of IEEE International Conference Image Processing*, Vol. 1. 1997.

51. Maes, F., Vandermeulen, D., and Suetens, P., Medical image registration using mutual information, *Proc. IEEE*, 91(10), 1699–1722, 2003.

52. Maintz, J. B., and Viergever, M., A survey of medical image registration, *Med. Image Anal.*, 2(1), 1–36, 1998.

53. Meyer, R., Boes, J. L., Kim, B., Bland, P. H., Zasadny, K. R., Kison, P. V., Koral, K. F., Frey, K. A., and Wahl, R. L., Demonstration of accuracy and clinical versatility of mutual information for automatic multimodality image fusion using affine and thin-plate spline warped geometric deformations, *Med. Image Anal.*, 1(3), 195–206, 1997, April.

54. Michel, O., Baraniuk, R., and Flandrin, P., Time-frequency based distance and divergence measures, pp. 64–67. In *IEEE International Time-Frequency and Time-Scale Analysis Symposium*, 1994, October.

55. Mount, D. M., and Arya, S., Approximate Nearest Neighbor Code. http://www.cs.umd.edu/~mount/ANN.

56. Narasimhan, G., Zhu, J., and Zachariasen, M., Experiments with computing geometric minimum spanning trees, pp. 183–196. In *Proceedings of Second Workshop on Algorithm Engineering and Experiments*, 2000.

57. NASA Visible Earth internet site.

58. Neemuchwala, H., Hero, A., and Carson, P., Image registration using entropic graph matching criteria. In *Proceedings of Asilomar Conference*, Monterey, CA, 2002, November.

59. Neemuchwala, H., Hero, A., Carson, P., and Meyer, C., Local feature matching using entropic graphs. In *Proceedings of the IEEE International Symposium on Biomedical Imaging*.

60. Neemuchwala, H., Hero, A. O., and Carson, P., Image matching using alpha-entropy measures and entropic graphs. *European Journal of Signal Processing, Special Issue on Content based image retrieval*, Accepted, 2004.

61. Neemuchwala, H., Hero, A. O., and Carson, P., Feature coincidence trees for registration of ultrasound breast images. In *IEEE International Conference on Image Processing*, Thessaloniki, Greece, 2001, October.

62. Nene, S. A., and Nayar, S. K., A simple algorithm for nearest neighbor search in high dimensions, *IEEE Trans. Pattern Anal. Mach. Intell.*, 19, 1997.

63. Nobel, A. B., and Olshen, R. A., Termination and continuity of greedy growing for tree-structured vector quantizers, *IEEE Trans. Inform. Theory*, IT-42(1), 191–205, 1996.

64. Olshausen, B. A., *Sparse Codes and Spikes*, MIT Press, Cambridge, 2001.

65. Park, H., and Meyer, C., Grid refinement in adaptive non-rigid registration, *Lecture Notes in Computer Sciences*, vol. 2879, pp. 796–803, 2003.

66. Penney, G. P., Weese, J., Little, J., Hill, D., and Hawkes, D., A comparison of similarity measures for used in 2-D-3-D medical image registration, *IEEE Trans. Med. Imag.*, 17(4), 586–595, 1998.

67. Prim, R. C., Shortest connection networks and some generalizations, *Bell Syst. Tech. Journ.*, 36, 1389–1401, 1957.

68. Project Atlanta.

69. Rangarajan, A., Hsiao, I.-T., and Gindi, G., Integrating anatomical priors in ect reconstruction via joint mixtures and mutual information. In *IEEE Medical Imaging Conference and Symposium on Nuclear Science*, Vol. III, 1998, October.

70. Redmond, C., and Yukich, J. E., Asymptotics for Euclidean functionals with power weighted edges, *Stochastic Processes and their Applications*, 6, 289–304, 1996.

71. R'enyi, A., On measures of entropy and information, pp. 547–561. In *Proceedings of 4th Berkeley Symposium on Mathematical Statistics and Probability*, Vol. 1, 1961.

72. Rohlfing, T., West, J., Beier, J., Liebig, T., Tachner, C., and Thornalc, U., Registration of functional and anatomical mri: accuracy, assessment and applications in navigated neurosurgery, *Comput. Aided Surg.*, 5(6), 414–425, 2000.

73. Rueckert, D., Clarkson, M. J., Hill, D. L. G., and Hawkes, D. J., Non-rigid registration using higher-order mutual information. In Proc. SPIE Medical Imaging 2000. Image Processing, San Diego, CA, 2000, pp. 438–447.

74. Srivastava, A., Lee, A. B., Simoncelli, E. P., and Zhu, S. C., On advances in statistical modeling of natural images, *J. Math. Imaging Vis.*, 18(1) 2003, January.

75. Steele, J. M., Probability theory and combinatorial optimization, CBMF-NSF regional conferences in applied mathematics, Vol. 69, Society for Industrial and Applied Mathematics (SIAM), 1997.

76. Stoica, R., Zerubia, J., and Francos, J. M., Image retrieval and indexing: A hierarchical approach in computing the distance between textured images. In *IEEE International Conference Image Processing*, Chicago, 1998, October.

77. Stoica, R., Zerubia, J., and Francos, J. M., The two-dimensional world decomposition for segmentation and indexing in image libraries. In *Proc. IEEE Int. Conf. Acoust., Speech, Sig. Proc.*, Seattle, May, 1998.

78. Stone, H., Le Moigne, J., and McGuire, M., The translation sensitivity of wavelet-based registration, *IEEE Trans. Pattern Anal. Mach. Intel.*, 21(10), 1074–1081, 1999.

79. Taneja, I. J., New developments in generalized information measures, *Adv. Imag. Elect. Phys.*, 91, 135 1995.
80. Toga, A., *Brain Warping*, Academic Press, ISBN: 0126925356, 1999.
81. Vasconcelos, N. and Lippman, A., A Bayesian framework for content-based indexing and retrieval. In *IEEE Compression Conference*, Snowbird, Utah, 1998; nuno.www.media.mit.edu/people/nuno/.
82. Vasconcelos, N. and Lippman, A., Bayesian representations and learning mechanisms for content based image retrieval. In *SPIE Storage and Retrieval for Media Databases 2000*, San Jose, CA, nuno.www.media.mit.edu/people/nuno/.
83. Vasicek, O., A test for normality based on sample entropy, *J. Roy. Stat. Soc., Ser. B*, 38, 54–59, 1976.
84. Viola, P. and Wells, W. M. III., Alignment by maximization of mutual information, pp. 16–23, In *Proceedings of IEEE International Conference on Computer Vision*, Los Alamitos, CA, 1995, June.
85. Wald, W., and Wolfowitz, J., On a test whether two samples are from the same population, *Ann. Math. Statist.*, 11, 147, 1940.
86. Williams, W. J., Brown, M. L., and Hero, A. O., Uncertainty, information, and time-frequency distributions, pp. 144–156. In *Proc. Meet. Intl. Soc. Opt. Eng. (SPIE)*, Vol. 1566, 1991.
87. Wu, Y., Kanade, T., Li, C., and Cohn, J., Image registration using wavelet-based motion model, *Int. J. Comput. Vis.*, 38(2), 129–152, 2000.
88. Yukich, J. E., Probability theory of classical Euclidean optimization, *Lecture Notes in Mathematics*, Vol. 1675, Springer-Verlag, Berlin, 1998.

A1. APPENDIX

Here we give a derivation of the entropic graph estimators of α-GA Equation 6.25 and α-MI Equation 6.26 estimators. The derivations are given for equal numbers m and n of features from the two images but are easily generalized to unequal m, n. The derivation is based on a heuristic and thus the convergence properties are, at present, unknown.

First consider estimating $\alpha D_{GA}(f,g) = (\alpha - 1)^{-1} \log I_{GA}(f,g)$, where $I_{GA}(f,g)$ is the integral in Equation 6.12, by $\widehat{\alpha D_{GA}} = (\alpha - 1)^{-1} \log \widehat{I_{GA}}$ where:

$$\widehat{I_{GA}} = \frac{1}{2n} \sum_{i=1}^{2n} \left(\frac{\hat{f}^p(z_i)\hat{g}^q(z_i)}{\hat{h}(z_i)} \right)^{1-\alpha} \tag{6.33}$$

Here $\hat{h}(z)$ is an estimate of the common probability density function $pf(z) + qg(z)$ of the i.i.d. pooled unordered sample $\{Z_i\}_{i=1}^{2n} = \{O_i, X_i\}_{i=1}^{n}$ and $\hat{f}$, $\hat{g}$ are estimates of the common densities f, g of the i.i.d. samples $\{O_i\}_{i=1}^{n}$ and $\{X_i\}_{i=1}^{n}$, respectively. We assume that the support set of f, g, h is contained in a bounded region S of R^d. If $\hat{f}$, $\hat{g}$, $\hat{h}$ are consistent, that is, they converge (a.s.) as $n \to \infty$ to f, g, h then by the strong law of large numbers $\widehat{I_{GA}}$ converges (a.s) to

$$E[\widehat{I_{GA}}] = E\left[\left(\frac{f^p(z_i)g^q(z_i)}{h(z_i)} \right)^{1-\alpha} \right] = \int_S \left(\frac{f^p(z)g^q(z)}{h(z)} \right)^{1-\alpha} h(z)\mathrm{d}z \tag{6.34}$$

Taking the log of expression Equation 6.34 and dividing by $\alpha - 1$, we obtain $\alpha D_{\mathrm{GA}}(f, g)$ in Equation 6.12 so that $\widehat{\alpha D}_{\mathrm{GA}}$ is asymptotically unbiased and its variance goes to zero.

Next divide the samples $\{Z_i\}_{i=1}^{2n}$ into two disjoint sets of samples Z_{train} and test samples Z_{test}. Using the training sample construct the Voronoi partition density estimators

$$\hat{h}(z) = \frac{\mu(\Pi_z(z))}{\lambda(\Pi_z(z))}, \qquad \hat{f}(z) = \frac{\mu(\Pi_o(z))}{\lambda(\Pi_o(z))}, \qquad \hat{g}(z) = \frac{\mu(\Pi_x(z))}{\lambda(\Pi_x(z))} \qquad (6.35)$$

where $\Pi_Z(z), \Pi_O(z), \Pi_X(z)$ are the cells of the Voronoi partition of $S \in \mathbb{R}^d$ containing the point $z \in \mathbb{R}^d$ and constructed from training samples $Z_{\mathrm{train}} \equiv \{O_{\mathrm{train}}, X_{\mathrm{train}}\}, O_{\mathrm{train}}$ and X_{train} respectively using K-means or other algorithm. Here μ and λ are the (normalized) counting measure and Lebesgue measure respectively, that is, $\mu(\Pi)$ is the number of points in the set Π divided by the total number of points and $\lambda(\Pi)$ is the volume of the set Π. Let $\{K_z, K_o, K_x\}$ be the number of cells in the partitions $\{\Pi_z, \Pi_o, \Pi_x\}$ respectively and let n_{train} be the number of training samples. The Voronoi partition density estimators are asymptotically consistent as $k, n_{\mathrm{train}} \to \infty$ and $k/n_{\mathrm{train}} \to 0$, for $k \in \{K_z, K_o, K_x\}$.[63]

Therefore, under these conditions and defining $Z_i = Z_{\mathrm{test}}(i)$,

$$\widehat{\alpha D}_{\mathrm{GA}} = \frac{1}{\alpha - 1} \log\left(\frac{1}{n_{\mathrm{test}}} \sum_{i=1}^{n_{\mathrm{test}}} \left(\frac{\hat{f}^p(\tilde{z}_i)\hat{g}^q(\tilde{z}_i)}{\hat{h}(\tilde{z}_i)} \right)^{1-\alpha} \right) \qquad (6.36)$$

is an asymptotically consistent estimator.

Next consider the following plug-in estimator of α-MI: $\widehat{\alpha\mathrm{MI}} = (\alpha - 1)\log \hat{I}_{\mathrm{MI}}$, where

$$\hat{I}_{\mathrm{MI}} = \frac{1}{n} \sum_{i=1}^{n} \left(\frac{\hat{f}_o(o_i)\hat{f}_x(x_i)}{\hat{f}_{ox}(o_i, x_i)} \right)^{1-\alpha} \qquad (6.37)$$

and $\hat{f}_{ox}$ is an estimate of the joint density of the $2d$ dimensional vector $(O_i, X_i) \in \mathbb{R}^{2d}$. $\hat{f}_o$ and $\hat{f}_x$ are estimates of the density of O_i and X_i, respectively. Again, if $\hat{f}_o, \hat{f}_x$ and $\hat{f}_{ox}$ are consistent then it is easily shown that $\hat{I}_{\mathrm{MI}}$ converges to the integral in the expression Equation 6.5 for α-MI:

$$\int_{S \times S} \int \left(\frac{\hat{f}_o(u)\hat{f}_x(v)}{\hat{f}_{ox}(u, v)} \right)^{1-\alpha} f_{ox}(u, v)\mathrm{d}u \, \mathrm{d}v \qquad (6.38)$$

where S is a bounded set containing the support of densities f_o and f_x. Similarly, separating $\{(O_i, X_i)\}_{i=1}^{n}$ into training and test samples, we obtain an asymptotically consistent estimator:

$$\widehat{\alpha\mathrm{MI}} = \frac{1}{\alpha - 1} \log\left(\frac{1}{n_{\mathrm{test}}} \sum_{i=1}^{n_{\mathrm{test}}} \left(\frac{\hat{f}_o(\tilde{o}_i)\hat{f}_x(\tilde{x}_i)}{\hat{f}_{ox}(\tilde{o}_i, \tilde{x}_i)} \right)^{1-\alpha} \right) \qquad (6.39)$$

The entropic graph estimators in Equation 6.25 and Equation 6.26 are obtained by specializing to the case $n_{train} = 0$, in which case $\mu(\Pi_Z(z)) = (2n)^{-1}$, $\mu(\Pi_O(z)) = \mu(\Pi_X(z)) = \mu(\Pi_{O\times X}(z)) = n^{-1}$, and using the Voronoi cell volume approximations

$$\lambda(\Pi_Z(z_i)) \asymp e_i^d(z) = \min\{e_i^d(o), e_i^d(x)\}, \qquad \lambda(\Pi_X(z_i)) \asymp e_i^d(x) \qquad (6.40)$$

$$\lambda(\Pi_Y(z_i)) \asymp e_i^d(o), \qquad \lambda(\Pi_{O\times X}(z_i)) \asymp e_i^{2d}(o \times x) \qquad (6.41)$$

where $\asymp$ denotes "proportional to" and $e_i(o)$, $e_i(x)$, $e_i(o \times x)$ are the nearest neighbor distances defined in Section III.B.

7 Fusion of Images from Different Electro-Optical Sensing Modalities for Surveillance and Navigation Tasks

Alexander Toet

CONTENTS

I. INTRODUCTION

Modern night-time cameras are designed to expand the conditions under which humans can operate. A functional piece of equipment must therefore provide an image that leads to good perceptual awareness in most environmental and operational conditions (to "own the weather," or "own the night"). The two most common night-time imaging systems either display emitted infrared (IR) radiation or reflected light, and thus provide complimentary information of the inspected scene. IR cameras have a history of decades of development. Although modern IR cameras function very well under most circumstances, they still have some inherent limitations. For instance, after a period of extensive cooling (e.g., after a long period of rain) the IR bands provide less detailed information due to low thermal contrast in the scene, whereas the visual bands may represent the background in great detail (vegetation or soil areas, texture). In this situation it can be hard or even impossible to distinguish the background of a target in the scene, using only the IR bands, whereas at the same time, the target itself may be highly detectable (when its temperature differs sufficiently from the mean temperature of its local background). On the other hand, a target that is well camouflaged for visual detection will be hard (or even impossible) to detect in the visual bands, whereas it can still be detectable in the thermal bands. A combination of visible and thermal imagery may then allow both the detection and the unambiguous localization of the target (represented in the thermal image) with respect to its background (represented in the visual image). A human operator using a suitably combined or fused representation of IR and (intensified) visual imagery may therefore be able to construct a more complete mental representation of the perceived scene, resulting in a larger degree of situational awareness.[1] In addition, a false color representation of fused night-time imagery that closely resembles a natural daylight color scheme[2] will help the observer by making scene interpretation more intuitive.

The rapid development of multiband IR and visual night vision systems has led to an increased interest in color fused ergonomic representations of multiple sensor signals.[3–15] Simply mapping multiple spectral bands of imagery into a three-dimensional color space already generates an immediate benefit, since the human eye can discern several thousand colors, while it can only distinguish about 100 shades of gray at any instance. Combining bands in color space therefore provides a method to increase the dynamic range of a sensor system.[16] Experiments have convincingly demonstrated that appropriately designed false color rendering of night-time imagery can significantly improve observer performance and reaction times in tasks that involve scene segmentation and classification.[6,10,17–20] However, inappropriate color mappings may hinder situational awareness.[10,19,21] One of the main reasons seems to be the counter intuitive appearance of scenes rendered in artificial color schemes and the lack of color constancy.[10] Hence, an ergonomic color scheme should produce night vision imagery with a natural appearance and with colors that are invariant for changes in the environmental conditions.[2]

In this chapter, we describe the results of several studies that we performed to test the complementarity of information, obtained from different types of night vision systems (IR and image intensifiers). We also tested the capability of several gray-level and color image fusion schemes to combine and convey information originating from different night vision imaging modalities, about both the global structure and the fine detail of scenes, for use in surveillance and navigation tasks performed by human observers. The structure of this chapter is as follows. First, we will present an experiment that was performed to test the effects of two gray-level and color image fusion schemes on the accuracy with which observers can localize a target while performing a military surveillance task. Second, we will discuss an experiment in which we investigated the merits of a gray-level image fusion method and two different color image fusion schemes for the recognition of detail and for situational awareness. Finally, we will discuss the general findings of these studies.

II. LOCALIZATION ACCURACY

A. IMAGERY

Visible light images were recorded using a sensitive CCD camera. IR images were obtained with a thermal focal plane array camera, operating in the 3 to 5 μm (midrange) band. The cameras were mounted on a common frame. The visual and thermal images were spatially registered as closely as possible, using a second order affine digital warping procedure to map corresponding points in the scene to corresponding pixels in the image plane. For full details about the equipment and the registration procedures we refer to Ref. 1.

The individual images used in this study correspond to successive frames of a time sequence. The time sequences represent three different scenarios, provided by the Royal Dutch Army. They simulate typical surveillance tasks and were chosen because of their military relevance.

Scenario I corresponds to the guarding of a UN camp, and involves monitoring a fence that encloses a military asset. To distinguish innocent passers-by from individuals planning to perform subversive actions, the guard must be able to determine the exact position of a person in the scene at any time. During the image acquisition period the fence is clearly visible in the visual (CCD) image. In the thermal (IR) image however, the fence is merely represented by a vague haze. A person (walking along the fence) is clearly visible in the IR image but can hardly be distinguished in the CCD image. In the fused images both the fence and the person are clearly visible. An observer's situational awareness can therefore be tested by asking the subject to report the position of the person relative to the fence.

Scenario II corresponds to guarding a temporary base. Only a small section of the dune-like terrain is visible, the rest is occluded by trees. The assignment of the guard is to detect and counter infiltration attempts in a very early stage. During the registration period the trees appear larger in the IR image than they really are because they have nearly the same temperature as their local background. In the CCD image however, the contours of the trees are correctly represented. A person (crossing the interval between the trees) is clearly visible in the IR image but is represented with low contrast in the CCD image. In the fused images both the outlines of the trees and the person are clearly visible. As a result, it is difficult to determine the position of the person relative to the trees using either the CCD or the IR images. The fused images correctly represent both the contours of the trees and the person. An observer's situational awareness can therefore be tested by asking the subject to report the position of the person relative to the midpoint of the interval delineated by the contours of the trees that are positioned on both sides of the person.

Scenario III corresponds to the surveillance of a large area. The scene represents a dune landscape, covered with semishrubs and sandy paths. The assignment of the guard is to detect any attempt to infiltrate a certain area. During the registration period the sandy paths in the dune area have nearly the same temperature as their local background, and are therefore represented with very low contrast in the IR image. In the CCD image however, the paths are depicted with high contrast. A person (walking along a trajectory that intersects the sandy path) is clearly visible in the IR image but is represented with less contrast in the CCD image. In the fused images both the outlines of the paths and the person are clearly visible. It is difficult (or even impossible) to determine the position of the person relative to the sandy path he is crossing from either the IR or the CCD images. An observer's situational awareness can therefore be tested by asking the subject to report the position of the person relative to the sandy path.

We used schematic (cartoon-like) reference images of the actual scenes to obtain a baseline performance and to register the observer responses. These schematic images were constructed from the visual images by:

1. Applying standard image processing techniques like histogram equalization and contrast stretching to enhance the representation of the reference contours in the original visual images,
2. Drawing the contours of the reference features (visually judged) on a graphical overlay on the contrast enhanced visual images, and
3. Filling the contours with a homogeneous gray-level value.

The images thus created represent segmented versions of the visual images.

B. Fusion Methods

The computational image fusion methodology was developed at the MIT Lincoln Laboratory[11–13,15,22,23] and derives from biological models of color vision[24–27] for fusion of visible light and IR radiation.[28,29]

In the case of color vision in monkeys and man, retinal cone sensitivities are broad and overlapping, but the images are contrast enhanced *within* bands by spatial opponent-processing (*via* cone–horizontal–bipolar cell interactions) creating both on and off center-surround response channels.[27] These signals are then contrast enhanced *between* bands *via* interactions among bipolar, sustained amacrine, and single-opponent color ganglion cells,[30,31] all within the retina.

Fusion of visible and thermal IR imagery has been observed in the optic tectum of rattlesnakes and pythons.[28,29] These neurons display interactions in which one modality (e.g., IR) can enhance or depress the response to the other sensing modality (e.g., visible) in a strongly nonlinear fashion. Such interactions resemble opponent-processing between bands as observed in primate retina.

For opaque surfaces in thermodynamic equilibrium, spectral reflectivity ρ and emissivity ϵ are linearly related at each wavelength λ: $\rho(\lambda) = 1 - \varepsilon(\lambda)$. This provides a rationale for the use of both on-center and off-center channels when treating IR imagery as characterized by thermal emissivity.

In the color image fusion methodology the individual input images are first enhanced by filtering them with a feedforward center-surround shunting neural network.[32] This operation serves to:

1. Enhance spatial contrast in the individual visible and IR bands,
2. Create both positive and negative polarity IR contrast images, and
3. Create two types of single-opponent color contrast images.

The resulting single-opponent color contrast images represent grayscale fused images that are analogous to the IR-depressed visual and IR-enhanced visual cells of the rattlesnake.[28,29]

To obtain a natural color representation of these single-opponent images (each being an 8-bit grayscale image) we have developed two alternative methodologies to choose from, based on the relative resolution of the visible and IR images.

In the case where the IR camera is of significantly lower resolution (i.e., half that of the visible camera), the enhanced visible is assigned to the green channel, the difference signal of the enhanced visible and IR images is assigned to the blue channel, and the sum of the visible and IR images is assigned to the red channel of an RGB display:[11–13,15,22]

$$\begin{pmatrix} R \\ G \\ B \end{pmatrix} = \begin{pmatrix} CCD^+ + IR^+ \\ CCD^+ \\ CCD^+ - IR^+ \end{pmatrix} \tag{7.1}$$

where the $(\ldots)^+$ indicates the center-surround operation. These channels correspond with our natural associations of warm (red) and cool (blue).

In the case where the IR and visible cameras are of comparable resolution, the enhanced sum of the enhanced visible and IR images is assigned to the green channel, the difference signal of the enhanced visible and IR images is assigned to the blue channel, and the difference of the enhanced IR and visible images is assigned to the red channel of an RGB display:[5]

$$\begin{pmatrix} R \\ G \\ B \end{pmatrix} = \begin{pmatrix} IR^+ - CCD^+ \\ (CCD^+ + IR^+)^+ \\ CCD^+ - IR^+ \end{pmatrix} \tag{7.2}$$

These transformations are followed by hue and saturation remapping. The result is a more naturally appearing color fused representation.

Because the resolution of the cameras used for our experiments are comparable, this is the methodology chosen to process the imagery used here.

The grayscale fused images are produced by taking the luminance component of the corresponding color fused images.

C. TEST METHODS

A computer was used to present the images on a CRT display. Each image presentation was followed by the presentation of a corresponding reference image. The subject's task was to assess from each presented image the position of the person in the scene relative to the reference features. The images were shown only briefly (for 1 sec). Viewing was binocular. The experiments were performed in a dimly lit room. A total of six participants, aged between 20 and 30 years, served in the experiments reported below. For full details on the experimental setup and procedures we refer to Ref. 1.

In Scenario I the reference features are the poles that support the fence. These poles are clearly visible in the CCD images (Figure 7.1a) but not represented in the IR images (Figure 7.1b) because they have almost the same temperature as the surrounding terrain. In the (gray-level and color) fused images (Figure 7.1c and Figure 7.1d) the poles are again clearly visible.

In Scenario II the outlines of the trees serve to delineate the reference interval. The contours of the trees are correctly represented in the CCD images (Figure 7.2a). However, in the IR images (Figure 7.2b) the trees appear larger than their actual physical size because they almost have the same temperature as the surrounding soil. As a result, the scene is incorrectly segmented after quantization and it is not possible to perceive the correct borders of the area between the trees. In the (gray-level and color) fused images (Figure 7.2c and d) the outlines of the trees are again correctly represented and clearly visible.

In Scenario III the area of the small and winding sandy path provides a reference contour for the task at hand. This path is represented at high contrast in the CCD images (Figure 7.3a), but it is not represented in the IR images (Figure 7.3b) because it has the same temperature as the surrounding soil. In the (gray-level and color) fused images (Figure 7.3c and d) the path is again clearly visible.

For each scenario a total of nine frames is used in the experiment. In each frame the person is at a different location relative to the reference features. These different locations are equally distributed relative to the reference features. Each stimulus is presented for 1 sec, centered on the midpoint of the screen, preceded by a blank screen with an awareness message, which is presented for 1 sec. A schematical representation of the reference features is shown immediately after each stimulus presentation. The position of the center of the reference image is randomly displaced around the center of the screen between presentations to ensure that subjects cannot use prior presentations as a frame of reference for detection and localization. A complete run consists of 135 presentations (5 image modalities $\times$ 3 scenarios $\times$ 9 frames per scenario), and typically lasts about 1 h.

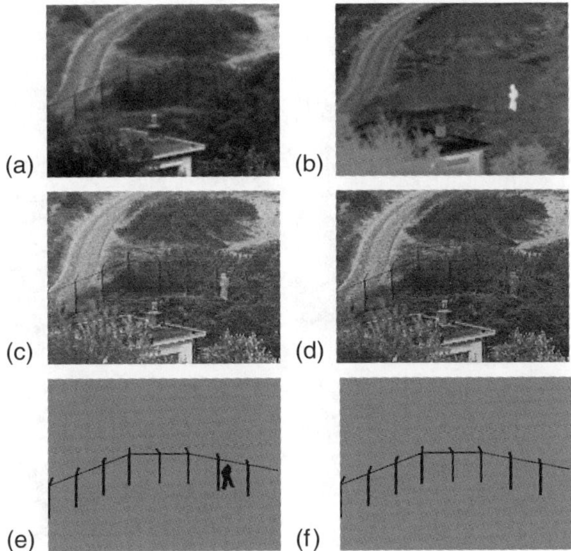

FIGURE 7.1 (See color insert following page 236) Original intensified visual image (II), original thermal image (IR), gray-level fused (GF) image, color fused (CF) image, baseline test image (baseline), and reference (reference) image of Scenario I.

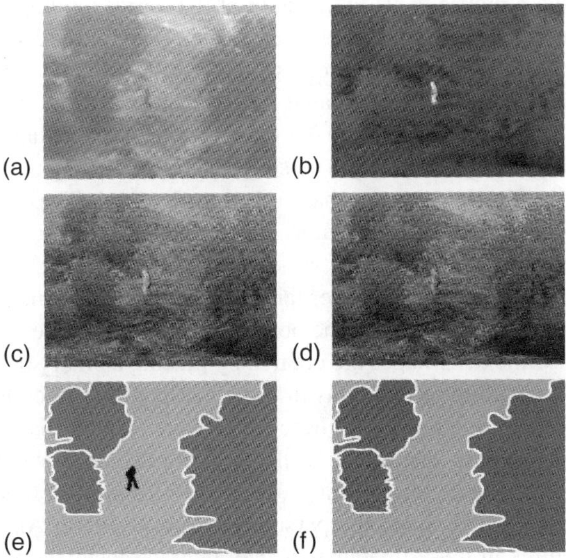

FIGURE 7.2 (See color insert) Original intensified visual image (II), original thermal image (IR), gray-level fused (GF) image, color fused (CF) image, baseline test image (baseline), and reference (reference) image for Scenario II.

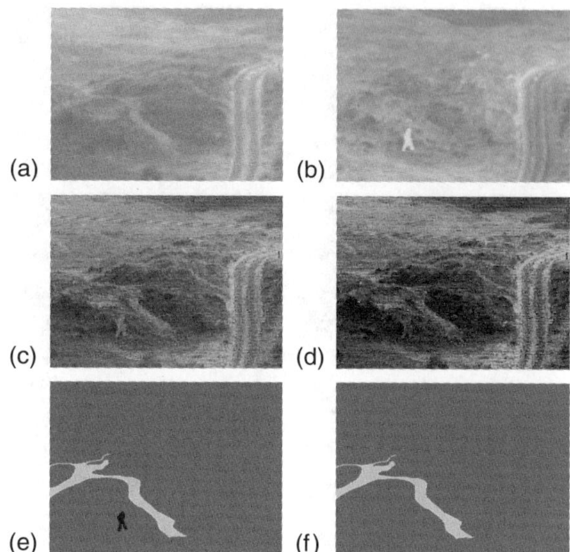

FIGURE 7.3 (See color insert) Original intensified visual image (II), original thermal image (IR), gray-level fused (GF) image, color fused (CF) image, baseline test image (baseline), and reference (reference) image for Scenario III.

The subject's task is to indicate the perceived location of the person in the scene by placing a mouse controlled cursor at the corresponding location in this schematical drawing. The subject has in principle unlimited time to reach a decision. When the left mouse button is pressed the computer registers the co-ordinates corresponding to the indicated image location (the mouse co-ordinates) and computes the distance in the image plane between the actual position of the person and the indicated location. The subject presses the right mouse button if the person in the displayed scene has not been detected. The subject can only perform the localization task by memorizing the perceived position of the person relative to the reference features.

The schematic reference images are also used to determine the optimal (baseline) localization accuracy of the observers. For each of the three scenarios a total of nine baseline test images are created by placing a binary (dark) image of a walking person at different locations in the corresponding reference scene. The different locations of the person in these images are equally distributed over the entire reference interval. The image of the walking person was extracted from a thresholded and inverted thermal image. In the resulting set of schematic images both the reference features and the person are highly visible. Also, there are no distracting features in these images that may degrade localization performance. Therefore, observer performance for these schematic test images should be optimal and may serve as a baseline to compare performance obtained with the other image modalities.

A complete run consists of 162 presentations (6 image modalities × 3 scenarios × 9 frames per scenario) in random order, and typically lasts about 1 h.

D. RESULTS

Figure 7.4 shows that subjects are uncertain about the location of the person in the scene for about 26% of the visual image presentations, and 23% of the thermal image presentations. The (gray-level and color) fused images result in a smaller fraction of about 13% "not sure" replies. The lowest number of "not sure" replies is obtained for the baseline reference images: only about 4%. This indicates that the increased amount of detail in fused imagery does indeed improve an observer's subjective situational awareness.

Figure 7.5 shows the mean weighted distance between the actual position of the person in each scene and the position indicated by the subjects (the perceived position), for the visual (CCD) and thermal (IR) images, and for the gray-level and color fusion schemes. This figure also shows the optimal (baseline) performance obtained for the schematic test images representing only the segmented reference features and the walking person. A low value of this mean weighted distance measure corresponds to a high observer-accuracy and a correctly perceived position of the person in the displayed scenes relative to the main reference features. High values correspond to a large discrepancy between the perceived position and the actual position of the person.

In all scenarios the person was at approximately 300 m distance from the viewing location. At this distance one pixel corresponds to 11.4 cm in the field.

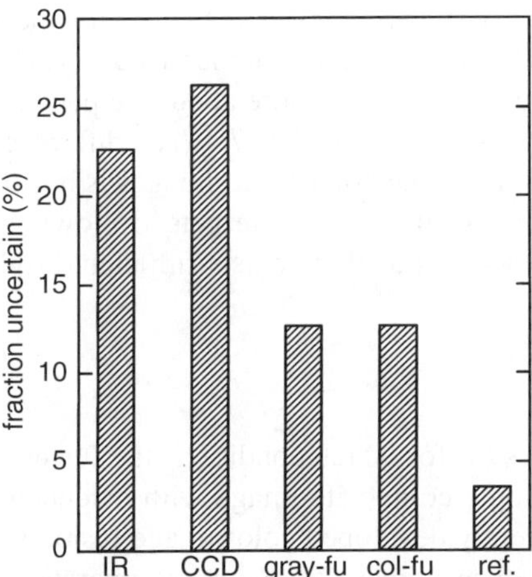

FIGURE 7.4 Percentage of image presentations in which observers are uncertain about the relative position of the person in the scene, for each of the five image modalities tested (IR, intensified CCD, gray-level fused, color fused, and schematical reference images).

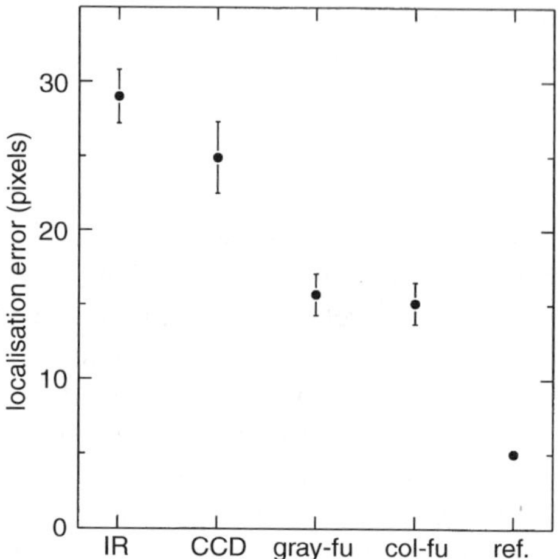

FIGURE 7.5 The mean weighted distance between the actual position of the person in the scene and the perceived position for each of the five image modalities tested (IR, intensified CCD, gray-level fused, color fused, and schematical reference images). The error bars indicate the size of the standard error in the perceived location.

Figure 7.5 shows that the localization error obtained with the fused images is significantly lower than the error obtained with the individual thermal and visual image modalities ($p = .0021$). The smallest errors in the relative spatial localization task are obtained for the schematic images. This result represents the baseline performance, since the images are optimal in the sense that they do not contain any distracting details and all the features that are essential to perform the task (i.e., the outlines of the reference features) are represented at high visual contrast. The lowest overall accuracy is achieved for the thermal images. The visual images appear to yield a slightly higher accuracy. However, this accuracy is misleading since observers are not sure about the person in a large percentage of the visual images, as shown by Figure 7.4. The difference between the results for the gray-level fused and the color fused images is not significant ($p = .134$), suggesting that spatial localization of targets (following detection) does not exploit color contrast as long as there exists sufficient brightness contrast in the gray fused imagery.

E. DISCUSSION

This study investigates (a) for which conditions the fusion of visual and thermal images results in a single composite image with extended information content, and (b) whether a recently developed color image fusion scheme[11,13,15,22,23] can enhance the situational awareness of observers operating under these specific conditions and using visual and thermal images.

Conditions in which fusion of visual and thermal imagery are most likely to result in images with increased information content occur around sunrise. At this time the contrast of both the visual and the thermal images is very low. One can construct other scenarios involving night operations in which both modalities are lacking in contrast. The visual contrast is low around sunrise because of the low luminance of the sky. However, contours of extended objects are still visible. After some image enhancement (e.g., center-surround shunting, histogram equalization, or contrast stretching) even an appreciable amount of detail can be perceived. Small objects with low reflectance, such as, a person wearing a dark suit or camouflage clothing, or objects that are partly obscured, are not represented in the visual image under these conditions, and can therefore not be detected. The thermal contrast is low around sunrise because most of the objects in the scene have about the same temperature after losing their excess heat by radiation during the night. As a result, the contours of extended objects are not at all or incorrectly represented in the thermal image. The fusion of images registered around sunrise should therefore result in images that represent both the context (the outlines of extended objects) and the details with a large thermal contrast (e.g., humans) in a single composite image. To test this hypothesis a large set of image sequences is captured around sunrise on different days. The scenes used in this study represent three somewhat different scenarios that were provided by the Royal Dutch Army. The images are fused using the recently developed MIT color fusion scheme.[11,13,15,22,23] Gray-level fused images are also produced by taking the luminance component of the color fused images. Visual inspection of the results shows that the fusion of thermal and visual images indeed results in composite images with an increased amount of information.

An observer experiment is performed to test if the increased amount of detail in the fused images can yield an improved observer performance in a task that requires a certain amount of situational awareness. The task that is devised involves the detection and localization of a person in the displayed scene, relative to some characteristic details that provide the spatial context. The person is optimally represented in the thermal imagery and the reference features are better represented in the visual imagery. The hypothesis is therefore that the fused images provide a better representation of the overall spatial structure of the depicted scene. To test this hypothesis, subjects perform a relative spatial localization task with a selection of thermal, visual, and (both gray-level and color) fused images representing the above-mentioned military scenarios. The results show that observers can indeed determine the relative location of a person in a scene with a significantly higher accuracy when they perform with fused images, compared to the individual image modalities.

This study shows no significant difference between the localization performance with color fused images and with their luminance components (the derived gray-level fused images). However, color fused images are easier to visually segment than gray-level fused images. As a result, color coding may greatly improve the speed and accuracy of information uptake,[33] and fewer fixation may be required to locate color coded targets.[34] Therefore, dynamic tasks like navigation and orienting, that depend upon a quick and correct scene segmentation, may benefit from a color fused image representation. Also, it is likely that the fusion of thermal and low-light level imagery may yield an even better observer performance over extended exposure times which often lead to exhaustion or distraction.

III. SITUATIONAL AWARENESS

A. IMAGERY

A variety of outdoor scenes, displaying several kinds of vegetation (grass, heather, semishrubs, trees), sky, water, sand, vehicles, roads, and humans, were registered at night with a recently developed dual-band visual intensified (DII) camera (see below), and with a state-of-the-art thermal middle wavelength band (3 to 5 μm) IR camera (Radiance HS). Both cameras had a field of view (FOV) of about 6 × 6 degrees. Some image examples are shown in Figure 7.6, Figure 7.7, Figure 7.8, Figure 7.9, and Figure 7.10.

The DII camera was developed by Thales Optronics and facilitated a two-color registration of the scene, applying two bands covering the part of the electromagnetic spectrum ranging from visual to near IR (400 to 900 nm). The crossover point between the bands of the DII camera lies approximately at 700 nm. The short (visual) wavelength part of the incoming spectrum is mapped to the R channel of an RGB color composite image. The long (near IR) wavelength band corresponds primarily to the spectral reflection characteristics of vegetation, and is therefore mapped to the G channel of an RGB color composite image. This approach utilizes the fact that the spectral reflection characteristics of plants are distinctly different from other (natural and artificial) materials in the visual and near IR range.[35] The spectral response of the long-wavelength channel ("G") roughly matches that of a Generation III image intensifier system. This channel is stored separately and used as an individual image modality (II).

Images were recorded at various times of the diurnal cycle under various atmospheric conditions (clear, rain, fog, etc.) and for various illumination levels (1 lx to 0.1 mlx). Object ranges up to several hundreds of meters were applied. The images were digitized on-site (using a Matrox Genesis frame grabber, using at least 1.8 times oversampling).

First, the recorded images were registered through an affine warping procedure, using fiducial registration points that were recorded at the beginning of each session. After warping, corresponding pixels in images taken with different cameras represent the same location in the recorded scene. Then, patches displaying different types of scenic elements were selected and cut out from corresponding images (i.e., images representing the same scene at the same instant in time, but taken with different cameras). These patches were deployed as stimuli in the pyschophysical tests. The signature of the target items (i.e., buildings, humans, vehicles, etc.) in the image test sets varied from highly distinct to hardly visible.

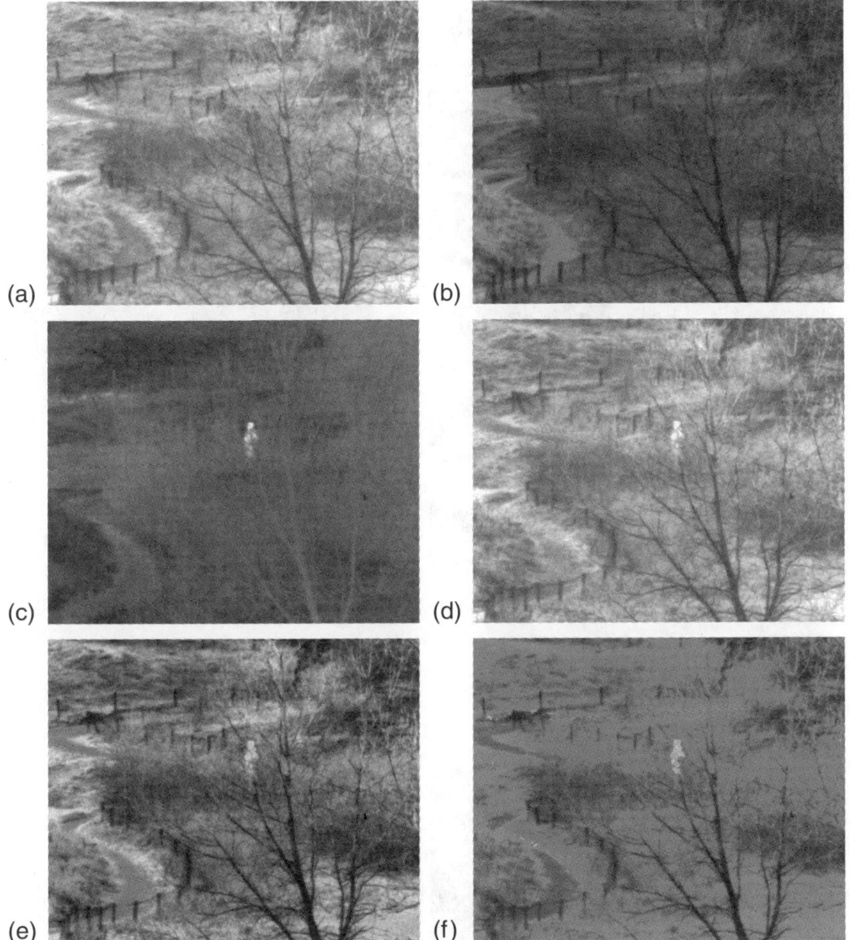

FIGURE 7.6 (See color insert) The different image modalities used in this study. II and DII: the long wavelength band and both bands of the false color intensified CCD image. IR: the thermal 3 to 5 μm IR image. GF: the gray-level fused image and CF1(2) and color fused images produced with Method 1(2). This image shows a scene of a person in terrain, behind a tree.

To test the *perception of detail*, patches were selected that displayed buildings, vehicles, water, roads, or humans. These patches are 280 × 280 pixels, corresponding to a FOV of 1.95 × 1.95°.

To investigate the *perception of global scene structure*, larger patches were selected, that represent either the horizon (to perform a horizon perception task), or a large amount of different terrain features (to enable the distinction between an image that is presented upright and one that is shown upside down). These patches are 575 × 475 pixels, corresponding to a FOV of 4.0 × 3.3°.

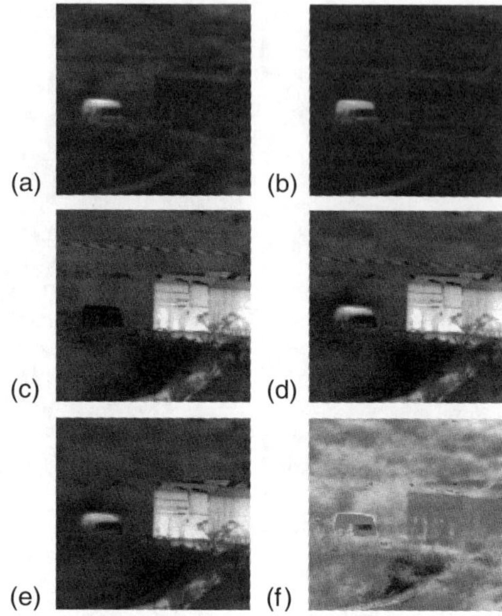

FIGURE 7.7 (See color insert) A scene displaying a road, a house, and a vehicle.

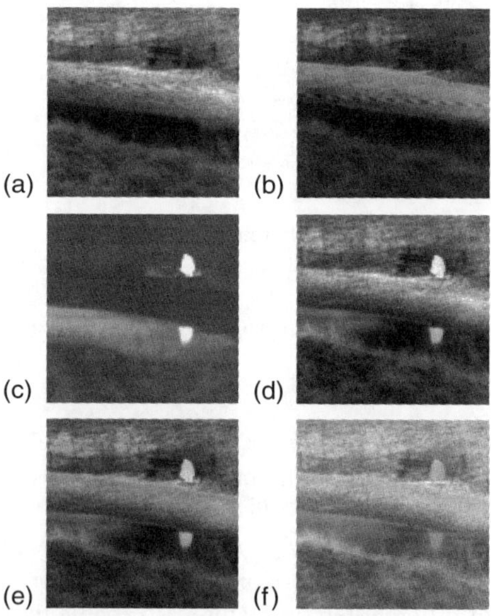

FIGURE 7.8 (See color insert) A scene displaying a person along a riverside. Notice the reflection of the person's silhouette on the water surface in the thermal image.

FIGURE 7.9 (See color insert) A scene displaying people on a road through the woods.

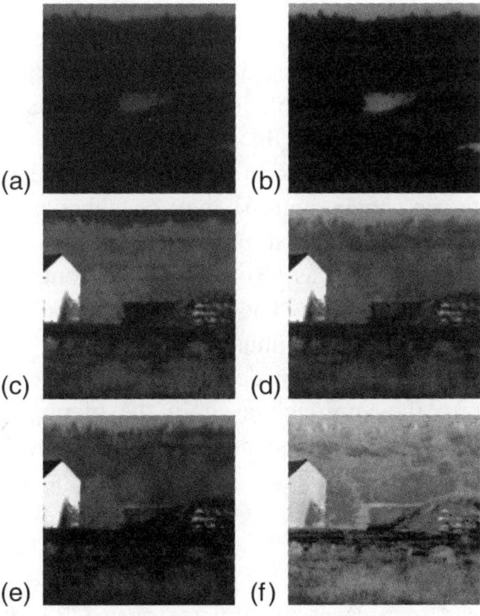

FIGURE 7.10 (See color insert) A scene displaying a house and trees.

B. FUSION METHODS

Grayscale fused (GF) images were produced by combining the IR and II images though a pyramidal image fusion scheme.[36–38] A seven-level Laplacian pyramid[36] was used, in combination with a maximum absolute contrast node (i.e., pattern element) selection rule.

Color fused imagery was produced by the following two methods:

– *Color fusion Method 1* (CF1): the short and long wavelength bands of the DII camera were, respectively, mapped to the R and G channels of an RGB color image. The resulting RGB color image was then converted to the YIQ (NTSC) color space. The luminance (Y) component was replaced by the corresponding aforementioned grayscale (II and IR) fused image, and the result was transformed back to the RGB color space (note that the input Y from combining the R and G channel is replaced by a Y which is created by fusing the G channel with the IR image). This color fusion method results in images in which grass, trees, and humans are displayed as greenish, and roads, buildings, and vehicles are brownish.
– *Color fusion Method 2* (CF2): first, an RGB color image was produced by assigning the IR image to the R channel, the long wavelength band of the DII image to the green channel (as in Method 1), and the short wavelength band of the DII image to the blue channel (instead of the red channel, as in Method 1). This color fusion method results in images in which vegetation is displayed as greenish, humans are reddish, buildings are red-brownish, vehicles are whitish/bluish, and the sky and roads are most often bluish.

The multiresolution grayscale image fusion scheme employed here, selects the perceptually most salient contrast details from both of the individual input image modalities, and fluently combines these pattern elements into a resulting (fused) image. As a side effect of this method, details in the resulting fused images can be displayed at higher contrast than they appear in the images from which they originate, i.e., their contrast may be enhanced.[39,40] To distinguish the perceptual effects from contrast enhancement from those of the fusion process, observer performance was also tested with contrast enhanced versions of the individual image modalities. The contrast in these images was enhanced by a multiresolution local contrast enhancement scheme. This scheme enhances the contrast of perceptually relevant details for a range of spatial scales, in a way that is similar to the approach used in the hierarchical fusion scheme. A detailed description of this enhancement method is given elsewhere.[39,40]

C. TEST METHODS

A computer was used to present the images on a CRT display, measure the response times, and collect the observer responses.

The perception of the global structure of a depicted scene was tested in two different ways. In the first test, scenes were presented that had been randomly mirrored along the horizontal, and the subjects were asked to distinguish the orientation of the displayed scenes (i.e., whether a scene was displayed right side up or upside down). In this test, each scene was presented twice: once upright and once upside down. In the second test, horizon views were presented together with short markers (55×4 pixels) on the left and right side of the image and on a virtual horizontal line. In this test, each scene was presented twice: once with the markers located at the true position (height) of the horizon, and once when the markers coincided with a horizontal structure that was opportunistically available (like a band of clouds) and that may be mistaken for the horizon. The task of the subjects was to judge whether the markers indicated the true position of the horizon. The perception of the global structure of a scene is likely to determine situational awareness.

The capability to discriminate fine detail was tested by asking the subjects to judge whether or not a presented scene contained an exemplar of a particular category of objects. The following categories were investigated: buildings, vehicles, water, roads, and humans. The perception of detail is relevant for tasks involving visual search, detection, and recognition.

The tests were blocked with respect to both (1) the imaging modality and (2) the task. This was done to minimize observer uncertainty, both with respect to the characteristics of the different image modalities, and with respect to the type of target.

Blocking by image modality yielded the following six classes of stimuli:

1. Grayscale images representing the thermal 3 to 5 μm IR camera signal.
2. Grayscale images representing the long-wavelength band (G-channel) of the DII images.
3. Color (R and G) images representing the two channels of the DII.
4. Grayscale images representing the IR and II signals fused by GF.
5. Color images representing the IR and DII signals fused by CF1.
6. Color images representing the IR and DII signals fused by CF2.

Blocking by task resulted in trial runs that tested the perception of global scene structure by asking the observers to judge whether:

- The horizon was veridically indicated
- The image was presented right side up

and the recognition of detail by asking the observers to judge whether the image contained an exemplar of one of the following categories:

- Building
- Person

- Road or path
- Fluid water (e.g., a ditch, a lake, a pond, or a puddle)
- Vehicle (e.g., a truck, car, or van)

The entire experiment consisted of 42 different trial runs (7 different image modalities × 7 different tasks). Each task was tested on 18 different scenes. The experiment therefore involved the presentation of 756 images in total. The order in which the image modalities and the tasks were tested was randomly distributed over the observers.

Before starting the actual experiment, the observers were shown examples of the different image modalities that were tested. They received verbal information, describing the characteristics of the particular image modality. It was explained how different types of targets are displayed in the different image modalities. This was done to familiarize the observers with the appearance of the scene content in the different image modalities, thereby minimizing their uncertainty.

Next, subjects were instructed that they were going to watch a sequence of briefly flashed images, and that they had to judge each image with respect to the task at hand. For a block of trials, testing the perception of detail, the task was to judge whether or not the image showed an exemplar of a particular category of targets (e.g., a building). For a block of trials, testing the perception of the overall structure of the scene, the task was to judge whether the scene was presented right side up, or whether the position of the horizon was indicated correctly. The subjects were instructed to respond as quickly as possible after the onset of a stimulus presentation, by pressing the appropriate one of two response keys.

Each stimulus was presented for 400 ms. This brief presentation duration, in combination with the small stimulus size, served to prevent scanning eye movements (which may differ among image modalities and target types), and to force subjects to make a decision based solely on the instantaneous percept aroused by the stimulus presentation. Immediately after the stimulus presentation interval, a random noise image was shown. This noise image remained visible for at least 500 ms. It served to erase any possible afterimages (reversed contrast images induced by, and lingering on after, the presentation of the stimulus, that may differ in quality for different image modalities and target types), thereby equating the processing time subjects can use to make their judgment. Upon each presentation, the random noise image was randomly left/right and up/down reversed. The noise images had the same dimensions as the preceding stimulus image, and consisted of randomly distributed subblocks of 5 × 5 pixels. For trial blocks testing the monochrome IR and II imaging modalities and grayscale fused imagery, the noise image subblocks were either black or mean gray. For trial blocks testing DII and color fused imagery, the noise image subblocks were randomly colored, using a color palette similar to that of the modality being tested. In all tests, subjects were asked to quickly indicate their visual judgment by pressing one of two response keys (corresponding to a yes/no response), immediately after the onset of a stimulus image presentation. Both the accuracy and the reaction time were registered.

D. RESULTS

For each visual discrimination task the numbers of hits (correct detections) and false alarms (fa) were recorded to calculate $d' = Z_{\text{hits}} - Z_{\text{fa}}$, an unbiased estimate of sensitivity.[41]

The effects of contrast enhancement on human visual performance are found to be similar for all tasks. Figure 7.11 shows that contrast enhancement significantly improves the sensitivity of human observers performing with II and DII imagery. However, for IR imagery, the average sensitivity decreases as a result of contrast enhancement. This is probably a result of the fact that the contrast enhancement method employed in this study increases the visibility of irrelevant detail and clutter in the scene. Note that this result does *not* indicate that (local) contrast enhancement in general should not be applied to IR images.

Figure 7.12 shows the results of all scene recognition and target detection tasks investigated here. As stated before, the ultimate goal of image fusion is to produce a combined image that displays more information than either of the original images. Figure 7.12 shows that this aim is only achieved for the following perceptual tasks and conditions:

- The detection of roads, where CF1 outperforms each of the input image modalities;
- The recognition of water, where CF1 yields the highest observer sensitivity; and
- The detection of vehicles, where three fusion methods tested perform significantly better than the original imagery.

These tasks are also the only ones in which CF1 performs better than CF2. An image fusion method that always performs at least as good as the best of

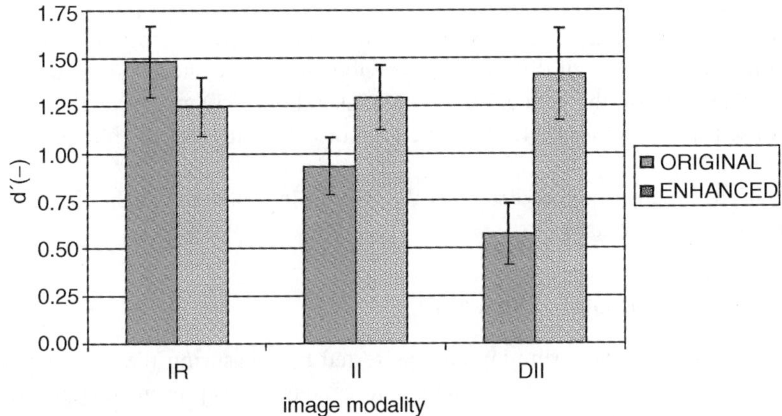

FIGURE 7.11 The effect of contrast enhancement on observer sensitivity d'.

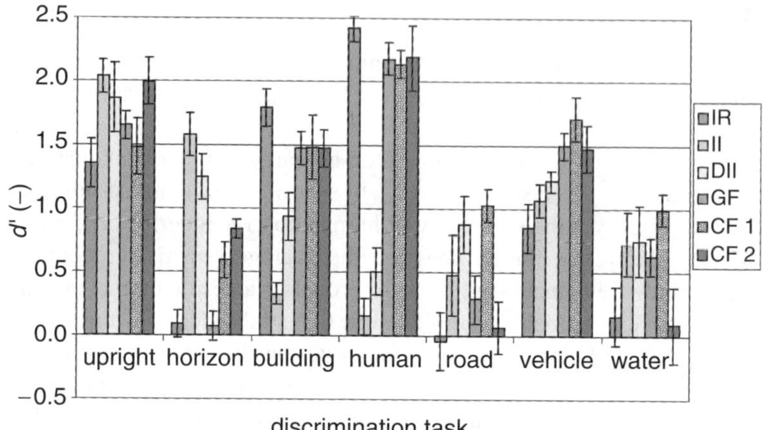

FIGURE 7.12 (See color insert) Observer sensitivity d' for discrimination of global layout (orientation and horizon) and local detail (buildings, humans, roads, vehicles, and water), for six different image modalities. These modalities are (in the order in which they appear in the labeled clusters above): infrared (IR), single-band or grayscale (II) and double-band or color (DII) intensified visual, grayscale (GF) and color fused (CF1, CF2) imagery.

the individual image modalities can be of great ergonomic value, since the observer can perform using only a single image. This result is obtained for the recognition of scene orientation from color fused imagery produced with CF2, where performance is similar to that with II and DII imagery. For the detection of buildings and humans in a scene, all three fusion methods perform equally well and slightly less than IR. CF1 significantly outperforms grayscale fusion for the detection of the horizon and the recognition of roads and water. CF2 outperforms grayscale fusion for both global scene recognition tasks (orientation and horizon detection). However, for CF2 observer sensitivity approaches zero for the recognition of roads and water.

Rather surprisingly, the response times (not shown here) did not differ significantly between all different image modalities. The shortest reaction times were obtained for the detection of humans (about 650 ms), and the longest response times were found for the detection of the position of the horizon (about 1000 ms).

The following section discusses the results in detail for each of the seven different perception tasks.

1. Perception of Global Structure

The perception of the scene layout was tested by measuring the accuracy with which observers were able to distinguish a scene that was presented right side up from one that was presented upside down, and perceive the position of the horizon.

The first group of bars in Figure 7.12 (labeled "upright") represents the results for the scene orientation perception task. For the original image modalities, the best results are obtained with the intensified imagery (the II performed slightly better than the DII). The IR imagery performs significantly worse. CF2 performs just as well as II, whereas CF1 performs similar to IR. Gray-level fusion is in between both color fusion methods. Observers remarked that they based their judgment mainly on the perceived orientation of trees and branches in the scene. CF2 displays trees with a larger color contrast (red–brown on a light greenish or bluish background) than CF1 (dark green trees on a somewhat lighter green background), resulting in a better orientation detection performance. Also, CF2 produces bright blue skies most of the time, which makes the task more intuitive.

The perception of the true position of the horizon, represented by the second group of bars in Figure 7.12, is best performed with II imagery, followed by the DII modality. Both intensified visual image modalities perform significantly better than IR or any kind of fused imagery. The low performance with the IR imagery is probably a result of the fact that a tree line and a band of clouds frequently have a similar appearance in this modality. The transposition of these "false horizons" into the fused image modalities significantly reduces observer performance. For gray-level fused imagery, the observer sensitivity is even reduced to a near-zero level, just as found for IR. Color fused imagery restores some of the information required to perform the task, especially CF2 that produces blue skies. However, the edges of the cloud bands are so strongly represented in the fused imagery that observer performance never attains the sensitivity level obtained for the intensified visual modalities alone (II and DII).

In both the orientation and horizon perception tasks subjects tend to confuse large bright areas (e.g., snow on the ground) with the sky.

2. Perception of Detail

The best score for the recognition of *buildings* is found for IR imagery. In this task, IR performs significantly better than II or DII. DII imagery performs significantly better than II, probably because of the color contrast between the buildings and the surrounding vegetation (red–brown walls on a green background, compared to gray walls on a gray background in case of the II imagery). The performance with fused imagery is slightly less than with IR, and independent of the fusion method.

The detection of *humans* is best performed with IR imagery, in which they are represented as white hot objects on a dark background. II imagery yields a very low sensitivity for this task; that is, humans are hardly ever noticed in intensified visual imagery. The sensitivity for the detection of humans in DII imagery is somewhat higher, but remains far below that found for IR. In this case, there is almost no additional information in the second wavelength band of the DII modality, and therefore almost no additional color contrast. As a result, most types of clothing are displayed as greenish, and are therefore hard to distinguish from vegetation. Performance with fused imagery is only slightly below that with IR. There is no significant difference between the different grayscale and color fusion types.

Roads cannot reliably be recognized from IR imagery (d' becomes even negative, meaning that more false alarms than correct detections are scored). DII performs best of the individual image modalities, and significantly higher than II because of the additional color contrast (DII displays roads as red–brown, on a green background). Grayscale fused imagery results in a performance that is significantly below that found for DII, and somewhat lower than that obtained for II imagery. This is probably a result of (1) the introduction of irrelevant luminance details from the IR imagery, and (2) the loss of color contrast as seen in the DII imagery. CF1 produces color fused imagery that yields a higher sensitivity than each of the original image modalities, although observer performance is not significantly better than with DII imagery. The additional improvement obtained with this combination scheme is probably caused by the contrast enhancement inherent in the fusion process. The sensitivity obtained for imagery produced by CF2 is near zero. This is probably a result of the fact that this method displays roads with a light blue color. These can therefore easily be mistaken for water or snow. This result demonstrates that the inappropriate use of color in image fusion severely degrades observer performance.

Image fusion clearly helps to recognize *vehicles* in a scene. They are best discriminated in color fused images produced with CF1, that displays vehicles in brown–yellow on a green background. CF2 (that shows vehicles as blue on a brown and green background) and grayscale fusion both result in equal and somewhat lower observer sensitivity. Fused imagery of all types performs significantly better than each of the original image modalities. The lowest recognition performance is obtained with IR imagery.

Water is best recognized in color fused imagery produced with CF1. This method displays water sometimes as brown-reddish, and sometimes as grayish. The II, DII and gray-level fusion schemes all yield a similar and slightly lower performance. CF2 results on a near zero observer sensitivity for this task. This method displays water sometimes as purple-reddish, thus giving it a very unnatural appearance, and sometimes as bluish, which may cause confusion with roads, that have the same color. These results again demonstrate that it is preferable not to use any color at all (grayscale), than to use an inappropriate color mapping scheme.

3. Summary

Table 7.1 summarizes the main findings of this study. IR has the lowest overall performance of all modalities tested. This results from a low performance for both large scale orientation tasks, and for the detection and recognition of roads, water, and vehicles. In contrast, intensified visual imagery performs best in both orientation tasks. The perception of the horizon is significantly better with II and DII imagery. IR imagery performs best for the perception and recognition of

TABLE 7.1
The Relative Performance of the Different Image Modalities for the Seven Perceptual Recognition Tasks

	IR	II	DII	GF	CF1	CF2
Upright	− 1	2	1			2
Horizon	− 1	2	1			
Building	2	− 1		1	1	1
Human	2	− 1		1	1	1
Road	− 1		1		2	
Vehicle	− 1			2	2	1
Water	− 1				2	
Overall	− 1	2	3	4	8	5

Rank orders − 1, 1, and 2 indicate, respectively, the worst, second best, and best performing image modality for a given task. The tasks involve the perception of the global layout (orientation and horizon) of a scene, and the recognition of local detail (buildings, humans, roads, vehicles, and water). The different image modalities are: infrared (IR), grayscale (II) and dual band false-color (DII) intensified visual, grayscale fused images (GF) and two different color fusion (CF1, CF2) schemes. The sum of the rank orders indicates the overall performance of the modalities.

buildings and humans — DII has the best overall performance of the individual image modalities. Thus, IR on one hand and (D)II images on the other hand contain *complementary* information, which makes each of these image modalities suited for performing different perception tasks.

CF1 has the best overall performance of the image fusion schemes tested here. The application of an appropriate color mapping scheme in the image fusion process can indeed significantly improve observer performance compared to grayscale fusion. In contrast, the use of an inappropriate color scheme can severely degrade observer sensitivity. Although the performance of CF1 for specific observation tasks is below that of the optimal individual sensor, for a combination of observation tasks (as will often be the case in operational scenarios) the CF1 fused images can be of great ergonomic value, since the observer can perform using only a single image.

E. Discussion

Night-time images recorded using an image intensified low-light CCD camera and a thermal middle wavelength band (3 to 5 μm) IR camera contain *complementary* information. This makes each of the individual image modalities only suited for specific observation task. However, the complementarity of the information of the image modalities can be exploited using image fusion, which would enable multiple observation tasks using a single night-time image representation.

Since there evidently exists no one-to-one mapping between the temperature contrast and the spectral reflectance of a material, the goal of producing a nighttime image, incorporating information from IR imagery, with an appearance similar to a color daytime image can never be fully achieved. The options are therefore (1) to settle for a single mapping that works satisfactory in a large number of conditions, or (2) to adapt (optimize) the color mapping to the situation at hand. However, the last option is not very attractive since a different color mapping for each task and situation tends to confuse observers.[21,42]

Multimodal image fusion schemes based on local contrast decomposition do not distinguish between material edges and temperature edges. For many tasks, material edges are the most important ones. Fused images frequently contain an abundance of contours that are irrelevant for the task that is to be performed. Fusion schemes incorporating some kind of contrast stretching enhance the visibility of all details in the scene, irrespective of their visual significance. The introduction of spurious or irrelevant contrast elements in a fused image may clutter the scene, distract the observer and lead to misinterpretation of perceived details. As a result, observer performance may degrade significantly. A useful image fusion scheme should therefore take into account the visual information content (meaning) of the edges in each of the individual image modalities, and combine them accordingly in the resulting image.

For most perceptual tasks investigated here (except for horizon and road detection), grayscale image fusion yields appreciable performance levels. When an appropriate color mapping scheme is applied, the addition of color to grayscale fused imagery can significantly increase observer sensitivity for a given condition and a certain task (e.g., color fusion Method 2 for orientation detection, both color fusion methods for horizon detection, color fusion Method 1 for road and water detection). However, inappropriate use of color can significantly decrease observer performance compared to straightforward grayscale image fusion (e.g., color fusion Method 2 for the detection of roads and water).

For the observation tasks and image examples tested here, optimal overall performance was obtained for images fused using color fusion Method 1. The overall performance was higher than for either of the individual image modalities. Note that in this fusion method, no color mapping is applied to the IR information. Instead, the IR information is blended into the image without changing the color.

IV. CONCLUSIONS

Intensified visual and thermal imagery represent complementary information. This is especially evident in imagery recorded around sunrise, as shown in the first part of this study.

The results of the first experiment show that observers performing a surveillance task can indeed localize a person in the scene significantly better with fused images than with each of the individual image modalities.

This indicates that the fused images provide a better representation of the spatial layout of the scene.

In the second experiment we tested the benefits of different gray-level and color image fusion schemes for global scene recognition (situational awareness) and the perception of details (target recognition). The results show that grayscale image fusion yields appreciable performance levels. When an appropriate color mapping scheme is applied, the addition of color to grayscale fused imagery can significantly increase observer sensitivity for a given condition and a certain task. However, inappropriate use of color can significantly decrease observer performance compared to straightforward grayscale image fusion. One of the main reasons seems to be the counter intuitive appearance of scenes rendered in artificial color schemes and the lack of color constancy.[10]

We conclude that image fusion can significantly enhance observer performance in surveillance and navigation tasks. When color is applied in image fusion schemes the color mapping should be such that (a) the resulting imagery has a natural appearance[2] and (b) the colors are invariant for changes in the environmental conditions (i.e., the image should always have more or less the same and intuitively correct appearance).

REFERENCES

1. Toet, A., IJspeert, J. K., Waxman, A. M., and Aguilar, M., Fusion of visible and thermal imagery improves situational awareness, *Displays*, 18, 85–95, 1997.
2. Toet, A., Natural colour mapping for multiband nightvision imagery, *Inf. Fusion*, 4(3), 155–166, 2003.
3. Aguilar, M., and Garret, A. L., Biologically based sensor fusion for medical imaging, In *Sensor Fusion: Architectures, Algorithms, and Applications V*, Vol. 4385, Dasarathy, B. V., Ed., The International Society for Optical Engineering, Bellingham, WA, pp. 149–158, 2001.
4. Aguilar, M., Fay, D. A., Ireland, D. B., Racamoto, J. P., Ross, W. D., and Waxman, A. M., Field evaluations of dual-band fusion for color night vision, In *Enhanced and Synthetic Vision 1999*, Vol. 3691, Verly, J. G., Ed., The International Society for Optical Engineering, Bellingham, WA, pp. 168–175, 1999.
5. Aguilar, M., Fay, D. A., Ross, W. D., Waxman, A. M., Ireland, D. B., and Racamoto, J. P., Real-time fusion of low-light CCD and uncooled IR imagery for color night vision, In *Enhanced and Synthetic Vision 1998*, Vol. 3364, Verly, J. G., Ed., The International Society for Optical Engineering, Bellingham, WA, pp. 124–135, 1998.
6. Essock, E. A., Sinai, M. J., McCarley, J. S., Krebs, W. K., and DeFord, J. K., Perceptual ability with real-world nighttime scenes: image-intensified, infrared, and fused-color imagery, *Hum. Factors*, 41(3), 438–452, 1999.
7. Fay, D. A., Waxman, A. M., Aguilar, M., Ireland, D. B., Racamato, J. P., Ross, W. D., Streilein, W., and Braun, M. I., Fusion of multi-sensor imagery for night vision: color visualization, target learning and search,

pp. TuD3-3–TuD3-10. In *Proceedings of the third International Conference on Information Fusion I.* ONERA, Paris, France, 2000.

8. Schuler, J., Howard, J. G., Warren, P., Scribner, D. A., Klien, R., Satyshur, M., and Kruer, M. R., Multiband E/O color fusion with consideration of noise and registration, In *Targets and Backgrounds VI: Characterization, Visualization, and the Detection Process*, Vol. 4029, Watkins, W. R., Clement, D., and Reynolds, W. R., Eds., The International Society for Optical Engineering, Bellingham, WA, pp. 32–40, 2000.

9. Scribner D. A., Warren P., and Schuler J., Extending color vision methods to bands beyond the visible, pp. 33–40. In *Proceedings of the IEEE Workshop on Computer Vision Beyond the Visible Spectrum: Methods and Applications, 1999.*

10. Varga, J. T., Evaluation of Operator Performance Using True Color and Artificial Color in Natural Scene Perception, Report AD-A363036, Naval Postgraduate School, Monterey, CA, 1999.

11. Waxman, A. M., Fay, D. A., Gove, A. N., Seibert, M. C., Racamato, J. P., Carrick, J. E., and Savoye, E. D., Color night vision: fusion of intensified visible and thermal IR imagery, In *Synthetic Vision for Vehicle Guidance and Control*, Vol. 2463, Verly, J.G., Ed., The International Society for Optical Engineering, Bellingham, WA, pp. 58–68, 1995.

12. Waxman, A. M., Gove, A. N., Fay, D. A., Racamoto, J. P., Carrick, J. E., Seibert, M. C., and Savoye, E. D., Color night vision: opponent processing in the fusion of visible and IR imagery, *Neural Netw.*, 10(1), 1–6, 1997.

13. Waxman, A. M., Carrick, J. E., Fay, D. A., Racamato, J. P., Augilar, M., and Savoye, E. D., Electronic imaging aids for night driving: low-light CCD, thermal IR, and color fused visible/IR. In *Proceedings of the SPIE Conference on Transportation Sensors and Controls*, Vol. 2902, The International Society for Optical Engineering, Bellingham, WA, 1996.

14. Waxman, A. M., Aguilar, M., Baxter, R. A., Fay, D. A., Ireland, D. B., Racamoto, J. P., and Ross, W. D., Opponent-color fusion of multi-sensor imagery: visible, IR and SAR, pp. 43–61. In *Proceedings of the 1998 Conference of the IRIS Specialty Group on Passive Sensors I*, 1998.

15. Waxman, A. M., Aguilar, M., Fay, D. A., Ireland, D. B., Racamoto, J. P., Ross, W. D., Carrick, J. E., Gove, A. N., Seibert, M. C., Savoye, E. D., Reich, R. K., Burke, B. E., McGonagle, W. H., and Craig, D. M., Solid-state color night vision: fusion of low-light visible and thermal infrared imagery, *MIT Lincoln Lab. J.*, 11, 41–60, 1999.

16. Driggers, R. G., Krapels, K. A., Vollmerhausen, R. H., Warren, P. R., Scribner, D. A., Howard, J. G., Tsou, B. H., and Krebs, W. K., Target detection threshold in noisy color imagery, In *Infrared Imaging Systems: Design, Analysis, Modeling, and Testing XII*, Vol. 4372, Holst, G. C., Ed., The International Society for Optical Engineering, Bellingham, WA, pp. 162–169, 2001.

17. Sinai, M. J., McCarley, J. S., Krebs, W. K., and Essock, E. A., Psychophysical comparisons of single- and dual-band fused imagery, In *Enhanced and Synthetic Vision 1999*, Vol. 3691, Verly, J.G., Ed., The International Society for Optical Engineering, Bellingham, WA, pp. 176–183, 1999.

18. Toet, A., IJspeert, J. K., Waxman, A. M., and Aguilar, M., Fusion of visible and thermal imagery improves situational awareness, In *Enhanced and Synthetic Vision 1997*, Vol. 3088, Verly, J. G., Ed., International Society for Optical Engineering, Bellingham, WA, pp. 177–188, 1997.

19. Toet, A., and IJspeert, J. K., Perceptual evaluation of different image fusion schemes, In *Signal Processing, Sensor Fusion, and Target Recognition X*, Vol. 4380, Kadar, I., Ed., The International Society for Optical Engineering, Bellingham, WA, pp. 436–441, 2001.

20. White, B. L., Evaluation of the Impact of Multispectral Image Fusion on Human Performance in Global Scene Processing, Report AD-A343639, Naval Postgraduate School, Monterey, CA, 1998.

21. Krebs, W. K., Scribner, D. A., Miller, G. M., Ogawa, J. S., and Schuler, J., Beyond third generation: a sensor-fusion targeting FLIR pod for the F/A-18, In *Sensor Fusion: Architectures, Algorithms, and Applications II*, Vol. 3376, Dasarathy, B. V., Ed., International Society for Optical Engineering, Bellingham, WA, USA, pp. 129–140, 1998.

22. Waxman, A. M., Gove, A. N., Seibert, M. C., Fay, D. A., Carrick, J. E., Racamato, J. P., Savoye, E. D., Burke, B. E., Reich, R. K., McGonagle, W. H., and Craig, D. M., Progress on color night vision: visible/IR fusion, perception and search, and low-light CCD imaging, In *Enhanced and Synthetic Vision 1996*, SPIE-2736, Verly, J. G., Ed., The International Society for Optical Engineering, Bellingham, WA, pp. 96–107, 1996.

23. Waxman, A. M., Gove, A. N., and Cunningham, R. K., Opponent-color visual processing applied to multispectral infrared imagery, pp. 247–262. In *Proceedings of 1996 Meeting of the IRIS Specialty Group on Passive Sensors II*. Infrared Information Analysis Center, ERIM, Ann Arbor, US, 1996.

24. Schiller, P. H., Central connections of the retinal ON and OFF pathways, *Nature*, 297(5867), 580–583, 1982.

25. Schiller, P. H., The connections of the retinal on and off pathways to the lateral geniculate nucleus of the monkey, *Vision Res.*, 24(9), 923–932, 1984.

26. Schiller, P. H., Sandell, J. H., and Maunsell, J. H., Functions of the ON and OFF channels of the visual system, *Nature*, 322(6082), 824–825, 1986.

27. Schiller, P. H., The ON and OFF channels of the visual system, *Trends Neurosci.*, 15(3), 86–92, 1992.

28. Newman, E. A., and Hartline, P. H., Integration of visual and infrared information in bimodal neurons of the rattlesnake optic tectum, *Science*, 213, 789–791, 1981.

29. Newman, E. A., and Hartline, P. H., The infrared "vision" of snakes, *Sci. Am.*, 246, 116–127, 1982.

30. Gouras, P., Color vision, In *Principles of Neural Science*, 3rd ed., Kandel, E. R., Schwartz, J. H., and Jessel, T. M., Eds., Elsevier, Oxford, UK, pp. 467–480, 1991.

31. Schiller, P. H., and Logothetis, N. K., The color-opponent and broad-band channels of the primate visual system, *Trends Neurosci.*, 13(10), 392–398, 1990.

32. Grossberg, S., *Neural Networks and Natural Intelligence*, MIT Press, Cambridge, MA, 1998.

33. Christ, R. E., Review and analysis of colour coding research for visual displays, *Hum. Factors*, 17, 542–570, 1975.

34. Hughes, P. K., and Creed, D. J., Eye movement behaviour viewing colour-coded and monochrome avionic displays, *Ergonomics*, 37, 1871–1884, 1994.

35. Onyango, C. M., and Marchant, J. A., Physics-based colour image segmentation for scenes containing vegetation and soil, *Image Vision Comput.*, 19(8), 523–538, 2001.

36. Burt, P. J., and Adelson, E. H., Merging images through pattern decomposition, In *Applications of Digital Image Processing VIII*, Vol. 575, Tescher, A.G., Ed.,

The International Society for Optical Engineering, Bellingham, WA, pp. 173–181, 1985.

37. Toet, A., van Ruyven, J. J., and Valeton, J. M., Merging thermal and visual images by a contrast pyramid, *Opt. Eng.*, 28(7), 789–792, 1989.
38. Toet, A., Hierarchical image fusion, *Machine Vision Appl.*, 3, 1–11, 1990.
39. Toet, A., Adaptive multi-scale contrast enhancement through non-linear pyramid recombination, *Pattern Recogn. Lett.*, 11, 735–742, 1990.
40. Toet, A., Multi-scale contrast enhancement with applications to image fusion, *Opt. Eng.*, 31(5), 1026–1031, 1992.
41. Macmillan, N. A., and Creelman, C. D., *Detection Theory: A User's Guide*, Cambridge University Press, Cambridge, MA, 1991.
42. Steele, P. M., and Perconti, P., Part task investigation of multispectral image fusion using gray scale and synthetic color night vision sensor imagery for helicopter pilotage, pp. 88–100. In *Proceedings of the SPIE Conference on Targets and Backgrounds, Characterization and Representation III*, Vol. 3062, Watkins, W., and Clement, D., Eds., International Society for Optical Engineering, Bellingham, WA, 1997.

8 A Statistical Signal Processing Approach to Image Fusion Using Hidden Markov Models*

Jinzhong Yang and Rick S. Blum

CONTENTS

---

* This material is based on work supported by the U.S. Army Research Office under grant number DAAD19-00-1-0431. The content of the information does not necessarily reflect the position or the policy of the federal government, and no official endorsement should be inferred.

I. INTRODUCTION

Image fusion methods based on multiscale transforms (MST) are a popular choice in recent research.[1] Figure 8.1 illustrates the block diagram of a generic image fusion scheme based on multiscale analysis. The basic idea is to perform a MST on each source image, then construct a composite multiscale representation from these. The fused image is obtained by taking an inverse multiscale transform (IMST).

Wavelet theory has emerged as a well developed yet rapidly expanding mathematical foundation for a class of multiscale representations. Some sophisticated image fusion approaches based on wavelet transforms have been proposed and studied.[1–9] However, the majority of fusion methods employing the wavelet transform have not attempted to model or capitalize on the correlations between wavelet coefficients, especially the correlations across the wavelet decomposition scales. The fusion was often performed separately on each wavelet subband.

Although the wavelet transform is sometimes interpreted as a "decorrelator" which attempts to make each wavelet coefficient statistically independent of all others, the research from Refs. 10–14 demonstrates that there are still some important dependencies between wavelet coefficients. The dependencies can be described using the statistical properties called *clustering* and *persistence*. *Clustering* is a property that states that if a particular wavelet coefficient is large (small) then adjacent coefficients are very likely also to be large (small). *Persistence* is a property that states that large (small) values of wavelet coefficients tend to propagate across scales. Recently, in Refs. 10,11, researchers have studied these properties and applied them in image coding, signal detection, and estimation. Based on the study of these properties (mainly persistence), a hidden Markov model (HMM)[15,16] approach was suggested. Here we employ an image formation model using a HMM to capture the correlations between wavelet coefficients across wavelet decomposition scales. Then, based on this image formation model, the expectation-maximization (EM)[17–21]

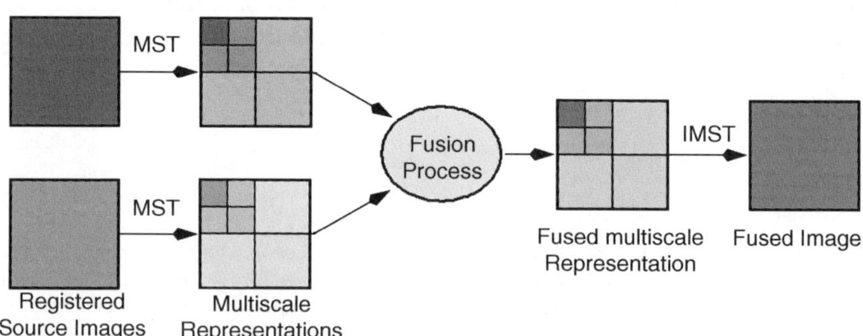

FIGURE 8.1 Block diagram of a generic image fusion scheme.

algorithm was used to estimate the model parameters and produce the fused image. We have applied this new image fusion approach to concealed weapon detection (CWD) cases and night vision applications with good fusion results.

II. THE IMAGE FORMATION MODEL BASED ON HMM

This research builds on the work in Ref. 22. The image formation model proposed in Ref. 22 assumed that each pyramid coefficient was statistically independent of all others. This approach was taken to promote simplicity. Here we propose a more realistic dependency model. A new image formation model has been created based on allowing correlations between the pyramid coefficients. In the new model, the sensor images are described as the true scene corrupted by additive non-Gaussian distortion, and a HMM describes the correlations between the wavelet coefficients in one sensor image.

A. Tree Structure of the Wavelet Coefficients

To describe the HMM for the wavelet coefficients, we need to use *graphs* and *trees*.[23] An undirected graph consists of a set of *nodes* $\{v_1, v_2, \ldots, v_N\}$ and a set of *edges* linking the nodes. A *path* is a set of edges connecting two nodes. A rooted *tree* is an undirected acyclic graph. All nodes that lie on a path from v_i to the root are called *ancestors* of v_i. All nodes that lie on paths from v_i going away from the root node are called *descendants* of v_i. A node is called the *parent* of v_i if it is the immediate ancestor of v_i. The parent of v_i is denoted by $v_{\rho(i)}$. A node is called the *child* of v_i if v_i is its parent. The children of v_i are denoted by $\{v_j\}_{j \in c(j)}$. Each node in the rooted tree has only one parent but may have several children. The root has no parent node. Nodes with no children are called *leaves* of the tree. In a rooted tree, if each node that is not a leaf has four children, this tree will be called a *quadtree*. A collection of quadtrees is called a *forest* of quadtrees.

Based on the persistence property of the wavelet coefficients, we organize the wavelet coefficients of a source image as a forest of quadtrees.[10,11] Each coefficient represents a node in one of the quadtrees. The trees are rooted at the wavelet coefficients in the high-frequency bands (HL, LH, HH bands) in the coarsest scale. The coefficients in the LL band are not included in the quadtrees and will be processed separately. Figure 8.2 illustrates the quadtrees. In a quadtree, each coefficient in a coarse scale subband has four child coefficients in the corresponding finer scale subband. The arrows in the figure point from the subband of the parents to the subband of the children. The number of quadtrees in the forest depends on the size of the image and the number of decomposition levels in the wavelet transform. For example, if we decompose an $N \times N$ image using a wavelet transform with L decomposition levels (L scales), then in each subband set (LH, HL or HH subband set), $(N \times 2^{-L})^2$ wavelet trees are obtained.

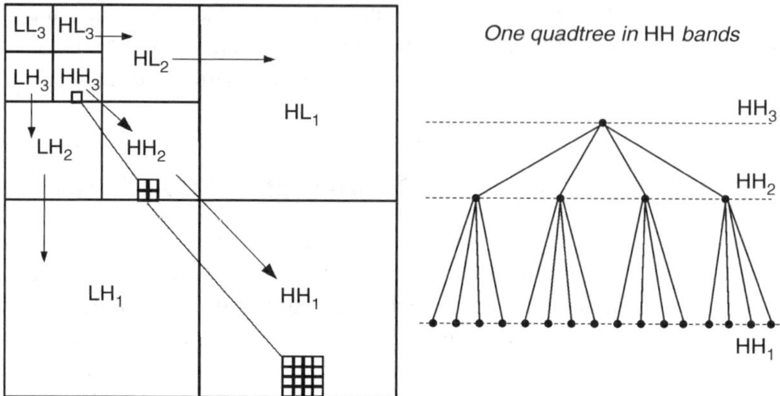

FIGURE 8.2 Forest of quadtrees of wavelet coefficients. A quadtree rooted in the HH$_3$ subband is shown.

B. HIDDEN MARKOV TREE MODEL

From the study in Ref. 11, a Gaussian mixture model appears to be able to closely fit the distribution of wavelet coefficient data. Hence, we can treat each wavelet coefficient from a given sensor image as a random variable with a Gaussian mixture probability density function (PDF). From the above discussion, it appears that a given coefficient may be statistically dependant on its parent and children. This dependency increases the complexity of the image formation model and the parameter estimation for the model. In order to solve this problem, a HMM is introduced. We associate a state variable with each wavelet coefficient. This random state variable will capture the parent–children relationship of wavelet coefficients. In a quadtree, we associate each coefficient node with a state node, as shown in Figure 8.3. Therefore, the wavelet coefficients are dependant on the associated state variable and independent of the other coefficients, when conditioned on the state variable. Therefore, the dependency of wavelet coefficients can be viewed as being "hidden" by the state variables. Further, we find that the state nodes in a quadtree form a Markov chain, since a node is statistically related only to its parent and children. The Markov model is fit on the tree structure (see Figure 8.2) of the wavelet coefficients. Hence we call it a hidden Markov tree (HMT) model.[11] Figure 8.3 gives an example of a HMT model for a particular quadtree. The definition of the state variable will be explained in more detail in Section II.C.

C. IMAGE FORMATION MODEL

The sensor images are described as the true scene corrupted by additive, and possibly non-Gaussian, distortion. Assume there are q sensor images to be fused. In the wavelet representation of each sensor image, suppose there are

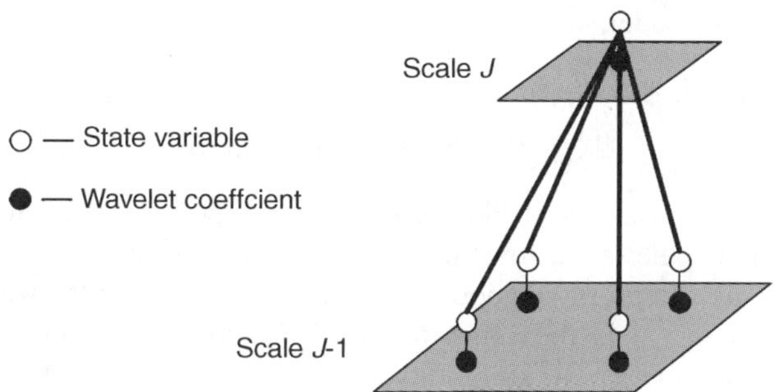

FIGURE 8.3 HMT for an image quadtree. The black nodes represent the wavelet coefficient and the white nodes represent the state variables.

K quadtrees. For each quadtree, there are P wavelet coefficients, or P nodes. Let $z_{i,k}(j)$ denote the wavelet coefficient at the jth node of the kth tree in the ith sensor image. Let $w_k(j)$ denote the wavelet coefficient at the jth node of the kth tree of the wavelet transform of the image describing the true scene. Then our image formation model is

$$z_{i,k}(j) = \beta_{i,k}(j)w_k(j) + \varepsilon_{i,k}(j)$$
$$i = 1,...,q; \quad k = 1,...,K; \quad j = 1,...,P$$

(8.1)

where $\beta_{i,k}(j)$ is the sensor selectivity factor and $\varepsilon_{i,k}(j)$ is the possibly non-Gaussian distortion.

We use a zero-mean, M-term, Gaussian mixture model to fit the distortion. Now we introduce a random state variable $S_{i,k}(j)$ with M states; $S_{i,k}(j)$ denotes the state variable for the jth node in the kth wavelet quadtree. Then, $p(S_{i,k}(j) = m)$ is the probability of being in state m. Now we associate state m ($m = 1,...,M$) of variable $S_{i,k}(j)$ with the mth term in the Gaussian mixture model for the distortion. Thus we model the distortion as being in M states. In each state, the distortion has a zero-mean Gaussian distribution. The state probability, denoted by $p_{S_{i,k}(j)}(m) = p(S_{i,k}(j) = m)$, is the probability that the distortion is in state m. To allow for efficient training, given only the source images, we assume that $p_{S_{i,k}(j)}(m)$ is the same for all i and k and we denote it by $p_{S(j)}(m)$ for simplicity. Since $\beta_{i,k}(j)w_k(j)$ is assumed to be deterministic, $z_{i,k}(j)$ also has a Gaussian mixture distribution. If the distortion is in state m, the coefficient $z_{i,k}(j)$ can also be thought to be in state m. Thus we will see that $z_{i,k}(j)$ has a distribution corresponding to the mth term in the mixture model for the distortion, but with a non-zero mean. Hence the state variables characterize the state (mixture term) of the sensor wavelet coefficients. As described previously, we associate each sensor coefficient node in a quadtree with a state node to obtain a HMT like the one shown in Figure 8.3. Let $S_{i,k}(1)$ denote the state variable of the root node and let $\rho(j)$ denote the parent node of node j in the kth tree of ith sensor image.

Due to the HMT model, we have

$$P\{S_{i,k}(j) = m | S_{i,k}(\rho(j)) = n_{\rho(j)}, \ldots, S_{i,k}(2) = n_2, S_{i,k}(1) = n_1\}$$
$$= P\{S_{i,k}(j) = m | S_{i,k}(\rho(j)) = n_{\rho(j)}\} \tag{8.2}$$

where $\rho(j), \ldots, 1$ represents all the nodes between $\rho(j)$ and root node 1 in the tree. We denote the state transition probability using $a^{m,n}_{j,\rho(j)} = P\{S_{i,k}(j) = m | S_{i,k}(\rho(j)) = n\}$. To promote good estimation based only on the source images, we assume that $a^{m,n}_{j,\rho(j)}$ is the same for all i, k and so we can write $a^{m,n}_{j,\rho(j)} = P\{S(j) = m | S(\rho(j)) = n\}$ for simplicity.

Given $S_{i,k}(j) = m$, the distribution of distortion $\varepsilon_{i,k}(j)$ is

$$f_{\varepsilon_{i,k}(j)}(\varepsilon_{i,k}(j) | S_{i,k}(j) = m) = \frac{1}{\sqrt{2\pi\sigma_m^2(j)}} \exp\left\{-\frac{\varepsilon_{i,k}(j)^2}{2\sigma_m^2(j)}\right\} \tag{8.3}$$

and using Equation 8.1, the distribution of $z_{i,k}(j)$ given $S_{i,k}(j) = m$ will be

$$f_{z_{i,k}(j)}(z_{i,k}(j) | S_{i,k}(j) = m) = \frac{1}{\sqrt{2\pi\sigma_m^2(j)}} \exp\left\{-\frac{(z_{i,k}(j) - \beta_{i,k}(j)w_k(j))^2}{2\sigma_m^2(j)}\right\}. \tag{8.4}$$

Thus removing the conditioning on $S_{i,k}(j)$, the distribution of $\varepsilon_{i,k}(j)$ is generally non-Gaussian as given by

$$f_{\varepsilon_{i,k}(j)}(\varepsilon_{i,k}(j)) = \sum_{m=1}^{M} \frac{p(S_{i,k}(j) = m)}{\sqrt{2\pi\sigma_m^2(j)}} \exp\left\{-\frac{\varepsilon_{i,k}(j)^2}{2\sigma_m^2(j)}\right\}. \tag{8.5}$$

If $\varepsilon_{i,k}(j)$ is Gaussian, then this is also modeled by Equation 8.5 with the proper choice of parameters.

The M-state Gaussian mixture model is used to fit the wavelet coefficient data of the sensor images. Generally speaking, for different coefficients the distribution of each $z_{i,k}(j)$ is different. However, they can be statistically related according to the statistical relationships of their state variable. The relationship is described by the HMT model. In this image formation model, the sensor selectivity factor $\beta_{i,k}(j)$ and true scene data $w_k(j)$ are generally unknown. The Gaussian variance of each state $\sigma_m^2(j)$ is also undetermined. These parameters explicitly appear in the image formation model. Two other variables, the state probability $p_{S(j)}(m)$, and the state transition probabilities $a^{m,n}_{j,\rho(j)}$, are also unknown and these will be needed to fully specify the image formation model as they describe the unconditioned distribution in Equation 8.5.

III. FUSION WITH THE EM ALGORITHM

The image formation model in Equation 8.1 and Equation 8.5 has been used in conjunction with the expectation-maximization (EM) algorithm[17−21] to develop a set of iterative equations to estimate the model parameters and to produce the fused image (the final true scene estimate). Approximate maximum likelihood estimates[18] are produced after using the EM algorithm. Since the image formation model in Equation 8.1 and Equation 8.5 is based on the coefficients in the high-frequency bands (HL, LH, HH bands), the iterative algorithm will be applied only to these bands. Fusion of coefficients in the LL band will be processed separately.

We assume the same distortion model for each wavelet quadtree. Then the iterative algorithm will be run over the set of wavelet representations of the sensor images to obtain the estimates of

$$\Phi = \left\{ p_{S(j)}(m), a_{j,\rho(j)}^{m,n}, \sigma_m^2(j), \beta_{i,k}(j), w_k(j) \right.$$
$$\left. | i = 1, \ldots q; \ k = 1, \ldots, K; \ j = 1, \ldots, P; \ m, n = 1, \ldots, M \right\} \tag{8.6}$$

We let

$$\Phi' = \left\{ p'_{S(j)}(m), a'^{m,n}_{j,\rho(j)}, \sigma'^2_m(j), \beta'_{i,k}(j), w'_k(j) \right.$$
$$\left. | i = 1, \ldots q; \ k = 1, \ldots, K; \ j = 1, \ldots, P; \ m, n = 1, \ldots, M \right\} \tag{8.7}$$

denote the updated values of Φ in each iteration of the iterative algorithm. Figure 8.4 gives the block diagram of the iterative fusion procedure. This iterative fusion procedure begins with the parameters' initialization. Reasonable initial values are given to Φ. Then, with the initial parameter values and observed data, the conditional probabilities list is computed using the upward–downward algorithm[11,24,25] based on the HMT model. The conditional probabilities list,

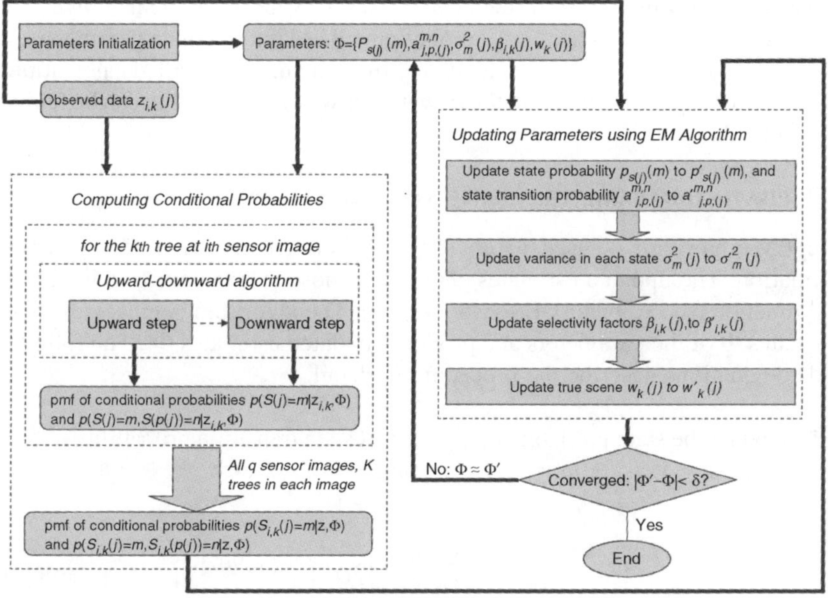

FIGURE 8.4 Block diagram of the iterative fusion procedure.

together with the initial parameters and observed data, are then used by the EM algorithm to update the parameters $\mathbf{\Phi}$ to $\mathbf{\Phi}'$. If the difference between $\mathbf{\Phi}$ and $\mathbf{\Phi}'$ is less than a given threshold δ, the final estimates of $\mathbf{\Phi}$ are set to $\mathbf{\Phi}'$ and we terminate the fusion procedure; otherwise, we update the current estimates using $\mathbf{\Phi}'$ and repeat the procedure until the iterative procedure converges. Each block in Figure 8.4 will be described in detail in the following.

A. COMPUTING CONDITIONAL PROBABILITIES

Let $\mathbf{z}$ denote the complete set of observations. From Equation 8.4 we know the wavelet coefficients of the sensor images are conditionally Gaussian given the hidden states. Hence, in order to carry out the iterative EM algorithm, we need to know the marginal probability mass functions (PMFs), $p(S_{i,k}(j) = m|\mathbf{z}, \mathbf{\Phi})$, and parent–child PMFs, $p(S_{i,k}(j) = m, \ S_{i,k}(\rho(j)) = n|\mathbf{z}, \mathbf{\Phi})$, for $i = 1, \dots, q$, $k = 1, \dots, K$ and $j = 1, \dots, P$. Since we assume that each quadtree is independent of the others, these conditional probabilities can be calculated independently for a given wavelet quadtree of a particular sensor image. Thus, we carry out these computations for each quadtree of each image. For computing efficiency in a particular tree, these probabilities are given by some intermediate parameters in our HMT model. The well-developed upward–downward algorithm[11,24,25] is used to produce these parameters. In order to determine the probabilities for the state variables, the state information must be propagated throughout the tree. The upward step of the algorithm calculates the parameters by transmitting information from fine-scale wavelet coefficients up to the states of the coarse-scale wavelet coefficients; the downward step of the algorithm calculates the parameters by propagating information from the coarse-scale wavelet coefficients down to the fine scale wavelet coefficients. After obtaining these intermediate parameters, the conditional probabilities can be calculated from them. The detailed calculation of the conditional probabilities and the upward–downward algorithm can be found in Appendix A.

B. UPDATING PARAMETERS USING THE EM ALGORITHM

The EM algorithm is used to develop the iterative equations for parameter estimation. The updated estimates chosen are those that maximize a likelihood-like function (see Appendix B for the details). The algorithm begins with current estimates $\mathbf{\Phi}$ of the parameters and produces updated estimates $\mathbf{\Phi}'$. The following update equations describe the procedure in detail.

1. Update the state probability $p_{S(j)}(m)$ and state transition probability $a_{j,\rho(j)}^{m,n}$ for all $j = 1, \dots, P$; $m = 1, \dots, M$ using

$$p'_{S(j)}(m) = \frac{\displaystyle\sum_{i=1}^{q}\sum_{k=1}^{K} p(S_{i,k}(j) = m|\mathbf{z}, \mathbf{\Phi})}{Kq} \tag{8.8}$$

$$a'^{m,n}_{j,\rho(j)} = \frac{\sum_{i=1}^{q} \sum_{k=1}^{K} p(S_{i,k}(j) = m, S_{i,k}(\rho(j)) = n | \mathbf{z}, \boldsymbol{\Phi})}{Kqp_{S(\rho(j))}(n)} \tag{8.9}$$

2. Update the Gaussian variance $\sigma_m^2(j)$ for all $j = 1, \ldots, P$; $m = 1, \ldots, M$ using

$$\sigma'^2_m(j) = \frac{\sum_{i=1}^{q} \sum_{k=1}^{K} (z_{i,k}(j) - \beta_{i,k}(j) w_k(j))^2 p(S_{i,k}(j) = m | \mathbf{z}, \boldsymbol{\Phi})}{Kqp_{S(j)}(m)} \tag{8.10}$$

3. To update $\beta_{i,k}(j)$ for $i = 1, \ldots, q$, $k = 1, \ldots, K$ and $j = 1, \ldots, P$, select $\beta'_{i,k}(j) = \pm 1, 0$ to maximize

$$\begin{aligned} Q = -\frac{1}{2} \sum_{m=1}^{M} & \left\{ \ln \sigma'^2_m(j) + \frac{(z_{i,k}(j) - \beta'_{i,k}(j) w_k(j))^2}{\sigma'^2_m(j)} \right\} \\ & \times p(S_{i,k}(j) = m | \mathbf{z}, \boldsymbol{\Phi}) \end{aligned} \tag{8.11}$$

4. Update the value of $w_k(j)$ for all $k = 1, \ldots, K$; $j = 1, \ldots, P$ using

$$w'_k(j) = \frac{\sum_{i=1}^{q} \sum_{m=1}^{M} \dfrac{\beta'_{i,k}(j) z_{i,k}(j)}{\sigma'^2_m(j)} p(S_{i,k}(j) = m | \mathbf{z}, \boldsymbol{\Phi})}{\sum_{i=1}^{q} \sum_{m=1}^{M} \dfrac{\beta'^2_{i,k}(j)}{\sigma'^2_m(j)} p(S_{i,k}(j) = m | \mathbf{z}, \boldsymbol{\Phi})} \tag{8.12}$$

The above update equations, Equation 8.8 to Equation 8.12, are derived from the SAGE version of the EM algorithm,[19] similar to the development in Ref. 22. The details of the derivation are presented in Appendix B.

C. INITIALIZATION OF THE FUSION ALGORITHM

Initial estimates are required for computing conditional probabilities and the iterative procedure (Equation 8.8 to Equation 8.12). We choose the initial estimates for the true scene $w_k(j)$ to come from the weighted average of the sensor images as per:

$$w_k(j) = \sum_{i=1}^{q} \lambda_{i,k}(j) z_{i,k}(j) \qquad k = 1, \ldots, K; \ j = 1, \ldots, P \tag{8.13}$$

where $\sum_{i=1}^{q} \lambda_{i,k}(j) = 1$. In order to determine the $\lambda_{i,k}(j)$ in Equation 8.13, we employ a salience measure that was discussed in Ref. 26. For coefficient $z_{i,k}(j)$, let (x, y) denote its co-ordinates in a subband. Then the salience measure for this coefficient is computed from the weighted average of the coefficients in a window around it, as per:

$$\Omega_{i,k}(j) = \sum_{x'=-2}^{2} \sum_{y'=-2}^{2} p(x',y') z_i^2(x+x',y+y') \quad i=1,\ldots,q \qquad (8.14)$$

where $p(x',y')$ is the weight for each coefficient around (x,y) and $z_i(x+x',y+y')$ denotes the coefficient value at co-ordinates $(x+x',y+y')$ in a subband. Here a five by five window of coefficients centered on (x,y) is used to calculate the salience measure of $z_{i,k}(j)$ using

$$\mathbf{p} = \begin{pmatrix} 1/48 & 1/48 & 1/48 & 1/48 & 1/48 \\ 1/48 & 1/24 & 1/24 & 1/24 & 1/48 \\ 1/48 & 1/24 & 1/3 & 1/24 & 1/48 \\ 1/48 & 1/24 & 1/24 & 1/24 & 1/48 \\ 1/48 & 1/48 & 1/48 & 1/48 & 1/48 \end{pmatrix} \qquad (8.15)$$

where $\mathbf{p}$ is the matrix notation of $p(x',y')$ in a five by five window. Then the $\lambda_{i,k}(j)$ for $i=1,\ldots,q$, are specified using:

$$\lambda_{i,k}(j) = \Omega_{i,k}(j) \bigg/ \sum_{l=1}^{q} \Omega_{l,k}(j) \quad i=1,\ldots,q \qquad (8.16)$$

A simple initialization for $\beta_{i,k}(j)$ is to assume that the true scene appears in each sensor image. Hence $\beta_{i,k}(j)=1$ for $i=1,\ldots,q; \; k=1,\ldots,K$ and $j=1,\ldots,P$. We assume the initial state probabilities $p_{S(j)}(m)=1/M$ for $m=1,\ldots,M$. At initialization, the parent–child transition probabilities are also assumed equal for all M states. Hence $a_{S(j),S(\rho(j))}^{m,n}=1/M$ for $m=1,\ldots,M$ and $n=1,\ldots,M$. In order to model the distortion in a robust way the distortion is initialized as impulsive.[20] Thus we set $\sigma_m^2(j)=\gamma\sigma_{m-1}^2(j)$ for $j=1,\ldots,P$ such that fixing γ and $\sigma_1^2(j)$ fixes $\sigma_m^2(j)$ for $m>1$. Then the value for $\sigma_1^2(j)$ is chosen so that the total variance of the mixture model, $\sigma^2(j)=\sum_{m=1}^{M}\sigma_m^2(j)/M$, matched the variance of the observations:

$$\sigma^2(j) = \sum_{i=1}^{q} \sum_{k=1}^{K} [z_{i,k}(j) - w_k(j)]^2 / qK \qquad (8.17)$$

We chose $\gamma=10$ so that the initial distortion model was fairly impulsive. This initialization scheme worked very well for the cases we have studied. We observed that the algorithm in our experiments generally converged in less than 15 iterations.

D. FUSION OF THE LL SUBBAND WAVELET COEFFICIENTS

Our HMT model applies to only the high-frequency subbands of the wavelet representations (see Figure 8.2) and so the above iterative estimation procedure produces only the fused high-frequency subbands. Therefore, the fused LL subband needs to be produced separately. We used the weighted average method in Section III.C to produce the fused LL subband. Let $w(x,y)$ denote the coefficient with co-ordinate (x,y) in the LL subband of fused image, and let $z_i(x,y)$ denote the coefficient with co-ordinate (x,y) in LL subband of sensor image i, with $i = 1, ..., q$. Then the coefficients in the LL subband are determined using:

$$w(x,y) = \sum_{i=1}^{q} \lambda_i(x,y)z_i(x,y) \tag{8.18}$$

where $\lambda_i(x,y)$ is the weight and $\sum_{i=1}^{q} \lambda_i(x,y) = 1$. We use the same method described in Equation 8.14, Equation 8.15, and Equation 8.16 to determine $\lambda_i(x,y)$ for $i = 1, ..., q$.

IV. EXPERIMENTAL RESULTS

We have applied this fusion algorithm to CWD applications. CWD is an increasingly important topic in the general area of law enforcement and image fusion has been identified as a key technology to enable progress on this topic.[7,27,28] With the increasing threat of terrorism, CWD is a very important technology. We also applied this algorithm to night vision applications.[29,30]

A. CWD WITH VISUAL AND MMW IMAGES

We used this algorithm to fuse the visual and MMW images[*] shown in Figure 8.5(a) and (b) for a CWD application. The order-3 Daubechies wavelet[31] is used to transform the visual and MMW images into multiscale representations with four decomposition levels. We created a two-state HMT model based on these wavelet representations and then use the algorithm described in Section III to fuse these two images. The fusion result is shown in Figure 8.5(c). We also used the EM-fusion algorithm presented at Ref. 22 to fuse these two images based on the wavelet representations. The fused result is shown in Figure 8.5(d). Figure 8.5(e) is the fused result obtained by using the same wavelet representations[†] and choosing the maximum sensor wavelet coefficients for the high-pass subband wavelet coefficients and averaging the sensor wavelet coefficients for the low-pass subband wavelet coefficients.[1] We call

---

[*] The source images were obtained from Thermotex Corporation
[†] We use the same wavelet representation for all wavelet methods used in each example we present in this chapter.

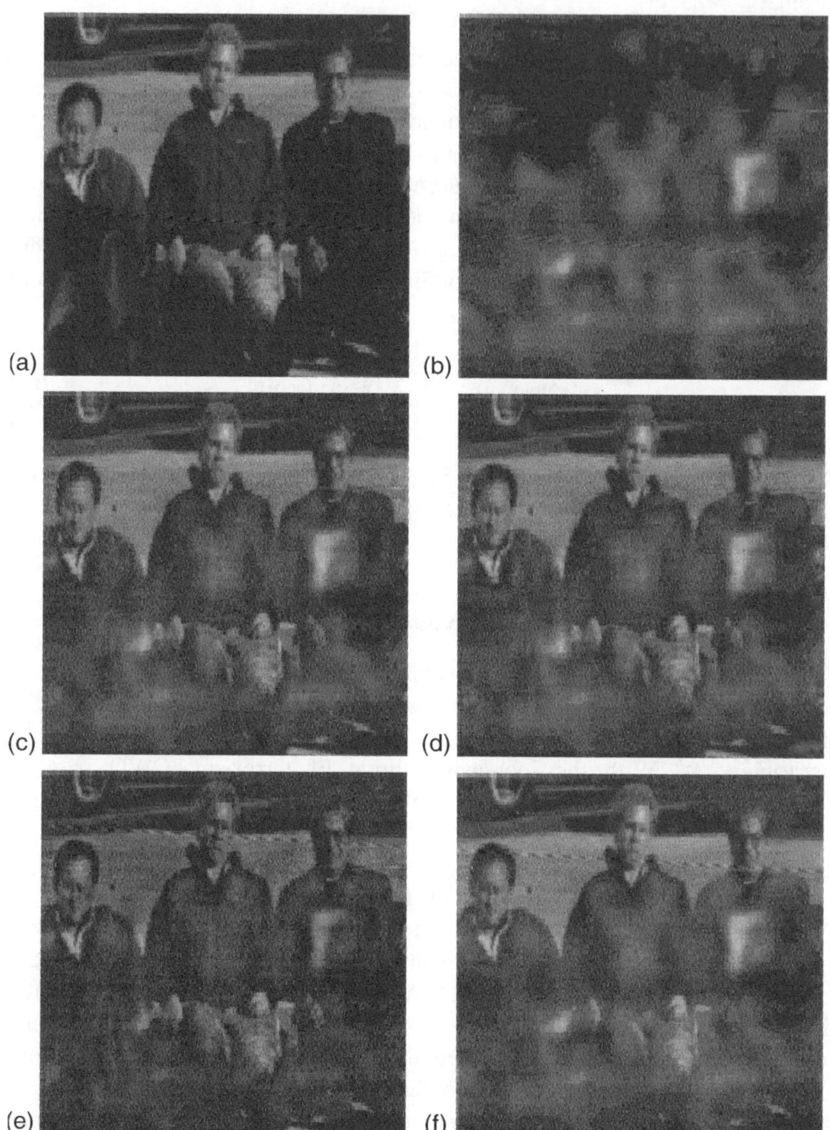

FIGURE 8.5 Visual and MMW images and fused result for CWD: (a) visual image, (b) MMW image, (c) HMT-based EM fusion, (d) EM fusion, (e) wavelet fusion, and (f) selecting maximum.

this wavelet fusion. Figure 8.5(f) is the fused result obtained by selecting the maximum pixel (no wavelet transform). We call this the selecting maximum algorithm. From the comparison, the HMT-based EM fusion algorithm performs better than the other three fusion methods. HMT-based EM fusion gives some improvement over the EM fusion method by considering the correlations of the coefficients at different scales. From the fused image, there is considerable evidence to suspect that the person on the right has a concealed gun beneath his clothes. This fused image may be very helpful to a police officer, for example, who must respond promptly.

B. CWD WITH VISUAL AND IR IMAGES

Another example considers a CWD application employing the visual and the IR images shown in Figure 8.6(a) and (b). The order-3 Daubechies wavelet[31] is used to transform the visual and IR images into wavelet representations employing six decomposition levels and a two-state HMT model is created. Figure 8.6(c) to (f) show the fused results obtained by the HMT-based EM fusion method, the EM fusion method in Ref. 22, the wavelet fusion method from Ref. 1, and the selecting maximum algorithm, respectively. In this case, the standard EM fusion method from Ref. 22 is slightly inferior to the wavelet fusion method from Ref. 1. However, after considering the correlations between different scale coefficients, the HMT-based EM fusion method seems to overcome any methodical limitations of Ref. 22. In comparison, the HMT-based EM fusion algorithm performs better than the EM fusion method from Ref. 22, the wavelet fusion method from Ref. 1, and the selecting maximum algorithm.

C. NIGHT VISION IMAGE FUSION APPLICATIONS

Some might suggest that in night vision applications, the most information is contained in the thermal images, with complementary information from the visual images. This is different from the daytime fusion system. In these cases, some might suggest that the most information is contained in the visual images. Hence, fusion methods which are good for daytime fusion systems may not work well for night vision applications. We applied the HMT-based EM fusion method for night vision applications with very good results.

Figure 8.7(a) and (b) show the IR and visual images. The order-3 Daubechies wavelet[31] is used to transform the IR and visual images into wavelet representations with five decomposition levels. A two-state HMT model is created. Figure 8.7(c) to (f) show the fused results obtained by the HMT-based EM fusion method, the EM fusion method in Ref. 22, the wavelet fusion method from Ref. 1, and selecting maximum algorithm, respectively. This example shows that the HMT-based EM fusion algorithm performs very well for night vision image fusion systems. In comparison, the HMT-based EM fusion algorithm performs better than the EM fusion from Ref. 22, the wavelet fusion method from Ref. 1, and the selecting maximum algorithm.

V. CONCLUSIONS

We have presented a new image fusion method based on a Gaussian mixture distortion model. In this method, we used a hidden Markov tree (HMT) model

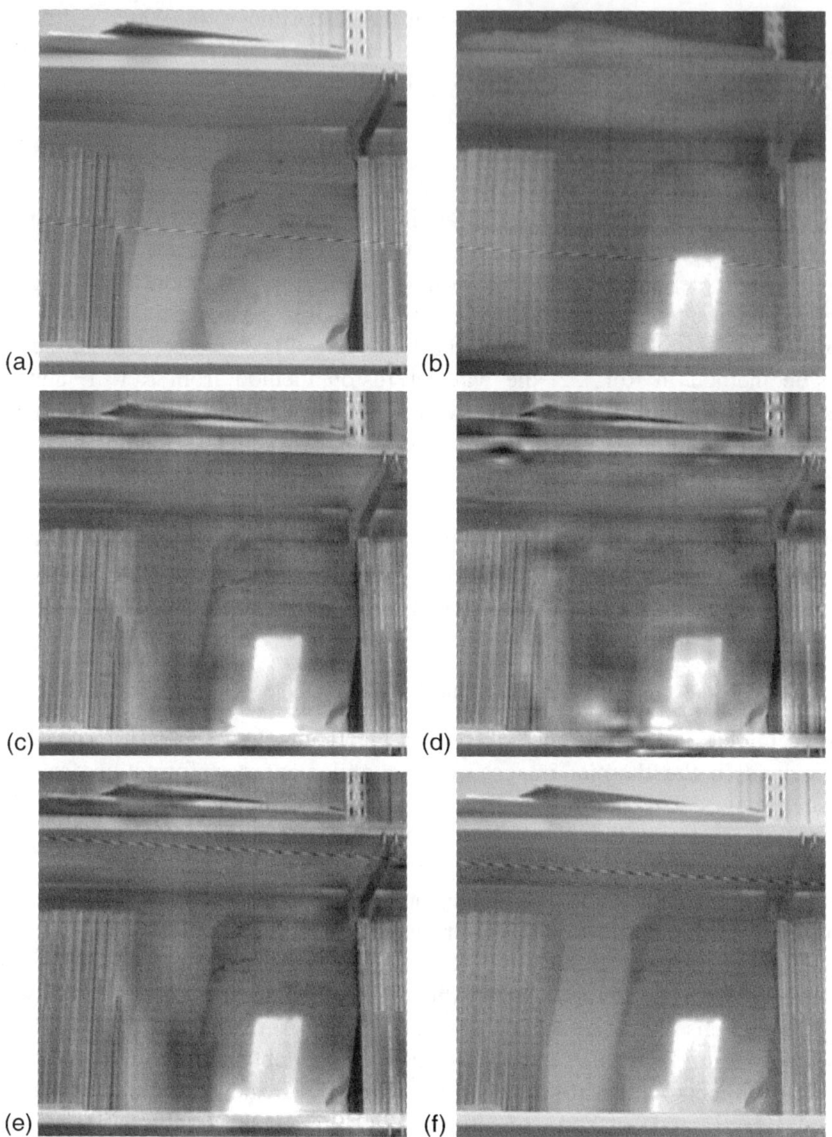

FIGURE 8.6 Visual and IR images and fused result for CWD:(a) visual image, (b) IR image, (c) HMT-based EM fusion, (d) EM fusion, (e) wavelet fusion, and (f) selecting maximum.

fitting to the wavelet multiscale representations of the sensor images so as to exploit the correlations between the wavelet coefficients across different scales. In our method, the fusion is performed over wavelet trees spanning the subbands of all wavelet scales. In most other wavelet-based fusion methods, the fusion is

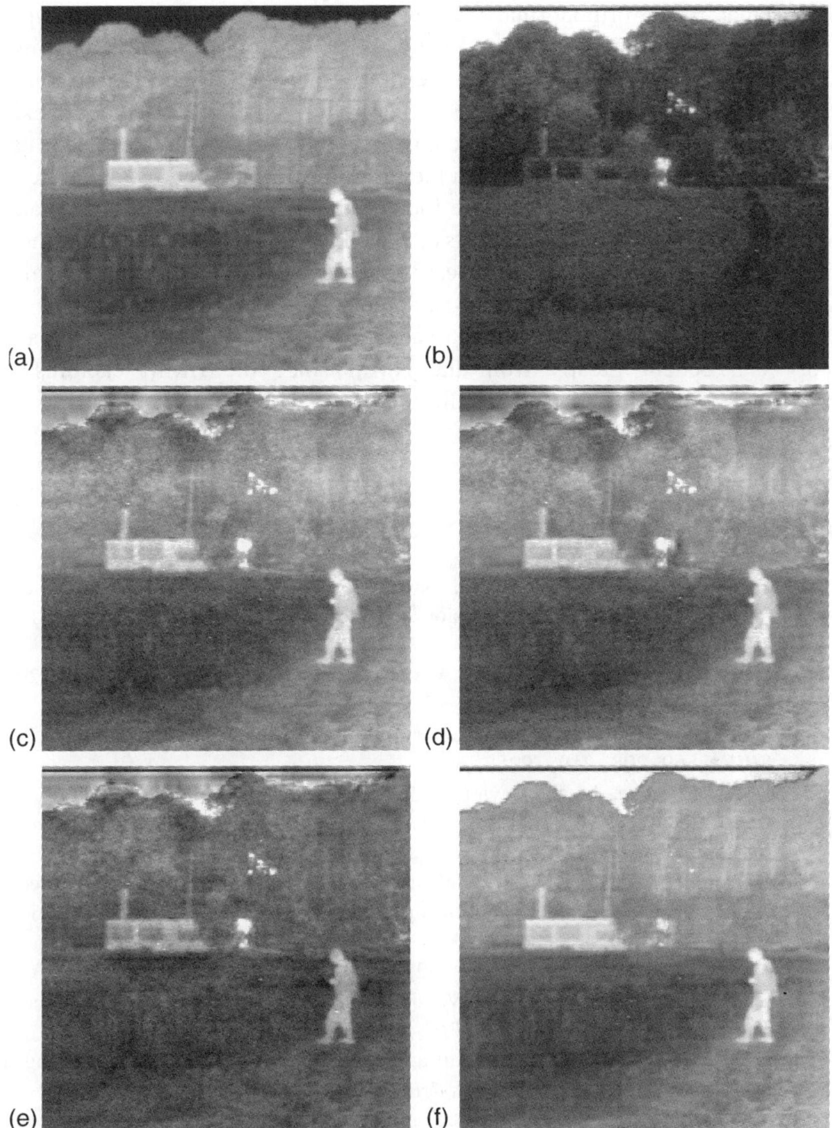

FIGURE 8.7 IR and night visual images and the fused results: (a) IR image, (b) visual image, (c) HMT-based EM fusion, (d) EM fusion, (e) wavelet fusion, and (f) selecting maximum.

performed subband by subband. We experimented with this method for CWD applications and night vision applications. The results showed the advantages of this HMT-based EM fusion approach. We also have studied the effect of the wavelet bases, number of wavelet decomposition levels, and the number of

HMT states. Some orthonormal wavelet bases[31] were used in our examples. When different orthonormal bases were used, we found there is not much difference between the fused results. Generally speaking, using more decomposition levels can be beneficial but this comes at the cost of increased complexity. In practice, we have found if a sufficient number of decomposition levels are used there is not much difference between the fused results obtained using a different number of decomposition levels. We found that a two-state HMT model can fit the distribution of wavelet coefficients data well. We also studied the HMT model with more than two states. In the cases we have studied, there is not much difference in the fused results for M-state models with $M > 2$.

We believe that wavelet-based image fusion algorithms using estimation theory can benefit greatly from the HMM since the HMM captures the correlations between subbands in different scales. The correlations are very helpful in estimation.

REFERENCES

1. Zhang, Z., and Blum, R. S., A categorization and study of multiscale-decomposition-based image fusion schemes, *Proc. IEEE*, 87(8), 1315–1328, 1999.
2. Huntsberger, T., and Jawerth, B., Wavelet based sensor fusion, *Proc. SPIE*, 2059, 488–498, 1993.
3. Chipman, L. J., Orr, T. M., and Graham, L. N., Wavelets and image fusion, *Proc. SPIE*, 2569, 208–219, 1995.
4. Koren, I., Laine, A., and Taylor, F., Image fusion using steerable dyadic wavelet transform, pp. 232–235. In *Proceedings of the IEEE International Conference on Image Processing*, 1995.
5. Wilson, T. A., Rogers, S. K., and Myers, L. R., Perceptual based hyperspectral image fusion using multiresolution analysis, *Opt. Eng.*, 34(11), 3154–3164, 1995.
6. Jiang, X., Zhou, L., and Gao, Z., Multispectral image fusion using wavelet transform, *Proc. SPIE*, 2898, 35–42, 1996.
7. Uner, M. K., Ramac, L. C., and Varshney, P. K., Concealed weapon detection: an image fusion approach, *Proc. SPIE*, 2942, 123–132, 1997.
8. Li, H., Manjunath, B. S., and Mitra, S. K., Multisensor image fusion using the wavelet transform, *Graphical Models Image Process.*, 57, 235–245, 1995.
9. Zhang, Z. and Blum, R. S., A region based image fusion scheme for concealed weapon detection, pp. 168–173. In *Proceedings of the 31th Annual Conference on Information Sciences and Systems*, 1997.
10. Shapiro, J., Embedded image coding using zerotrees of wavelet coefficients, *IEEE Trans. Signal Process.*, 41, 3445–3462, 1993.
11. Crouse, M., Nowak, R. D., and Baraniuk, R. G., Wavelet-based statistical signal processing using hidden Markov models, *IEEE Trans. Signal Process.*, 46(2), 886–902, 1998.
12. Ramchandran, K., and Orchard, M. T., An investigation of wavelet-based image coding using an entropy-constrained quantization framework, *IEEE Trans. Signal Process.*, 46(2), 342–353, 1998.

13. Mallat, S., and Zhong, S., Characterization of signals from multiscale edges, *IEEE Trans. Pattern Anal. Machine Intell.*, 14, 710–732, 1992.

14. Mallat, S., and Hwang, W., Singularity detection and processing with wavelets, *IEEE Trans. Inf. Theory*, 38, 617–643, 1992.

15. Rabiner, L., A tutorial on hidden Markov models and selected applications in speech recognition, *Proc. IEEE*, 77, 257–285, 1989.

16. Smyth, P., Heckerman, D., and Jordan, M., Probabilistic independence networks for hidden Markov probability models, *Neural Comput.*, 9(2), 227–269, 1997.

17. Dempster, A. P., Laird, N. M., and Rubin, D. B., Maximum likelihood from incomplete data via the EM algorithm, *J. Royal Stat. Soc. B*, 39(1), 1–38, 1977.

18. Redner, R. A., and Walker, H. F., Mixture densities, maximum likelihood and the EM algorithm, *SIAM Rev.*, 26(2), 195–239, 1984.

19. Fessler, J. A., and Hero, A. O., Space-alternating generalized expectation-maximization algorithm, *IEEE Trans. Signal Process.*, 42(10), 2664–2677, 1994.

20. Blum, R. S., Kozick, R. J., and Sadler, B. M., An adaptive spatial diversity receiver for non-Gaussian interference and noise, *IEEE Trans. Signal Process.*, 47(8), 2100–2111, 1999.

21. Blum, R. S., Kozick, R. J., and Sadler, B. M., EM-based approaches to adaptive signal detection on fading channels with impulsive noise, pp. 112–117. In *31th Annual Conference on Information Sciences and Systems,* Baltimore, MD, 1997.

22. Yang, J., and Blum, R. S., A statistical signal processing approach to image fusion for concealed weapon detection, pp. I-513–I-516. In *IEEE International Conference on Image Processing,* Rochester, NY, 2002.

23. Thulasiraman, K., and Swamy, M. N. S., *Graphs: Theory and Algorithms*, Wiley, New York, pp. 1–54, 1992.

24. Pearl, J., *Probabilistic Reasoning in Intelligent Systems: Networks of Plausible Inference*, Morgan Kaufmann, San Francisco, CA, pp. 143–238, 1988.

25. Ronen, O., Rohlicek, J., and Ostendorf, M., Parameter estimation of dependence tree models using the EM algorithm, *IEEE Signal Process. Lett.*, 2, 157–159, 1995.

26. Burt, P. J., and Kolczynski, R. J., Enhanced image capture through fusion, pp. 173–182. In *Proceedings of the 4th International Conference on Computer Vision*, 1993.

27. Ferris, D. D. Jr., McMillan, R. W., Currie, N. C., Wicks, M. C., and Slamani, M. A., Sensors for military special operations and law enforcement applications, *Proc. SPIE*, 3062, 173–180, 1997.

28. Slamani, M. A., Ramac, L., Uner, M., Varshney, P. K., Weiner, D. D., Alford, M. G., Ferris, D. D. Jr., and Vannicola, V. C., Enhancement and fusion of data for concealed weapons detection, *Proc. SPIE*, 3068, 8–19, 1997.

29. Reese, C. E., and Bender, E. J., Multispectral/image fused head tracked vision system (HTVS) for driving applications, *Proc. SPIE*, 4361, 1–11, 2001.

30. Bender, E. J., Reese, C. E., and VanDerWal, G. S., Comparison of additive image fusion vs. feature-level image fusion techniques for enhanced night driving, *Proc. SPIE*, 4796, 140–151, 2003.

31. Daubechies, I., Orthonormal bases of compactly supported wavelets, *Commun. Pure Appl. Math.*, 41, 909–996, 1998.

APPENDIX A OUTLINE OF COMPUTING THE CONDITIONAL PROBABILITIES

A.1. DERIVATION OF CONDITIONAL PROBABILITIES

Let $\mathbf{z}_{i,k}$ denote the vector consisting of the wavelet coefficients for the kth tree in the ith image. Consider the computations of $p(S(j) = m|\mathbf{z}_{i,k}, \mathbf{\Phi}), j = 1,...,P$ and $p(S(j) = m, S(\rho(j)) = n|\mathbf{z}_{i,k}, \mathbf{\Phi}), j = 1,...,P$, where $S(j)$ is the shorthand notation for $S_{i,k}(j)$ since we focus on the kth tree in ith image.

We now focus on the kth tree to simplify notation. For the kth tree, we define T_j to be the subtree of observed wavelet coefficients with a root at node j, so that the subtree T_j contains coefficient $z(j)$ and all of its descendants. If T_l is a subtree of T_j (i.e., $z(l)$ and all its descendants are members of T_j), then we define $T_{j\setminus l}$ to be the set of wavelet coefficients obtained by removing the subtree T_l from T_j. Let $z(1)$ denote the root node of the kth tree. Thus, T_1 denotes the kth tree and we have $T_1 = \mathbf{z}_{i,k}$.

To simplify the following discussion, we define further notation. Let $f(\cdot)$ denote the PDF of its argument (excusing the abuse of notation for its brevity). For each subtree T_j, we define the conditional likelihoods

$$B_j(m) = f(T_j|S(j) = m, \mathbf{\Phi}) \tag{8.19}$$

$$B_{j,\rho(j)}(m) = f(T_j|S(\rho(j)) = m, \mathbf{\Phi}) \tag{8.20}$$

$$B_{\rho(j)\setminus j}(m) = f(T_{\rho(j)\setminus j}|S(\rho(j)) = m, \mathbf{\Phi}) \tag{8.21}$$

We note that a PDF times the appropriate differential can be thought of as the probability that the random variable or vector is in some incrementally small region. Thus, we express the joint probability that a discrete random variable $S(j)$ equals a particular value m and that a continuous random variable is in some incrementally small region as

$$A_j(m) = p(S(j) = m, T_{1\setminus j}|\mathbf{\Phi}) \tag{8.22}$$

after multiplication by the appropriate differential.

From the HMT properties, we know T_j and $T_{1\setminus j}$ are independent given the state variable $S(j)$. Therefore we obtain:

$$p(S(j) = m, T_1|\mathbf{\Phi}) = A_j(m)B_j(m) \tag{8.23}$$

$$p(S(j)=m,S(\rho(j))=n,T_1|\mathbf{\Phi})=B_j(m)B_{\rho(j)\setminus j}(n)a_{j,\rho(j)}^{m,n}A_{\rho(j)}(n) \tag{8.24}$$

The likelihood of $\mathbf{z}_{i,k}$ can be obtained from

$$f(\mathbf{z}_{i,k}|\mathbf{\Phi})=f(T_1|\mathbf{\Phi})=\sum_{m=1}^{M} p(S(j)=m,T_1|\mathbf{\Phi})=\sum_{m=1}^{M} B_j(m)A_j(m) \tag{8.25}$$

The marginal state PMF is obtained from Equation 8.23 and Equation 8.25 as

$$p(S(j)=m|\mathbf{z}_{i,k},\Phi)=\frac{A_j(m)B_j(m)}{\displaystyle\sum_{n=1}^{M}A_j(n)B_j(n)} \tag{8.26}$$

and the parent–child PMF is obtained from Equation 8.24 and Equation 8.25 as

$$p(S(j)=m,S(\rho(j))=n|\mathbf{z}_{i,k},\Phi)=\frac{B_j(m)a_{j,\rho(j)}^{m,n}A_{\rho(j)}(n)B_{\rho(j)\backslash j}(n)}{\displaystyle\sum_{l=1}^{M}A_j(l)B_j(l)} \tag{8.27}$$

A.2. Upward–Downward Algorithm

The upward–downward algorithm[11,24,25] is used to produce the quantities needed for calculating the conditional probabilities in Equation 8.26 and Equation 8.27. The upward step in the algorithm produces the B coefficients and the downward step produces the A coefficients. We assume the wavelet transform uses L decomposition levels and that $J=1$ is the index for the decomposition level with the finest scale, while $J=L$ is the index for the decomposition level with the coarsest scale. Recall that $\rho(j)$ denotes the parent of node j and $c(j)$ denotes the children of node j. Define the shorthand notation

$$g(z;\beta,w,\sigma^2)=\frac{1}{\sqrt{2\pi\sigma^2}}\exp\left\{-\frac{(z-\beta w)^2}{2\sigma^2}\right\} \tag{8.28}$$

The upward–downward algorithm is performed over one wavelet quadtree in the following two steps (for simplification, we omit the subscripts in $z_{i,k}(j)$, $\beta_{i,k}(j)$ and $w_k(j)$ in the upward and downward steps since this algorithm is applied to a given quadtree for a particular image thus a given (i,k).

A.2.1. Upward Step

Initialize: For all state variables $S(j)$ at the finest scale $J=1$, calculate for $m=1,...,M$:

$$B_j(m)=g(z(j);\beta(j),w(j),\sigma_m^2(j)) \tag{8.29}$$

1. For all state variables $S(j)$ at scale J, compute for $m = 1, \dots, M$:

$$B_{j,\rho(j)}(m) = \sum_{n=1}^{M} a_{j,\rho(j)}^{n,m} B_j(n) \tag{8.30}$$

$$B_{\rho(j)}(m) = \ g(z(\rho(j)); \beta(\rho(j)), w(\rho(j)), \ \sigma_m^2(\rho(j))) \cdot \prod_{j \in c(\rho(j))} B_{j,\rho(j)}(m) \tag{8.31}$$

$$B_{\rho(j)\backslash j}(m) = \frac{B_{\rho(j)}(m)}{B_{j,\rho(j)}(m)} \tag{8.32}$$

2. Set $J = J + 1$, move up the tree one scale.
3. If $J = L$, stop; else return to Step 1.

A.2.2. Downward Step

Initialize: For state variable $S(1)$ at the coarsest scale $J = L$, set for $m = 1, \dots, M$;

$$A_1(m) = p_{S(1)}(m) \tag{8.33}$$

1. Set $J = J - 1$, move down the tree one scale.
2. For all state variables $S(j)$ at scale J, compute for $m = 1, \dots, M$:

$$A_j(m) = \sum_{n-1}^{M} a_{j,\rho(j)}^{m,n} A_{\rho(j)}(n) B_{\rho(j)\backslash j}(n) \tag{8.34}$$

3. If $J = 1$, stop; else return to Step 1.

APPENDIX B OUTLINE OF THE DERIVATION OF THE UPDATE EQUATIONS

The EM algorithm provides a general approach to the iterative computation of maximum-likelihood estimates from incomplete data.[17] Let $\mathbf{X}$ denote the incomplete data set and let $\mathbf{X}_c$ denote the complete data set. From the image formation model (Equation 8.1 and Equation 8.5) we have

$$\mathbf{X} = \left\{ z_{i,k}(j) : i = 1, \dots, q; \ k = 1, \dots, K; \ j = 1, \dots, P \right\} \tag{8.35}$$

and

$$\mathbf{X}_c = \left\{ (z_{i,k}(j), p_{S(j)}(m)) : i = 1, \dots, q; \ l = 1, \dots L; \ j = 1, \dots, P \right\} \tag{8.36}$$

Here we assume that each quadtree employs the same distortion model as shown in Equation 8.5. Further, we assume $p_{S(j)}(m)$ identifies the probability that $\varepsilon_{i,k}(j)$

from Equation 8.1 comes from a particular term m in the Gaussian mixture PDF (Equation 8.5) which models $\varepsilon_{i,k}(j)$, the additive distortion in the observation $z_{i,k}(j)$. The common parameter set is $\mathbf{\Phi}$ defined in Equation 8.6. The conditional probabilities used in the iterative EM algorithm, $p(S_{i,k}(j) = m|z, \mathbf{\Phi})$ and $p(S_{i,k}(j) = m, S_{i,k}(\rho(j)) = n|\mathbf{z}, \mathbf{\Phi})$, are derived in Appendix A.

Let $\mathbf{S} = \{S_{i,k}(j) : \ i = 1, ..., q; \ k = 1, ..., K; \ j = 1, ..., P\}$ denote the complete set of state variables. Each iteration of the EM algorithm involves two steps, the expectation step (E-step) and the maximization step (M-step).[17] The E-step of the EM algorithm performs an average over the complete data, conditioned upon the incomplete data to produce the cost function.

$$Q(\mathbf{\Phi}'|\mathbf{\Phi}) = E_S\{\ln f_z(\mathbf{z}, \mathbf{S}|\mathbf{\Phi}')|\mathbf{z}, \mathbf{\Phi}\}$$

$$= E_S\{\ln p_S(\mathbf{S}|\mathbf{\Phi}') + \ln f_z(\mathbf{z}|\mathbf{S}, \mathbf{\Phi}')|\mathbf{z}, \mathbf{\Phi}\}$$

$$= E_S\left\{\ln\left[\prod_{i=1}^{q}\prod_{k=1}^{K} p(S_{i,k}(1) = m|\mathbf{\Phi}')\right]\bigg|\mathbf{z}, \mathbf{\Phi}\right\}$$

$$+ E_S\left\{\ln\left[\prod_{i=1}^{q}\prod_{k=1}^{K}\prod_{j=2}^{P} p(S_{i,k}(j) = m|S_{i,k}(\rho(j)) = n, \mathbf{\Phi}')\right]\bigg|\mathbf{z}, \mathbf{\Phi}\right\}$$

$$+ E_S\left\{\ln\left[\prod_{i=1}^{q}\prod_{k=1}^{K}\prod_{j=1}^{P} f_{z_{i,k}(j)}(z_{i,k}(j)|S_{i,k}(j) = m, \mathbf{\Phi}')\right]\bigg|\mathbf{z}, \mathbf{\Phi}\right\}$$

$$= E_S\left\{\sum_{i=1}^{q}\sum_{k=1}^{K} \ln p(S(1) = m|\mathbf{\Phi}')|\mathbf{z}, \mathbf{\Phi}\right\}$$

$$+ E_S\left\{\sum_{i=1}^{q}\sum_{k=1}^{K}\sum_{j=2}^{P} p(S(j) = m|S(\rho(j)) = n, \mathbf{\Phi}')|\mathbf{z}, \mathbf{\Phi}\right\}$$

$$+ E_S\left\{\sum_{i=1}^{q}\sum_{k=1}^{K}\sum_{j=1}^{P} \ln f_{z_{i,k}(j)}(z_{i,k}(j)|S_{i,k}(j) = m, \mathbf{\Phi}')|\mathbf{z}, \mathbf{\Phi}\right\}$$

$$= F_1 + F_2 + F_3 \tag{8.37}$$

Here, E_S implies an average over $S_{1,1}(1), ..., S_{q,K}(1)$, and $S_{1,1}(2), ..., S_{q,K}(P)$. Given the conditioning, we use the current (old) parameters in the distribution used in this averaging. Note that in the final result in (8.37), we have assumed the probabilities $p(S_{i,k}(j) = m|S_{i,k}(\rho(j)) = n, \mathbf{\Phi}')$ and $p(S_{i,k}(1) = m|\mathbf{\Phi}')$ are independent of i and k. Hence, we denote $p(S_{i,k}(1) = m|\mathbf{\Phi}')$ by $p(S(1) = m|\mathbf{\Phi}')$ in

Equation 8.37 to remind the reader. Likewise we write $p(S_{i,k}(j) = m | S_{i,k}(\rho(j)) = n, \mathbf{\Phi}')$ as $p(S(j) = m | S(\rho(j)) = n, \mathbf{\Phi}')$ in Equation 8.37. The EM algorithm updates the parameter estimates to new values $\mathbf{\Phi}'$, defined by Equation 8.7, that maximize $Q(\mathbf{\Phi}'|\mathbf{\Phi})$ in Equation 8.37. This is called the M-step of the EM algorithm.

In order to maximize $Q(\mathbf{\Phi}'|\mathbf{\Phi})$ analytically, we update each parameter one at a time. We write out the first term in Equation 8.37 as

$$F_1 = E_S\left\{ \sum_{i=1}^{q} \sum_{k=1}^{K} \ln p_{S(1)}(S(1) = m | \mathbf{\Phi}') | \mathbf{z}, \mathbf{\Phi} \right\}$$

$$= \sum_{i=1}^{q} \sum_{k=1}^{K} \sum_{m=1}^{M} \ln p'(S(1) = m) \cdot p(S_{i,k}(1) = m | \mathbf{z}, \mathbf{\Phi})$$

$$= \sum_{i=1}^{q} \sum_{k=1}^{K} \sum_{m=1}^{M} \ln p'_{S(1)}(m) \cdot p(S_{i,k}(1) = m | \mathbf{z}, \mathbf{\Phi}) \tag{8.38}$$

where $p'_{S(1)}(m) = p(S(1) = m)$. The update estimate for $p_{S(1)}(m)$ is obtained from maximizing Equation 8.38 analytically by solving $\partial F_1/\partial p'_{S(1)}(m) = 0$ subject to $\sum_{m=1}^{M} p'_{S(1)}(m) = 1$, for $m = 1, \ldots, M$. The update estimate for $p_{S(1)}(m)$ is

$$p'_{S(1)}(m) = \frac{\displaystyle\sum_{i=1}^{q} \sum_{k=1}^{K} p(S_{i,k}(1) = m | \mathbf{z}, \mathbf{\Phi})}{Kq} \tag{8.39}$$

This is the result of Equation 8.8 for $j = 1$.

The second term in Equation 8.37 can be written as

$$F_2 = E_S\left\{ \sum_{i=1}^{q} \sum_{k=1}^{K} \sum_{j=2}^{P} p(S(j) = m | S(\rho(j)) = n, \mathbf{\Phi}') | \mathbf{z}, \mathbf{\Phi} \right\}$$

$$= \sum_{i=1}^{q} \sum_{k=1}^{K} \sum_{j=2}^{P} \sum_{m=1}^{M} \sum_{n=1}^{M} \ln a'^{m,n}_{j,\rho(j)} \cdot p(S_{i,k}(j) = m, S_{i,k}(\rho(j)) = n | \mathbf{z}, \mathbf{\Phi}) \tag{8.40}$$

where we use $a'^{m,n}_{j,\rho(j)} = P'\{S(j) = m | S(\rho(j)) = n\}$ as discussed after 8.2. The updated estimate for $a^{m,n}_{j,\rho(j)}$ in Equation 8.9 is obtained from maximizing Equation 8.40 analytically by solving $\partial F_2/\partial a'^{m,n}_{j,\rho(j)} = 0$, subject to $\sum_{m=1}^{M} \sum_{n=1}^{M} a'^{m,n}_{j,\rho(j)} p_{S(\rho(j))}(n) = 1$, for $j = 1, \ldots, P; m = 1, \ldots, M$.

Now consider the derivation of Equation 8.8 for $j = 2, \ldots, P$. For any node $j = 2, \ldots, P$, there is the relationship between it and its parent node as

$$p'_{S(j)}(m) = \sum_{n=1}^{M} p'_{S(\rho(j))}(n) a'^{m,n}_{j,\rho(j)} \tag{8.41}$$

Plugging Equation 8.9 into Equation 8.41 and computing the sum over n, we can obtain Equation 8.8 for $j = 2, \ldots, P$ in Section III.B.

The third term in Equation 8.37 can be written as

$$F_3 = E_S\left\{\sum_{i=1}^{q}\sum_{k=1}^{K}\sum_{j=2}^{P}\ln f_{z_{i,k}(j)}(z_{i,k}(j)|S_{i,k}(j)=m,\mathbf{\Phi}')|\mathbf{z},\mathbf{\Phi}\right\}$$

$$=\sum_{i=1}^{q}\sum_{k=1}^{K}\sum_{j=1}^{P}\sum_{m=1}^{M}\left\{\ln\frac{1}{\sqrt{2\pi\sigma_m'^2(j)}}-\frac{(z_{i,k}(j)-\beta'_{i,k}(j)w'_k(j))^2}{2\sigma_m'^2(j)}\right\}$$

$$\times\, p(S_{i,k}(j)=m|\mathbf{z},\mathbf{\Phi})$$

$$=B-\frac{1}{2}\sum_{i=1}^{q}\sum_{k=1}^{K}\sum_{j=1}^{P}\sum_{m=1}^{M}\left\{\ln\sigma_m'^2(j)+\frac{(z_{i,k}(j)-\beta'_{i,k}(j)w'_k(j))^2}{\sigma_m'^2(j)}\right\}$$

$$\times\, p(S_{i,k}(j)=m|\mathbf{z},\mathbf{\Phi}) \tag{8.42}$$

where B is a term that is independent of $\mathbf{\Phi}'$. To simplify we use the SAGE version of EM[19] which allows us to update one parameter at a time. The updated estimate for $\sigma_m^2(j)$ in Equation 8.10 is obtained from maximizing Equation 8.42 analytically by solving $\partial F_3/\partial\sigma_m'^2(j)=0$ for $j=1,...,P$; $m=1,...,M$, and then using the updated $p'_{S(j)}(m)$, $a'^{m,n}_{j,\rho(j)}$, and other old parameter values. Because $\beta_{i,k}(j)$ is discrete, $\beta'_{i,k}(j)$ is updated to have the value from the set $\{0,-1,+1\}$ that maximizes Equation 8.42 with $p'_{S(j)}(m)$, $a'^{m,n}_{j,\rho(j)}$, $\sigma_m^2(j)$, and all the other parameters set at their old values. The updated estimate for $w_k(j)$ in Equation 8.12 is obtained from maximizing Equation 8.42 analytically by solving $\partial F_3/\partial w'_k(j)=0$ for $k=1,...,K$; $j=1,...,P$, using the updated parameters where possible.

9 Multimodal Human Recognition Systems

Arun Ross and Anil K. Jain

CONTENTS

I. INTRODUCTION

Establishing the identity of a person is becoming critical in our vastly interconnected society. Questions like "Is she really who she claims to be?", "Is this person authorized to use this facility?", or "Is he on the watch list posted by the government?" are routinely being posed in a variety of scenarios ranging from the issuance of driver licenses to gaining access to a secure facility. The need for reliable user authentication techniques has increased in the wake of heightened concerns about security and rapid advancements in networking, communication, and mobility. Traditionally, passwords (knowledge-based security) and ID cards (token-based security) have been used to establish identity and moderate access to secure systems. However, security can be easily breached in such systems when a password is divulged to an unauthorized user or a card is stolen by an impostor. Furthermore, these methods cannot be reliably used to counter repudiation claims wherein a user accesses a privilege and later denies using it.

Biometrics, described as the science of recognizing an individual based on physiological or behavioral traits, is beginning to gain acceptance as a legitimate method of determining an individual's identity.[1] Biometric systems have now been deployed in various commercial, civilian, and forensic applications as a means of establishing identity. These systems rely on the evidence of fingerprints, hand geometry, iris, retina, face, hand vein, facial thermogram, signature, voice, and so on, to either validate or determine an identity (Figure 9.1). The Schiphol Privium scheme at the Amsterdam airport, for example, employs iris scan cards to speed up the passport and visa control procedures. Passengers enrolled in this scheme insert their card at the gate and look into a camera; the camera acquires the eye image of the traveler which is then processed to locate the iris and compute the iris code;[2] the computed iris code is compared with the data residing in the card to complete user verification. A similar scheme is also being used to verify the identity of Schiphol airport employees working in high-security areas. Thus, biometric systems can be used to enhance user convenience while improving security.

A generic biometric system has four important modules: (1) the sensor module which captures the *trait* in the form of raw *data*; (2) the feature extraction module which processes the data to extract a *feature set* that is a compact representation of the trait; (3) the matching module which employs a classifier to compare the extracted feature set with the *templates* residing in the database to generate match scores; (4) the decision module which uses the matching scores to either determine an identity or validate a claimed identity. Most biometric systems deployed in real-world applications are unimodal, that is, they rely on the evidence of a single source of biometric information for authentication

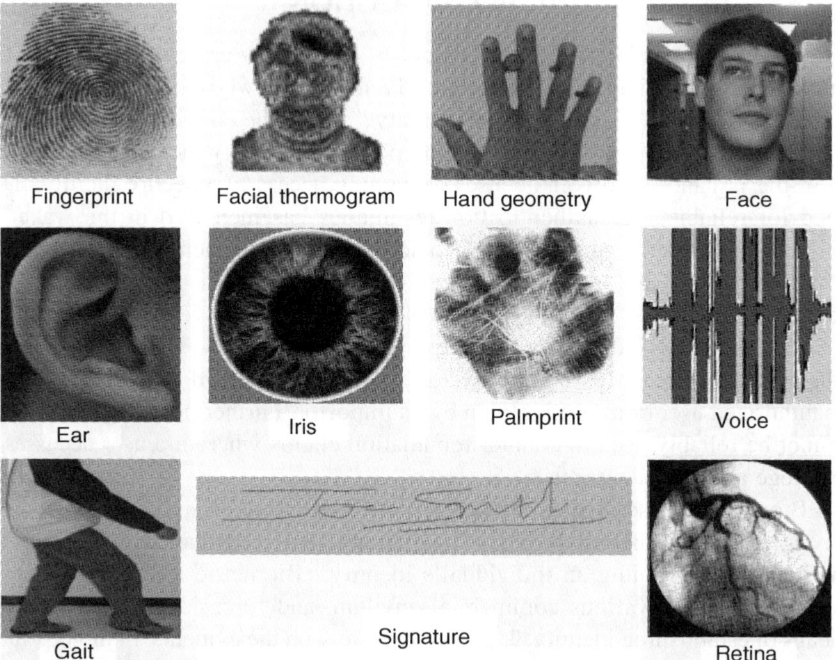

FIGURE 9.1 Examples of some of the biometric traits that can be used for authenticating an individual.

(e.g., a single fingerprint or face). Besides having high error rates (see Table 9.1), these systems have to contend with a variety of problems as indicated below:

1. *Noise in sensed data*: Some examples of noisy data are a fingerprint image with a scar or a voice sample altered by a cold. Noisy data could also result from defective or improperly maintained sensors (e.g., accumulation of dirt on a fingerprint sensor) or unfavorable ambient conditions (e.g., poor illumination of a user's face in a face recognition system). Noisy biometric data may be incorrectly matched with templates in the database, resulting in a user being incorrectly rejected.

2. *Intraclass variations*: These variations are typically caused by a user who is incorrectly interacting with the sensor (e.g., incorrect facial pose; see Figure 9.2), or due to changes in the biometric characteristics of a user over a period of time (e.g., change in width and length of the hand). Intraclass variations are typically handled by storing multiple templates for every user and updating these templates over time.

3. *Interclass similarities*: In a biometric system comprising a large number of users, there may be interclass similarities (overlap) in the feature space of multiple users. Golfarelli et al.[3] state that the number of distinguishable patterns in two of the most commonly used representations of hand geometry and face are only of the order of 10^5 and 10^3, respectively.

4. *Nonuniversality*: The biometric system may not be able to acquire meaningful biometric data from a subset of users. A fingerprint biometric system, for example, may extract incorrect minutiae features from the fingerprints of certain individuals, due to the poor quality of

---

TABLE 9.1
Error Rates Associated with Fingerprint, Face, and Voice Biometric Systems

	Test	Test Parameter	False Reject Rate (%)	False Accept Rate (%)
Fingerprint	FVC 2002[32]	Users mostly in age group 20–39	0.2	0.2
Face	FRVT 2002[33]	Enrolment and test images were collected in indoor environment and could be on different days	10	1
Voice	NIST 2000[34]	Text dependent	10–20	2–5

The accuracy estimates of biometric systems depend on a number of test conditions.

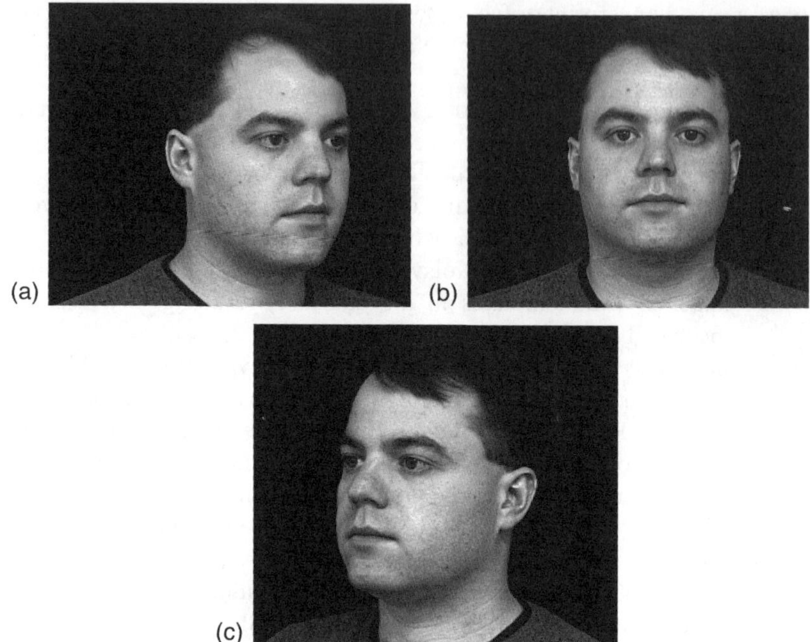

FIGURE 9.2 Intraclass variation associated with an individual's face image. Due to change in pose, an appearance-based face recognition system will not be able to match these three images successfully, although they belong to the same individual.[35]

the ridges (Figure 9.3). Thus, there is a failure to enroll (FTE) rate associated with using a single biometric trait.

5. *Spoof attacks*: This type of attack is especially relevant when behavioral traits such as signature[4] and voice[5] are used. However, physical traits such as fingerprints are also susceptible to spoof attacks.[6]

Some of the limitations imposed by unimodal biometric systems can be overcome by including multiple sources of biometric information for establishing identity.[7] Such systems, known as *multimodal biometric systems*, are expected to be more reliable due to the presence of multiple, (fairly) independent pieces of evidence.[8] These systems are able to meet the stringent performance requirements imposed by high-security applications such as accessing a nuclear facility. They address the problem of nonuniversality, since the use of multiple traits ensures that more users can be accommodated in the system (i.e., it increases population coverage). Multimodal systems also deter spoofing since it would be difficult for an impostor to spoof multiple biometric traits of a genuine user simultaneously. Furthermore, they can support a challenge–response type of mechanism by requesting the user present a random subset of biometric traits thereby ensuring that a "live" user is indeed present at the point of data acquisition.

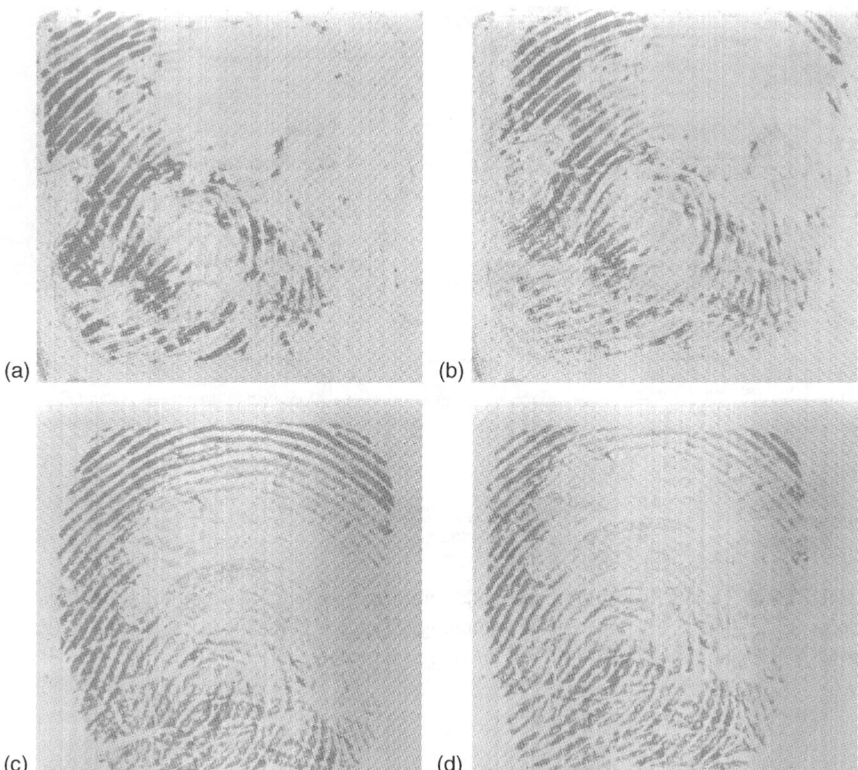

FIGURE 9.3 Non-universality of fingerprints. The four impressions of a user's fingerprint shown here cannot be enrolled in some fingerprint systems, due to the poor image quality of the ridges. Consequently, alternate biometrics must be adopted in order to include this user in an identity management system.

In this chapter we examine the levels of fusion and the various scenarios that are possible in a multimodal biometric system. We also discuss the different modes of operation, the integration strategies adopted to consolidate evidence, and issues related to the design and deployment of these systems.

II. LEVELS OF FUSION

In a multimodal biometric system information reconciliation can occur in any one of the modules constituting the biometric system (see Figure 9.4).

1. *Fusion at the data or feature level*: In this level, either the raw data or the feature sets originating from multiple sensors/sources are fused.
2. *Fusion at the match score level*: Each classifier outputs a match score indicating the proximity of the two feature sets that are

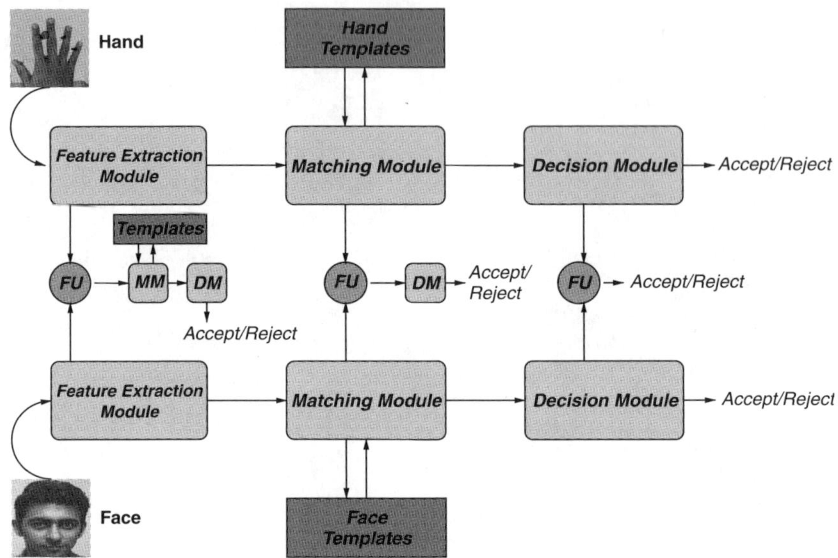

FIGURE 9.4 Levels of fusion in a bimodal biometric system operating in the verification mode; FU: Fusion Module, MM: Matching Module, DM: Decision Module. (From Ross, A. and Jain, A. K., Information fusion in biometrics, *Pattern Recognit Lett.*, 24(13), 2115–2125, 2003.)

being compared. The scores generated by the classifiers pertaining to different modalities are combined.

3. *Fusion at the decision level*: The final outputs[*] of multiple classifiers are consolidated *via* techniques such as majority voting.[9]

Biometric systems that integrate information at an early stage of processing are believed to be more effective than those systems which perform integration at a later stage. Since the feature set contains richer information about the input biometric data than the matching score or the output decision of a matcher, fusion at the feature level is expected to provide better recognition results. However, fusion at this level is difficult to achieve in practice because (1) the feature sets of the various modalities may not be compatible (e.g., eigen-coefficients of the face and minutiae sets of fingers), and (2) most commercial biometric systems do not provide access to the feature sets (nor the raw data) which they use in their products. Fusion at the decision level is considered to be rigid due to the availability of limited information. Thus, fusion at the match score level is

---

[*] In a verification system the output is an "accept" or a "reject" while in an identification system the output is the identity label of an enrolled user.

usually preferred, as it is relatively easy to access and combine the scores presented by the different modalities.

III. FUSION SCENARIOS

Depending on the number of traits, sensors, and feature sets used, a variety of scenarios are possible in a multimodal biometric system.

1. *Single biometric trait, multiple sensors*: Here, multiple sensors are used to record the same biometric trait. Thus, raw biometric data pertaining to different sensors are obtained. Chang et al.[10] acquire both two and three dimensional images of the face and combine them at the data level as well as the match score level to improve the performance of a face recognition system. Kumar et al.[11] describe a hand-based verification system that combines the geometric features of the hand with palm prints at the feature and match score levels. Interestingly, in their experiments, fusion at the match score level results in better performance than fusion at the feature level. This could be due to the high dimensionality of the fused feature set (the "curse-of-dimensionality" problem) and, therefore, the application of a feature reduction technique may have been appropriate (see Section V).

2. *Single biometric trait, multiple classifiers*: Unlike the previous scenario, only a single sensor is employed to obtain raw data; this data is then used by multiple classifiers. Each of these classifiers either operates on the same feature set extracted from the data or generates its own feature sets. Jain et al.[12] use the logistic function to integrate the matching scores obtained from three different fingerprint matchers operating on the same minutiae sets (also see Ref. 13). Ross et al.[14] combine the matching score of a minutiae-based fingerprint matcher with that of a texture-based matcher to improve matching performance. Lu et al. extract three different types of feature sets from the face image of a subject (using Principal Component Analysis (PCA), Linear Discriminant Analysis (LDA), and Independent Component Analysis (ICA)) and integrate the output of the corresponding classifiers at the match score level.[15]

3. *Single biometric trait, multiple units*: In the case of fingerprints (or irises), it is possible to integrate information presented by two or more fingers (or both the irises) of a single user. This is an inexpensive way of improving system performance since this does not entail deploying multiple sensors, incorporating additional feature extraction, or matching modules.

4. *Multiple biometric traits*: Here, multiple biometric traits of an individual are used to establish the identity. Such systems employ multiple sensors to acquire data pertaining to different traits. The independence of the traits ensures that a significant improvement in performance is obtained. Brunelli et al.[16] use the face and voice traits of an individual for identification. A HyperBF network is used to combine the normalized scores of five different classifiers operating on the voice and face feature sets of a user. Bigun et al. develop a statistical framework based on Bayesian statistics to integrate the speech (text dependent) and face data of a user.[17] The estimated biases of each classifier are taken into account during the fusion process. Hong and Jain associate different confidence measures with the individual matchers when integrating the face and fingerprint traits of a user.[18] They also suggest an indexing mechanism wherein face information is first used to retrieve a set of possible identities and the fingerprint information is then used to select a single identity. A commercial product called BioID[19] uses the voice, lip motion, and face features of a user simultaneously in order to verify identity.

IV. MODES OF OPERATION

A multimodal system can operate in one of three different modes: serial mode, parallel mode, or hierarchical mode. In the serial mode of operation, the output of one modality is typically used to narrow down the number of possible identities before the next modality is used.[18] Therefore, multiple sources of information (e.g., multiple traits) do *not* have to be acquired simultaneously. Further, a decision could be made before acquiring *all* the traits. This can reduce the overall recognition time. In the parallel mode of operation, information from multiple modalities is used simultaneously in order to perform recognition. In the hierarchical scheme, individual classifiers are combined in a treelike structure. This mode is relevant when the number of classifiers is large.

V. INTEGRATION STRATEGIES

The strategy adopted for integration depends on the level at which fusion is performed. Fusion at the feature level can be accomplished by concatenating two *compatible* feature sets. Feature selection/reduction techniques may be employed to identify a subset of feature attributes that result in good performance. Fusion at the match score level has been well studied in the literature.[20,21] Robust and efficient normalization techniques are necessary to transform the scores of multiple matchers into a common domain prior to consolidating them.[22] In the

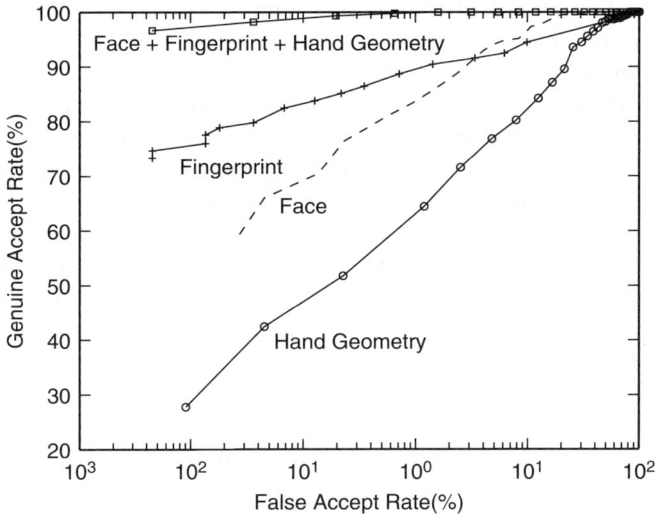

FIGURE 9.5 Performance gain obtained using the sum rule. (From Ross, A., and Jain, A. K., Information fusion in biometrics, *Pattern Recognit. Lett.*, 24(13), 2115–2125, 2003.)

context of verification, two distinct strategies exist for fusion at this level. In the first approach, the fusion is viewed as a classification problem where a feature vector is constructed using the matching scores output by the individual matchers; this feature vector is then classified into one of two classes: "accept" (genuine user) or "reject" (impostor).[23] In the second approach, the fusion is viewed as a combination problem where the individual matching scores are combined to generate a single scalar score which is then used to make the final decision.[24,25] General strategies for combining multiple classifiers have been suggested in Refs. 26,27. Ross and Jain have shown[7] that the simple sum rule is sufficient to obtain a significant improvement in the matching performance of a multimodal biometric system (Figure 9.5). They also suggest a technique to incorporate user-specific weights to further improve the system performance.[28] Fusion strategies at the decision level include majority voting,[9] behavior knowledge space method,[29] weighted voting based on the Dempster–Shafer theory of evidence,[30] AND/OR rules,[31] and so on.

VI. DESIGN ISSUES

A variety of factors should be considered when designing a multimodal biometric system. These include (1) the choice and number of biometric traits, (2) the level in the biometric system at which information provided by multiple traits should be integrated, (3) the methodology adopted to integrate the information, and (4) the cost versus matching performance trade-off. The choice and number of biometric traits is largely driven by the nature of the application, the overhead introduced by multiple traits (computational demands and cost, for example), and the correlation between the traits considered (negatively correlated information is preferred since the performance improvement is more pronounced in this case). In a cell phone that is equipped with a camera, it might be easier to combine the face and voice traits of a user, while in an ATM application it might be easier to combine the fingerprint and face traits of the user. In identification systems comprising of a large number of users (in the order of millions), an indexing mechanism may be facilitated using a multimodal approach.[18] Researchers are currently studying the performance gain that can be obtained using state-of-the-art commercial off-the-shelf fingerprint and face systems on a large population of individuals.[22]

VII. SUMMARY AND CONCLUSIONS

Multimodal biometric systems elegantly address several of the problems present in unimodal systems. By combining multiple sources of information, these systems improve matching performance, increase population coverage, deter spoofing, and facilitate indexing. Various fusion levels and scenarios are possible in multimodal systems. Fusion at the match score level is the most popular due to the ease in accessing and consolidating matching scores. Performance gain is pronounced when negatively correlated information are used in a multimodal system. Incorporating user-specific parameters can further improve the performance of these systems. With the widespread deployment of biometric systems in several civilian and government applications, it is only a

FIGURE 9.6 A prototype multimodal biometric login system. The three sensors used for acquiring the biometric data are seen in the figure.

matter of time before multimodal biometric systems begin to impact the way in which identity is established in the 21st century (Figure 9.6).

REFERENCES

1. Jain, A. K., Ross, A., and Prabhakar, S., An introduction to biometric recognition, *IEEE Trans. Circuits Syst. Video Technol.*, 14(1), 4–20, 2004.
2. Daugman, J., Recognizing persons by their iris patterns, In *Biometrics Personal Identification in a Networked Society*, Jain, A. K., Bolle, R., and Pankanti, S., Eds., Kluwer Academic Publishers, Dordrecht, pp. 103–121, 1999.
3. Golfarelli, M., Maio, D., and Maltoni, D., On the error-reject tradeoff in biometric verification systems, *IEEE Trans. Pattern Anal. Mach. Intell.*, 19(7), 786–796, 1997.
4. Harrison, W., *Suspect Documents, Their Scientific Examination*, Nelson-Hall Publishers, 1981.
5. Eriksson, A., and Wretling, P., How flexible is the human voice? A case study of mimicry, pp. 1043–1046. In *Proceedings of the European Conference on Speech Technology*, Rhodes, 1997.
6. Matsumoto, T., Matsumoto, H., Yamada, K., and Hoshino, S., Impact of artificial gummy fingers on fingerprint systems, pp. 275–289. In *Proceedings SPIE.* Vol. 4677, San Jose, USA, 2002.
7. Ross, A., and Jain, A. K., Information fusion in biometrics, *Pattern Recognit. Lett.*, 24(13), 2115–2125, 2003.
8. Kuncheva, L. I., Whitaker, C. J., Shipp, C. A., and Duin, R. P. W., Is independence good for combining classifiers? pp. 168–171. In *Proceedings of International Conference on Pattern Recognition (ICPR)*, Vol. 2, Barcelona, Spain, 2001.
9. Zuev, Y., and Ivanon, S., The voting as a way to increase the decision reliability, pp. 206–210. In *Foundations of Information/Decision Fusion with Applications to Engineering Problems.* Washington, DC, USA, 1996.
10. Chang, K. I., Bowyer, K. W., and Flynn, P. J., Face recognition using 2D and 3D facial data, pp. 25–32. In *Proceedings of Workshop on Multimodal User Authentication.* Santa Barbara, CA, 2003.
11. Kumar, A., Wong, D. C. M., Shen, H. C., and Jain, A. K., Personal verification using palmprint and hand geometry biometric, pp. 668–678. In *Proceedings of fourth International Conference on Audio and Video-based Biometric Person Authentication (AVBPA).* Guildford, UK, 2003.
12. Jain, A. K., Prabhakar, S., and Chen, S., Combining multiple matchers for a high security fingerprint verification system, *Pattern Recognit. Lett.*, 20, 1371–1379, 1999.
13. Marcialis, G. L., and Roli, F., Experimental results on fusion of multiple fingerprint matchers, pp. 814–820. In *Proceedings of fourth International Conference on Audio and Video-based Biometric Person Authentication (AVBPA).* Guildford, UK, 2003.
14. Ross, A., Jain, A. K., and Reisman, J., A hybrid fingerprint matcher, *Pattern Recognit.*, 36(7), 1661–1673, 2003.

15. Lu, X., Wang, Y., and Jain, Y. K., Combining classifiers for face recognition, pp. 13–16. In *Proceedings of IEEE International Conference on Multimedia and Expo (ICME)*. Vol. 3. Baltimore, MD, 2003.

16. Brunelli, R., and Falavigna, D., Person identification using multiple cues, *IEEE Trans. Pattern Anal. Mach. Intell.*, 12(10), 955–966, 1995.

17. Bigun, E., Bigun, J., Duc, B., and Fischer, S., Expert conciliation for multi-modal person authentication systems using Bayesian Statistics, pp. 291–300. In *Proceedings of first International Conference on Audio and Video-based Biometric Person Authentication (AVBPA)*. Crans-Montana, Switzerland, 1997.

18. Hong, L., and Jain, A. K., Integrating faces and fingerprints for personal identification, *IEEE Trans. Pattern Anal. Mach. Intell.*, 20(12), 1295–1307, 1998.

19. Frischholz, R. W., and Dieckmann, U., Bioid: a multimodal biometric identification system, *IEEE Comput.*, 33(2), 64–68, 2000.

20. Duin, R. P. W., and Tax, D. M. J., Experiments with classifier combining rules, pp. 16–29. In *Proceedings of first Workshop on Multiple Classifier Systems*. Vol. LNCS 1857. Springer, Cagliari, Italy, 2000.

21. Tax, D. M. J., Breukelen, M. V., Duin, R. P. W., and Kittler, J., Combining multiple classifiers by averaging or by multiplying?, *Pattern Recognit.*, 33(9), 1475–1485, 2000.

22. Indovina, M., Uludag, U., Snelick, R., Mink, A., and Jain, A. K., Multimodal biometric authentication methods: A COTS approach. pp. 99–106. In *Proceedings of Workshop on Multimodal User Authentication*. Santa Barbara, CA, 2003.

23. Verlinde, P., and Cholet, G., Comparing decision fusion paradigms using k-NN based classifiers, decision trees and logistic regression in a multi-modal identity verification application, pp. 188–193. In *Proceedings of second International Conference on Audio- and Video-based Person Authentication*. Washington, DC, USA, 1999.

24. Dieckmann, U., Plankensteiner, P., and Wagner, T., Sesam: a biometric person identification system using sensor fusion, *Pattern Recognit. Lett.*, 18(9), 827–833, 1997.

25. Ben-Yacoub, S., Abdeljaoued, Y., and Mayoraz, E., Fusion of face and speech data for person identity verification, *IEEE Trans. Neural Networks*, 10, 1065–1074, 1999.

26. Ho, T. K., Hull, J. J., and Srihari, S. N., Decision combination in multiple classifier systems, *IEEE Trans. Pattern Anal. Mach. Intell.*, 16(1), 66–75, 1994.

27. Kittler, J., Hatef, M., Duin, R. P., and Matas, J. G., On combining classifiers, *IEEE Trans. Pattern Anal. Mach. Intell.*, 20(3), 226–239, 1998.

28. Jain, A. K., and Ross, A., Learning user-specific parameters in a multibiometric system, pp. 57–60. In *Proceedings of the International Conference on Image Processing (ICIP)*. Rochester, USA, 2002.

29. Lam, L., and Suen, C. Y., Optimal combination of pattern classifiers, *Pattern Recognit. Lett.*, 16(9), 945–954, 1995.

30. Xu, L., Krzyzak, A., and Suen, C., Methods of combining multiple classifiers and their applications to handwriting recognition, *IEEE Trans. Syst. Man Cybern.*, 22(3), 418–435, 1992.

31. Daugman, J., Combining multiple biometrics, http://www.cl.cam.ac.uk/users/jgd1000/combine/.

32. Maio, D., Maltoni, D., Cappelli, R., Wayman, J. L., and Jain, A. K., FVC2002: Fingerprint verification competition, pp. 744–747. In *Proceedings of the International Conference on Pattern Recognition (ICPR)*. Quebec City, Canada, 2002.
33. Philips, P. J., Grother, P., Micheals, R. J., Blackburn, D. M., Tabassi, E., and Bone, M., FRVT 2002: Overview and summary. Technical Report available at http://www.frvt.org/frvt2002/documents.htm.
34. Martin, A., and Przybocki, M., The NIST 1999 speaker recognition evaluation — an overview, *Digit. Signal Process.*, 10, 1–18, 2000.
35. Hsu, R. L., Face detection and modeling for recognition, Ph.D. Thesis, Michigan State University, 2002.

10 Change Detection/ Interpretation with Evidential Fusion of Contextual Attributes — Application to Multipass RADARSAT-1 Data

*Alexandre Jouan, Yannick Allard,
and Yves Marcoz*

CONTENTS

I. INTRODUCTION

Lockheed Martin Canada has, over the years, developed various applications in multisensor data fusion. The research and development effort was initially focused on developing a multisensor data fusion test-bed for the Canadian Patrol Frigate[1] integrating the five levels of fusion as proposed by the US Joint Directors of Laboratories (JDL) group (Level 0: preprocessing; Level 1: single object refinement; Level 2: situation refinement; Level 3: threat refinement; Level 4: resources management). Then, a second generation of the multisensor data fusion test-bed has been developed to simulate a typical mission management system for a surveillance aircraft. To achieve this objective, several databases (platforms, geopolitical data, and emitters) of the test-bed were reconfigured to include target related features extracted from the processing of synthetic aperture radar (SAR) imagery and other sources of information provided by nonimaging sensors.[2-4]

These two test-beds have the same objective: studying the conditions and confidence level under which a data fusion system could provide as complete a description of the scene as possible, and identifying the algorithms yielding the most precise identification of the targets in the surrounding environment, while being robust to confusion.

These projects were successful in demonstrating how the performance of target detection/target recognition (TD/TR) algorithms could significantly be improved by incorporating contextual information.

Several studies have already shown that RADARSAT-1 would provide valuable information for coastal surveillance missions and activity monitoring. Based on the expertise gained on the multisensor data fusion projects, the development was proposed of a methodology for change and activity monitoring based on the fusion of contextual features extracted from multipass imagery. This chapter describes the structure and the first results obtained with the proposed system prototype. It is organized in the four sections:

1. Section II describes a preprocessing step needed to make use of incomplete Geographical Information Systems (GIS) databases as *a priori* information.
2. Section III describes the data fusion system designed for land-use mapping and activity monitoring using evidential fusion.
3. Section IV shows the results obtained using the data set provided by the Canadian Space Agency as part of the second installment of the Application Development and Research Opportunity (ADRO) program.
4. Section V concludes the work and presents related on-going activities.

II. *A PRIORI* INFORMATION

The study site is located around the airport of Stephenville, Newfoundland, Canada.

TABLE 10.1
RADARSAT-1 Acquisitions

Scene #	Dataset	Date	Beam	Orbit	Proj.	Corr.	Pixel/Line Spacing (m)
Scn1	C0016960	1998/06/14	F1	Asc	UTM–WGS84	SSG	6.25/6.25
Scn2	C0016820	1997/11/20	F2F	Asc	UTM–WGS84	SSG	6.25/6.25
Scn3	C0016809	1998/05/14	F2	Asc	UTM–WGS84	SSG	6.25/6.25
Scn4	C0016811	1998/05/21	F1	Asc	UTM–WGS84	SSG	6.25/6.25
Scn5	C0017335	2000/08/31	F2	Asc	UTM–WGS84	SSG	6.25/6.25
Scn6	C0017338	2000/09/01	F4	Desc	UTM–NAD83	SSG	6.25/6.25
Scn7	C0019852	2001/04/28	F2	Asc	UTM–WGS84	SSG	6.25/6.25

SSG, Synthetic Aperture Radar Systematically Geocoded.

Table 10.1 presents acquisition data related to each of the scenes provided by the Canadian Space Agency.

In addition to remotely sensed imagery, GIS themes were provided by Geomatics Canada and weather information was provided by Environment Canada. These additional sources of information are critical for the interpretation and understanding of the backscattered signal from the ground.

Table 10.2 shows the data provided by Environment Canada and the Canadian Hydrographic Service for each of the acquired scenes.

A. COMPLETING THE GIS FILES

1. Methodology

The GIS themes provided by the National Topographic Database (NTDB) were built from data that were not updated recently and showed incompleteness. Section II.A describes a preprocessing step combining the original GIS themes and the imagery as an attempt to improve the classification from GIS data.

In order to get an estimate for the class label in areas with incomplete GIS information, a stochastic labeling followed by a relaxation step using the Iterated Conditional Mode (ICM) strategy was employed.

The information used in this classification process consists in:

1. The incomplete ground cover map from the GIS files.
2. A set of textural features derived from the gray level co-occurrence matrix (GLCM),
3. The radar backscattered intensity.

In the first step of the process, the areas with no labels (no information available in GIS files) were stochastically labeled using backscattered information from the first available radar image without allowing the existing cluster to change.

TABLE 10.2

Environmental Conditions Provided by the Atlantic Climate Center of Environment Canada and the Canadian Hydrographic Services (Bedford Institute of Oceanography) for Port Harmon

Scene #	Dataset	Date	Tide (m)	Wind (km/h)	RH (%)	Temp (C)	Total Daily Rainfall	Total Daily Snowfall	Snow on Ground (cm)
Scn2	C0016820	1997/11/20	—	9	70	0.1	0	0	3
Scn3	C0016809	1998/05/14	0.982	4	53	10.0	0	0	0
Scn4	C0016811	1998/05/21	0.881	7	69	13.7	0	0	0
Scn1	C0016960	1998/06/14	0.607	9	80	11.7	0	0	0
Scn5	C0017335	2000/08/31	1.001	13	87	17.1	0	0	0
Scn6	C0017338	2000/09/01	1.0	—	—	—	—	—	—
Scn7	C0019852	2001/04/28	0.471	9	82	−1.2	34.8	1.2	0

A pseudo–random stochastic labeling process was used to constrain the random initialization of the class labels to some classes. As an example, the label for the class "airport" could only be randomly selected for pixels located in the vicinity of the airport according to a distance map computed from the GIS files.

The image data model used here is very simple as it is based on the gray level of the image and can therefore be applied to any kind of images (SAR, electro-optical (EO), etc.). We applied the Maximum *a posteriori* (MAP) estimator of the label space, maximizing the *a posteriori* probability:

$$P(\omega|F) = \frac{1}{P(F)} P(F|\omega)P(\omega)$$

Where F is the observed image and ω is the set of all possible labels, $P(F)$ only depends on the observations and is estimated as:

$$P(F|\omega) = \prod_{s \in S} P(f_s|\omega_s)$$

where f_s is the observed pixel at site s and ω_s is its label. The equivalence between Markov random fields and Gibbs distributions (Hammersley–Clifford theorem) is invoked to write the *a priori* probability distribution as:

$$P = \frac{1}{Z} \exp\{- U\}$$

where U is the energy function and Z is the normalization constant.

The energy function to be minimized for the stack of the A layers F^a made of the backscattered data and the extracted textural features can be written as:

$$U = \sum_{a=1}^{A} U_{F^a} + U_{\text{spatial, map}}$$

where

$$U_{F^a} = \sum_{s \in S} \left(\ln(\sqrt{2\pi}\sigma_{\omega_s}) + \frac{(f_s^a - \mu_{\omega_s})^2}{2\sigma_{\omega_s}^2} \right)$$

and

$$U_{\text{spatial, map}} = U_{\text{spatial}} + U_{\text{map}}$$

with

$$U_{\text{spatial}} = - \sum_{\{s,r\} \in C} \beta\delta(\omega_s, \omega_r)$$

and the parameter β controls the homogeneity of the regions.

U_{map} stands for the map transition model. It controls the transition from one class to another using the class conditional transition probabilities (CCTP) $t(\omega_s|\omega_r)$ and is defined as:

$$U_{\text{map}} = -\beta_{\text{map}} \sum_{\{s,r\} \in C} t(\omega_s|\omega_r)$$

$t(\omega_s | \omega_r)$ is the probability of transition from the class ω_s to class ω_r between the time of creation of the map and the current time.

2. Choosing the Clique Topology and the Optimization Algorithm

Two different Markov random field cliques configurations were tested, namely the classical eight-connectivity and the eight-connectivity "with holes", as shown on Figure 10.1.

Four different stochastic optimization algorithms were tested for the energy minimization task:

- the Metropolis algorithm,[5]
- the Gibbs sampler,[6]
- the modified metropolis dynamics (MMD),[7]
- the Polya urn sampling model (super urn case).[8]

The hyperparameter, which controls the homogeneity of the segmentation, has been fixed at 0.45 for the classical topology and 0.3 for the topology with holes to account for the increase in the number of neighbors involved. Table 10.3 shows the results of a simulation conducted on a test image (Figure 10.2) obtained from a simulated ground truth (Figure 10.3).

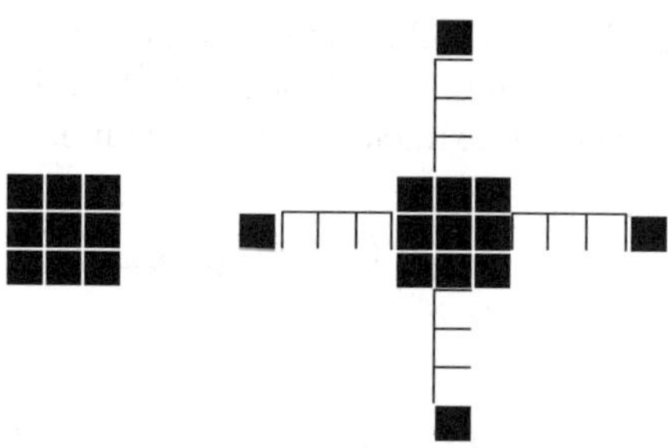

FIGURE 10.1 Classical topology and topology with holes.

TABLE 10.3
Optimization Results

	Classical		With Holes	
	Time/Iteration	%	Time/Iteration	%
Metropolis	1.89	92.9	1.93	96.5
MMD	1.35	92.8	1.39	96
Gibbs	2.71	97.1	2.78	97.5
Polya	5.9	59.6	6.9	60.1

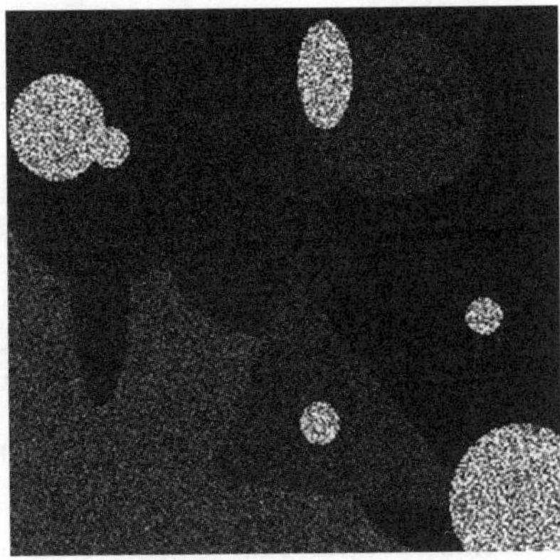

FIGURE 10.2 Noisy version of the test image.

Table 10.3 clearly shows that the neighborhood topology with holes performs better than the classical one. As stated by Pony,[9] this topology destroys stable but undesirable small structures. Despite a little increase in the computing time, the topology with holes has been selected to define the Markov Random Fields (MRF) cliques used for this work.

With respect to the optimization algorithm, we observed that the Polya urn-sampling scheme is not working well in a very noisy environment for it naturally emphasizes the majority labels in a given neighborhood. The computation time required for the Gibbs sampler is about twice that of the other tested simulated annealing-type algorithms. This may prevent its use in an operational mode. The MMD algorithm will be used for the rest of the study since it provides the best compromise between precision of the resulting segmentation and the required computation time.

FIGURE 10.3 Test image.

A deterministic temporal relaxation labeling is performed with the ICM algorithm.[10] This algorithm is relatively fast and performs quite well on the maps produced by the MMD relaxation.

B. RESULTS ON RADARSAT-1 IMAGERY

Figure 10.4 shows the selected study site. The city of Stephenville is visible at the upper left side of the figure; Stephenville airport is located at the center of the figure and Port Harmon harbor is at the bottom of it.

Figure 10.5 shows the color-coding for six class labels, namely: airport, water, wetland, vegetation, grass, and town.

Figure 10.6 shows the classification map extracted directly from the GIS files provided by Geomatics Canada. The black areas correspond to regions with no GIS information.

Figures 10.7 to 10.12 show results obtained with the algorithm described in Section II.A.

Figure 10.7 shows the map resulting from the pseudo−random initialization step. Figure 10.8 shows the result of the stochastic labeling on the areas with incomplete GIS information (blackened area in Figure 10.6). Figure 10.9 shows the result of the temporal relaxation labeling with the ICM.

Classification maps obtained by applying the ICM at time T are used for the initialization of the relaxation process, classifying the structures of the image taken at time $T + 1$. This approach provides a classification map for each available image. Pixel-based class transitions between two successive maps provide useful information for change detection.

According to the successive classification maps we can see that some buildings (characterized by strong scatterers) are misclassified as vegetation.

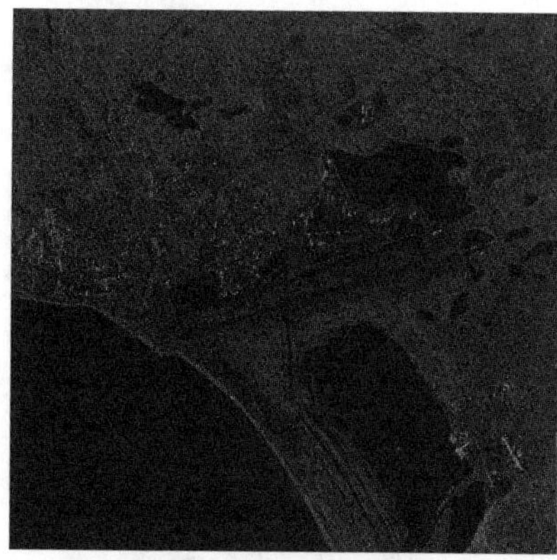

FIGURE 10.4 RADARSAT-1 scene of Stephenville (Newfoundland, Canada). (© 1998 Canadian Space Agency, l'Agence spatiale canadienne. With permission.)

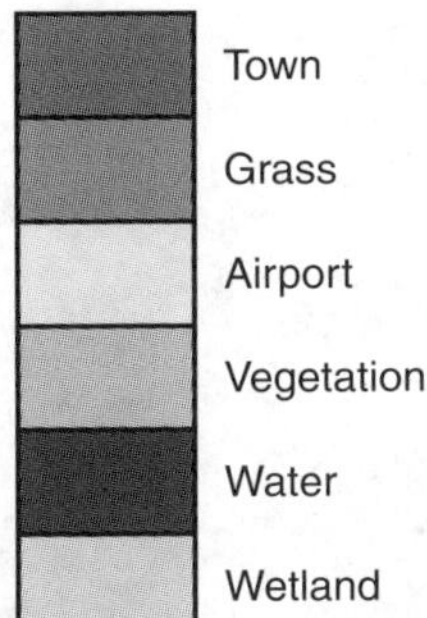

FIGURE 10.5 **(See color insert following page 236)** Classification labels.

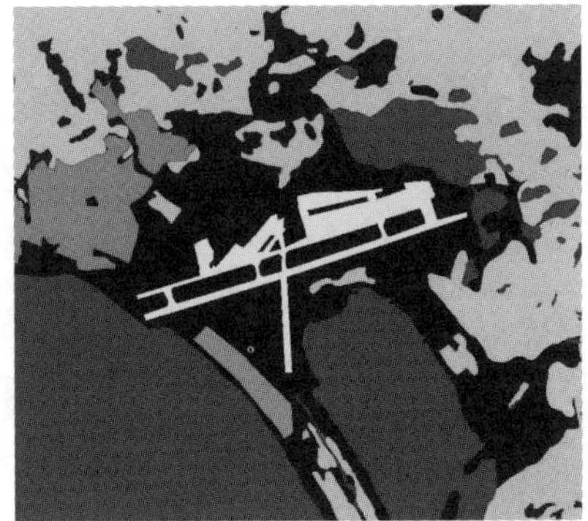

FIGURE 10.6 **(See color insert)** Incomplete land-use map generated from GIS files.

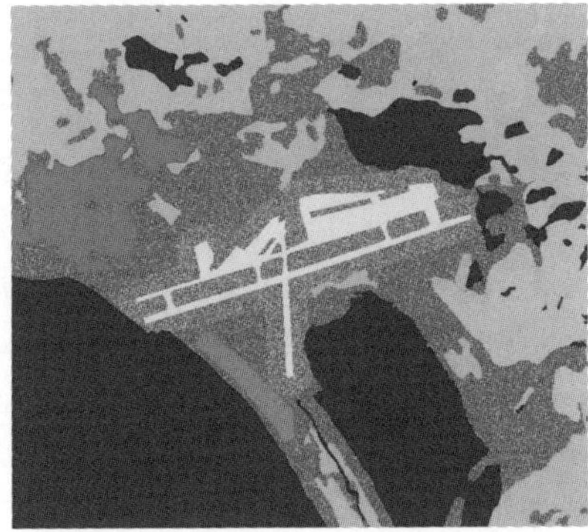

FIGURE 10.7 **(See color insert)** Pseudo-random initialization.

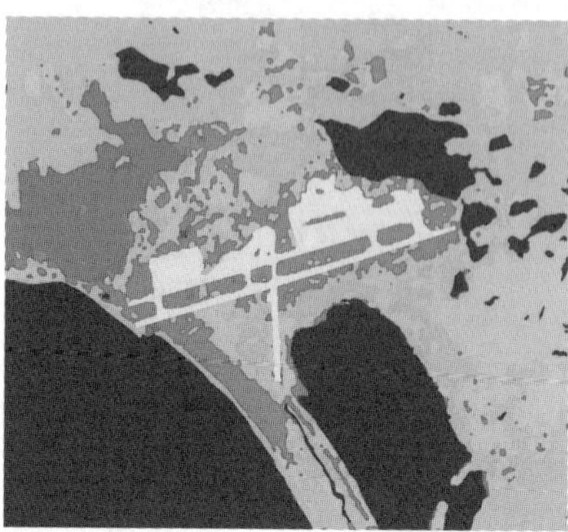

FIGURE 10.8 (See color insert) Stochastic labeling of the regions with no GIS data.

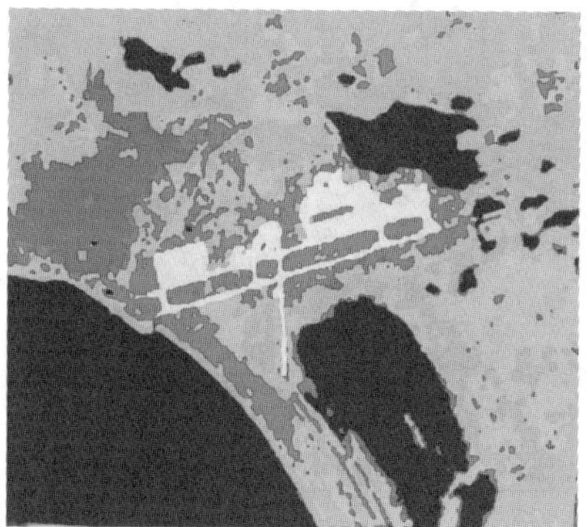

FIGURE 10.9 (See color insert) Result of the first deterministic relaxation (without CCTP).

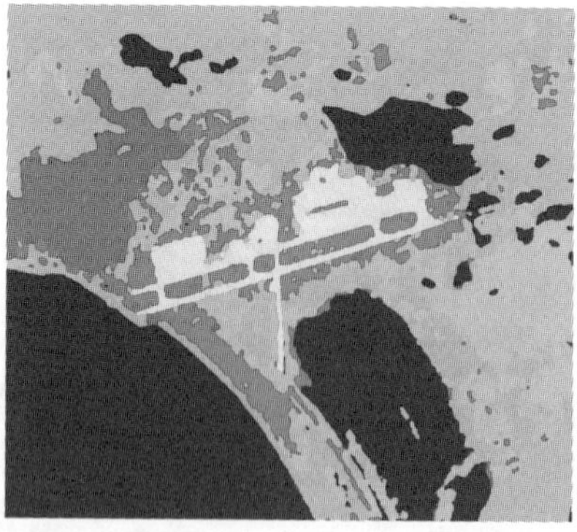

FIGURE 10.10 (See color insert) Deterministic relaxation T1 (with CCTP).

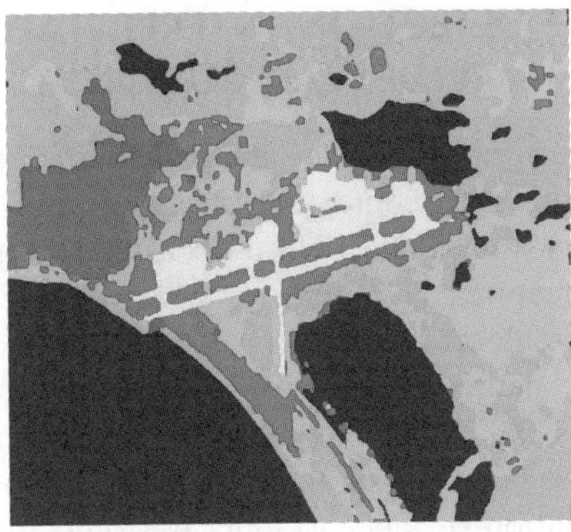

FIGURE 10.11 (See color insert) Deterministic relaxation T2 (with CCTP).

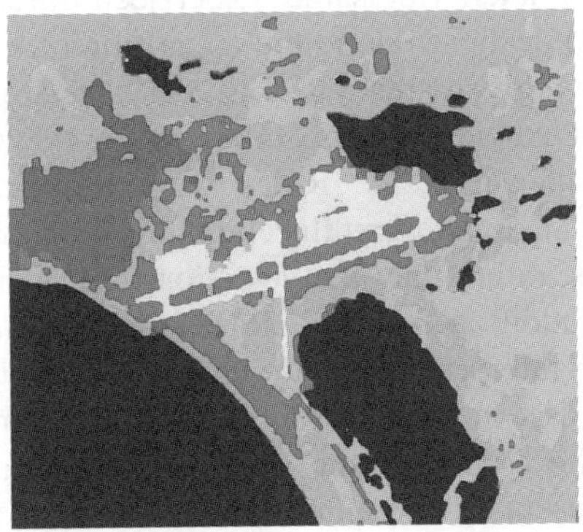

FIGURE 10.12 (See color insert) Deterministic relaxation T3 (with CCTP).

This is due to an inaccurate estimation of the parameters of this class obtained from the GIS layer of built-up areas. This layer also includes grass, trees, and buildings, as well as roads and a lot of dark structures. The estimated mean of the class "building" is lower than it should be and the variance is greater. The estimated parameters of the class "building" are then much closer to those of the class vegetation. Several options exist to correct for this misclassification. A subclass of the GIS layer "building" called "man-made structure" could be defined that would provide a more accurate estimation of the corresponding parameters. Another approach would be to fuse the result of a specular reflector detector with the previous result to correct for the initial mislabeling and generate more consistent parameters.

III. CHANGE DETECTION/INTERPRETATION USING EVIDENTIAL FUSION

Textural analysis is performed with a neural network (multilayer perceptron) trained to classify pixels into contextual classes from a set of eight textural features computed from the gray level co-occurrence matrix.[11] The type (and number) of textural features is optimized according to their discrimination properties using genetic algorithms.

The output of the neural classifier is a pixel-based distribution of label weights. These weights can be used to generate a classification map by labeling each pixel with the class label having the maximum weight value. Alternatively, the pixel-based distribution of label weights is stored for further combination with the output of the textural classifier executed on another acquisition. This combination is performed according to the Dempster–Shafer theory.[12,13] Rule-based reasoning and a taxonomy tree were used to interpret class transitions (compound hypotheses generated from fusion) and discriminate between changes due to acquisition parameters (the same object seen differently due to different acquisition conditions) or those due to the apparition of new structures.

As an example, in a given area where two bridges are present, both bridges were detectable on a given view but only one on a second view that was taken at a slightly different incidence angle. Rules were implemented coupling GIS information and metadata to interpret the class transition as the result of different acquisition parameters, rather than the detection of an unknown object. On the other hand, when no additional information can explain a class transition, the system presents its own interpretation of the change (compound hypothesis). As an example, pixels labeled as "man made object" when previously labeled as "land" on an area known as the Trans-Canada Highway are presented to the operator as pixels potentially representing a mobile structure (vehicle, mobile building).

Rules were also implemented using environmental data to interpret other types of compound hypotheses that may be generated by fusion. As an example, a "shallow water" or "high bank" hypothesis would typically result from the fusion of an area that could alternatively be classified as "water", "land", or "high bank", depending on tide levels.

A. HIERARCHICAL CLASSIFIER

The selected classifier is a multilayer perceptron with 8 neurons on the input layer corresponding to the height of textural features that have been selected for the study, 14 neurons on the hidden layer, and 4 neurons on the output layer corresponding to the desired set of classes.

The textural features are computed from the GLCM using a 9×9 pixel neighborhood window. The number of gray levels of the SAR image was reduced to 11.

Given the dataset available for this study, the neural network was trained to classify textures in five different classes, namely: "water", "wetland/short grass", "vegetation", "man-made objects", and "unclassified". This step is critical for good performance because of the high dependency of the classifier on the training phase. We selected our training data from Scene 2 (see Table 10.1) labeling the vectors according to various GIS themes. Incomplete GIS themes were updated according to the method described in Section II. Even with the help of updated GIS themes, a visual inspection of the data used to build the training vectors is necessary to get rid of time-varying information that may corrupt the training process (presence of ships in ocean or harbor, vehicles on parking lot, aircraft on tarmac, etc.).

The output of the classifier consists of five values, one for each class plus the "unclassified" class, ranging from zero to one. These values are normalized so that they can be interpreted as levels of confidence that the pixel truly belongs to the classes (one meaning it belongs to this class and zero meaning it does not).

The output of the textural classifier can lead to the generation of two different products:

- A *thematic map* can be built by labeling pixels according to the highest output of the classifier. Figure 10.1 shows a typical result of such thematic mapping.
- Another approach is to build a *pixel-based declaration* (in the Dempster–Shafer sense) from the whole set of confidence values. Temporal monitoring can thus be performed through the fusion of the declarations produced by the classifier at different times. In order to increase or decrease the level of confidence associated to the classification, as well as to improve the interpretation of changes over time, the temporal declarations are fused according to the Dempster–Shafer combination rules.

B. Selection of GLCM Features Using Genetic Algorithms

Eight GLCM textural features[14] were selected amongst the 14 proposed by Haralick.[11] These are the maximum (MAX), the first-order difference moment (FDM), the second-order difference moment or inertia (SDM), the first-order inverse difference moment (FIDM), the second-order inverse difference moment (SIDM), the entropy (ENT), the uniformity or angular second moment (UNI), and the cluster shade (CLSH).

The selection of these features was the result of a comparison between the classification performances obtained with the height GLCM textural features and those obtained using the 11 Gaussian Markov random field (GMRF) parameters defined in a typical eight-neighborhood.[15,16] On airborne SAR imagery, this comparison study showed that the GLCM features were performing better than the GMRF features, and that even though some features are not necessary, they do not decrease the performance either. These results need to be confirmed for space-borne imagery. The method followed to obtain those results is described in the following sections.

1. Fitness Value

In genetic algorithms, the fitness value is a measure of the performance of a chromosome. Our objective is to improve the texture classification rate. The classification map is computed using the features selected by each chromosome. The classifier is a multilayer perceptron with 18 neurons on the hidden layer trained using selected regions in the image. The percentage of good classification has been evaluated using the GIS data as ground truth. A linear transformation of this percentage is used as the fitness value for the chromosome. The main disadvantage of this type of classifier is the randomness of the initialization. Two training runs with the same training base will not necessarily give the same percentage of error (about 0.5% difference maximum) so it is difficult to compare two sets of features having close fitness values.

2. Genetic Operations

A population of 20 chromosomes was randomly generated and several genetic operations were applied to them until a selected maximum number of generations was reached.

The genetic operations are crossover, mutation, and reproduction.

The crossover operation consists of selecting pairs in the population, and mixing the genes of the chromosomes of the pair. We decided to choose the two points' crossover for our application. This crossover consists in slicing each parent chromosome at 2 random points, and to replace the middle part of one chromosome by the middle part of the second chromosome. The crossover rate was set at 90%.

The mutation operation is an inversion of a gene. Each gene has a 1 in 50 probability of being mutated.

The reproduction step uses the biased roulette wheel. Each chromosome has its fitness value computed, and is given a probability of being selected proportional to its fitness value ($p_i = f_i / \sum_i f_i$, where f_i is the fitness value of the ith chromosome).

Twenty new chromosomes were then selected among the offspring of the previous population.

3. Results

The algorithm was stopped after 1000 generations. As expected, the algorithm did not converge to a single set of features, but to several similar sets. Twenty chromosomes were selected among those having the lowest classification error rate. The features (or pattern of features) having the highest frequency of occurrence among these chromosomes were selected.

Table 10.4 presents the 20 chromosomes along with their classification error rate.

This preliminary study showed that among the height GLCM features, five are collectively more efficient in providing a good classification rate. These five

TABLE 10.4
Best Chromosomes Selected by the GA and the Associated Error Percentage

Error Percentage	GMRF Features	GLCM Features
12.78	10010001101	00101111
12.77	10110011101	00111111
12.76	11110011100	10101101
12.76	10111011001	00101101
12.74	10110011001	10111111
12.73	10011011001	00111111
12.72	10110011011	00111111
12.70	10111001011	10101111
12.67	10011001101	00101111
12.65	00111001011	00101111
12.65	10110011001	00101111
12.64	10111011001	00101111
12.63	00111011011	00101111
12.59	11110011010	01101111
12.58	10110011101	00101111
12.56	00111011000	10101111
12.56	10011011001	00101111
12.55	11110011101	00101111
12.53	10011011101	00101111
12.34	10111011001	00111111

GLCM features are respectively the SDM, SIDM, ENT, UNI, and CLSH. As a matter of fact, this result is consistent with Schistad Solberg and Jain's study[14] where SDM, SIDM, and CLSH were identified as the set of features providing superior classification scores for space-borne SAR images. As per the GMRF parameters, the study shows that the 1st, 3rd, 4th, 5th, 8th, 9th, and 11th parameters appear to play a collectively more significant role for texture classification than the other parameters.

IV. RESULTS

This section contains some of the first results obtained with the proposed system prototype. Its capability to perform change detection/identification has been evaluated on a set of seven RADARSAT-1 scenes acquired over the city of Stephenville and its vicinity. Some of the scenes from this dataset were part of the archives of the Canadian Space Agency, others were ordered specifically for the ADRO220 project in order to complete the seasonal sampling of the area.

Figure 10.13a represents the southern area of Saint-George Bay (corresponding to Scene 1 of Table 10.1). Figure 10.13b and Figure 10.13c are classification

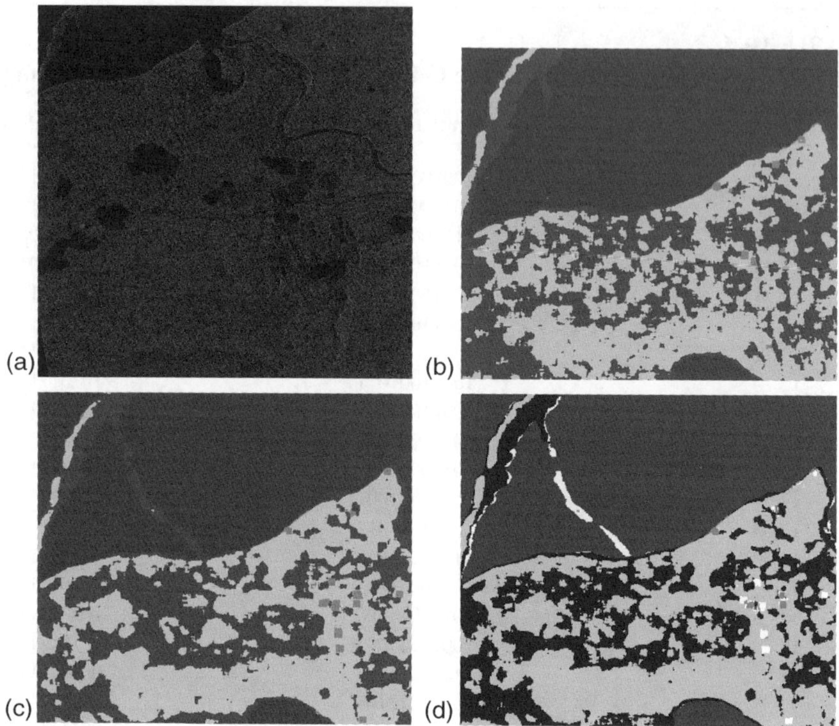

FIGURE 10.13 (See color insert) Change monitoring using evidential fusion. (© 1998 Canadian Space Agency, l'Agence spatiale canadienne. With permission.)

maps obtained from a close-up area of Scene 1 and Scene 4. Figure 10.13d represents the fusion map.

A high bank located between the peninsula and the land is not visible in Scene 1, so its corresponding pixels are classified as "water" (blue). However, due to a lower water level on Scene 4, the high bank is clearly visible on Figure 10.13c and its corresponding pixels have been classified as "wetland" (light gray). This area is classified as "shallow water" (sand) in the fused map.

Figure 10.14a and Figure 10.14b show a region located at the east of Stephenville (respectively, corresponding to Scenes 1 and 4 of Table 10.1). The Trans-Canada Highway is clearly visible and crosses the image vertically. Figure 10.14c shows the fused classification map. Some pixels in this map were classified as land on the first scene and classified as man made object (MMO) on the second one. For these pixels the fusion process generates a hypothesis supporting the possible presence of a vehicle (white). Figure 10.14d shows profiles along the highway centered on the detection of the vehicle. The blue (respectively green) profile corresponds to the Figure 10.14a (respectively, Figure 10.14b). The presence of the vehicle is highlighted in the difference of the two profiles in Figure 10.14e.

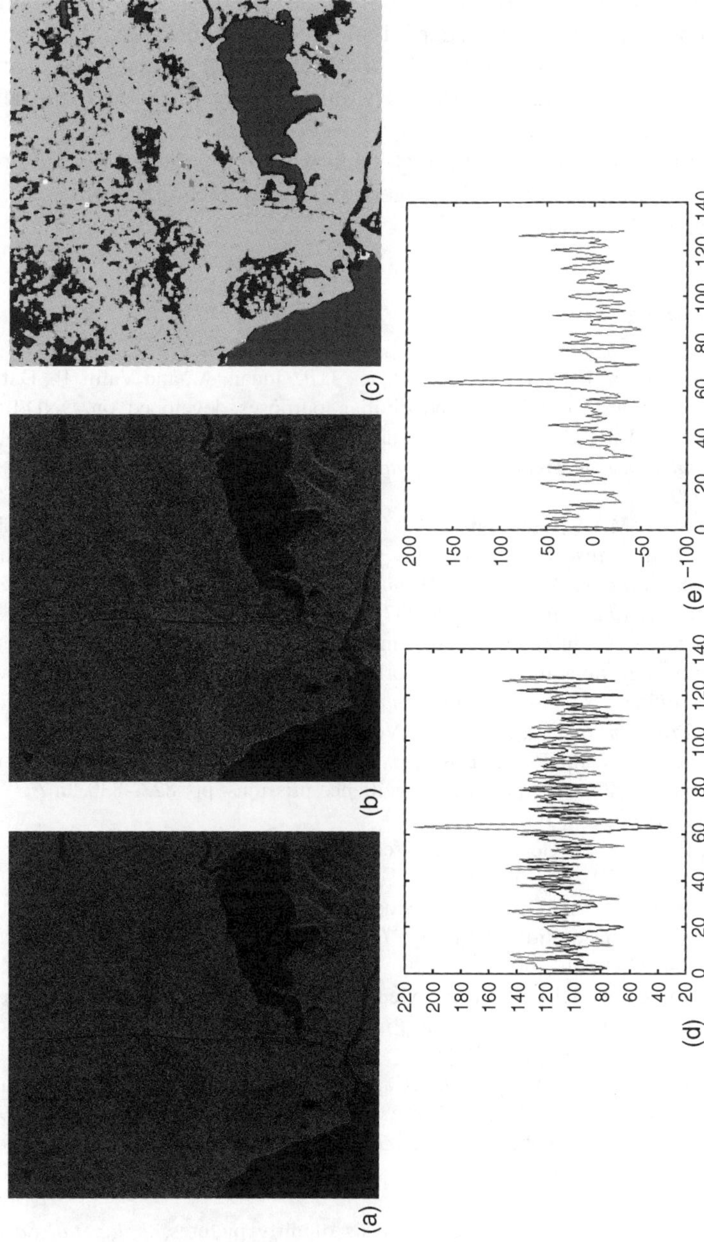

FIGURE 10.14 (See color insert) Detection of a vehicle on the Trans-Canada highway. (© 1998 Canadian Space Agency, l'Agence spatiale canadienne. With permission.)

V. CONCLUSION

We presented an application of coastal monitoring/target detection using evidential fusion. The results obtained so far are very promising. The methodology can easily be adapted to other monitoring applications using multidate imagery by selecting the appropriate set of features, the appropriate frame of discernment, and the necessary training process leading to the definition of the mass functions.

A similar approach is currently underway using the evidential fusion of polarimetric features from SAR imagery and spectral features from multispectral and hyperspectral data cubes.[17,18]

REFERENCES

1. Shahbazian, E., Bergeron, P., Duquet, J. P., Jouan, A., and Valin, P., Data fusion applications for military and civilian purposes developed on DND/Lockheed Martin (LM) Canada decision support testbed, pp. 420–424. In *33rd Asilomar Conference on Signal, Systems and Computers*, Pacific Grove CA, Oct. 24–27, 1999.

2. Simard, M. A., and Jouan, A., Platform database and evidential combination method to fuse ESM detections with the attribute data from other sensors. In *Proceedings NATO–AGARD on Electronic Warfare Integration for Ships, Aircraft, and Land Vehicles*, 1997.

3. Jouan, A., Gagnon, L., Shahbazian, E., and Valin, P., Fusion of imagery attributes with non imaging sensor reports by truncated Dempster–Shafer evidential reasoning, pp. 549–556. In *International Conference on Multisource–Multisensor Information Fusion, Fusion'98*, Vol. II. Las Vegas, July 6–9, 1998.

4. Jouan, A., Valin, P., and Bossé, E., Testbed for fusion of imaging and non-imaging sensor attributes in airborne surveillance missions, pp. 823–830. In *FUSION'99*, Sunnyvale, CA, July 6–8, 1999.

5. Kirkpatrick, S., Gelatt, C., and Vecchi, M., Optimization by simulated annealing, *Science*, 220, 671–680, 1983.

6. Geman, S., and Geman, D., Stochastic relaxation, Gibbs distribution and the Bayesian restoration of images, *IEEE Trans. Pattern Anal. Mach. Intell.*, 6(6), 452–472, 1984.

7. Kato, Z., Modélisation markoviennes multirésolutions en vision par ordinateur, Application a la segmentation d'images SPOT. Ph.D. dissertation, Université de Nice-Sophia Antipolis, 1994.

8. Banerjee, A., Burline, P., and Alajaji, F., Image segmentation and labeling using the polya urn model, *IEEE Trans. Image Process.*, 8(9), 1243–1253, 1999.

9. Pony, O., Descombes, X., and Zerubia, J., Classification d'images satellitaires hyperspectrales en zone rurale et periurbaine. Research Report No. 4008. INRIA, France, 2000.

10. Besag, J., On the statistical analysis of dirty pictures, *J. R. Stat. Soc.*, 48(3), 259–302, 1986.

11. Haralick, M., Shanmugam, K., and Dinstein, I., Textural features for image classification, *IEEE Trans. Syst. Man Cybern.*, 3(6), 610–621, 1973.

12. Dempster, A. P., A generalization of Bayesian inference, *J. R. Stat. Soc.*, 30-B, 1968.
13. Shafer, G., *A Mathematical Theory of Evidence*, Princeton University Press, Princeton, NJ, 1976.
14. Schistad Solberg, A. H., and Jain, A. K., Texture Analysis of SAR Images: A Comparative Study. Technical Report, Norwegian computing center, 1997.
15. Li, S. Z., *Markov Random Field in Computer Vision*, Springer, Tokyo, 1995.
16. Chellappa, R., and Chatterjee, S., Classification of textures using Gaussian Markov random fields, *IEEE Trans. Acoustics, Speech Signal Process.*, 33(4), 959–963, 1984.
17. Jouan, A., Allard, Y., Secker, J. Beaudoin, A., and Shahbazian, E., Intelligent data fusion system (IDFS) for Airborne/Spaceborne hyperspectral and SAR data analysis. In *Proceedings of the Workshop on Multi/Hyperspectral Technology and Applications*. Redstone Arsenal, Huntsville, Alabama, February 5–7, 2002.
18. Jouan, A., and Allard, Y., Land use mapping with evidential fusion of features extracted from polarimetric synthetic aperture radar and hyperspectral imagery, *Inform. Fusion*, 5, 251–267, 2004.

11 Multisensor Registration for Earth Remotely Sensed Imagery

Jacqueline Le Moigne and Roger D. Eastman

CONTENTS

I. INTRODUCTION

A. DEFINITION AND IMPORTANCE OF REMOTELY SENSED IMAGE REGISTRATION

For many applications of image processing, such as medical imagery, robotics, visual inspection and remotely sensed data processing, digital image registration is one of the first major processing steps. For all of these applications, image registration is defined as the process which determines the most accurate match between two or more images acquired at the same or different times by different or identical sensors. Registration provides the "relative" orientation of two images (or one image and other sources, e.g., a map), with respect to each other, from which the absolute orientation into an absolute reference system can be derived. As an illustration of this definition, Figure 11.1(a) shows an image extracted from a Landsat Thematic Mapper (TM) scene acquired over the Pacific Northwest. A corresponding image transformed by a rotation of 10° and a shift of 20 pixels horizontally and 60 pixels vertically is shown in Figure 11.1(b).

Satellite remote sensing systems provide large amounts of global and repetitive measurements. Multiple satellite platforms carry sensors that provide complementary information, and the combination of all these data at various resolutions — spatial, radiometric and temporal — enables a better understanding of Earth and space science phenomena. Utilized for global change analysis, validation of new instruments, or for new data analysis, accurate image registration (or geo-registration) is also very useful for real-time applications such as those related to direct readout data, or on-board navigation and processing systems. A common example of the utility of image registration and integration can be found in land cover applications, where the combination of coarse-resolution viewing systems for large area surveys and finer resolution sensors for more

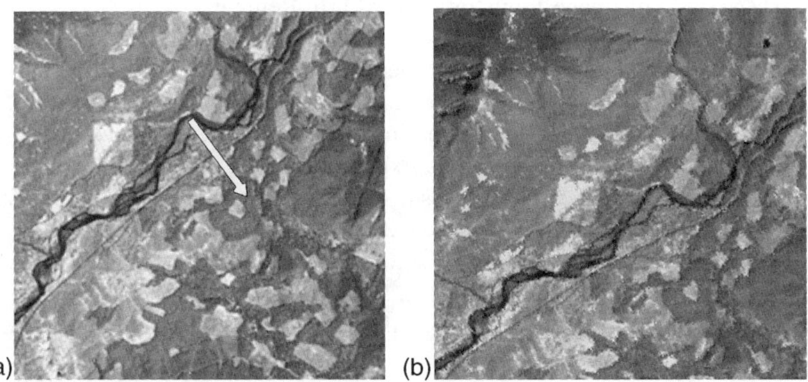

FIGURE 11.1 (a) and (b) Reference and transformed Landsat images deformation by 10° rotation and (20,60) pixels shift shown by the white arrow (Image Courtesy: Compton Tucker, NASA Goddard Space Flight Center).

detailed studies provide high accuracy assessment of the extent area of important land transformations. High-resolution sensors are very good for monitoring vegetation changes, for example, changes in forest cover, when landscape features are local in scale. However, studies at a global or continental scale at high spatial and temporal resolutions would require the processing of very large volumes of data. It is therefore necessary to combine information from both types of sensors to conduct feasible, accurate studies. Another application of remote sensing image registration deals with constellation, formation flying, and web sensor systems, when a fleet of sensors will simultaneously gather complementary information and will make decisions, on-board and in real-time, based on this information.

More generally, the different applications of geo-registration can be summarized as:

1. *Multimodal registration* which enables the integration of complementary information from different sensors,
2. *Multitemporal registration* when observations over several days, whether consecutive or over a longer period, can be integrated,
3. *Viewpoint registration* which integrates information from one moving platform or multiple platforms flying together into one three-dimensional models; and
4. *Template registration* which looks for the correspondence between newly sensed data and a previously developed model or dataset.

For all of these applications, the main requirements of an image geo-registration system are speed, accuracy, consistency and a high level of autonomy that will facilitate the processing of large amounts of data in real-time. Traditionally, because automated procedures do not offer the needed reliability and accuracy, remote sensing scientists have preferred a manual selection of the control points (e.g., from maps) to determine the transformation between maps and imagery. But such a point selection represents a repetitive labor- and time-intensive task that becomes prohibitive for large amounts of data. Also, since the interactive choice of control points in satellite images is sometimes difficult, too few points, inaccurate points, or ill-distributed points might be chosen thus leading to large registration errors. Such errors lead to data analysis errors as shown in two studies reported in Refs. 1,2. The two studies demonstrate that a small error in registration may have a large impact on the accuracy of global change measurements, and Ref. 1 shows that even a small error in registration may have a large impact on the accuracy of global change measurements. For example, when looking at simulated 250 m spatial resolution moderate resolution imaging spectrometer (MODIS) data, a one pixel misregistration can produce 50% error in the computation of the normalized difference vegetation index (NDVI). It also shows that a registration accuracy of less than 0.2 pixels is required to achieve change detection errors of less than 10%.

So, automatic Earth remote sensing imagery registration is an important processing step that promises to ease the workload, speed up the processing, and improve the accuracy in image registration.

B. SPECIAL ISSUES IN MULTISENSOR REMOTE SENSING REGISTRATION

For most sensors directed at the Earth, the distortions that should be considered are defined by the combined effects of sensor operation, the Earth's rotation, orbit and attitude anomalies, and atmospheric and terrain effects. These distortions, which are noticeable when dealing with only one sensor, are even larger when considering multiple sensors carried by multiple platforms on different orbits. Refer to Refs. 3,4 for a good description of distortion sources and distortion categories.

Most of these distortions are usually taken into account and corrected by the navigation model, a process called "systematic correction" that utilizes ephemeris data collected on-board the satellite. Ephemeris data can be defined as the information that describes the type, orientation, and shape of a satellite orbit, and enables one to determine where a satellite is located at any given moment. Image registration refers to "precision correction", and while systematic correction is model-based, precision correction is feature-based. The results of the systematic correction are usually accurate within a few or a few tens of pixels. Recent navigation systems that utilize Global Positioning System (GPS) information are usually accurate within a few pixels, while older sensor data can have errors up to tens or hundreds of pixels. Nevertheless, even for these accurate navigation systems errors may occur, for example, during a maneuver of the spacecraft. Therefore, for both cases, and because of the importance of reaching subpixel accuracy, selected image features or control points are utilized to refine this geo-location accuracy within one pixel or a subpixel. Currently, there are a large number of potential image registration methods developed for aerial or medical images that are applicable to remote sensing images.[5-8] However, there are no systematic studies that enable one to select the most appropriate method for a remote sensing application and predict its applicability as a function of the specific characteristics of the data, the acquisition conditions, the accuracy needed, the predicted accuracy of the navigation system, and the computational resources available.

Specific to Earth remote sensing image registration are issues related to the data acquisition conditions, the size of the data, and the lack of a "reference model" or fiducial points such as those utilized in medical image registration. The issues presented below show how image registration applied to remote sensing data needs to take into account these very specific data characteristics.

1. Data Acquisition Issues

Many conditions other than the image distortions related to satellite and sensor operations described earlier can affect the quality of the acquired data, and most of these are presented below. The design of satellite instruments is based on the principle that targets of interest can be identified based on their spectral characteristics. Each Earth surface feature, such as vegetation or water, presents very distinctive reflectance or emittance curves that are a function of the energy wavelength. These curves are called the "spectral signatures" of the objects being observed. Although these curves are extremely representative of each feature and can help identify them, they do not correspond to unique and absolute responses. For different reasons, such as atmospheric interactions or temporal and location variations, the response curves of a given object observed under different conditions might vary. For this reason, these curves are often called "spectral response patterns" instead of "spectral signatures".

a.   Atmospheric and Weather Effects

The effect of the atmosphere depends on the distance the radiation travels through the atmosphere (usually called "path length"), and on the magnitude of the energy signal. The two main atmospheric effects are known as "scattering" and "absorption". Scattering is the unpredictable redirection of radiation from particles suspended in the atmosphere and mainly depends on the size of these particles, but also on the wavelength of the radiation and the atmospheric path length. Atmospheric absorption occurs in specific wavelengths at which gases, such as water vapor, carbon dioxide, and ozone, absorb the energy of solar radiations instead of transmitting it. "Atmospheric windows" are defined as the intervals of the electromagnetic spectrum outside of these wavelengths, and Earth remote sensors usually concentrate their observations within the atmospheric windows: this is where the spectral "bands" (or "channels") are defined for each sensor. Most of these and other atmospheric effects, such as atmospheric humidity and the concentration of atmospheric particles, are corrected by utilizing physical models. But information related to these models, as well as altitudinal variation of this information, might affect data in a way that cannot be corrected with these models.

Related to the atmospheric effects is the effect of local and temporal weather when acquiring the data. In many areas, only very limited time windows are ideal for acquiring remote sensing data, and although atmospheric scientists prefer observations later in the morning to allow for cloud formation, researchers performing land studies prefer earlier morning observations to minimize cloud cover. For geo-registration purposes, clouds usually introduce unreliable features that need to be discarded unless, for calibration purposes, interchannel registration of only one sensor is performed, and clouds are known to appear similar in the channels to be registered and therefore can be utilized as features.

b.   Temporal Changes

Other radiometric and spatial variations are temporal, and correspond to variations in illumination due to different times of the day, different seasons or ground changes due to nature or human intervention (e.g., forest fires or human-induced deforestation). In this case, features may or may not be visible, and may or may not exist. As an illustration of cloud cover and radiometric variations due to time of day, Figure 11.2 shows the same area of Baja California observed by the GOES satellite, Channel 1 (visible band), at different times of the day and with a moving cloud coverage.

Another example, shown in Figure 11.3, presents the same features observed by the MODIS sensor in April and May 2002, during a flood of the Mississippi River. This example shows how the same geographic area leads to very different image features in a time interval of just one month. In this case, a successful image registration algorithm needs to rely only on features that have not changed through time.

c. *Terrain Relief*

Another issue related to satellite remote sensing is linked to the topography of the terrain. If a topographic surface is illuminated at an angle, variations in image

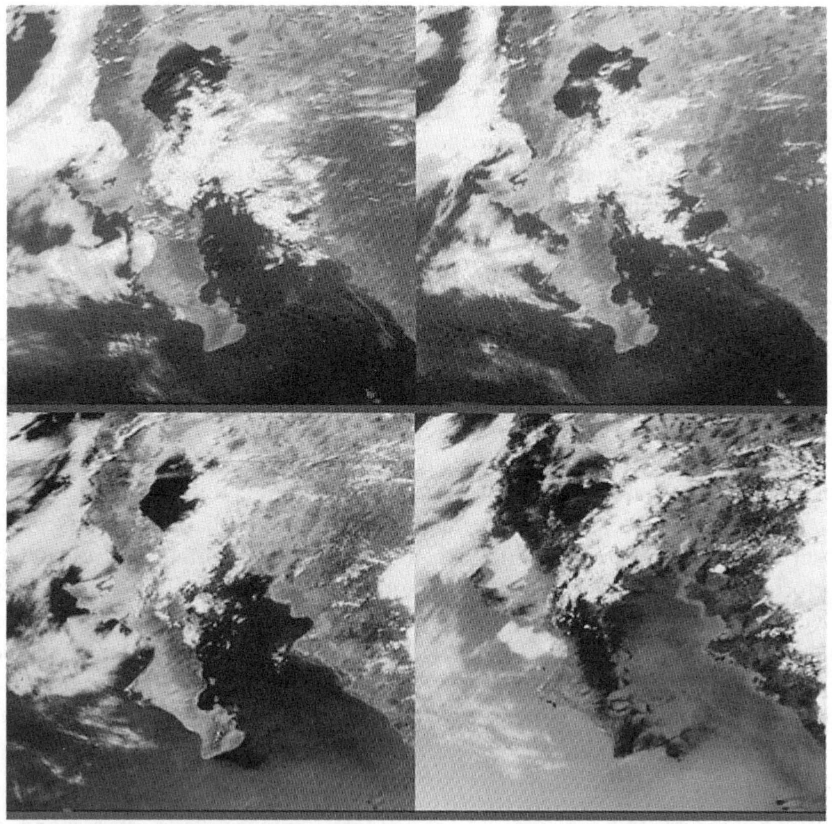

FIGURE 11.2 Baja Peninsula, California, observed by goes8-channel 1 at four different times of the day (Image Courtesy: Dennis Chesters, NASA Goddard Space Flight Center).

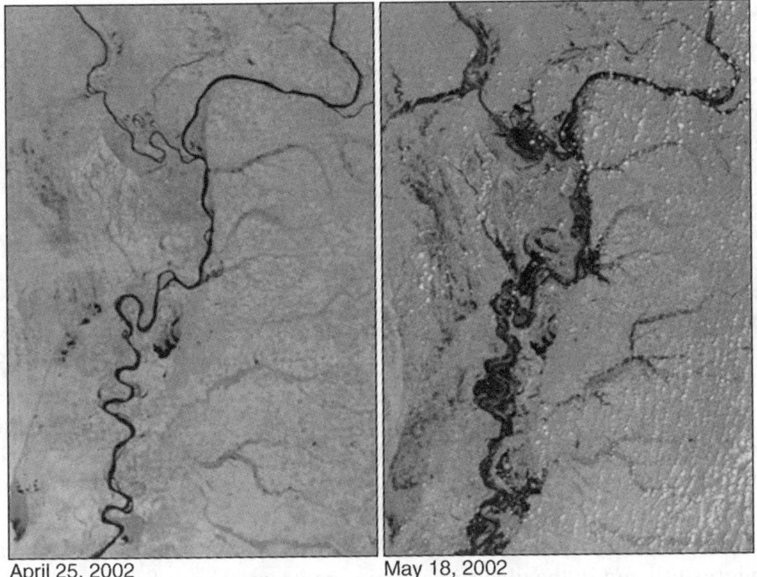

FIGURE 11.3 Mississipi and Ohio Rivers observed by the MODIS sensor on the NASA Terra spacecraft at two different dates showing the floods of spring 2002 (Image Courtesy: Jacques Descloitres, NASA Goddard Space Flight Center).

brightness will carry information concerning the slopes of individual pixels on the ground. But, depending on the slope of the geographic relief, the spatial resolution of the sensor and the orbit characteristics of the satellite, terrain relief effects will or will not be observed in the sensed data. Large areas of topographic effect are usually corrected as part of the systematic correction but local effects may remain in the image and might affect the effectiveness and the accuracy of a feature-based registration. Figure 11.4 illustrates this problem, with a SAR and a Landsat-TM image of the Lope area in Gabon, Africa, where the topography shown on the left side of the SAR image affects the radiometries of the corresponding vegetation features as seen in the Landsat image, through a shadow effect.

d. Multisensor Issues

When a new sensor is being designed the type of features to be observed, and the accuracy with which they will be mapped, define which wavelengths are of interest, the widths of the wavelength intervals to be used, what is the accuracy to be achieved in these bandwidths and what is the "smallest" or "faintest" feature which might be detected by the sensor. These requirements correspond to the "resolutions" of the sensor by which it is usually identified — spectral, radiometric, spatial, and temporal resolutions. The term "resolution" is usually employed to define the smallest unit of measurement or granularity that can be

FIGURE 11.4 Lope Area in Gabon, Africa observed by the SAR and the Landsat-TM Sensors topography effect is shown in the left side of the images (Image Courtesy: Nadine Laporte, formerly University of Maryland, Currently Woods Hole Research Center).

recorded in the observed data. The spectral resolution of a sensor is defined by the bandwidths utilized in the electromagnetic spectrum. The radiometric resolution defines the number of "bits" that are used to record a given energy corresponding to a given wavelength. The spatial resolution corresponds to the area covered on the Earth's surface to compute one measurement (or one picture element, "pixel") of the sensor. The temporal resolution (or frequency of observation), defined by the orbit of the satellite and the scanning of the sensor, describes how often a given Earth location is recorded by the sensor. Another characteristic of each sensor is also its number of bands or channels. In general, most Earth remote sensors are multispectral; that is, they utilize several bands to capture the energy emitted or reflected from Earth features. The addition of panchromatic imagery, which usually has a much better spatial resolution than multispectral imagery in the visible part of the spectrum, provides higher quality detail information. Multispectral and panchromatic data, usually acquired simultaneously, are coregistered and can be easily merged to obtain high spatial and spectral resolution.

Ideally, if a sensor had an infinite number of spectral channels (or bands), each observed area on the ground (or pixel) could be represented by a continuous spectrum and then identified from a database of known spectral response patterns. But, due to recent advances in solid state detector technology, it has only recently been possible to increase the number of bands without decreasing the signal-to-noise ratio, thus, the rise of new "hyperspectral" sensors. Although the boundary between multispectral and hyperspectral sensors is sometimes defined as low as 10 bands, hyperspectral imaging usually refers to the simultaneous detection in hundreds to thousands of spectral channels.

Finally, another difference between sensors comes from their scanning systems that sweep over the Earth to produce a two dimensional image of the terrain.[9,10] Most Earth remote sensors utilize either across-track scanning or along-track scanning systems. Both types of scanning systems record the data in scan lines that are perpendicular to the direction of the spacecraft motion. But cross-track scanners use a scanning mirror that rotates to acquire the data, while along-track scanners utilize a linear array of detectors that move along (or are "pushed along") with the platform: these systems are also known as "pushbroom scanners". Both types of scanning mechanisms introduce geometric distortions that can be modeled and systematically corrected, but these corrections also introduce errors in the areas and the radiometries of the features that may be used for registration.

Therefore, multisensor image registration must address the following issues:

1. Define transformation spaces that address multiple spatial resolutions.
2. Select image features that are invariant for radiometric differences due to multiple spectral and temporal resolutions.
3. Perform band-to-band registration by choosing the relevant corresponding channels that approximately cover the same portion of the electromagnetic spectrum.
4. Address the consistency of the algorithms in terms of various sensor geometry.

2.　Data Handling Issues

a.　Data Size

Compared to other domains, the registration of remote sensing imagery must deal with very large amounts of data. As an example, each Landsat scene is about 7000×7000 pixels. Therefore, handling such amounts of data in real-time must take into account computational requirements in terms of speed as well as memory requirements. It will be particularly important when designing on-board computing capabilities. This may be done either by bounding the number of features used in the registration process, or by extracting windows of interest in the images, performing several local registrations and then merging local registrations together into a global transformation. The implementation of these methods on parallel, distributed or reconfigurable architectures must also be considered in order to process these large amounts of data in real-time.

b.　Lack of Ground Truth

In medical imagery, new image registration algorithms are usually validated using either fiducial points as "ground truth" points or a database of test data or both. In the area of remote sensing, no such points can be easily introduced in the data, and ground truth usually refers to semimanual registration performed by a human operator. But, contrary to fiducial ground truth points, manual registration includes errors due to uncertainty about the exact location of control points (also called Ground Control Points (GCPs)) when geographic landmarks are chosen, and due to positioning error by the human eye, especially in nonvisible wavelengths.

Other ways to obtain validation of remote sensing image registration algorithms are to use "synthetic" data (where real data are artificially transformed), to compute algorithm consistency (e.g., by performing "circular" registrations between three or more images, requiring that computed transformations are mutually consistent), or to use higher resolution data degraded to lower resolution in order to assess the algorithm accuracy.

II. OVERVIEW OF AUTOMATIC REMOTE SENSING IMAGE REGISTRATION

Despite years of research on automatic image registration for geo-registration and other applications, no single algorithm or method has become universally accepted. The breadth of existing approaches for image registration is well documented in a number of literature reviews including Brown,[5] Maurer and Fitzpatrick,[11] Fonseca and Manjunath,[6] Maintz and Viergever,[12] and Zitová and Flusser.[8] The reader is referred to these articles for a thorough survey of registration approaches. In this article we will focus on methods used in image registration for remote sensing fusion, and on methods that form a basis for our own research with selected example references.

A. CHARACTERISTICS OF IMAGE REGISTRATION METHODS FOR REMOTE SENSING

Given the breadth of remotely sensing imagery, and the wide number of proposed automatic algorithms with competing claims to effectiveness, we can define several characteristics that relate the nature of the imagery to be registered and the algorithms used.

1. *The class of image registration problems to be solved.* Here we ask what transformations between images the algorithm is intended to be effective at finding, such as:
   A Geometric transformations
   B Radiometric transformations
   C Temporal changes
2. *Characteristics of the algorithm designed to solve the problem.* Here we ask about the main steps in the algorithm including:
   A Features used in the matching process
   B Similarity metrics for comparison of features
   C Search strategy for optimizing the metric

We have left off a commonly mentioned fourth step, the transformation and resampling of one image to the co-ordinate system of the second, because for remote sensing the resampling step may vary based on the needs of the application. Some Earth scientists use nearest neighbor interpolation or data indexing (in data indexing new absolute co-ordinates (such as latitude and longitude) of chosen points (such as corners of the scene) are recorded with the data file but no resampling is performed), to preserve the radiometry of each pixel. Since these methods do not introduce new intensity levels visualization specialists require smoother interpolation methods, such as bicubic or spline, that result in better looking images for animations or display. We view the transformation and resampling as a postprocessing step that should be under the control of the user once a registration algorithm finds an optimal geometric transformation.

We will expand our definition of each of the characteristics in turn.

1A. *Geometric transformations.* The geometric transformation f gives a co-ordinate transformation for converting a corresponding point in one image into the second, as in the equation $I_2(x, y) = I_1(f(x, y))$ (as formulated in Ref. 5). Geometric transformations include: (1) global rigid body translation, rotation and scale, (2) global affine transformations, (3) global polynomial transformations, including perspective transforms, (4) piece-wise affine or polynomial transformation, or (5) pixel-wise motion model that accounts for terrain elevation.

1B. *Radiometric transformations.* The radiometric transformation g gives a relationship between the sensed gray level in one image and the sensed gray level at the corresponding point in the second image, as in the equation $I_2(x, y) = g(I_1(f(x, y)))$. In the case of NASA satellite data, validation studies often include efforts to relate the spectral response of a new instrument to the well-understood response of an existing instrument.[10] In the case of multisensor fusion this transformation should be accounted for in algorithm design, but the relationship can be complex since this transformation can be compounded with the spectral signature of the objects and their temporal changes being sensed. The radiometric transformation is usually not stated explicitly in the derivation of a registration algorithm but instead implicitly accounted for in the similarity measure. Following the categories used in Ref. 13, common radiometric functions used include: (1) identity relationship, or radiometric invariance, in which $I_2 = I_1$, (2) the affine or linear relationship in which $I_2 = aI_1 + b$, (3) a general functional relationship $I_2 = g(I_1)$, which may be parametric as the gamma correction equation used in Ref. 14 or (4) a statistical relationship in which g may be discontinuous or nonmonotonic or both.

1C. *Temporal changes and noise.* The temporal changes possible between two satellite images have been noted in Section I.B.1.6 and can be taken into consideration in designing the robustness of an algorithm for the nature and scale of temporal changes, as well the possible sources of image noise.

2A. *Features used in the matching process.* An algorithm typically matches a set of features deemed reasonably invariant under the assumed geometric and radiometric transformations. These can include: (1) raw image intensities that may not be optimal when the radiometric transformation is complex but, for those cases in which similar spectral bands are matched, may be the least modified and most accurate, (2) continuous functions of image intensities that are typically the result of an edge filter or filters applied to the images, (3) point features computed by an interest operator that selects high-energy points that can be localized, such as local peaks, corners or intersections,[15] (4) contour and line features, and their invariants, that are extracted from edge images or as region boundaries from segmentation,[16] (5) regions, that result from segmentation or the extraction of compact objects,[17] (6) transform domain features, including the coefficients in Fourier or Wavelet transforms and derived computations.[18]

2B. *Similarity metrics for comparison of features.* To select the best matching features, an algorithm uses a similarity metric to produce a numeric value for the quality of a specific feature correspondence or transform parameters. These include: (1) L1 norm, which assumes radiometric invariance,[19] (2) L2 norm and correlation coefficient, which assume a linear radiometric function,[20] (3) correlation ratio, which assumes a functional relationship,[21] (4) mutual information, which assumes a weak functional relationship,[22-24] (5) consistency of spatial arrangements for feature points, such as the Hausdorff distance,[25] (6) contour and region differences, such as moment invariants.[16]

2C. *Search strategy for optimizing the measure.* The time efficiency of an algorithm is critically based on the strategy used to find the optimal transform. These include: (1) brute force, or an exhaustive search over all possible transform parameter combinations, which is rarely efficient enough to be used in operational algorithms but can be useful for small images or for testing and visualizing a similarity measure, (2) multiresolution, a general strategy combination with other search methods that uses an image pyramid to compute the final transformation in a coarse to fine approach, saving computation time by using smaller levels of the pyramid, (3) numerical search approaches, such as gradient descent[26-28,42] or Marquardt–Levenberg,[14] (4) dynamic programming and other discrete search methods.[28-30]

B. Image Registration Algorithms

The characteristics of image registration algorithms defined in the previous section often appear in particular combinations. For example, a similarity measure that is expensive to compute is usually combined with an efficient search strategy to compensate. In this section we describe some of the combinations that serve as background for our current research.

1. Intensity, Area-Based Algorithms

In traditional area-based registration two images, or selected regions, are matched under the assumption of radiometric invariance or linear relationship and raw image intensities are used as features. These can find the transformation that minimizes an L1 norm, an L2 norm, a related norm such as the normalized cross-correlation function or the correlation coefficient.

However, these measures can be expensive to compute with a brute force strategy and even more expensive if computed to subpixel accuracy. An option is to use a numerical optimization scheme, such as gradient descent from the least squares minimization of the L2 norm.[26–28] An alternative derivation uses the Marquardt–Levenberg optimization scheme and thereby adds a tuning parameter to control the rate of convergence based on recent performance.[14] The advantage of the method is subpixel accuracy and reasonable search times, even in higher parameter spaces, which can include translation, rotation and scale, but the strong radiometric assumptions limit its use in cross-band, cross-instrument fusion. In the latter case, the method can be used under the assumption of a parametric radiometric function[14,31] or under weaker radiometric assumptions with derived image features such as gradient information[32] or wavelets.[22]

2. Fourier Domain Algorithms

Matching in the Fourier domain offers performance, and other, advantages. Fourier algorithms can implement correlation efficiently using the fast Fourier transform (FFT) to compute the Fourier transforms of the images, and then using the product of the transforms to compute the correlation. Phase correlation algorithms use the cross-power spectrum of the Fourier transform to estimate the translation parameters.[18] One advantage is robustness against highly correlated noise and low-frequency luminance variation. The basic approach can be adapted for subpixel registration[33] and, with careful elimination of aliased and unreliable frequencies, subpixel accuracy can be quite high.[34]

3. Feature Point Algorithms

Feature point, or landmark, matching algorithms extract a set of distinct locations in each image and determine an optimal point-to-point correspondence. The feature points can be computed a large number of ways, from corner detection, wavelets, road intersections and many others (see Refs. 5,8 for reviews). Feature point matching has the advantage in image fusion of using distinctive neighborhoods with high gradients or variance that are relatively invariant to geometric and radiometric transformations (corners being a good example), and of focusing the computations on critical areas in what can be very large satellite images. If the feature points are few and distinctive, then a sequential matching processes based on point-wise comparison can be effective;[15] if the feature points are dense and less unique, then more global, point set methods are appropriate.[25]

4. Mutual Information Algorithms

In cases that cross-band and cross-instrument pairings without a functional radiometric relationship, mutual information offers an information-theoretic similarity measure that maximizes the information redundancy between the registered images.[22–24] A significant advantage for cross-band and cross-instrument registration is the weak radiometric assumption that while image intensities match between the images, the functional relationship can be discontinuous. A disadvantage is the computational expense, so algorithms based on mutual information use sophisticated search methods such as Marquardt–Levenberg,[24] Brent and Powell,[23] and Spall's.[22]

III. A STUDY OF IMAGE REGISTRATION FOR EARTH REMOTE SENSING IMAGE DATA

The goal of our project is to develop and assess image registration methodologies that will enable the required accurate multisource Earth remote sensing data integration. As described in Section II, image registration can be defined by three main steps:

1. Extraction of features to be used in the matching process,
2. Computation of the similarity measure
3. Search strategy for optimizing the measure

As seen in Section II, many choices are available for each step, and this section evaluates some potential choices for each step. Possible features for Step 1 are investigated in the context of first correlation-based methods, and then optimization-based methods. Once features have been evaluated, Step 2 and Step 3 are investigated by considering different similarity measures and matching methods.

A. CORRELATION-BASED EXPERIMENTS

Using correlation as a similarity metric our first experiments focused on feature assessment. As expected, this work showed that edges, or edge-like features like wavelets, are more robust to noise and local intensity variations than raw image values. Wavelet features considered as potential registration features are either low-pass features, which provide a compressed version of the original data and some texture information, or high-pass features, which provide detailed edge-like information. Comparing edges and wavelets, we observed that orthogonal wavelet-based registration was usually faster although not always as accurate as a full-resolution, edge-based registration.[35] This was obtained by exploiting the multiresolution nature of wavelets, where an approximation of the transformation is computed at very low-spatial resolution, and then iteratively refined at higher and higher resolutions. But because of this decimation, orthogonal wavelets lose the invariance to translation since features can migrate between frequency subbands. By lack of translation (respective rotation) invariance, we mean that the wavelet transform does not commute with the translation (respective rotation) operator. To study the effects of translation, we first conducted a study[36] that quantitatively assessed the use of orthogonal wavelet subbands as a function of features' sizes. The results showed that high-pass subbands are more sensitive to translation than low-pass subbands which are relatively insensitive, provided that the features of interest have an extent at least twice the size of the wavelet filters. A second study[37] investigated the use of an overcomplete frame representation, the "steerable pyramid".[38] It was shown that, as expected and due to their translation- and rotation- invariance, Simoncelli's steerable filters perform better than Daubechies' filters. Rotation errors obtained with steerable filters were minimal, independent of rotation size or noise amount. Noise studies also reinforced the results that steerable filters show a greater robustness to larger amounts of noise than do orthogonal filters. Another result of this study is that Simoncelli band-pass features are more robust, but less accurate than low-pass features.

B. OPTIMIZATION-BASED EXPERIMENTS

The earlier work focusing on correlation-based methods used exhaustive search. One of the main drawbacks of this method is the prohibitive computation times when the number of transformation parameters increases (e.g., affine transformation vs. shift-only), or when the size of the data increases (full-size scenes vs. small portions, multiband processing vs. mono-band). To answer this concern, we looked at different features using an optimization-based method.

In these experiments, we chose an optimization based on a gradient descent method using an L2 norm as similarity metrics. Using this matching methodology, we compared features obtained from two different multi-resolution decompositions; the Simoncelli steerable pyramid and the spline decompositon.[38,14] While the Simoncelli steerable pyramid produces low-pass and band-pass features, the spline pyramid only produces low-pass features. These experiments were performed on synthetic data shown in Figure 11.5 and created using a 512×512 section of a Landsat-TM (band 4) of a Pacific Northwest scene. This reference image is transformed by using combinations of scaling, rotation, and translation, and adding various amounts of Gaussian noise. Figure 11.5 shows the reference data with three examples of transformed images. Both rotation and scaling are applied with respect to the center of the image. After the transformation is applied, the 256×256 centers of the transformed images are extracted and registered to the 256×256 center of the original reference image.

Results showed that for a gradient optimization matching, Simoncelli low-pass features have a better radius of convergence, while Simoncelli band-pass features are the best in terms of accuracy and consistency, but that when they converge, the spline features present the best accuracy.[40]

C. SIMILARITY MEASURES EXPERIMENTS

Using Simoncelli band-pass features as registration features, exhaustive searches, as well as a stochastic gradient optimization matching strategy (Spall's algorithm),[22] were utilized to compare two similarity measures: correlation and

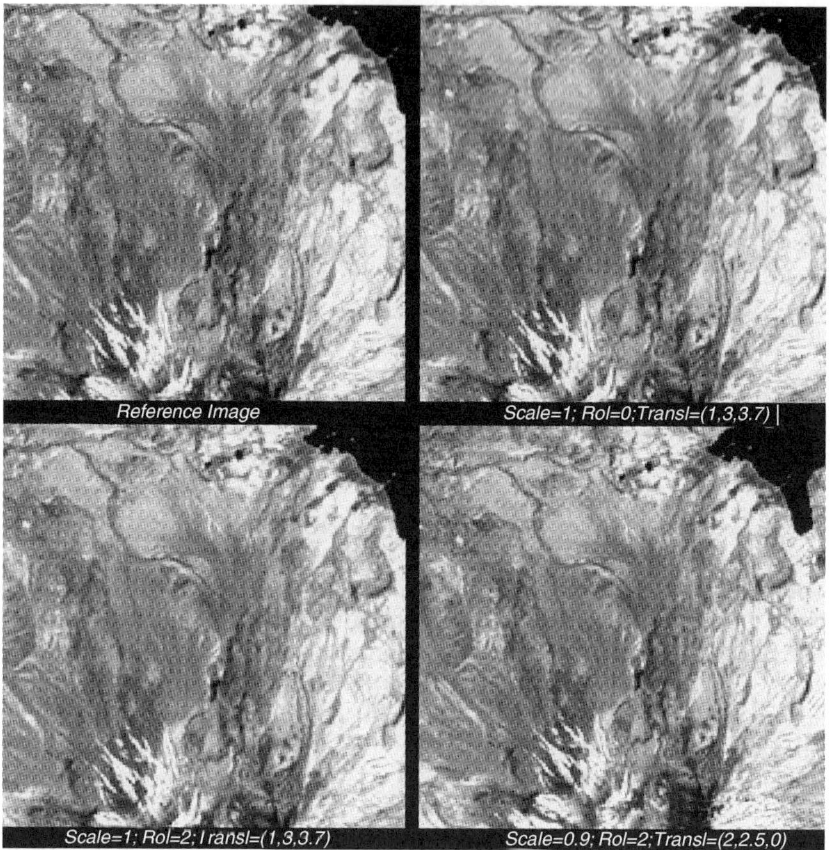

FIGURE 11.5 Synthetic test data — reference image and three transformed input images.

mutual information. The results show that similarity peaks obtained with mutual information are sharper than those obtained with correlation which can be important to reach subpixel registration accuracy. Furthermore, when using mutual information with a stochastic gradient, results show an accuracy of 0.01 pixels on synthetic test data, 0.64 pixels on multitemporal (cloudy) data, and 0.34 pixels on multisensor data.

D. COMBINATION ALGORITHMS EXPERIMENTS

Following those preliminary experiments focusing on one or another of the components of the registration process, the next step is then to combine these components in different ways and assess them on well-chosen test data. Before performing these comparisons in a systematic fashion, an early set of these combination algorithms was tested on three different datasets, using transformations composed of a rotation, a translation and an isometric scaling (i.e., scale factor similar in both x- and y-directions). By combining the previous features, similarity metrics, and matching components five algorithms were developed:

Method 1: Gray levels matched by fast Fourier correlation.[41]

Method 2: Gray levels matched by gradient descent[42] using a least squares criterion.

Method 3: Spline or Simoncelli (band- or low-pass) pyramid features matched by optimization using Marquardt–Levenberg algorithm.[40]

Method 4: Simoncelli wavelet features matched by optimization of the mutual information criterion using Spall algorithm.[22]

Method 5: Simoncelli wavelet features matched using a robust feature matching algorithm and a generalized Hausdorff distance.[25,43]

For some of the methods (Method 1 and Method 5), registration is computed on individual subimages and then integrated by computing a global transformation. For the others (Method 2, Method 3, and Method 4), registration is computed on the entire images but iteratively, using the pyramid decompositions. Three different datasets were utilized for the study and results are presented below.

1. Synthetic Dataset Results

Using the process described in Section III.B, seven different artificially transformed images are created using scales in the range [0.9, 1.1], rotations varying between 0° and 3°, and translations between 0 and 4 pixels in each direction. Method 1, Method 2, Method 3, and Method 4 were applied on this dataset. The lowest error was obtained with Method 3 using Simoncelli low-pass features, and the results showed that most results are within at most 1/3 pixel of the "truth transformation".

2. Multitemporal Dataset

The multitemporal dataset was acquired over two areas, central Virginia and the Washington DC/Baltimore area. For each area, one reference scene is chosen and six to eight reference chips (of size 256×256) are extracted. Also for each area, the dataset includes four input scenes known from their universal transverse Mercator projection (UTM) co-ordinates. From these co-ordinates, windows corresponding to each reference chip of that area are computed and extracted, and local chip/window registrations are performed using Method 5.[43] Figure 11.6 shows a few examples of chip/window pairs. After all local registrations have been performed, a global registration is computed for each pair of scenes with a generalized least mean squares method that combines all previous local registrations. Compared to manual registration, these

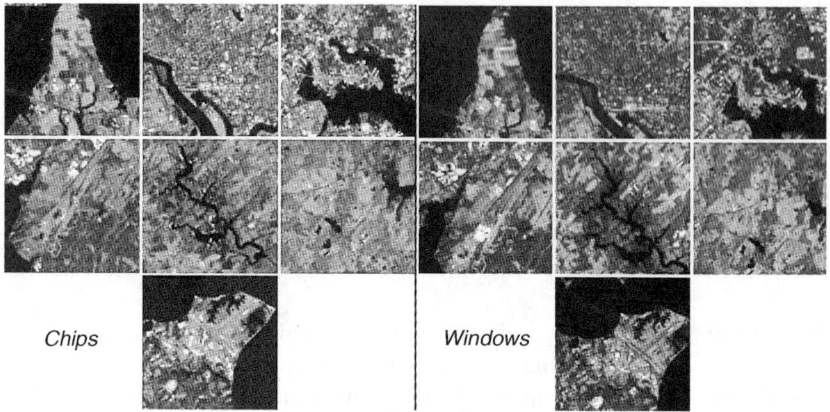

FIGURE 11.6 Examples of chip/window pairs for the DC/Baltimore area (Image Courtesy: Jeffrey Masek, NASA Goddard Space Flight Center).

multitemporal experiments produced registration accuracies included between 0.21 and 0.59 pixels.

3. Multisensor Dataset

The third dataset used in this study represents multisensor data acquired by two different sensors over four of the MODIS Validation Core Sites. The four sites represent four different types of terrain in the United States:

1. Coast reserve area represented by the Virgina site, data acquired in October 2001.
2. Agricultural area represented by the Konza Prairie in the state of Kansas, data acquired July to August 2001.
3. Mountainous area represented by the CASCADES site, data acquired in September 2000.
4. Urban area represented by the USDA, Greenbelt, Maryland, site, data acquired in May 2001. Figure 11.7 shows an example of these data.

The two sensors and their respective bands and spatial resolutions involved in this study are:

1. IKONOS Bands 3 (red) and 4 (near-infrared), spatial resolution of 4 m per pixel.
2. Landsat-7/ETM + Bands 3 (red) and 4 (near-infrared), spatial resolution of 30 m per pixel.

In this study, wavelet decomposition was utilized not only to compute registration features, but also to bring various spatial resolution data to similar resolutions, by performing recursive decimation by two. For example, after three

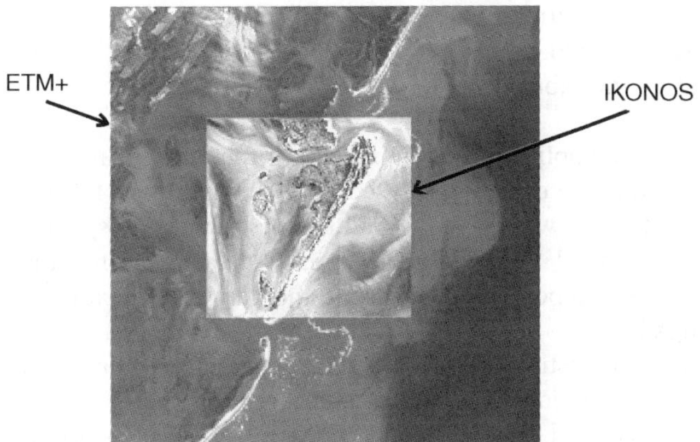

FIGURE 11.7 ETM and IKONOS data of the Virginia Coastal area (Image Courtesy: EOS Land Validation Core Sites, NASA Goddard Space Flight Center, Http://landval. gsfc.nasa.gov/MODIS/).

levels of wavelet decomposition, the IKONOS spatial resolution is brought to 32 m that, compared to the Landsat spatial resolution, corresponds to a scaling of 1.07. This is the scaling expected when registering IKONOS to Landsat data. Method 1, Method 2, Method 3, and Method 4 were applied to the registration of the two sites.

Since no exact ground truth is available in this case, we can validate the results by expecting the multimodal intrasensor registrations to be scale $= 1$, rotation $= 0$, translation $= (0,0)$, and a scale of 1.07 for the IKONOS to Landsat registrations. As expected, the results showed that the registrations based on gray levels are less reliable on interband registrations than those based on edge-like features, but, when reliable, these results are more accurate. Another evaluation of the algorithms was also performed by measuring consistency between algorithms:[44] most results were within 1/4 to 1/3 pixels of each other. Self-consistency of two of the methods was also checked, by performing circular registrations: for example if three images A, B, and C are considered then registrations of pairs (A,B), (B,C) and (A,C) are compared. In this experiment, we observed that the tested algorithms were self-consistent within 1/8 of a pixel.

IV. CONCLUSION AND FUTURE WORK

Earth remote sensing data registration is a necessary preprocessing step that will enable multisource data to be integrated with a high level of accuracy. Such integration has applications in many Earth Science applications, global change detection and assisted decision making in emergency situations such as rapid fire response, hurricane management, or invasive species, to name a few. But fast, accurate and automatic image registrations will also have applications in future distributed and reconfigurable spacecraft systems, such as sensor webs.

Keeping these applications in mind, Earth remote sensing data registration must be adapted to the design of the sensors, the conditions, locations and times of data acquisition, as well as to the multiple resolutions — spatial, spectral and temporal — exhibited by the multiple sensor data to be integrated. Subpixel accuracy is a very important requirement and various validation methods have been investigated.

This chapter investigated various multisensor and multiresolution precision correction or registration components and algorithms, and showed on several Earth Science datasets that subpixel accuracy, self-consistency and interalgorithm consistency can be achieved. Future work will involve an extended synthetic dataset with noise added, radiometric transformation applied, and a larger number of transformations, as well as a larger number of sensors for multisensor data registration. In order to enable new registration components and algorithms to be tested in a rigorous fashion, a modular image registration framework is also currently being developed.[45] The concept guiding this framework is that various components of the registration process can be combined in several ways in order to reach optimum registration on a given type of data and under given circumstances. Thereby, the purpose of this framework will be double-fold: (1) it will represent a testing framework to assess various combinations of components as a function of the applications, as well as to assess a new registration component compared to other known ones; (2) it will be the basis of a registration tool where a user will be able to "schedule" a combination of components as a function of the application at hand, the available computational resources and the required registration accuracy.

From all these tests, it will be possible to categorize the different algorithms in terms of their accuracy but also in terms of their sensitivity to initial (i.e., navigation) conditions, their computational and memory requirements, and their implementations on high-performance and reconfigurable implementations, for real-time and on-board application.

ACKNOWLEDGMENTS

The authors would like to acknowledge the support of the NASA Intelligent Systems/Intelligent Data Understanding Program. All the results reported in Section III were obtained under this project by the NASA Goddard Image registration group and the authors would like to thank the following researchers for allowing them to present their results in this section, especially Arlene Cole-Rhodes, Kisha Johnson, Nathan Netanyahu, Jeffrey Morisette, Harold Stone and Ilya Zavorin.

REFERENCES

1. Townshend, J., Justice, C. O., Gurney, C., and McManus, J., The impact of misregistration on change detection, *IEEE Trans. Geosci. Remote Sens.*, 30(5), 1992.

2. Dai, X., and Khorram, S., The effects of image misregistration on the accuracy of remotely sensed change detection, *IEEE Trans. Geosci. Remote Sens.*, 36(5), 1998.

3. Van Wie, P., and Stein, M., A Landsat digital image rectification system, *IEEE Trans. Geosci. Electron.*, GE-15(3), 130–137, 1977.

4. Richards, J. A., *Remote Sensing Digital Image Analysis: An Introduction*, 2nd ed., Springer, New York, 1993.

5. Brown, L., A survey of image registration techniques, *ACM Comput. Surveys*, 24(4), 1992.

6. Fonseca, L. M. G., and Manjunath, B. S., Registration techniques for multisensor sensed imagery, *Photogrammetr. Eng. Remote Sens. J.*, 62(9), 1049–1056, 1996.

7. Le Moigne, J., Campbell, W. J., and Cromp, R. F., An automated parallel image registration technique of multiple source remote sensing data, *IEEE Trans. Geosci. Remote Sens.*, 40(8), 1849–1864, 2002.

8. Zitova, B., and Flusser, I., Image registration methods: a survey, *Image Vision Comput.*, 21, 977–1000, 2003.

9. Multi-spectral scanning. In *Fundamentals of Remote Sensing*, chap. 2.8; http://www.ccrs.nrcan.gc.ca/ccrs/learn/tutorials/fundam/chapter2/chapter2_8_e.html; Canada Center for Remote Sensing, 2003.

10. Campbell, J. B., *Introduction to Remote Sensing*, 2nd ed., Guilford Press, New York, 1996.

11. Maurer, C. R., and Fitzpatrick, J. M., A review of medical image registration, In *Interactive Image Guided Neurosurgery*, Maciunas, R. J., Ed., American Association of Neurological Surgeons, Parkridge, IL, pp. 17–44, 1993.

12. Maintz, J. B., and Viergever, M. A., A survey of medical image registration, *Med. Image Anal.*, 2(1), 1998.

13. Roche, A., Malandain, G., and Ayache, N., Unifying maximum likelihood approaches in medical image registration, *Int. J. Imaging Syst. Technol.*, 11(1), 71–80, 2000, Special Issue on 3D Imaging.

14. Thévenaz, P., Ruttimann, U., and Unser, M., A pyramid approach to sub-pixel registration based on intensity, *IEEE Trans. Image Process.*, 7(1) 1998.

15. Stewart, C. V., Tsai, C.-L., and Roysam, B., The dual-bootstrap iterative closest point algorithm with application to retinal image registration, *IEEE Trans. Med. Imaging*, 22(11), 1379–1394, 2003.

16. Li, H., Manjunath, B., and Mitra, S. K., A contour-based approach to multi-sensor image registration, *IEEE Trans. Image Process.*, 4(3) 1995.

17. Goshtasby, A., and Stockman, G. C., A region-based approach to digital image registration with subpixel accuracy, *IEEE Trans. Geosci. Remote Sens.*, 24(3), 390–399, 1986.

18. Kuglin, C. D., and Hines, D. C., The phase correlation image alignment method, pp. 163–165. In *IEEE Conference on Cybernetics and Society*, New York, 1975.

19. Barnea, D. I., and Silverman, H. F., A class of algorithms for fast digital image registration, *IEEE Trans. Comput.*, 21(2), 179–186, 1972.

20. Rosenfeld, A., and Kak, A. C., *Digital Picture Processing*, 2nd ed., Academic Press, Orlando, 1982.

21. Roche, A., Malandain, G., Pennec, X., and Ayache, N., The correlation ratio as a new similarity measure for multimodal image registration, pp. 1115–1124. In *Proceeding of First International Conference on Medical Image Computing and Computer-Assisted Intervention (MICCAI'98)*, Vol. 1496 of LNCS, Springer, Cambridge, 1998.

22. Cole-Rhodes, A., Johnson, K., Le Moigne, J., and Zavorin, I., Multiresolution registration of remote sensing imagery by optimization of mutual information using a stochastic gradient, *IEEE Trans. Image Process.*, 12(12), 1495–1511, 2003.

23. Maes, F., Vandermeulen, D., Marchal, G., and Suetens, P., Fast multimodality image registration using multiresolution gradient-based maximization of mutual information, pp. 191–200. *Image Registration Workshop*, NASA Goddard Space Flight Center, 1997.

24. Thévenaz, P., and Unser, M., Optimization of mutual information for multi-resolution image registration, *IEEE Trans. Image Process.*, 9(12), 2083–2099, 2000.

25. Mount, D. M., Netanyahu, N. S., and Le Moigne, J., Efficient algorithms for robust feature matching, *Pattern Recogn.*, 32(1), 17–38, 1999, Special Issue on Image Registration.

26. Lucas, B. D., and Kanade, T., An iterative image registration technique with an application to stereo vision. *DARPA Image Understanding Workshop*, April 1981.

27. Keren, D., Peleg, S., and Brada, R., Image sequence enhancement using sub-pixel displacement, *Comput. Vision Pattern Recognit., Santa Barbara*, 742–746, 1988.

28. Irani, M., and Peleg, S., Improving resolution by image registration, *CVGIP: Graph. Models Image Process.*, 53(3), 231–239, 1991.

29. Maitre, H., and Wu, Y., Improving dynamic programming to solve image registration, *Pattern Recognit.*, 20(4), 1997.

30. Maitre, H., and Yifeng, W., A dynamic programming algorithm for elastic registration of distorted pictures based on autoregressive model, *IEEE Trans. Acoustics, Speech, Signal Process.*, 37(2) 1989.

31. Jianchao, Y., and Chern, C. T., The practice of automatic satellite image registration, pp. 221–226. In *Proceedings of 22nd Asian Conference on Remote Sensing*, Singapore, 2001.

32. Irani, M., and Anandan, P., Robust multi-image sensor image alignment, pp. 959–966. *International Conference on Computer Vision*, Bombay, 1998.

33. Foroosh (Shekarforoush), H., Zerubia, J., and Berthod, M., Extension of phase correlation to sub-pixel registration, *IEEE Trans. Image Process.*, 11(3), 188–200, 2002.

34. Stone, H. S., Orchard, M., Chang, E., and Martucci, S., A fast direct Fourier-based algorithm for subpixel registration of images, *IEEE Trans. Geosci. Remote Sens.*, 39(10), 2235–2243, 2001.

35. Le Moigne, J., Xia, W., Tilton, J., El-Ghazawi, T., Mareboyana, M., Netanyahu, N., Campbell, W., and Cromp, R., First evaluation of automatic image registration methods, pp. 315–317. *International Geoscience and Remote Sensing Symposium*, Seattle, 1998.

36. Stone, H. S., Le Moigne, J., and McGuire, M., The translation sensitivity of wavelet-based registration, *IEEE Trans. Pattern Anal. Mach. Intell.*, 21(10), 1074–1081, 1999.

37. Le Moigne, J., and Zavorin, I., Use of wavelets for image registration, *SPIE Aerosense Conference, Wavelet Applications*, Orlando, April 2000.

38. Simoncelli, E., and Freeman, W., The steerable pyramid: a flexible architecture for multi-scale derivative computation, *IEEE International Conference on Image Processing*, 1995.

39. Zavorin, I., Stone, H. S., and Le Moigne, J., Evaluating performance of automatic techniques for sub-pixel registration of remotely sensed imagery, *SPIE Electronic Imaging, Image Processing: Algorithms and Systems II Conference*, Santa Clara, 2003.

40. Zavorin, I., and Le Moigne, J., On the use of wavelets for image registration, *IEEE Transactions on Image Processing*, to appear in June 2005.

41. Stone, H. S., Progressive wavelet correlation using Fourier methods, *IEEE Trans. Signal Process.*, 47(1), 97–107, 1999.

42. Eastman, R. D., and Le Moigne, J., Gradient-descent techniques for multi-temporal and multi-sensor image registration of remotely sensed imagery. *Fourth International Conference Information Fusion*, Montreal, 2001.

43. Netanyahu, N., Le Moigne, J., and Masek, J., Geo-registration of Landsat data by robust matching of wavelet features, *IEEE Transactions on Geosciences and Remote Sensing*, Volume 42, No. 7, pp. 1586–1600, July 2004.

44. Le Moigne, J., Cole-Rhodes, A., Eastman, R., Johnson, K., Morisette, J., Netanyahu, N., Stone, H. S., and Zavorin, I., Multi-sensor image registration for on-the-ground or on-board science data processing, *Science Data Processing Workshop*, Greenbelt, 2002, pp. 9b1–9b6.

45. Le Moigne, J., Stone, H. S., Cole-Rhodes, A., Eastman, R., Jain, P., Johnson, K., Morisette, J., Netanyahu, N., and Zavorin, I., A study of the sensitivity of automatic image registration algorithms to initial conditions. In *Proceedings of IEEE International Geoscience and Remote Sensing Symposium*, Anchorage, 2004.

12 System and Model-Based Approaches to Data Fusion for NDE Applications

Lalita Udpa, Satish Udpa,
and Antonello Tamburrino

CONTENTS

I. INTRODUCTION

Data fusion can be defined as the synergistic integration of information from multiple sensors to infer more accurate and comprehensive information about a system.[1] Data fusion algorithms are of interest in a number of disciplines including defense, geophysical exploration, medicine, and nondestructive evaluation (NDE). In particular, NDE is an application field that lends itself naturally to data fusion largely because it involves the use of multiple sensors. For example, NDE methods involve the application of a suitable form of energy to the specimen under test. A snapshot of the material/energy interaction process is then taken via an appropriate transducer and analyzed to ascertain the state of the specimen. The nature of the energy applied to interrogate the specimen depends on the physical properties of the specimen, the kind of information desired, and the prevailing test conditions. Therefore, a wide variety of testing methods have evolved in response to differing industrial needs. Each method offers different "views" of the information depending on the type of excitation energy and its characteristics, including amplitude, frequency and wavelength. It can, therefore, be argued that a system capable of extracting complementary segments of information from data collected from multiple NDE tests can offer additional insight relative to that obtained using a single NDE test. For instance, ultrasonic methods offer good penetration into the material and resolution, but the signal suffers from poor signal to noise ratio (SNR), whereas electromagnetic methods offer good sensitivity and SNR but poor resolution. Data fusion algorithms exploit both the redundancy and complementarity of information in multiple inspection techniques to enhance the resulting image. The redundant information is used to improve the SNR of the data and the complementary information is used to augment the overall information content, which in turn increases the accuracy of the characterization.

The concept of data fusion is not new. Human beings and a number of other organisms fuse information from a variety of sensors routinely for making decisions. Newman and Hartline[2] discuss how information from infrared signals generated by the pit organ is fused with visual information provided by the eyes in the optic tectum of rattlesnakes. Sensors can be homogeneous (e.g., two eyes used in stereo vision) or heterogeneous (e.g., vision and touch) and, correspondingly, provide similar or dissimilar pieces of information. Methods for fusing data should therefore take into account these differences.

Data fusion algorithms in NDE can be broadly classified as phenomenological or nonphenomenological. Phenomenological algorithms utilize the knowledge of the underlying physical process as a basis for deriving the data fusion procedure. Nonphenomenological approaches, in contrast, tend to ignore the underlying physics and attempt to model the NDE processes by linear, time invariant systems, thereby allowing the use of well-established techniques from systems theory for data fusion. These techniques fuse multisensor data using probabilistic models or least squares techniques. A number of strategies using a variety of signal and image processing techniques have been proposed over the years. Mitiche and Aggarwal[3] present methods using image-processing techniques. A more recent survey of approaches that includes methods for fusing data at the signal, pixel, feature, and symbol levels can be found in Ref. 4. These methods range from classical Kalman filter and statistical techniques, to supervised learning and fuzzy logic based procedures. These techniques, while computationally efficient, require a large database to either estimate the parameters required by the algorithm or train the system to accomplish the task.

This chapter presents both nonphenomenological (system based) and phenomenological (model based) algorithms for NDE data fusion. In particular, the section will focus on two types of sensors, namely, eddy current and ultrasonic transducers. The eddy current technique based on electromagnetic induction is governed by a physical process that is diffusive in nature, whereas the ultrasonic method relies on the propagation and reflection of an ultrasonic wave into the test specimen.[5] Ultrasonic imaging techniques offer excellent resolution, but the method is sensitive to a wide variety of measurement conditions, including surface roughness and coupling, which result in corruption of the signal with excessive noise. In contrast, eddy current techniques do not require a couplant and are relatively insensitive to surface roughness conditions. The disadvantages associated with the eddy current method lie in its poor resolution capabilities. Although eddy current methods offer excellent flaw detection capabilities, they are not an effective method for characterizing closely located multiple flaws due to their poor resolution characteristics. Fusion of ultrasonic and eddy current data will therefore provide superior resolution as well as improved flaw detectability and SNR.

II. SYSTEM-BASED METHODS

System-based methods for data fusion can be classified into four categories, namely, signal level, pixel level, feature level, and symbol level. Signal level fusion methods are usually applicable when the sensors have identical or similar characteristics, or when the relationship between the signals from different sensors is explicitly known. Pixel level fusion methods can be applied when sensors are used to generate data in the form of images. Statistical characteristics of images, combined with information concerning the relationship between sensors, are used to develop the fusion strategy. Feature level fusion implies fusion of a reduced set of data representing the signal, called features, and symbol level fusion calls for extracting abstract elements of information called symbols.

A. LINEAR MINIMUM MEAN SQUARE ERROR FILTER

An optimal approach for the system-based fusion of multiple images derived from a heterogeneous sensor environment can be obtained using a linear minimum mean square error (LMMSE) filter.[6] Assume that the signals from N sensors are linearly degraded, as shown in Figure 12.1, where both signal and noise are random processes, and that we have knowledge of their spectral characteristics. We also assume that the signals are registered. Under these assumptions we can derive a filter that minimizes the mean-square error (MSE) of the resulting image. The linear system used for fusion consists of N filters whose outputs are added together to generate the fused signal $r(t)$, as shown in Figure 12.1. We also assume that the system and signals are real and stationary.

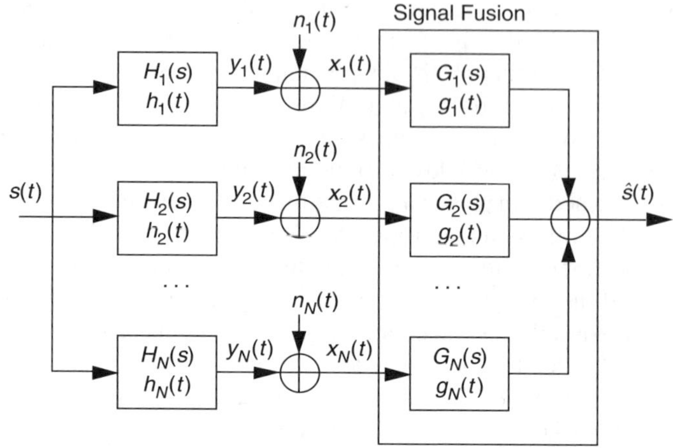

FIGURE 12.1 Systems model for linear signal fusion.

Notation

$s(t)$	*original signal.*
$\hat{s}(t)$	restored signal.
$h_j(t)$	impulse response associated with jth degradation stage, $1 \le j \le N$.
$H_j(s)$	transfer function associated with jth degradation stage, $1 \le j \le N$.
$y_j(t)$	degraded signal corresponding to the jth stage, $1 \le j \le N$.
$n_j(t)$	additive noise at the input of the jth stage restoration filter, $1 \le j \le N$.
$x_j(t)$	input to the filter at the jth stage, $1 \le j \le N$.
$g_j(t)$	restoration filter at the jth stage, $1 \le j \le N$.

1. LMMSE Filter — System with *N* Inputs without Degradation

Consider an *N*-input LMMSE filter for one-dimensional signals with additive noise only. Figure 12.2 presents the corresponding block diagram for the overall approach. The input $x_j(t)$ for the *j*th stage is

$$x_j(t) = s(t) + n_j(t) \tag{12.1}$$

The restored signal $\hat{s}(t)$ is given by

$$\hat{s}(t) = \sum_{i=1}^{N} g_i(t) * x_i(t) = \sum_{i=1}^{N} \int_{-\infty}^{\infty} g_i(t - \lambda)x_i(\lambda)d\lambda \tag{12.2}$$

We design the filter by minimizing the MSE defined by

$$\text{MSE} \equiv E\big[(s(t) - \hat{s}(t))^2\big] \tag{12.3}$$

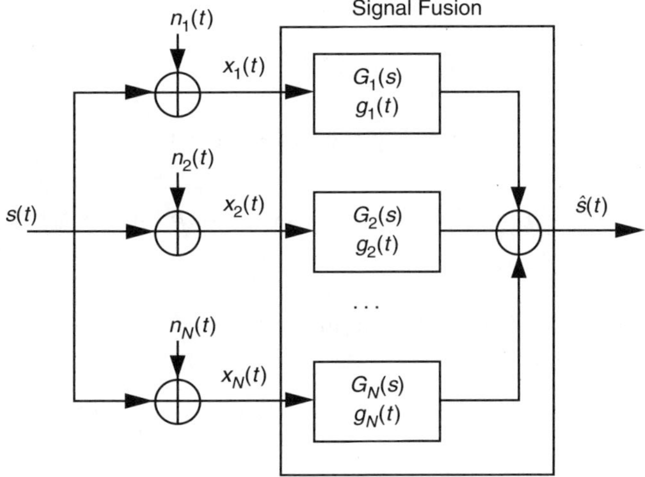

FIGURE 12.2 Multiple input LMMSE — signal without degradation.

The MSE is minimized if $g_j(1 \leq j \leq N)$ satisfies the orthogonality condition.[6]

$$E[(s(t) - \hat{s}(t))x_j(t')] = 0, \quad \text{for } t' \in R \quad \text{and} \quad \forall j, \ 1 \leq j \leq N \qquad (12.4)$$

$$E\left[\left(s(t) - \sum_{i=1}^{N}\left\{\int_{-\infty}^{\infty} g_i(\lambda)x_i(t - \lambda)\right\}\right)x_j(t')\right] = 0,$$

$$\text{for } t' \in R \quad \text{and} \quad \forall j, \ 1 \leq j \leq N \qquad (12.5)$$

Let the correlation function be defined as follows for a real and stationary signal.

$$R_{x,y}(\tau) = E[x(t + \tau)y(t)] \qquad (12.6)$$

where $E[\cdot]$ denotes the expectation operator. Equation 12.5 can then be expressed as

$$\sum_{i=1}^{N}\left\{\int_{-\infty}^{\infty} g_i(\lambda)R_{x_i,x_j}(t - \lambda - t')\right\} = R_{s,x_j}(t - t'),$$

$$\text{for } t' \in R \quad \text{and} \quad \forall j, \ 1 \leq j \leq N \qquad (12.7)$$

Let $\tau = t - t'$

$$\sum_{i=1}^{N}\left\{\int_{-\infty}^{\infty} [g_i(\lambda)R_{x_i,x_j}(\tau - \lambda)]d\lambda\right\} = R_{s,x_j}(\tau), \quad \text{for } \forall j, \ 1 \leq j \leq N$$

$$(12.8)$$

If $s(t)$, $n_i(t)$ and $n_j(t)$ are uncorrelated, then

$$R_{x_i,x_j}(t) = R_{s+n_i,s+n_j}(t) = R_s(t) + R_{n_i,n_j}(t), \qquad R_{s,x_j}(t) = R_{s,s+n_j}(t) = R_s(t) \tag{12.9}$$

Substituting Equation 12.9 into Equation 12.8, we get

$$\sum_{i=1}^{N} \int_{-\infty}^{\infty} [g_i(\lambda)\{R_s(\tau - \lambda) + R_{n_i,n_j}(\tau - \lambda)\}]d\lambda = R_s(\tau), \qquad \forall j, \ 1 \leq j \leq N \tag{12.10}$$

Taking the two-sided Laplace transform, we have

$$\sum_{i=1}^{N} [G_i(s)S_s(s)] + G_j(s)S_{n_j}(s) = S_s(s), \quad \forall j, \ 1 \leq j \leq N \tag{12.11}$$

If

$$\sum_{i=1}^{N} G_i(s) = \Gamma$$

then by rearranging Equation 12.11 we obtain

$$G_j(s) = \frac{S_s(s)}{S_{n_j}(s)}(1 - \Gamma), \quad \forall j, \ 1 \leq j \leq N \tag{12.12}$$

Summing Equation 12.12 over all channels, we get

$$\Gamma = \sum_{j=1}^{N} G_j(s) = \sum_{j=1}^{N} \left\{ \frac{S_s(s)}{S_{n_j}(s)}(1 - \Gamma) \right\} \tag{12.13}$$

Solving for Γ, we obtain

$$\Gamma = \frac{S_s(s) \sum_{i=1}^{N} \dfrac{1}{S_{n_i}(s)}}{1 + S_s(s) \sum_{i=1}^{N} \dfrac{1}{S_{n_i}(s)}} \tag{12.14}$$

or

$$\Gamma = 1 - \frac{1}{1 + S_s(s) \sum_{i=1}^{N} \dfrac{1}{S_{n_i}(s)}} \tag{12.15}$$

Substituting Γ into the Equation 12.12 for $G_j(s)$ we have

$$G_j(s) = \frac{S_s(s)}{S_{n_j}(s)} \frac{1}{1 + S_s(s) \sum_{i=1}^{N} \frac{1}{S_{n_i}(s)}}, \quad \forall j, \; 1 \le j \le N \qquad (12.16)$$

From Equation 12.11, we can establish the following relationship between the different restoration filters $G_j(s)$.

$$G_i(s)S_{n_i}(s) = G_j(s)S_{n_j}(s), \quad \forall j, \; 1 \le j \le N \qquad (12.17)$$

For the special case where $N = 2$, the two input LMMSE filter can be rearranged in the following format.

$$G_1(s) = \frac{1}{1 + \dfrac{S_{n_1}(s)}{S_s(s)} + \dfrac{S_{n_1}(s)}{S_{n_2}(s)}}, \qquad G_2(s) = \frac{1}{1 + \dfrac{S_{n_2}(s)}{S_s(s)} + \dfrac{S_{n_2}(s)}{S_{n_1}(s)}} \qquad (12.18)$$

2. LMMSE Filter — System with *N* Inputs Degraded by Additive Noise

Expanding the previous result to a situation where the original signal $s(t)$ undergoes N different degradation processes, $h_j(t)$, to generate $y_j(t)$, $1 < j < N$ as shown in Figure 12.1

$$y_j(t) = h_j(t) * s(t) \quad \text{for } 1 \le j \le N$$

The input to each restoration filter is given by

$$x_j(t) = y_j(t) + n_j(t) = h_j(t) * s(t) + n_j(t) \quad \text{for } 1 \le j \le N \qquad (12.19)$$

The restored signal $\hat{s}(t)$ is then expressed as

$$\hat{s}(t) = \sum_{i=1}^{N} g_i(t) * x_i(t) = \sum_{i=1}^{N} \int_{-\infty}^{\infty} g_i(t - \lambda)x_i(\lambda)d\lambda \qquad (12.20)$$

The MSE as defined by Equation 12.3 is minimized if $g_i(1 \le i \le N)$ satisfies the orthogonality condition in Equation 12.4.

$$E\left[\left(s(t) - \sum_{i=1}^{N}\left\{\int_{-\infty}^{\infty} g_i(\lambda)x_i(t - \lambda)d\lambda\right\}\right)x_j(t')\right] = 0,$$
$$\text{for } t' \in R \quad \text{and} \quad \forall j, \; 1 \le j \le N \qquad (12.21)$$

Using the same procedure as before, we have

$$\sum_{i=1}^{N}\left\{\int_{-\infty}^{\infty}[g_i(\lambda)R_{x_i,x_j}(\tau - \lambda)]d\lambda\right\} = R_{s,x_j}(\tau), \qquad \forall j, \; 1 \le j \le N \qquad (12.22)$$

If $s(t)$, $n_i(t)(1 \leq i \leq N)$ are uncorrelated, then

$$R_{x_i,x_j}(t) = R_{y_i+n,y_j+n_j}(t) = R_{y_i,y_j}(t) + R_{n_i,n_j}(t),$$

$$R_{s,x_j}(t) = R_{s,y_j+n_j}(t) = R_{s,y_j}(t)$$

$$(12.23)$$

Substituting Equation 12.23 into Equation 12.22, we have

$$\sum_{i=1}^{N} \int_{-\infty}^{\infty} \left[g_i(\lambda)\{R_{y_i,y_j}(\tau - \lambda) + R_{n_i,n_j}(\tau - \lambda)\} \right] d\lambda = R_{s,y_j}(\tau), \quad \forall j, \; 1 \leq j \leq N$$

$$(12.24)$$

Using the relation shown in Figure 12.3

$$y_1(t) = \int_{-\infty}^{\infty} s(t - \lambda)h_1(\lambda)d\lambda \qquad y_2(t) = \int_{-\infty}^{\infty} s(t - \lambda)h_2(\lambda)d\lambda \qquad (12.25)$$

By multiplying the first equation by $y_2(t - \tau)$ and the second equation by $s(t + \tau)$, and using the result obtained for the multiple input system, we have

$$y_1(t)y_2(t - \tau) = \int_{-\infty}^{\infty} s(t - \lambda)y_2(t - \tau)h_1(\lambda)d\lambda$$

$$s(t + \tau)y_2(t) = \int_{-\infty}^{\infty} s(t + \tau)s(t - \lambda)h_2(\lambda)d\lambda$$

$$(12.26)$$

Computing the expected value of both sides, we have the two equations

$$R_{y_1,y_2}(\tau) = \int_{-\infty}^{\infty} R_{s,y_2}(\tau - \lambda)h_1(\lambda)d\lambda = R_{s,y_2}(\tau) * h_2(\tau)$$

$$R_{s,y_2}(\tau) = \int_{-\infty}^{\infty} R_s(\tau + \lambda)h_2(\lambda)d\lambda = R_s(\tau) * h_2(-\tau)$$

$$(12.27)$$

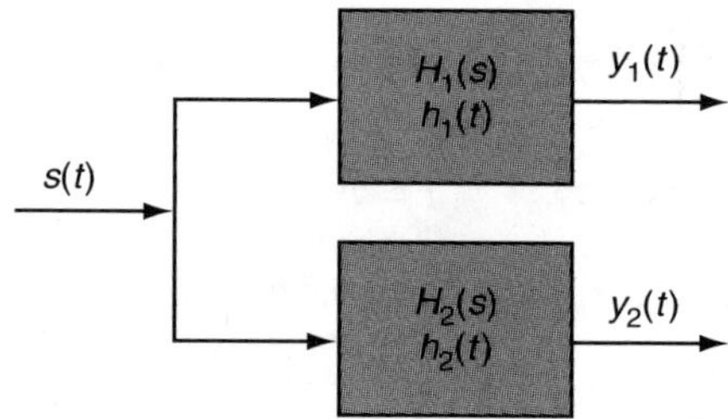

FIGURE 12.3 Model for computing the cross-correlation for multiple terminal outputs.

The Laplace transform of Equation 12.27 is given by

$$S_{y_1, y_2}(s) = S_{s, y_2}(s)H_1(s)$$

$$S_{s, y_2}(s) = S_s(s)H_2(-s) \tag{12.28}$$

Therefore, the expression for the cross correlation becomes

$$S_{y_1, y_2}(s) = S_s(s)H_1(s)H_2(-s) \tag{12.29}$$

Using this expression for the cross-correlation and taking the two-sided Laplace transform of both sides of Equation 12.24, we obtain

$$\left[\sum_{i=1}^{N} G_i(s)S_{y_i, y_j}(s)\right] + G_j(s)S_{n_j}(s) = S_{s, y_j}(s), \qquad \forall j, \ 1 \le j \le N \tag{12.30}$$

$$\left[\sum_{i=1}^{N} G_i(s)H_i(s)H_j(-s)S_s(s)\right] + G_j(s)S_{n_j}(s) = H_j(-s)S_s(s), \quad \forall j, \ 1 \le j \le N \tag{12.31}$$

$$\left[\sum_{i=1}^{N} G_i(s)H_i(s)H_j(-s)\right] + G_j(s)\frac{S_{n_j}(s)}{S_s(s)} = H_j(-s), \qquad \forall j, \ 1 \le j \le N \tag{12.32}$$

We now have an expression for the jth restoration filter $G_j(s)$.

$$G_j(s)\frac{S_{n_j}(s)}{S_s(s)} = H_j(-s)\left(1 - \sum_{i=1}^{N} G_i(s)H_i(s)\right), \qquad \forall j, \ 1 \le j \le N \tag{12.33}$$

Let $\Gamma(s) = \sum_{i=1}^{N} G_i(s)H_i(s)$. Rewriting Equation 12.33, we obtain

$$G_j(s) = \frac{S_s(s)H_j(-s)}{S_{n_j}(s)}(1 - \Gamma(s)) \quad \text{for } 1 \le j \le N \tag{12.34}$$

Multiplying both sides by $H_j(s)$ and summing over all filters we get

$$\Gamma(s) = \sum_{i=1}^{N} G_i(s)H_i(s) = (1 - \Gamma(s))\sum_{i=1}^{N}\frac{S_s(s)}{S_{n_i}(s)}|H_i(s)|^2 \tag{12.35}$$

We can solve Equation 12.35 to obtain $\Gamma(s)$.

$$\Gamma(s) = 1 - \frac{1}{1 + \sum_{i=1}^{N}\dfrac{S_s(s)}{S_{n_i}(s)}|H_i(s)|^2} \tag{12.36}$$

Substituting $\Gamma(s)$ in the equation for $G_j(s)$, we have

$$G_j(s) = \frac{S_s(s)}{S_{n_j}(s)} \frac{H_j(-s)}{1 + \displaystyle\sum_{i=1}^{N} \frac{S_s(s)}{S_{n_i}(s)} |H_i(s)|^2} \quad \text{for } 1 \leq j \leq N \qquad (12.37)$$

From Equation 12.32 we can derive the following relationship between the different restoration filters.

$$\frac{G_i(s)S_{n_i}(s)}{H_i(-s)} = \frac{G_j(s)S_{n_j}(s)}{H_j(-s)}, \quad \forall i,j \ \ 1 \leq i, j \leq N \qquad (12.38)$$

In the special case where $H_i(s) = 1$, for all $i(1 \leq i \leq N)$ Equation 12.38 reduces to Equation 12.17. If $H_i(s)H_j(-s) = 0$, for $i \neq j$, we can derive the reduced expression from Equation 12.34 as

$$G_j(s) = \frac{H_j(-s)}{|H_i(s)|^2 + \dfrac{S_{n_j}(s)}{S_s(s)}} \quad \forall j, \ 1 \leq j \leq N \qquad (12.39)$$

which degenerates to the case of N independent LMMSE filters.

In matrix notation, Equation 12.32 can be written as

$$H(s) = \begin{bmatrix} H_1(s) \\ H_2(s) \\ \vdots \\ H_N(s) \end{bmatrix}, \quad G(s) = \begin{bmatrix} G_1(s) \\ G_2(s) \\ \vdots \\ G_N(s) \end{bmatrix},$$

$$S_{\text{NS}}(s) = \frac{1}{S_s(s)} \times \begin{bmatrix} S_{n_1}(s) & & & 0 \\ & S_{n_2}(s) & & \\ & & \ddots & \\ 0 & & & S_{n_x}(s) \end{bmatrix} \qquad (12.40)$$

$$\times \{H(s)H^{\mathrm{T}}(-s) + S_{\text{NS}}(s)\}G(s) = H(-s)$$

where $H^{\mathrm{T}}(-s)$ is the transpose of $H(-s)$.

$$G(s) = (H(s)H^{\mathrm{T}}(-s) + S_{\text{NS}}(s))^{-1}H(-s) \qquad (12.41)$$

As a simple example, if $N = 2$, the two input LMMSE filter can be rewritten as

$$\begin{bmatrix} |H_1(s)|^2 S_s(s) + S_{n_1}(s) & S_s(s)H_1(s)H_2(-s) \\ S_s(s)H_1(-s)H_2(s) & |H_2(s)|^2 S_s(s) + S_{n_2}(s) \end{bmatrix} \begin{bmatrix} G_1(s) \\ G_2(s) \end{bmatrix} = \begin{bmatrix} H_1(-s)S_s(s) \\ H_2(-s)S_s(s) \end{bmatrix}$$

$$(12.42)$$

$$G_1(s) = \frac{H_1(-s)S_{n_2}(s)}{|H_1(s)|^2 S_{n_2}(s) + |H_2(s)|^2 S_{n_1}(s) + \dfrac{S_{n_1}(s)S_{n_2}(s)}{S_s(s)}} \qquad (12.43)$$

$$G_2(s) = \frac{H_2(-s)S_{n_1}(s)}{|H_1(s)|^2 S_{n_2}(s) + |H_2(s)|^2 S_{n_1}(s) + \dfrac{S_{n_1}(s)S_{n_2}(s)}{S_s(s)}} \qquad (12.44)$$

Again, as a special case, if the noise spectra are all identical, for example, $S_{n_j} = S_n$, $\forall j$, $1 \le j \le N$, the filter simplifies to

$$G_j(s) = \frac{H_j(-s)}{\displaystyle\sum_{i=1}^{N} |H_i(s)|^2 + \dfrac{S_n(s)}{S_s(s)}} \qquad (12.45)$$

When the noise and signal spectra are not available, which is common in practice, we can approximate S_n/S_s suitably with a constant $K = S_n/S_s$

$$G_j(s) = \frac{H_j(-s)}{\displaystyle\sum_{i=1}^{N} |H_i(s)|^2 + K} \qquad \text{for } 1 \le j \le N \qquad (12.46)$$

If the signal is noise free, for example, $S_{n_j} = 0$, $\forall j$, $1 \le j \le N$, the filter reduces to

$$G_j(s) = \frac{H_j(-s)}{\displaystyle\sum_{i=1}^{N} |H_i(s)|^2} \qquad (12.47)$$

Often the transfer function representing the degradation is not available. Under these circumstances it is possible to use the spectra of the acquired signals. We will derive a filter that does not require a transfer function representing the degradation, specifically

$$S_{x_i}(s) = |H_i(s)|^2 S_s(s) + S_{n_i}(s) \qquad (12.48)$$

$$|H_i(s)|^2 = \frac{S_{x_i}(s) - S_{n_i}(s)}{S_s(s)} \tag{12.49}$$

If we assume that the system has a symmetric spectrum $H(s) = H(-s)$ which is common in practice,

$$H_i(s) = H_i(-s) = \sqrt{\frac{S_{x_i}(s) - S_{n_i}(s)}{S_s(s)}} \tag{12.50}$$

In this case, by substituting Equation 12.50 into Equation 12.37 we have

$$G_j(s) = \frac{S_s(s)}{S_{n_j}(s)} \frac{\sqrt{\dfrac{S_{x_j}(s) - S_{n_j}(s)}{S_s(s)}}}{1 + \displaystyle\sum_{i=1}^{N} \dfrac{S_{x_i}(s) - S_{n_i}(s)}{S_{n_i}(s)}} \quad \text{for } 1 \leq j \leq N \tag{12.51}$$

or:

$$G_j(s) = \frac{\sqrt{S_s(s) S_{x_j}(s) - S_{n_j}(s)}}{(1 - N)S_{n_j}(s) + S_{n_j}(s) \displaystyle\sum_{i=1}^{N} \dfrac{S_{x_i}(s)}{S_{n_i}(s)}} \quad \text{for } 1 \leq j \leq N \tag{12.52}$$

Using Equation 12.52, we can design a multiple input LMMSE filter using the spectra of the degraded signals together with the original signal and noise spectra.

When the noise spectra are much smaller than the signal spectrum and are all identical, for example, $S_{n_j} = S_n$ and $S_{x_j} \gg S_n$, $1 \leq j \leq N$, Equation 12.52 becomes

$$G_j(s) = \frac{\sqrt{S_s(s) S_{x_j}(s)}}{\displaystyle\sum_{i=1}^{N} S_{x_i}(s)} \quad \text{for } 1 \leq j \leq N \tag{12.53}$$

When the spectrum of the original signal is not available, we can use the approximation.

$$G_j(s) = K \frac{\sqrt{S_{x_j}(s)}}{\displaystyle\sum_{i=1}^{N} S_{x_i}(s)} \quad \text{for } 1 \leq j \leq N \tag{12.54}$$

3. Multiple Input LMMSE Filter for the Two-Dimensional Discrete Case

For the two-dimensional discrete signal case the multiple input LMMSE is modified using z-transform to

$$G_j(z_1,z_2) = \frac{S_s(z_1,z_2)}{S_{n_j}(z_1,z_2)} \frac{H_j(z_1^{-1},z_2^{-1})}{1 + \sum_{i=1}^{N} \frac{S_s(z_1,z_2)}{S_{n_i}(z_1,z_2)} |H_i(z_1,z_2)|^2} \qquad \text{for } 1 \le j \le N \quad (12.55)$$

$G_j(z_1,z_2)$ represents the jth LMMSE filter in the filter bank for the multiple input image system.

Assuming that the spectra are much smaller than the signal spectrum and are all identical, for example, $S_{n_j} = S_n$ and $S_{x_j} \gg S_n$, for $1 \le j \le N$, we get

$$G_j(z_1,z_2) = \frac{\sqrt{S_s(z_1,z_2)S_{x_j}(z_1,z_2)}}{\sum_{i=1}^{N} S_{x_i}(z_1,z_2)} \qquad \text{for } 1 \le j \le N \qquad (12.56)$$

4. Experimental Results

In order to validate the approach, the image shown in Figure 12.4(a) is used as a reference image. Degraded versions of the image are generated by low and high pass filtering the image and superimposing uniformly distributed noise. The degraded images are shown in Figure 12.4(b) and Figure 12.4(c). Figure 12.4(d) illustrates the reconstructed image using the two degraded images. To facilitate comparison, the results of applying a single input LMMSE filter for the two degraded images are shown in Figure 12.5(a) and 12.5(b). It is clear that the results are not satisfactory due to the additive noise components. The average of these two reconstructed images, as shown in Figure 12.5(c), is also much worse than the result obtained using the two input LMMSE filters shown in Figure 12.4(d). The use of the cross-correlation parameters of the two images allows the two-input LMMSE filters to outperform the traditional LMMSE filter. The performance improvements are apparent from the expression derived for the mean square error as well as from the experimental results.

In real applications, we only have the acquired image data rather than the true spectrum of the data. To overcome this problem, we can assume that the images are ergodic, and estimate the spectrum by partitioning the image, and taking the average spectra of each of the blocks. Further improvements can be obtained if we have explicit knowledge of the degradation transfer function, and estimates of the noise and signal spectra.

B. MORPHOLOGICAL PROCESSING APPROACH TO FUSION

A second system-based algorithm for data fusion is based on morphological methods.[7] In this approach, the noise reduced ultrasonic image is used to derive

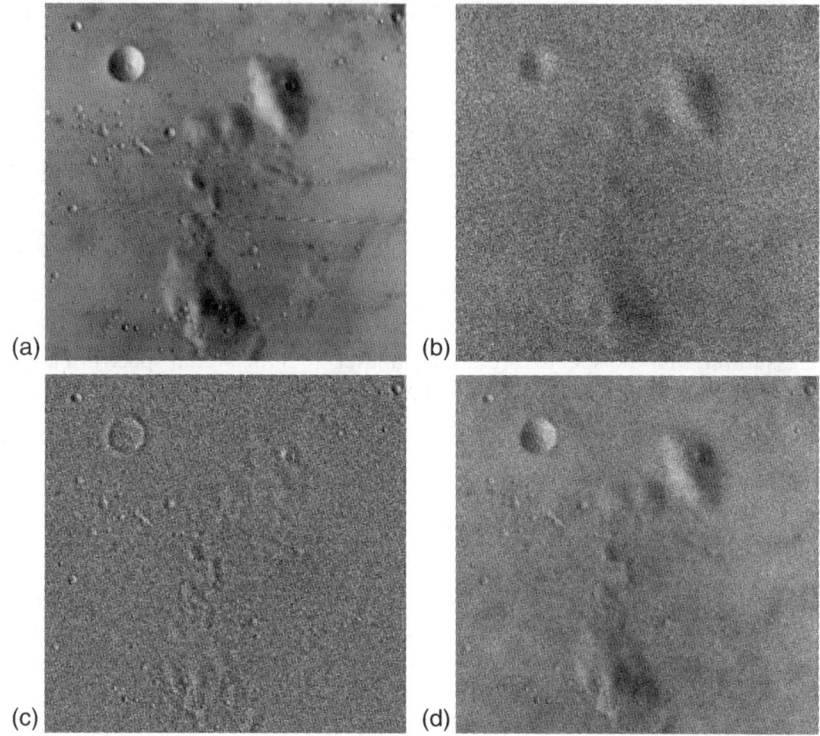

FIGURE 12.4 Multiple input LMMSE filtered image (a) original image, (b) degraded image 1 with additive noise, (c) degraded image 2 with additive noise, and (d) multiple input LMMSE filtered image.

information relating to the locations and boundaries of the flaw using a morphological closing operation. The gray levels of the eddy current image are then superimposed within the defect regions using an AND operation to obtain the fused image. The major challenge associated with this step lies in identifying a structuring element that is optimal in size. This is accomplished by computing the granulometric size distribution[8] using iterative closing operations with a sequence of structuring element sizes.

Notation

$Y(m, n)$	noise reduced ultrasonic image
$F_i(m, n)$	fused image
$S_k(m, n)$	kth structuring element in a family of structuring elements of different sizes

The preprocessed ultrasonic image containing the outline of the defect is thresholded to obtain a binary image where the defect boundary pixels have a gray level "one" and the background or nondefect outline pixels are tagged "zero". It is observed that single C-scan acoustic microscopy images offer little

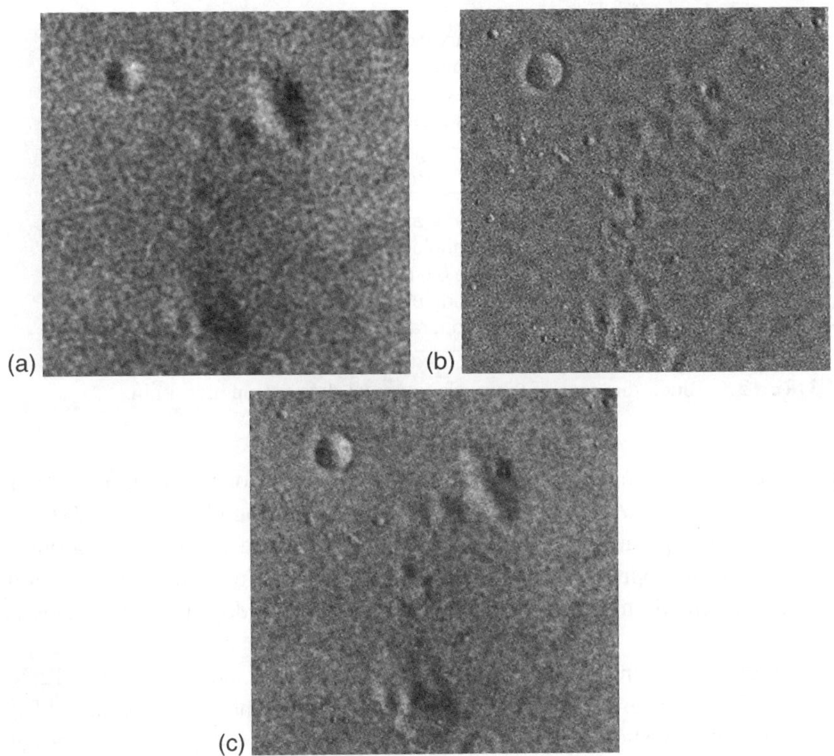

FIGURE 12.5 Comparison with one input LMMSE filter. (a) 1-input filtered image of Figure 12.4(b), (b) 1-input filtered image of Figure 12.4(c), and (c) average of images in (a) and (b).

information for inferring the depth of narrow cracks. Consequently, such images are utilized largely to identify the boundary of the defect.

The morphological closing operation is capable of filling "holes" that are generally smaller than the predetermined structuring element. The closing operation can be used, therefore, to merge the unidentified defect regions. The challenge is to determine the size of the structuring element that is necessary to merge such regions effectively. One approach for estimating the structuring element size is by examining the granulometric size (density) distribution obtained by using iterative closing operations. Consider a sequence of structuring elements $\{S_k | k = 0, 1, 2, \ldots\}$, where S_{k+1} is a superset of S_k. Since the closing operation has the extensivity property, the closed image $C(B, S_{k+1})$ is also a superset of $C(B, S_k)$. Thus, if we define $N'(k)$ as the number of pixels activated in $C(B, S_k)$, then $N(k) = N'(k)/M^2$, called the granulometric size distribution, is a nondecreasing function of k, where M^2 is the total number of pixels in the image. Here, $N(0)$ is the fraction of pixels activated in the binary image B itself. The granulometric size density can be computed using $p(k) = \{N(k + 1) - N(k)\}$. The optimal size of the structuring

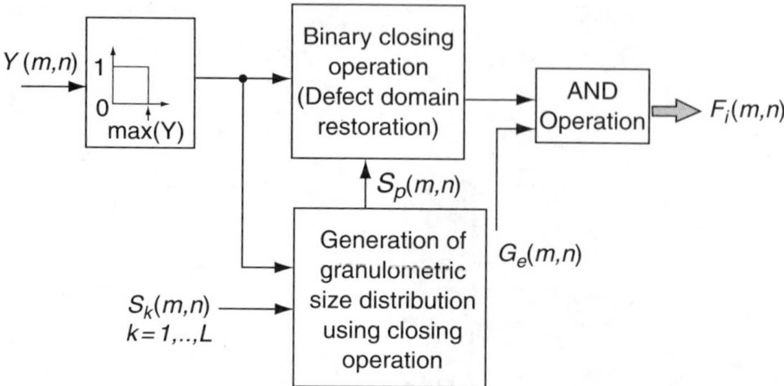

FIGURE 12.6 Block diagram of the morphological data fusion algorithm.

element can be determined by looking for abrupt transitions in the granulometric size density. After identifying the appropriate structuring element size $S_p(m, n)$, binary closing operations can be performed to restore the defect regions. A schematic of this approach is illustrated in Figure 12.6. Some initial results of the implementation of this scheme using ultrasonic and eddy current images are presented next.[9]

The test specimen used for generating the images is shown in Figure 12.7(a) and the original ultrasonic and eddy current measurement images obtained from the specimen are shown in Figure 12.7(b) and Figure 12.7(c), respectively. Figure 12.8(a) and Figure 12.8(b) show examples of granulometric size $N(k)$ and density functions $p(k)$ obtained from the image shown in Figure 12.7(b), respectively. The variable k along the horizontal axes in Figure 12.8(a) and Figure 12.8(b) indicate that the size of the structuring element is $(2k + 1) \times (2k + 1)$ pixels, $k = 1, 2, \dots$. In Figure 12.8(a), the values of vertical axis represent the normalized numbers of pixels which are merged using binary closing operations employing structuring elements of size $(2k + 1) \times (2k + 1)$ pixels, $k = 1, 2, \dots$. In this example, a structuring element of size 5×5 pixels was determined to be optimal based on the granulometric size density function. A closing operation using this structuring element is performed to merge the unidentified defect regions in the binary image to obtain the result shown in Figure 12.8(c). The gray levels in the eddy current image shown in Figure 12.7(c) are then superimposed on the defect regions using an AND operation. The highest gray level in the original eddy current image was chosen as the background level in the fused image in simulation. Figure 12.9(a) and Figure 12.9(b) show sample line scans of the original ultrasonic and eddy current images shown in Figure 12.7(b) and Figure 12.7(c), respectively. Figure 12.10 shows the fused image and line scans obtained using the proposed algorithm and Figure 12.11 shows the corresponding fused results obtained using the LMMSE technique. A line scan of the fused image shown in Figure 12.9(b) reveals the gray level transitions in the defect region. The fused image obtained using the proposed algorithm reveals the defect locations and

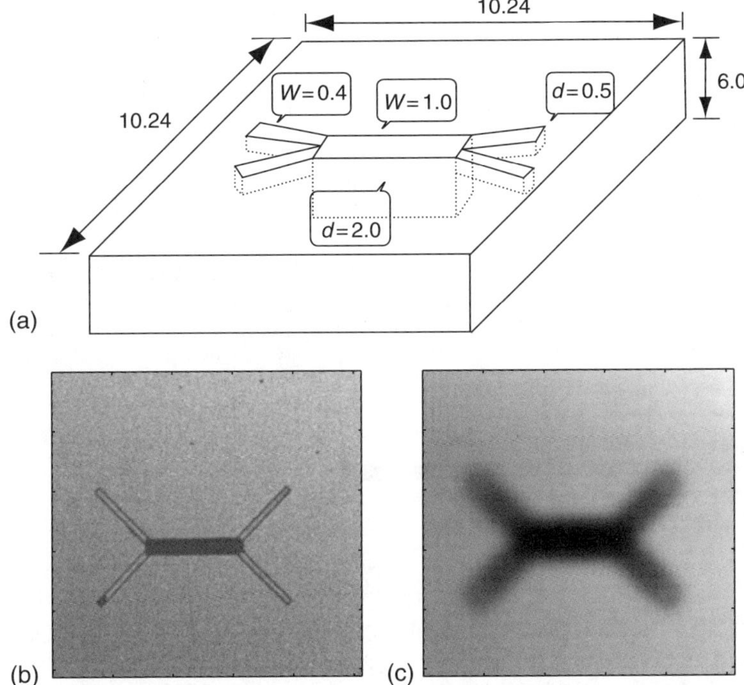

FIGURE 12.7 (a) Geometry of test specimen with machined defect, (b) ultrasonic, and (c) eddy current measurement data.

their gray levels clearly demonstrate the effectiveness of the approach compared to that obtained using the LMMSE filters.

III. MODEL-BASED DATA FUSION

This class of approaches for data fusion utilizes the underlying physical model of the multiple sensor data. The characteristics and nature of data generated by sensors are typically different, and consequently, information must first be mapped to a common "format". For instance, for fusing data from diffusion and wave propagation sensors, the information has to be appropriately transformed and processed before it is integrated. In addition, data registration issues have to be addressed before fusing the data. Data fusion algorithms that build on models of the associated physical processes take into account the fact that eddy current techniques based on electromagnetic induction are governed by a physical process that is diffusive in nature, whereas the ultrasonic methods rely on the propagation and reflection of an ultrasonic wave into the test specimen. An innovative approach for mapping data from the two processes is via the use of Q-transform.[10] Q-transform can be employed to map propagating wave fields

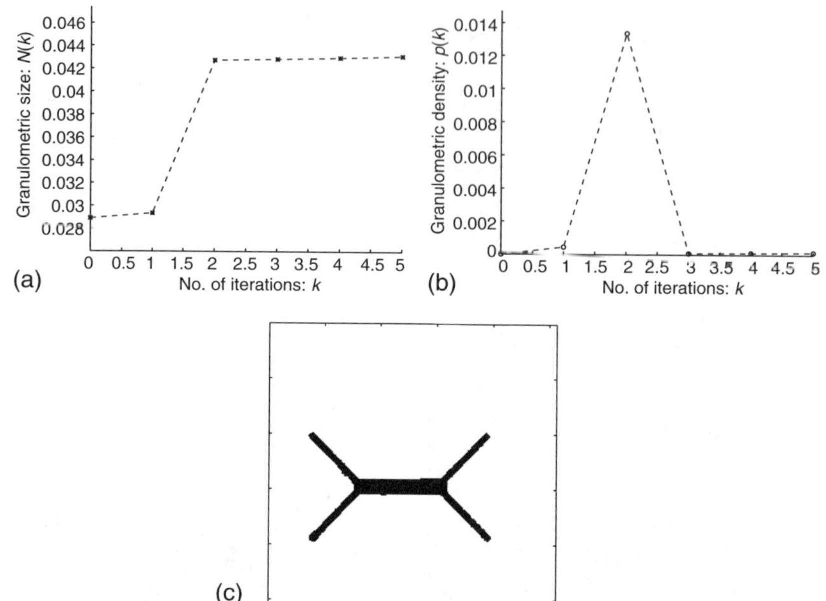

FIGURE 12.8 Results obtained using the algorithm shown in Figure 12.6. (a) Granulometric size distribution, (b) density for the image shown in Figure 12.7(b), and (c) restored binary image obtained after using a closing operation with a structuring element of size 5×5 pixels.

generated by sensors that rely on wave phenomena such as ultrasonic NDT sensors onto equivalent diffusive fields, such as those generated by eddy current probes. Once the data have been mapped onto a common format, system based methods can be used for fusing the data. The fusion algorithm then combines overlapping or redundant segments of information to enhance the signal-to-noise ratio. Complementary segments are combined to augment the overall information content of the composite signal.

The challenges associated with implementing the approach are formidable. The Q-transform is a compact operator. Therefore, the inverse Q-transform is inherently unstable, and consequently, algorithms for computing Q^{-1} in a robust manner need to be developed. However, the computation of the Q-transform is stable. Nevertheless, a time-shift in the data may alter its Q-transform.

A detailed description of the approach based on Q-transform for model based data fusion is discussed in the following sections.

A. Q-TRANSFORM

The Q-transform is an integral transformation that can be used to map solutions of hyperbolic problems onto solutions of parabolic problems. This is the key idea

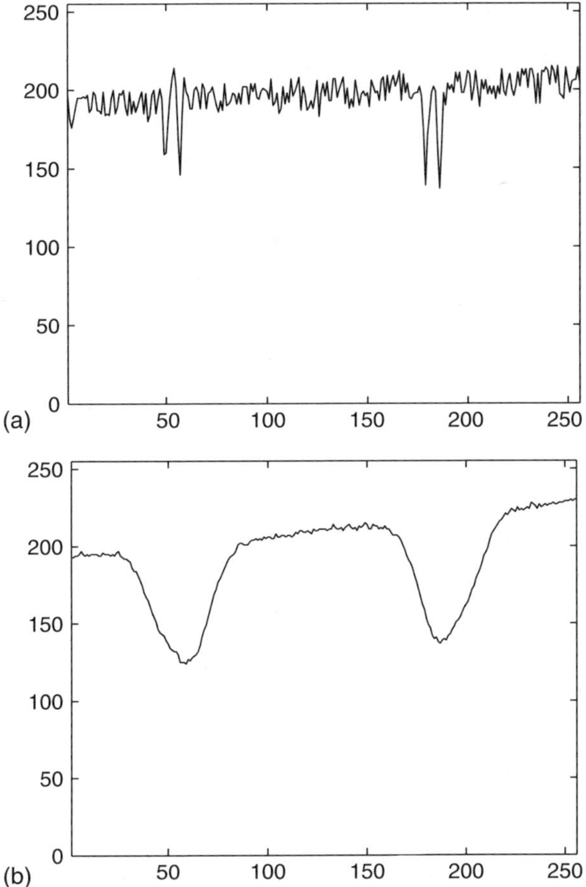

FIGURE 12.9 Line scans obtained from the original (a) ultrasonic, and (b) eddy current images.

for mapping the data provided by eddy current and ultrasonic sensors onto a common format.

The mathematical properties of the Q-transform have been studied systematically and exploited to develop a theoretical framework that enables robust transformation to and from the wave domain.

Mathematicians, geophysicists and others have studied the concept of transforming diffusive fields to wave propagation fields and *vice versa*. Bragg and Dettman[11] as well as Filippi and Frisch[12] were among the earliest to establish a connection between diffusion and wave equations. Others who have worked on establishing such linkages include Isaev and Filatov.[13] Pierce,[14] inspired by the work of Symes,[15] proposed an approach to developing a method for inverting diffusion data to solve wave propagation problems. In the process, he also establishes parallels between the two cases. Romanov[16] and Reznitskaya[17] derive

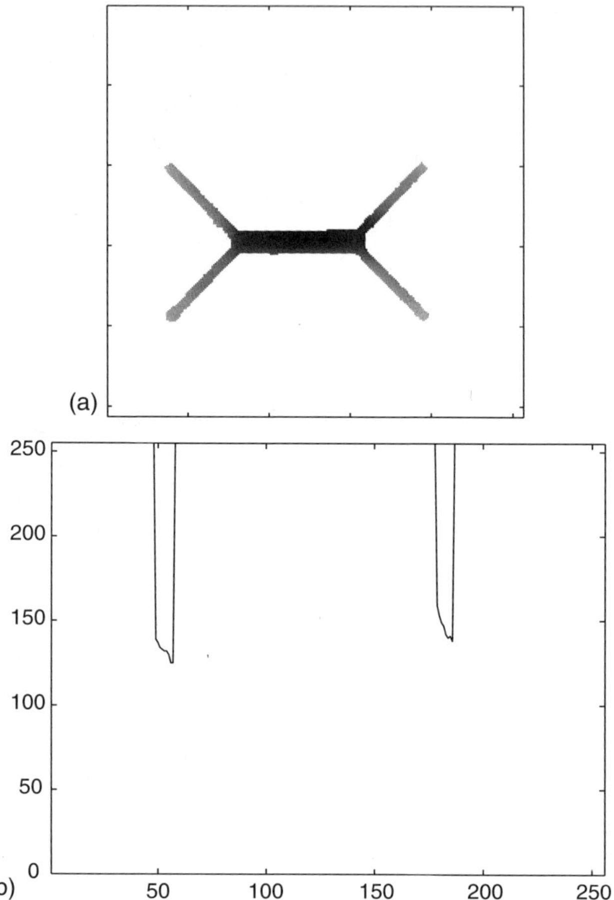

FIGURE 12.10 (a) Fused image, and (b) line scan obtained from the fused image using the morphological processing algorithm.

a transform linking diffusive and wave fields. Lee et al.[18] extended the work to vector electromagnetic fields and the corresponding wave fields. Lee et al.[19] use the inverse Q-transform to map diffusive fields to their corresponding wave field and use ray tomography to estimate the slowness and hence the conductivity map. In the same line, Ross et al.[20] use the inverse Q-transform for estimating the thickness of a conductive slab. Sun et al.[21] analyzed the Q-transform from the data fusion perspective. Gibert et al.[22] and Lee et al.[23] use the Q-transform to parameterize the waveform of eddy current data in terms of time-of-flight. Tamburrino et al.[24,25] use proper waveforms for the driving signal to relate the peak position in the measured waveform to the time-of-flight from the probe to a defect. A systematic study of properties and tables of Q-transform can be found in Ref. 26.

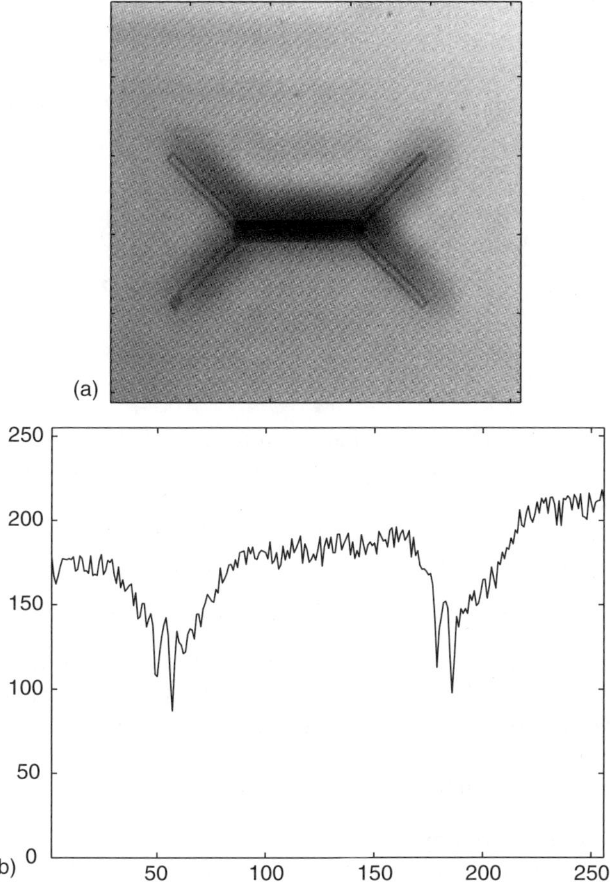

FIGURE 12.11 (a) Fused image, and (b) line scan obtained from the fused image using the LMMSE filters.

B. DEFINITION AND MAPPING PROPERTY

The Q-transform is an integral operator (acting only on the time co-ordinate) defined as

$$Q : u(x, q) \rightarrow v(x, t) \triangleq \int_0^{+\infty} \frac{q}{2\sqrt{\pi t^3}} \exp\left(-\frac{q^2}{4t}\right) u(x, q) dq \qquad (12.57)$$

where the variable q has the dimension of square root of time. The Q-transform in terms of Laplace transform is given by $V(x, s) = Q[U(x, \sqrt{s})]$ where V and U are the Laplace transforms of v and u, respectively.

The main property of this transformation is the ability to map solutions of hyperbolic problems to solutions of parabolic problems. To be more specific, let us consider the following two problems:

$$\nabla^2 \phi(x, t) - c(x)\partial_t \phi(x, t) = F(x, t), \quad x \in \Omega, \ t \geq 0 \qquad (12.58)$$

$$\alpha(x)\phi(x, t) + \beta(x)\partial_n \phi(x, t) = G(x, t), \quad x \in \partial\Omega, \ t \geq 0 \qquad (12.59)$$

$$\phi(x, 0) = h(x), \quad x \in \Omega \qquad (12.60)$$

and:

$$\nabla^2 \psi(x, q) - c(x)\partial_{qq} \psi(x, q) = f(x, q), \quad x \in \Omega, \ q \geq 0 \qquad (12.61)$$

$$\alpha(x)\psi(x, q) + \beta(x)\partial_n \psi(x, q) = g(x, q), \quad x \in \partial\Omega, \ q \geq 0 \qquad (12.62)$$

$$\psi(x, 0) = 0, \quad x \in \Omega \qquad (12.63)$$

$$\partial_q \psi(x, 0) = h(x), \quad x \in \Omega \qquad (12.64)$$

where $\Omega \subseteq \Re^N$ ($N \geq 1$), c is a real and non-negative function, ∂_n represents the normal derivative on the boundary and the two scalar functions α, β are given. The Q-transform maps the solution of Equation 12.61, Equation 12.62 and Equation 12.64 to the solution of Equation 12.58, Equation 12.59 and Equation 12.60 in the following manner:

$$\phi = Q[\psi] \quad \text{if} \ \ F = Q[f] \ \ \text{and} \ \ G = Q[g] \qquad (12.65)$$

We notice that the initial condition h appearing in Equation 12.60 is assigned to $\partial_q \psi(x, 0)$ in Equation 12.64 whereas $\psi(x, 0) = 0$. It is also possible to define an alternative Q-transform where h is assigned to $\psi(x, 0)$ and $\partial_q \psi(x, 0) = 0$.

C. Numerical Computation of the Q-Transform

The Q-transform is given by an improper integral. However, its kernel rapidly decays to zero for large q due to the exponential factor $\exp(-q^2/4t)$. The Q-transform of a function u can be written in the following manner

$$Q[u] = \frac{1}{t\sqrt{2\pi e}} \int_0^{+\infty} w\left(\frac{q}{\sqrt{2t}}\right) u(x, q)\,dq \qquad (12.66)$$

where the weight function w is given by

$$w(x) = x \exp\left(-\frac{x^2}{2}\right)\sqrt{e} \qquad (12.67)$$

The plot of the weight function is shown in Figure 12.12.

Its maximum value (equal to 1) is achieved at $x = 1$. Outside the interval $6.07 \times 10^{-6} \leq x \leq 5.228$ the function w is less than 10^{-5}. Therefore, apart from particular waveforms u, the integral in Equation 12.66 can be truncated to the range $8.58 \times 10^{-6}\sqrt{t} \leq q \leq 7.39\sqrt{t}$ or in general, to the range

$$k_{\min}\sqrt{t} \leq q \leq k_{\max}\sqrt{t} \qquad (12.68)$$

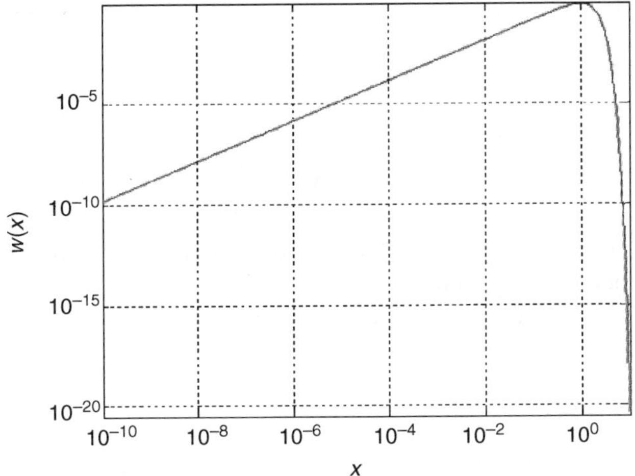

FIGURE 12.12 Plot of the weight function w. Notice the extremely rapid decay for x greater than 1.

To develop a numerical method for computing the Q-transform,[26] we recall that the Q-transform is a compact operator (smoothing operator), and is therefore not sensitive to the fine details of the waveform u. Hence we can replace u with the approximation given by

$$u_a(q) = \sum_{k=1}^{N} u_k p_k(q) \tag{12.69}$$

where u_k are samples of u and $p_k(q)$ are the interpolating functions having a stable Q-transform. Taking into account Equation 12.68 and assuming that $0 < t_{min} \leq t \leq t_{max}$, u_a should properly approximate u for $k_{min}\sqrt{t_{min}} \leq q \leq k_{max}\sqrt{t_{max}}$.

For a piecewise constant approximation where $u_k = u(q_k)$,

$$p_k(q) = \begin{cases} 1 & \text{for } q_k \leq q < q_{k+1} \\ 0 & \text{otherwise} \end{cases} \tag{12.70}$$

and $0 \leq q_1 < \ldots < q_{N+1}$ it turns out that

$$Q(u_a) = \sum_{k=1}^{N+1} (u_k - u_{k-1})F_k(t) \tag{12.71}$$

where $F_k(t) = (\pi t)^{-1/2}\exp(-q_k^2/4t)$ and u_0 and u_{N+1} are set to zero. We notice that F_k is bounded for q_k greater than zero and achieves its maximum value (equal to $q_k^{-1}\sqrt{2/\pi e}$) at $t = q_k^2/2$. If q_1 (the smallest among the q_k's) is equal to zero, then $F_1(t) = 1/\sqrt{\pi t}$ and any noise affecting sample u_1 may give rise to large errors, particularly for small t.

For a linear piecewise approximation, such as $u_a(q_k) = u_k$ and $u_a(q) = 0$ for $q < q_1$ and $q > q_N$, we have

$$Q(u_a) = (\pi t)^{-1/2}[q_1 \exp(-q_1^2/4t) - q_N \exp(-q_N^2/4t)]$$
$$+ \sum_{k=1}^{N+1} (u_k' - u_{k-1}')\mathrm{erfc}\left(q_k/2\sqrt{t}\right) \tag{12.72}$$

where $u_k' = (u_{k+1} - u_k)/(q_{k+1} - q_k)$, u_0 and u_{N+1} are set to zero and erfc is the complementary error function.

Properties of the Q-transform and tables of transform pairs can be found in Ref. 26.

1. Signal Level Data Fusion

The Q-transform relates solutions of hyperbolic (e.g., Equation 12.61 to Equation 12.64) and parabolic problems (e.g., Equation 12.58 to Equation 12.60). In principle we can convert measurements from eddy current sensors to the "ultrasonic format" and, similarly, measurements from ultrasonic sensors to "eddy current format". However, there is a substantial difference between the two alternatives. Specifically, recalling that the Q-transform is a compact operator (see Ref. 24 for its singular values decomposition), the inverse Q-transform magnifies the unavoidable noise affecting the eddy current test (ECT) data. Consequently, it is easier to convert ultrasonic data into the "eddy current format".

This is illustrated with reference to a simple example of diffusion and wave propagation in homogeneous media.[27] We consider a defect in an infinite homogeneous media (copper, $\sigma_{Cu} = 3.54 \times 10^7$, $\mu_r = 1$). The conductivity of the defect σ_D is represented by the contrast function $\chi_D = \sigma_D/\sigma_{Cu} - 1$, which, in this example, is equal to -0.25. This means* that $c(x) = \mu_0\sigma(x)$, where $\sigma(x)$ is either σ_{Cu} or σ_D depending on x, in Equation 12.58 and Equation 12.61. The defect is a sphere of radius 0.1 mm; the distance between its center and the driving source, assumed to be a point source, is 10 mm. The functional value of the point source for the hyperbolic problem (Equation 12.61) is given by

$$f(q) = (q - q_i)^2 \quad \text{for } q \geq 0 \tag{12.73}$$

where q_i is a non negative arbitrary constant. The corresponding driving signal associated with the diffusion problem Equation 12.58 is (see Equation 12.65):

$$F(t) = 2q_i\left[\frac{2}{q_i}\sqrt{\frac{t}{\pi}}\exp\left(-\frac{q_i^2}{4t}\right) - \mathrm{erfc}\left(\frac{q_i}{2\sqrt{t}}\right)\right], \quad t \geq 0 \tag{12.74}$$

---

*In addition, $\Omega = \Re^3$ and the boundary conditions 12.59 and 12.62 have to be replaced with the requirement that the fields are vanishing at infinity.

Figure 12.13 shows the plot of the solution of the diffusive problem and the Q-transform of the solution of the associated wave propagation problem ($q_i = 0$ in Equation 12.73). We assume that the solution is the reaction field (the field due to the defect) measured at the origin of the co-ordinate system. Specifically, $v_s(x, t)$ and $u_s(x, t)$ are the reaction fields for the diffusion and wave propagation problems, respectively.

In practice, the ultrasonic and eddy current measurements are governed by partial differential equations having different coefficient functions. When the coefficients are constant, a modified version of the Q-transform can be used:[23]

$$Q : u(x, q) \rightarrow v(t) \triangleq \int_0^{+\infty} \frac{\lambda q}{2\sqrt{\pi t^3}} \exp\left(-\frac{\lambda^2 q^2}{4t}\right) u(x, q) dq \qquad (12.75)$$

This Q-transform relates the solutions of problems Equation 12.58 to Equation 12.64 where Equation 12.58 and Equation 12.64 have been replaced by

$$\nabla^2 \phi(x, t) - \lambda^2 c(x) \partial_t \phi(x, t) = F(x, t), \quad x \in \Omega, \ t \geq 0 \qquad (12.76)$$

$$\partial_q \psi(x, 0) = \lambda^2 h(x), \quad x \in \Omega \qquad (12.77)$$

We notice that with this new definition the dimension of the independent variable q is $T^{1/2}[\lambda]^{-1/2}$. When Equation 12.76 describes an ultrasonic test (UT), q has the dimension of a time.

Sun et al.[23] have studied the effect of a time shift in the data on the Q-transform. Specifically, it is possible to prove that

$$Q\{u(q - q_0)\} = h_{q_0} * Q\{u\} \qquad (12.78)$$

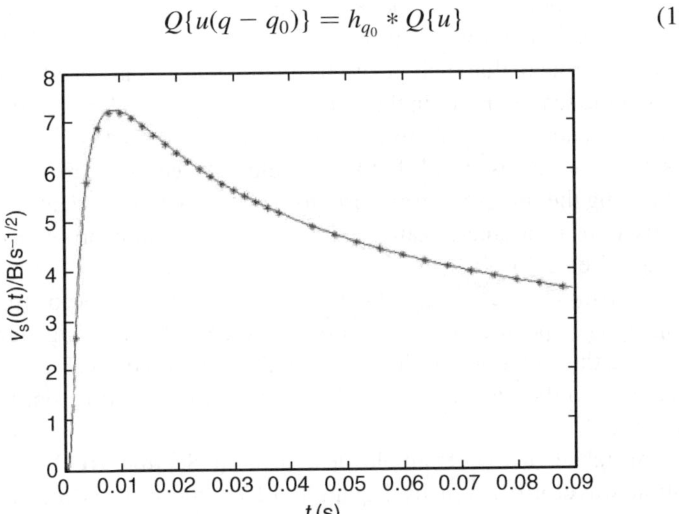

FIGURE 12.13 Plot of $v_s(0, t)$ (solid) together with the plot of $Q\{u_s(0, t)\}$ ($*$). B is normalization constant.

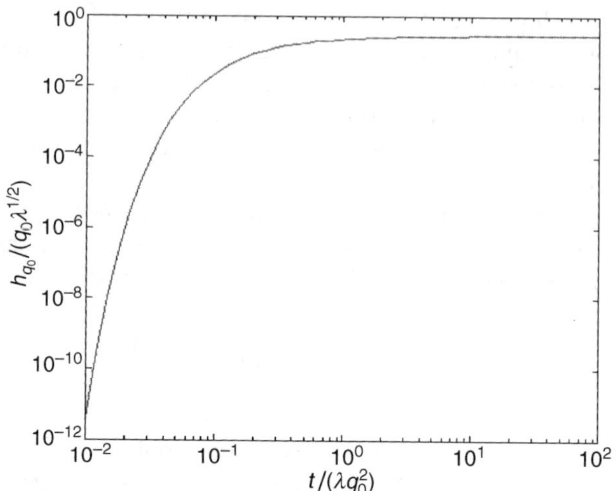

FIGURE 12.14 Normalized plot of h_{q_0}.

where $h_{q_0}(t) = \sqrt{\lambda/4\pi}\,\exp(-\lambda q_0^2/4t)q_0$. The effect of a time shift is therefore to smear the transform $Q(u)$ by the kernel h_{q_0} (see Figure 12.14).

2. Feature Level Data Fusion

Q-transform can also be used for feature level fusion. For example, let us consider again the problem of defect detection. In ultrasonic tests the received signal contains echoes of the waveform used to energize the material. These echoes have a central role in that their positions on the time axis are used to determine the location of the scatterer within the test sample. In order to fuse the signals from an ultrasonic test and ECT at feature level, we need to extract from ECT signals the same set of features used in UT, for example, the "echoes" from defects. This is done by using the inverse Q-transform to convert a parameterization for diffusive problems into a parameterization for wave propagation problems (see Gibert et al.[24] and Lee et al.[25]).

Tamburrino et al.,[25,26] also take into account the degree of freedom of the probing signal. Specifically, under proper choice of the probing signal, it can be shown that the position of the peak of the measured ECT signal is related algebraically to the distance of the defect (or interface etc.) from the surface of the specimen.

For instance, in the example described in Section III.C.1, assuming the excitation waveform given by Equation 12.74, the peak is located very close (within 2.3%, see Figure 12.14) to the theoretical value of 0.009 s predicted by $t_{\text{peak}} = q_f^2/2$ where $q_f = q_i + 2L/c_0$, L is the distance between probe and defect, and $c_0 = 1/\sqrt{\mu_0\sigma_0}$. In this way, we can extract from the feature (peak position) the location L of the defect,

$$L = c_0\left(\sqrt{2t_{\text{peak}}} - q_i\right)\big/2 \tag{12.79}$$

In the case of UT measurement, the defect location from the probe (which we call feature t_{fe} [position of the first echo]) is related to L through $L = t_{\text{fe}}v_0/2$, where v_0 is the velocity of the sound wave in the sample.

Thus the Q-transform based approach allows us to parameterize ECT waveform in terms of the UT features. We can also work in the opposite direction, that is, parameterize the UT waveform in terms of ECT features, but in this case, the relationship between features and symbols is more involved.

In conclusion, the Q-transform is a feasible approach for preprocessing data from multiple sensors when the underlying physical processes are governed by wave and diffusion equations. The final step of fusing the corresponding mapped data can then be performed using the system-based methods described earlier.

REFERENCES

1. Abidi, A. A., and Gonzalez, R. C., *Data Fusion in Robotics and Machine Intelligence*, Academic Press, London, 1992.
2. Newman, E. A., and Hartline, P. H., The infrared "vision" of snakes, *Sci. Am.*, 246(3), 116–127, 1982.
3. Mitchie, A., and Aggarwal, J. K., Multiple sensor integration/fusion through image processing: A review, *Opt. Eng.*, 25(3), 380–386, 1986.
4. Luo, R. C., and Kay, M. G., Data fusion and sensor integration: state-of-the-art 1990s, In *Data Fusion in Robotics and Machine Intelligence*, Abidi, M. A., and Gonzalez, R. C., Eds., Academic Press, San Diego, CA, pp. 7–135, 1992.
5. Udpa, L., and Lord, W., A discussion of the inverse problem in electromagnetic NDT, In *Review of Progress in Quantitative NDE*, Vol. 5, Thompson, D. O., and Chimenti, D. E., Eds., Plenum, New York, NY, pp. 375–382, 1985.
6. Yim, J., Image Fusion Using Multi-Resolution Decomposition and LMMSE Filter, Ph.D. dissertation, Iowa State University, Ames, IA, 1995.
7. Dougherty, E. R., *An Introduction to Morphological Image Processing*, SPIE Optical Engineering Press, Bellingham, WA, 1992.
8. Zhuang, X., and Haralick, R. M., Morphological structuring element decomposition *Comput. Vis. Graph. Image Process.*, 35(3), 370–382, 1986.
9. Songs, Y., NDE Data Fusion Using Morphological Approaches, Ph.D. dissertation, Iowa State University, Ames, IA, 1997.
10. Sun, K., NDE Data Fusion Using Phenomenological Approaches, Ph.D. dissertation, Iowa State University, Ames, IA, 1995.
11. Bragg, L. R., and Dettman, J. W., Related partial differential equations and their applications, *SIAM J. Appl. Math.*, 16, 459–467, 1968.
12. Filippi, P., and Frisch, U., Equations aux derives partielles. Relation entre l'equation de la chaleur et l'equation des ondes de Helmholtz, *C.R. Acad. Sc. Paris*, 268(Ser A), 804–807, 1969.

13. Isaev, G. A., and Filatov, V. V., Physiomathematical principles of visualization of nonstationary electromagnetic fields, *Geol. I Geofiz. (Sov. Geol. Geophys.)*, 22, 89–95, 1981.

14. Pierce, A., Wave methods for an inverse problem in diffusion, *Inverse Prob.*, 2, 205–217, 1986.

15. Symes, W. W., Inverse boundary problems and a theorem of Gel'fand and Levitan, *J. Math. Anal. Appl.*, 71, 379–402, 1979.

16. Romanov, V. G., *Inverse Problems of Mathematical Physics*, VNU Science Press BV, 1987.

17. Reznitskaya, K. G., The connection between solutions of the Cauchy problem for equations of different types and inverse problems, *Mat. Problemy Geofiz. Vyp.*, 5(part 1), 55–62, 1974.

18. Lee, K. K., Liu, G., and Morrison, H. F., A new approach to modeling the electromagnetic response of conductive media, *Geophysics*, 54(9), 1180–1192, 1989.

19. Lee, K. H., and Xie, G., A new approach to imaging with low frequency electromagnetic fields, *Geophysics*, 58(6), 786–796, 1993.

20. Ross, S., Lusk, M., and Lord, W., Application of a diffusion-to-wave transformation for inverting eddy current nondestructive evaluation data, *IEEE Trans. Magnet.*, 32, 535–546, 1996.

21. Sun, K., Udpa, S. S., Xue, T., and Lord, W., Registration issues in the fusion of eddy current and ultrasound NDE data using Q-transform, In *Review of Progress in Quantitative Nondestructive Evaluation*, Vol. 15, Thompson, D. O., and Chimenti, D. E., Eds., Plenum Press, New York, pp. 813–820, 1996.

22. Gibert, D., Tournerie, B., and Virieux, J., High-resolution electromagnetic imaging of the conductive earth interior, *Inverse Probl.*, 10, 341–351, 1994.

23. Lee, T. J., Suh, J. H., Kim, H. J., Song, Y., and Lee, K. H., Electromagnetic traveltime tomography using an approximate wavefield transform, *Geophysics*, 67, 68–76, 2002.

24. Tamburrino, A., and Udpa, S. S., Solution of inverse problems for parabolic equations using the Q-transform: time domain analysis. NASA Performance Report, 2002, June.

25. Tian, Y., Tamburrino, A., Udpa, S. S., and Udpa, L., Time-of-flight measurements from eddy current tests, In *Review of Progress in Quantitative Nondestructive Evaluation*, Thompson, D. O., and Chimenti, D. E., Eds., American Institute of Physics, pp. 593–600, 2003.

26. Tamburrino, A., Fresa, R., Udpa, S. S., and Tian, Y., Three-dimensional defect localization from time-of-flight/eddy current testing data, COMPUMAG Conference, Saratoga Spring, USA, 2003.

13 Fusion of Multimodal NDI Images for Aircraft Corrosion Detection and Quantification

Zheng Liu, David S. Forsyth, and Jerzy P. Komorowski

CONTENTS

I. INTRODUCTION

This chapter will focus on the fusion of multifrequency eddy current and pulsed eddy current (P-ET) data (lift-off-intersection scan) for the detection and quantification of hidden corrosion on aircraft structures. In order to perform risk assessments on aircraft structure, nondestructive inspection (NDI) must be able to provide quantitative assessments of fatigue and corrosion damage.[1,2] Often, in the case of complex structures or complex modes of degradation such as corrosion and fatigue and their interactions, more than one NDI technique is required. For example, the techniques used to measure thickness loss, corrosion pillowing, and fatigue cracks in aircraft fuselage lap joints are significantly different. The studies presented in this chapter illustrate how the data fusion technique can improve the NDI of corrosion damage in aircraft.

One of the earliest works in NDI data fusion was carried out by Jain et al.[3] In their work, a complete map of the defect region in a fiber-reinforced composite material was derived from the inspections of x-ray, ultrasonic C-scan, and acoustic emission. The region maps were first derived from multimodal images by applying an adaptive threshold algorithm. The detective regions were integrated through a region matching process. Data fusion techniques have since become more widespread in the NDI community. Pioneering work may be found in the publication of annual NDI conferences, for example, *The Review of Progress in Quantitative Nondestructive Evaluation*, and Gros' two monographs.[4,5] Generally speaking, the data fusion techniques can provide a useful tool to improve the classification and combination results in NDI. The certainty and reliability of the classification is increased by fusing the observations and data from multiple NDI sensors or techniques. The signal enhancement or quantification can be achieved in a combination process. As far as the methodology is concerned, there are four typical categories for NDI data fusion techniques: optimization, multiresolution analysis, heuristic, and probabilistic methods as listed in Table 13.1. Readers are referred to the corresponding references for detailed information. The requirements of a specific application largely determine the choice of multiple NDI techniques as well as the fusion algorithms. Therefore, the initial step is to clarify what output is expected.

In this chapter, the approaches for combination and classification are investigated. Both eddy current (ET) and P-ET images are registered with a post-teardown x-ray image so that the fused results of ET and P-ET images can be compared with the ground truth. Two algorithms were tried using a combination approach; in the first, the fusion is implemented at the pixel level by using multiresolution analysis methods. The second combination approach used a generalized additive model to combine the multiple NDI data under the regression mechanism. For the classification approach, the classification operation is carried out on a feature vector derived from multimodal NDI data. Each pixel is classified into predefined corrosion types of different percentages of material loss.

This study will clarify what the fusion algorithms contribute to the corrosion quantification compared with the traditional calibration method and what is the most appropriate combination of NDI data for characterizing the corrosion located at different layers in the multilayer aircraft fuselage joint example. The two fusion approaches are restricted by the capability of the inspection methodologies themselves. Extensive experimental results indicate the efficiency

TABLE 13.1
Data Fusion Approaches for Nondestructive Inspection and Evaluation

Fusion Algorithms		NDI Modalities	Achievements	References
Optimization method	Linear minimum mean square error (LMMSE)	(1) Eddy current image: real and imaginary parts	Image enhancement	6,7
		(2) Multifrequency eddy current data		
Multiresolution analysis	Wavelet	(1) X-ray, ultrasonic C-scan and acoustic emission	Image enhancement	8,10
		(2) Multifrequency eddy current data		
	Image pyramid	Eddy current C-scan and Infrared thermography	Defect characterization	9
Heuristic	Radial basis function neural network	(1) Multifrequency eddy current data	Image enhancement	11,12
	Multilayer perceptron neural network	(2) Ultrasonic and eddy current data	Image enhancement, defect classification	11–13
	Wavelet neural network	Eddy current and pulsed eddy current	Corrosion quantification	14

Continued

TABLE 13.1
Continued

Fusion Algorithms		NDI Modalities	Achievements	References
Probabilistic	Bayesian inference	(1) ET and infrared thermo-graphical inspection of composite material	(1) Defect characterization	9,15–17
		(2) X-ray image system operated at tow different conditions	(2) Improve dynamic range of x-ray image	
		(3) ET and ultrasonic testing of pressure tubes	(3) Increase measurement reliability	
	Dempster–Shafer theory	(1) ET and infrared inspection	(1) Defect localization and characterization	15,18,19
		(2) Ultrasonic and x-ray inspection of mock-up from energy and chemical industry	(2) Signal interpretation and defect quantification	
		(3) Ultrasonic testing and x-ray inspection		
	Fuzzy set theory	(1) Multiple features from ultrasonic testing	(1) Flaw classification	20,21
		(2) Stress, sound wave, temperature, and ultrasonic sensors	(2) Crack detection	

of the fusion algorithms for the problem of corrosion quantification. Possible directions for future work are also discussed.

II. MULTIMODAL NDI FOR CORROSION DETECTION

A. CORROSION PROBLEMS IN TRANSPORT AIRCRAFT

The design of transport aircraft is relatively mature, and there are common elements across many manufacturers and models. The outer shell of the fuselage is generally built up from circumferential and longitudinal stringers, analogous to a skeleton, and aluminum skins are fastened to these stringers. This chapter focuses on fuselage splice joints. These are the joints of the individual sections of the outer skin, and are called lap or butt joints, depending on the configuration. Lap joints, where the two pieces are overlapped and riveted, are very common on Boeing and Airbus designs and, thus, of interest to many aircraft operators. In both lap and butt joints (where the two sheets are butted up against each other) there are often additional reinforcing layers added to the joint area. Given this configuration of multiple sheets of aluminum, one on top of the other, crevice corrosion can occur on the interior, hidden (known as "faying") surfaces of these sheets if sealants and corrosion protection systems break down.[22]

Lifing analysis of fuselage splice joints based on holistic structural integrity design models has recently been shown to be very effective.[23] One available commercial tool for these risk assessments, the ECLIPSE model developed by AP/ES Incorporated,[23] requires estimation of general thickness loss on each of the joint layers, pillowing deformation, and local topography on the faying surfaces. Cracks due to fatigue or intergranular corrosion should also be known. Given the NDI estimates of thickness for the individual layers, the local corrosion topography can be estimated using empirically based relationships.[24] Corrosion pillowing can be measured directly[25] or it can also be calculated from models which relate the pillowing deformation to the average thickness loss and the particular joint geometry under study.[26]

B. NDI TECHNIQUES FOR CORROSION DETECTION

The effects of corrosion on structural integrity are evaluated by a number of corrosion metrics.[27] To obtain these metrics for structure integrity analysis, multiple NDI techniques can be involved. Potentially useful techniques include eddy current, ultrasound, optical methods, thermography, radiography, and microwave techniques.[27,28] In this chapter, the ET and P-ET techniques, which belong to the eddy current category, are applied to detect hidden corrosion in a multilayer lap joint from a Boeing 727 aircraft.

ET inspections are the most commonly used NDI techniques in aircraft maintenance. This technique is based on inducing electrical currents in the material being inspected and measuring the interaction between these currents and the material. A brief description of the theory of ET follows; for more details

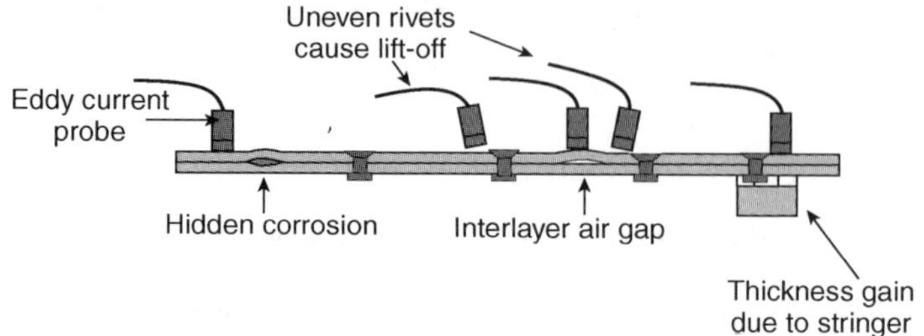

FIGURE 13.1 Typical noise factors encountered in the inspection.

see Ref. 29. When a periodical magnetic field intersects an electrical conductor, eddy currents are induced as described by Faraday's and Ohm's Laws. The induced currents generate their own magnetic field, which opposes the excitation field. The equilibrium field is reduced resulting in a change of coil impedance. By monitoring coil impedance, the electrical, magnetic, and geometrical properties of the sample can be measured. The penetration depth into a material depends on its electrical resistivity, its magnetic permeability, and on test frequency. This value is proportional to the inverse of the root of excitation frequency; consequently, applying multiple frequencies simultaneously provides more information at different depths. The different depths of penetration enable multifrequency eddy current testing (MF-ET) techniques to locate corrosion damage in multiple-layered structures in aircraft. A quantitative relationship between the signals representing the changes of impedance in the specimen and the corrosion damages is established by further data processing.

In contrast to the conventional ET technique, the P-ET technique employs a broadband pulse instead of a continuous wave as the excitation signal. Therefore, the P-ET measurement is equivalent to multiple conventional eddy current measurements as a function of frequency, where the information at different depths is available. However, as illustrated in Figure 13.1, in the presence of anomalies such as variations in interlayer gap and the probe-specimen separation (known as lift-off), the quantitative interpretation of ET or P-ET inspection data for hidden corrosion characterization raises a challenging problem.

C. Test Component

The specimen used in this experiment comes from a 30-year-old service-retired Boeing 727 aircraft. It is a two-layer lap joint cut out from below the cargo floor, near the belly of the fuselage. This simple joint is a common structural element in transport aircraft fuselage design, and is shown in Figure 13.2. The picture of the front and back view of the specimen is shown in Figure 13.3. The material is aluminum 2024-T3, except for the stringer, which is aluminum 7075-T6. The thickness of each layer is 0.045 in. (1.14 mm).

Following inspections, the specimen was dissembled and cleaned of all corrosion products. Then, thickness measurement was carried out using a digital

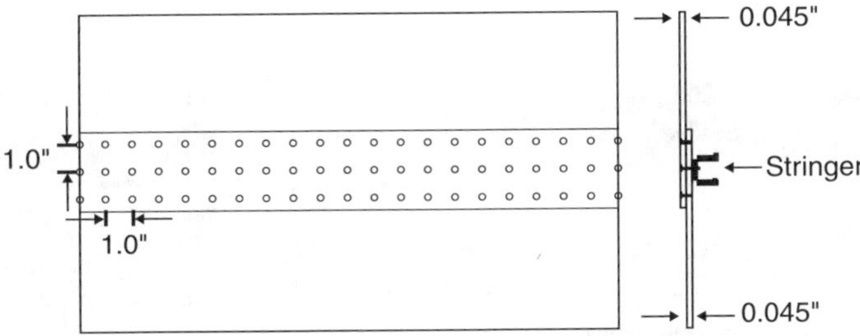

FIGURE 13.2 A schematic drawing of the Boeing 727 lap joint specimen used in this study.

x-ray mapping technique on each layer. The x-ray thickness map serves as the ground truth. The lap joint specimen was cut into seven sections labeled from A to G, resulting in 14 pieces in total including both the top and the bottom plates. Data from Section C, Section D, and Section F are selected for use in this study because there are salient corroded regions in these three sections.

The inspection data is given in Figure 13.4. As stated previously, the ET detects the hidden corrosion through the changes in impedance. ET techniques are usually designed to have 90° phase separation (or as close as possible to 90°) between impedance changes due to lift-off and those due to the signal of interest. By convention, the phase is adjusted so that the lift-off signal is the real component and thus for this analysis, only the imaginary component of the ET signal is used.

The P-ET signal is a series of (voltage, time) data pairs called an "A-scan" as shown in Figure 13.5, where a point named lift-off intersection (LOI) is presented. The signal amplitude at this point in time is insensitive to variations

FIGURE 13.3 Boeing 727 lap joint: (a) front view and (b) back view.

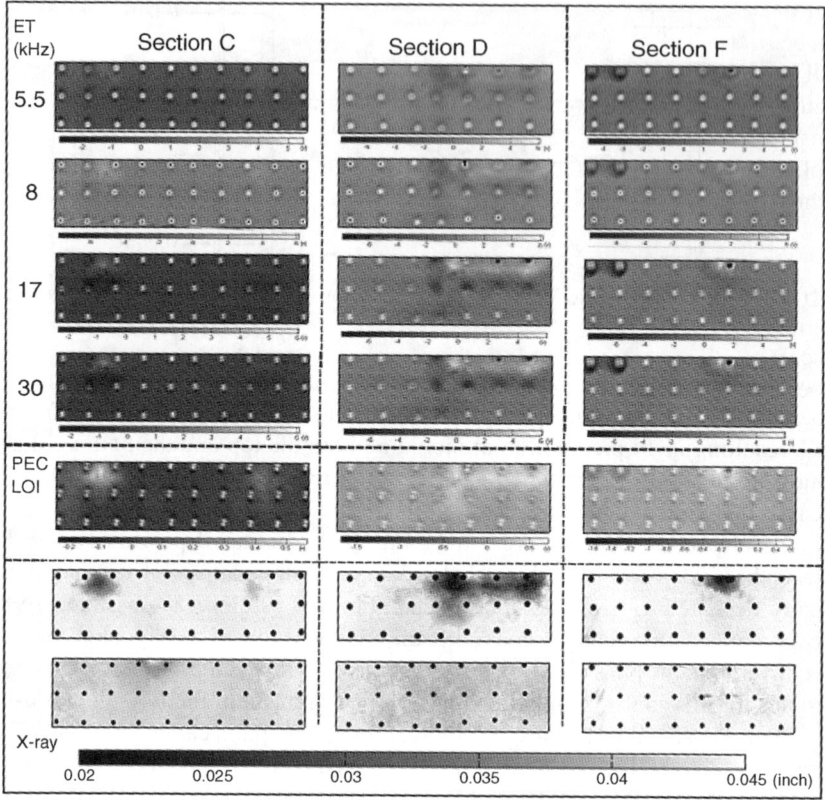

FIGURE 13.4 MF-ET and P-ET inspection of the lap joint from a Boeing 727 aircraft.

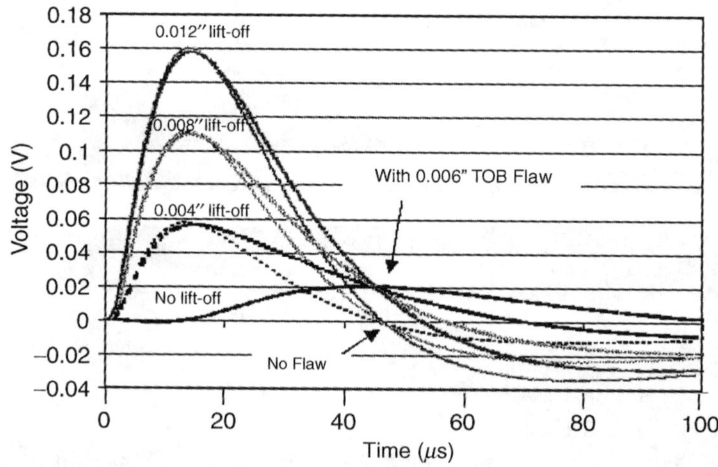

FIGURE 13.5 Typical A-scan signals of P-ET inspection and the concept of lift-off intersection.

in lift-off, and depends on the probe and conductivity of the specimen.[30] The LOI can be found experimentally on the time axis. The extracted image is called P-ET LOI scan. The typical noise factors encountered in the inspection are illustrated in Figure 13.1. These factors introduce uncertainties to the inspection, which are not considered or compensated for in the calibration process. The data fusion is applied to improve the reliability of the inspection through fusing multiple NDI techniques.

D. QUANTIFICATION OF NDI RESULTS

A straightforward way to quantify the inspection results is through the calibration process. The calibration specimen is prepared and inspected with the same experimental setup. A linear regression curve is derived from the experimental results. The quantitative value for material loss can be found from the corresponding measurement value on the calibration curve. The calibration results for Section C, Section D, and Section F are listed in Table 13.2.

It was found upon disassembly of the test specimen that there was little corrosion damage on the second layer, making calibration of the NDI results to second layer corrosion effectively impossible. It should be noted that, even if there was second layer corrosion damage for calibration, the NDI data is not unique for damage on different layers, so simple calibration cannot be used to determine where damage occurs. If the damage occurs on the same layer as the calibration, then the interpretation of the NDI results may be accurate.

TABLE 13.2
Evaluation of Calibration Results for the Bottom-of-Top Layer Corrosion

Section	Frequency	RMSE (E-3)	CORR2	PSNR (dB)	$\mu \times 10^{-3}$	σ
C	5.5 kHz	3.0343	0.9917	17.115	−2.100	0.0029
	8 kHz	2.8824	0.9948	17.8266	−2.500	0.0022
	17 kHz	1.4906	0.9977	14.6323	0.723	0.0016
	30 kHz	1.5338	0.9975	14.6417	0.697	0.0017
	P-ET LOI	2.4801	0.9928	16.0033	−1.100	0.0027
D	5.5 kHz	5.8684	0.9561	18.0669	−1.300	0.0071
	8 kHz	5.2297	0.9680	19.55	−2.200	0.006
	17 kHz	4.6639	0.9756	14.6149	1.200	0.0056
	30 kHz	4.8639	0.9733	13.8218	1.100	0.0059
	P-ET LOI	4.4633	0.9821	21.8191	−3.400	0.0043
F	5.5 kHz	3.8023	0.9873	22.97	−2.500	0.0038
	8 kHz	3.3997	0.9920	24.64	−2.800	0.0030
	17 kHz	2.2987	0.9938	14.75	0.321	0.0027
	30 kHz	2.3814	0.9934	13.93	0.286	0.0028
	P-ET LOI	1.4557	0.9975	28.04	−0.320	0.0017

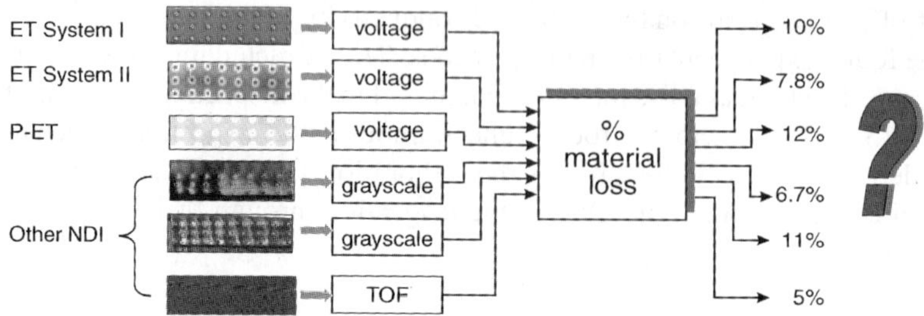

FIGURE 13.6 Multimodal NDI and the need for data fusion technique.

As shown in Figure 13.6, each NDI technique may provide a quantification result with the uncertainty introduced by the inspection methodology. The question is which technique is more reliable and how to achieve a more reliable result. Data fusion techniques may give an answer to these questions.

III. FUSION OF MULTIPLE NDI IMAGES

A. DATA ALIGNMENT AND REGISTRATION

Registration is a prerequisite step for the fusion process and can be implemented at two different levels: the physical or the geometrical level. The data are generated by heterogeneous physical interaction. Prior to the comparison or combination, the data or images should be converted to a format with the same physical meaning; for example thickness, length, or weight. An example in NDI is the registration of the ultrasonic and eddy current data presented by Sun and Udpa,[31] where the ultrasonic method relies on wave propagation and the eddy current relies on the energy diffusion. Averaging the magnitudes of an ultrasonic testing and an eddy current testing does not make any sense. A so-called Q-transform is applied to convert the eddy current data to the same format as the ultrasonic data so that the time-of-flight information can be extracted from the eddy current data.[31,32] Such transformation is necessary for signal-level fusion. Although the physical registration provides an ideal solution, there are implementation difficulties. However, if inspection data are converted to higher-level descriptions, for example, the probabilities of being defective or nondefective, the probability formulas can be applied to these values with further transformation. Even though the fusion is implemented at higher level, the geometrical registration still needs to be carried out.

The geometrical registration associates each pixel to a physical point on the specimen. A general procedure is shown in Figure 13.7. The start and end position of the scanning should be recorded for each inspection. The corresponding common region can be found from these parameters. However, the resolution of the scanned images is usually different due to the heterogeneous characteristics of each sensor. The images can be converted to the same resolution with a bilinear interpolation approach. The images are initially aligned. However, this process does not assure a precise matching

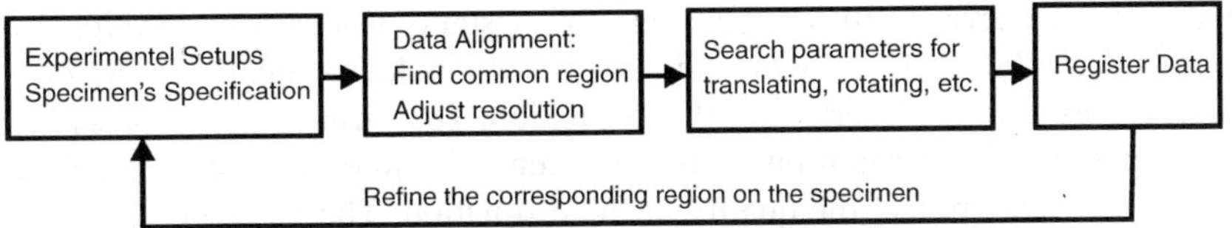

FIGURE 13.7 Procedure for registering multiple NDI images.

between images, where translation and rotation errors may still exist. The authors proposed a multiresolution-based approach for an automatic registration of multiple NDI images. The input images are decomposed into subimage components by the steerable pyramid and the registration parameters are searched and refined by matching the band-pass image component of the decomposition through a coarse-to-fine procedure. Readers are referred to Ref. 33 for the details of the algorithm. The crucial step in the registration is feature extraction; the finding and matching of common features in different images. After the input images are fully registered, the corresponding region on the specimen needs to be refined. Therefore, each pixel can be mapped to a corresponding physical point on the specimen.

Another preprocessing step is carried out to remove the regions of rivets from NDI images in this study. It is observed that the pixels at the rivet reach the maximum or the minimum value of the whole measurement. To get rid of their effects on the evaluation, these parts are removed by using the registered x-ray image. It is easy to set a threshold value for the x-ray image to remove the rivet. Together with a predefined rivet size, the rivet regions are set to zero in all NDI images, especially for those used for training.

B. VERIFICATION AND EVALUATION

The assessment and evaluation of the fused results is a crucial issue in image fusion. When the perfect reference is available, a comparison with the reference is carried out. If the reference is a thematic map, that is, in a classification-based application, the metric could be the classification rate represented by a confusion matrix and a receiver-operating characteristic (ROC) curves. In a combination-based application, the metrics include: root mean square error (RMSE), normalized least-square error (NLSE), peak signal-to-noise ratio (PSNR), cross-correlation, difference entropy, and mutual information (MI). Unfortunately, in most practical applications the perfect reference is not always available. A blind evaluation is thus preferred in these cases. Potential solutions were proposed by Qu,[34] Xydeas,[35] and Piella[36] in their publications. However, extensive experiments need to be carried out to verify the efficiency of those algorithms.

To evaluate the efficiency of fusion algorithms for ET and the P-ET LOI image, the fused results are verified with the ground truth x-ray thickness map, which is obtained by teardown inspection. Such destructive verification procedure is carried out at the stage of system design and development. To compare the fused results and the actual thickness map, the RMSE is employed as a major criterion for the evaluation. The mean (μ) and variance (σ) of the probability distribution of the difference between the fused result and ground truth also reflect the quality of the estimation. The variance can then be used to estimate the effects of NDI uncertainty in risk assessment procedures.[37]

C. STRATEGIES FOR NDI IMAGE FUSION

1. Pixel-Level Fusion

Pixel-level fusion implements an image-in–image-out fusion as shown in Figure 13.8. As stated in Section II.B, through the calibration process, the measurements are transformed to the same format or physical meaning, that is, thickness measure. Thus, operations like maximum or minimum selection, averaging, and weighted averaging can be applied directly. An efficient approach is the multiresolution image fusion (MRIF). The idea of MRIF originated from Burt and Adelson's work on image pyramids.[38] The multiresolution algorithms represent an image in a way in which features of an image at different resolution or scale can be easily accessed and modified. The multiresolution algorithms for image fusion basically include two categories: image pyramids and wavelets. Images to be fused are first decomposed into subimages or coefficients with multiresolution algorithms. The subimages or coefficients in the transform domain are combined by a fusion rule. The reconstruction generates a composite image that possesses most of the salient features of the inputs. The features of salience may include lines, edges, and region boundaries. A detailed review of this technique can be found in the first chapter of this book and Ref. 39.

In this study, a steerable image pyramid proposed by Simoncelli et al.[40] is employed to represent the input ET and P-ET LOI images. The basic rule to fuse the subimages is averaging the low-pass component and retaining the pixels with larger absolute value in the other frequency bands. More sophisticated rules consider the effect of a local region and corresponding regions at different resolutions. The translation- and rotation-invariant capabilities of the steerable pyramid allow it to represent the positional and orientational structures in

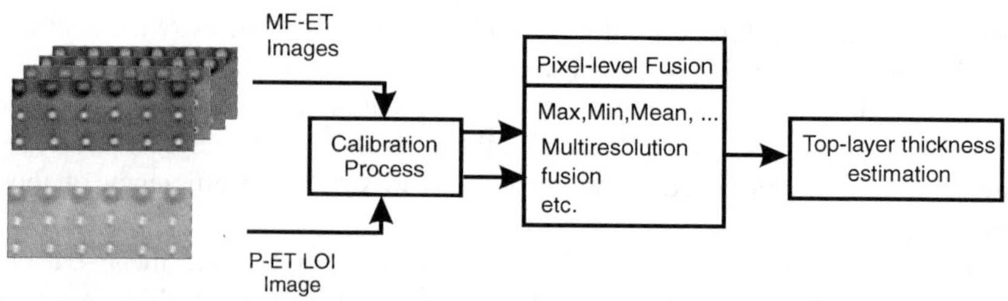

FIGURE 13.8 Pixel-level fusion of multiple NDI images.

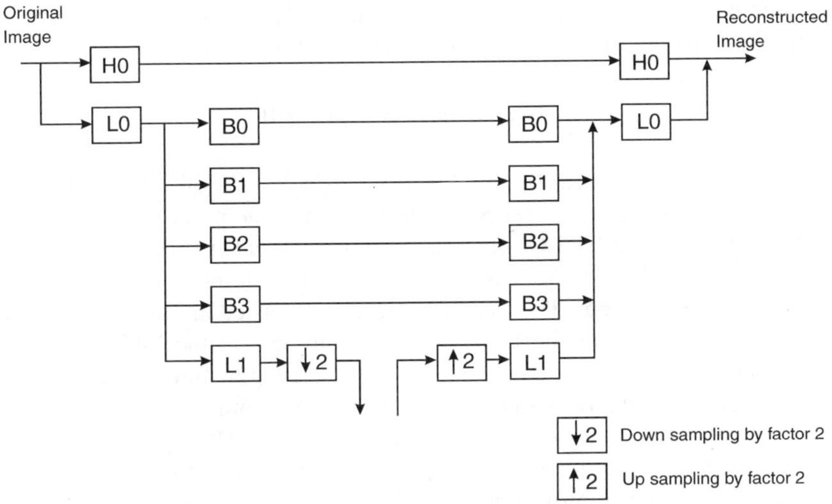

FIGURE 13.9 Structure of the steerable pyramid with one-level decomposition.

an image. This is the advantage of the steerable pyramid. The schematic of the steerable pyramid structure is given in Figure 13.9, where first level decomposition is presented and four band-pass filters oriented at 0, 45, 90, and 135° are employed. One level representation consists of the outputs of band-pass filter B_k (here $k = 0, 1, 2, 3$) and the output of low-pass filter L_1 down-sampled by a factor of two. This image can be further decomposed to obtain the lower level representations. The relationship between the input image $X(\vec{\omega})$ and the reconstructed image $\hat{X}(\vec{\omega})$ in frequency domain is

$$\hat{X}(\vec{\omega}) = \left[|H_0(\vec{\omega})|^2 + |L_0(\vec{\omega})|^2 \left(|L_1(\vec{\omega})|^2 + \sum_{k=0}^{n} |B_k(\vec{\omega})|^2 \right) \right] X(\vec{\omega}) \tag{13.1}$$

Here, $H_0(\vec{\omega})$ is a nonoriented high-pass filter while $L_1(\vec{\omega})$ is a narrowband low-pass filter. $L_0(\vec{\omega})$ is another low-pass kernel. $B_k(\vec{\omega})$ refer to the band-pass filters ($k=0,1,\ldots,n$) and the total number of band-pass filters is $n+1$. In order to avoid aliasing in the band-pass portion, the band-pass components are not sampled. To eliminate aliasing, avoid amplitude distortion, and cascade the system recursively, there are constraints as follows in Equation 13.2 to Equation 13.4

$$L_1(\vec{\omega}) = 0 \quad \text{for} \quad |\vec{\omega}| > \frac{\pi}{2} \tag{13.2}$$

$$|H_0(\vec{\omega})|^2 + |L_0(\vec{\omega})|^2 = 1 \tag{13.3}$$

$$|L_1(\vec{\omega})|^2 + \sum_{k=0}^{n} |B_k(\vec{\omega})|^2 = 1 \tag{13.4}$$

The angular constraint on the band-pass filters $B_k(\vec{\omega})$ is determined by the condition of steerability and can be expressed as in Ref. 40

$$B_k(\vec{\omega}) = B(\vec{\omega})[-j\cos(\theta - \theta_k)]^n \qquad (13.5)$$

where $\theta = \arg(\vec{\omega})$, $\theta_k = k\pi/(n+1)$ for $B(\vec{\omega}) = \sqrt{\sum_{k=0}^{n}|B_k(\vec{\omega})|^2}$. For our application, only one band-pass filter $B_0(\vec{\omega})$ is involved in the pyramid, that is, $k = 0$.

The MF-ET and P-ET LOI scanning data of Section D are firstly fused at the pixel level. The averaging, choosing minimum and multiresolution fusion operations are carried out. The evaluation results are listed in Table 13.3. Again, the fusion results are for the BOT corrosion only due to the limitation of the calibration process. For comparison, the Laplacian pyramid and Daubechies wavelet four are employed with the same fusion rule to combine the input images. For the RMSE criterion, the average value of ET at 17 kHz and P-ET LOI scan achieves the best estimation result compared with the other approaches, although all the schemes gain improvement over the individual data sets. Only the high frequency data from MF-ET inspection is considered because the responses for the high frequencies provide more information for the top layer.

2. Classification-Based Approach

Another way to quantify the hidden corrosion is through classifying the material loss into the predefined types. The corrosion types are defined based on the percentage of material loss in advance, where each type gives a range of percentage of material loss as shown in Table 13.4. The actual thickness is 0.045 in. by layer, that is, 45 thousandths. To do the classification, the Bayesian inference and Dempster–Shafer theory can be applied. The success of these approaches largely depends on how to model the measurement data, i.e., defining the conditional probability or probability mass function. Extra attention must be paid to defining the *a priori* probability when the Bayesian inference is involved. An equal averaged value is suggested due to the lack of information for estimating the priori probability in most cases. However, such a solution does not always assure a good result. A preliminary study on the corrosion quantification problem with the Bayesian inference and Dempster–Shafer theory can be found in Refs. 41–43. In this part of the work, we focus on using a classifier to quantify the hidden corrosion. As illustrated in Figure 13.10, the idea of the classification-based approach is to construct a feature vector from multiple inputs. The classification is carried out on the input feature vectors. The goal is to characterize the material loss by layer. With the knowledge of relationship between the frequency in ET and the penetration depth, the lower frequency data will be used for the estimation of second layer material loss and the data acquired at higher frequency is usually used for the first-layer corrosion estimation.

The data from Section D is used to train five classifiers. The one which achieved the best results is then applied to Section C and Section F for testing. The classifiers include a normal densities based linear discriminant

TABLE 13.3
Evaluation of the Pixel-Level Fusion Results for the Bottom-of-Top Layer Corrosion (Section D)

Pixel-level Fusion of MF-ET and P-ET-LOI Images (Section D)

Section D	Average				Minimum			Multiresolution Image Fusion		
								Laplacian Pyramid	Daubechies Wavelet	Steerable Pyramid
	8 & 17	17 & 30	17 & LOI	30 & LOI	17 & 30	17 & LOI	30 & LOI	17 & LOI	17 & LOI	17 & LOI
RMSE ($\times 10^{-3}$)	4.1138	4.7617	3.1347	3.2376	5.5516	5.0206	5.1861	4.1055	3.2508	3.4174
CORR2	0.9784	0.9745	0.9881	0.9874	0.9659	0.9776	0.9756	0.9821	0.9872	0.9858
PSNR (dB)	20.6041	14.2031	23.2110	22.9908	15.0784	15.7809	14.7815	22.0392	20.9767	20.9675
$\mu \times 10^{-3}$	-0.504	1.200	-1.100	-1.200	-3.100	-4.000	-4.100	1.800	-1.100	-1.200
σ	0.0050	0.0057	0.0037	0.0038	0.0060	0.0047	0.0048	0.0047	0.0038	0.0040

TABLE 13.4
Definition of Corrosion Types

Corrosion Type	Percentage of Material Loss (%)			Actual Thickness (Thousandths)	
	Up Bound	Down Bound	Average	Up Bound	Down Bound
1	0.6883	0.6425	0.6654	14.0	16.1
2	0.6425	0.5966	0.6195	16.1	18.2
3	0.5966	0.5507	0.5736	18.2	20.2
4	0.5507	0.5048	0.5277	20.2	22.3
5	0.5408	0.4589	0.4818	22.3	24.3
6	0.4589	0.4130	0.4360	24.3	26.4
7	0.4130	0.3671	0.3901	26.4	28.5
8	0.3671	0.3212	0.3442	28.5	30.5
9	0.3212	0.2753	0.2983	30.5	32.6
10	0.2753	0.2294	0.2524	32.6	34.7
11	0.2294	0.1836	0.2065	34.7	36.7
12	0.1836	0.1377	0.1606	36.7	38.8
13	0.1377	0.0918	0.1147	38.8	40.9
14	0.0918	0.0459	0.0688	40.9	42.9
15	0.0459	0	0.0229	42.9	45.0

classifier (LDC), a Fisher's least square linear classifier (FisherC), a linear classifier using linear perceptron (PERLC), a logistic linear classifier (LOGLC), and a nearest mean classifier (NMC).[44] Table 13.5 shows the comparison of the classification error for these classifiers. The normal densities based linear classifier is based on the normal distribution

$$p(\mathbf{x}|\omega_i) = \frac{1}{(2\pi)^{p/2}|\mathbf{\Sigma}_i|^{1/2}} \exp\left\{-\frac{1}{2}(\mathbf{x} - \mathbf{\mu}_i)^T \mathbf{\Sigma}_i^{-1}(\mathbf{x} - \mathbf{\mu}_i)\right\} \qquad (13.6)$$

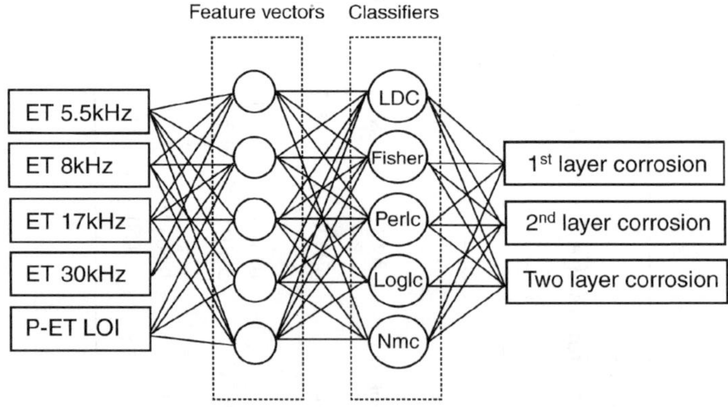

FIGURE 13.10 Classification of corrosion types.

TABLE 13.5
Classification Error of NDI Images from Section C

Section C		MF-ET				P-ET LOI
		5.5 kHz	8 kHz	17 kHz	30 kHz	
First layer estimation	LDC	0.3495	0.3525	0.3397	0.3398	0.3208
	Fisher	0.3530	0.3530	0.3530	0.3530	0.3518
	PERLC	0.3512	0.9586	0.3315	0.3382	0.6185
	LOGLC	0.3525	0.3530	0.3529	0.3529	0.3364
	NMC	0.6086	0.9550	0.8129	0.8107	0.4213
Second layer estimation	LDC	0.4026	0.4129	0.4174	0.4174	0.4173
	Fisher	0.4045	0.4163	0.4165	0.4165	0.4165
	PERLC	0.5395	0.4162	0.4225	0.4230	0.4197
	LOGLC	0.4026	0.4132	0.4165	0.4165	0.4169
	NMC	0.8201	0.6124	0.7241	0.7201	0.7453
Tow layer estimation	LDC	0.5767	0.5819	0.5677	0.5678	0.5343
	Fisher	0.5820	0.5821	0.5821	0.5821	0.5801
	PERLC	0.6828	0.9625	0.8447	0.8364	0.9188
	LOGLC	0.5799	0.5821	0.5781	0.5781	0.5413
	NMC	0.9489	0.9203	0.7448	0.7471	0.6115

where $\mathbf{x}$: observation or variable, $\mathbf{\Sigma}$: covariance matrix, $\mathbf{\mu}$: mean value, and $p(\mathbf{x}|\omega_i)$: multivariate normal density.

Classification is achieved by assigning a pattern to a class for which the posterior probability, $p(\omega_i|\mathbf{x})$, or equivalently $\log(p(\omega_i|\mathbf{x}))$, is the greatest.[44] By applying Bayes' rule, there is:

$$\log(p(\omega_i|\mathbf{x})) = -\frac{1}{2}(\mathbf{x} - \mathbf{\mu}_i)^T \mathbf{\Sigma}_i^{-1}(\mathbf{x} - \mathbf{\mu}_i) - \frac{1}{2}\log(|\mathbf{\Sigma}_i|)$$
$$- \frac{p}{2}\log(2\pi) + \log(p(\omega_i)) - \log(p(\mathbf{x})) \tag{13.7}$$

Omitting the constant items, a linear discriminant function is defined as

$$g_i(\mathbf{x}) = \log(p(\omega_i)) - \frac{1}{2}\log(|\mathbf{\Sigma}_i|) - \frac{1}{2}(\mathbf{x} - \mathbf{\mu}_i)^T \mathbf{\Sigma}_i^{-1}(\mathbf{x} - \mathbf{\mu}_i) \tag{13.8}$$

In the above equation, $\mathbf{u}_i$ and $\mathbf{\Sigma}$ are estimated from the training data set. Suppose there are in total C classes, $p(\omega_i)$ can be simply assigned a value $1/C$ or $n_i/\sum_j^C n_j$, where n_i is the number in class ω_i. With the assumption of equal covariance matrices, Equation 13.8 becomes

$$g_i(\mathbf{x}) = \log(p(\omega_i)) - \frac{1}{2}\mathbf{m}_i^T \mathbf{S}_W^{-1}\mathbf{m}_i + \mathbf{x}^T \mathbf{S}_W^{-1}\mathbf{m}_i \tag{13.9}$$

where there are

$$\mathbf{S}_W = \sum_{i=1}^{C} \frac{n_i}{n} \hat{\mathbf{\Sigma}}_i, \quad \hat{\mathbf{\Sigma}} = \frac{1}{n} \sum_{i=1}^{n} (\mathbf{x}_i - \mathbf{m})(\mathbf{x}_i - \mathbf{m})^T, \quad \text{and} \quad \mathbf{m} = \frac{1}{n} \sum_{j=1}^{n} \mathbf{x}_j$$

$\mathbf{S}_W$ is the common group covariance matrix. When the matrix $\mathbf{S}_W$ is taken to be identity and the class priors $p(\omega_i)$ are set as equal, the discriminant function turns out to be the nearest mean classifier, that is,

$$g_i(\mathbf{x}) = \mathbf{m}_i^T \mathbf{m}_i - 2\mathbf{x}^T \mathbf{m}_i \qquad (13.10)$$

In Fisher's discrimination function, a criterion to be optimized in a multiclass case becomes

$$J_F(\mathbf{a}) = \frac{\mathbf{a}^T \mathbf{S}_B \mathbf{a}}{\mathbf{a}^T \mathbf{S}_W \mathbf{a}} \qquad (13.11)$$

where there are

$$\mathbf{S}_B = \sum_{i=1}^{n} \frac{n_i}{n} (\mathbf{m}_i - \mathbf{m})(\mathbf{m}_i - \mathbf{m})^T \quad \text{and} \quad \mathbf{S}_W = \sum_{i=1}^{C} \frac{n_i}{n} \hat{\mathbf{\Sigma}}_i$$

$\mathbf{m}$ is the sample means and $\mathbf{m}_i$ and $\hat{\mathbf{\Sigma}}_i$ are the sample means and covariance matrices of each class. Vectors $\mathbf{a}$ are the solution for maximizing the criterion in Equation 13.11 and can be obtained from

$$\mathbf{S}_B \mathbf{A} = \mathbf{S}_W \mathbf{A} \mathbf{\Lambda} \qquad (13.12)$$

where $\mathbf{A}$ is the matrix whose column are the $\mathbf{a}_i$ and $\mathbf{\Lambda}$ is the diagonal matrix of eigenvalues.

The perceptron-based linear classifier is expressed as

$$g_i(\mathbf{x}) = \mathbf{v}_i^T \mathbf{z} \qquad (13.13)$$

where $\mathbf{z}$ is the augmented data vector, $\mathbf{z}^T = (1, \mathbf{x}^T)$. $\mathbf{v}$ is a $(p+1)$-dimensional vector. A generalized error-correction procedure is used to train the classifier. For the LOGLC, the assumption for C classes is that the log-likelihood ration is linear for any pair of likelihoods.

$$\log\left(\frac{p(\mathbf{x}|\omega_s)}{p(\mathbf{x}|\omega_C)}\right) = \beta_{s0} + \mathbf{\beta}_s^T \mathbf{x}, \qquad s = 1, \ldots, C-1, \qquad (13.14)$$

where the posterior probabilities are

$$p(\omega_s|x) = \frac{\exp(\beta'_{s0} + \mathbf{\beta}_s^T \mathbf{x})}{1 + \sum_{s=1}^{C-1} \exp(\beta'_{s0} + \mathbf{\beta}_s^T \mathbf{x})} \qquad (13.15)$$

$$p(\omega_C|x) = \frac{1}{1 + \sum_{s=1}^{C-1} \exp(\beta'_{s0} + \boldsymbol{\beta}_s^T \mathbf{x})} \tag{13.16}$$

and there is $\beta'_{s0} = \beta_{s0} + \log(p(\omega_s)/p(\omega_C))$.

The decision rule will assign $\mathbf{x}$ to class ω_j, if there is max $(\beta'_{s0} + \boldsymbol{\beta}_s^T \mathbf{x}) = \beta'_{j0} + \boldsymbol{\beta}_j^T \mathbf{x} > \mathbf{0}$; otherwise assign $\mathbf{x}$ to class ω_C.

The algorithms are implemented in the pattern recognition toolbox by Duin.[45] In Table 13.6, the percentage of data that is incorrectly classified by the classifier is listed. The lower value indicates a better classification result. The classified results are further quantified by assigning each type the averaged material loss value of the corresponding range. The results for Section C and Section F are given in Figure 13.11 and Figure 13.12, respectively. See Table 13.7 for the numerical evaluation of the classification results.

3. Estimation with a General Additive Model

The quantification process is typically an inverse problem. From the measurement value, the material loss of the specimen is characterized. A straightforward implementation is the linear or nonlinear regression of the

TABLE 13.6
Classification Error of Multiple NDI Inputs of Section D

Section D		Data Combination				
		5.5 & 8 & 17 & 30	5.5 & 8 & 17 & 30 & LOI	5.5 & 17 & LOI	5.5 & 8 & LOI	5.5 & 8 & 17 & LOI
First layer	LDC	0.2851	0.2742	0.2801	0.2843	0.2745
estimation	Fisher	0.3480	0.3473	0.3513	0.3504	0.3475
	PERLC	0.3373	0.3404	0.3221	0.3268	0.3225
	LOGLC	0.3005	0.2994	0.3021	0.3090	0.2993
	NMC	0.5504	0.4790	0.4700	0.5424	0.4727
Second layer	LDC	0.4082	0.4008	0.4068	0.4053	0.4011
estimation	Fisher	0.4072	0.4037	0.4060	0.4061	0.4037
	PERLC	0.4157	0.4093	0.4079	0.4102	0.4082
	LOGLC	0.3931	0.3671	0.3880	0.4012	0.3671
	NMC	0.7360	0.7367	0.7649	0.6500	0.7416
Tow layer	LDC	0.4710	0.4480	0.4753	0.4870	0.4481
estimation	Fisher	0.5734	0.5708	0.5737	0.5761	0.5711
	PERLC	0.5551	0.5555	0.5635	0.5604	0.5601
	LOGLC	0.4753	0.4439	0.4673	0.4911	0.4443
	NMC	0.7136	0.7029	0.7107	0.6762	0.7008

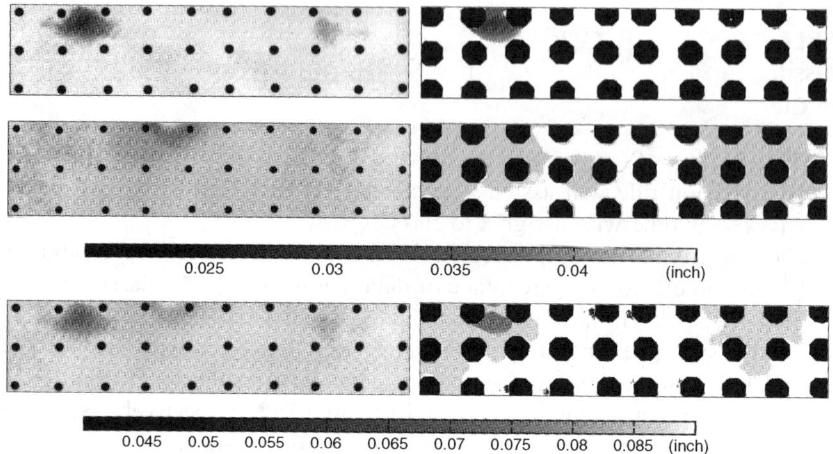

FIGURE 13.11 Right column shows the results for Section C by the classification approach (from top to bottom): first layer, second layer, and two layers. The corresponding x-ray thickness maps are given on the left column.

calibration process where a quantitative relationship between the observed variable and a response is established. A linear or nonlinear regression can be used for single input prediction. Therefore, we can see how the additive model can improve the estimation. The classification process entails placing an object in a certain category based on the measured features. The regression process yields

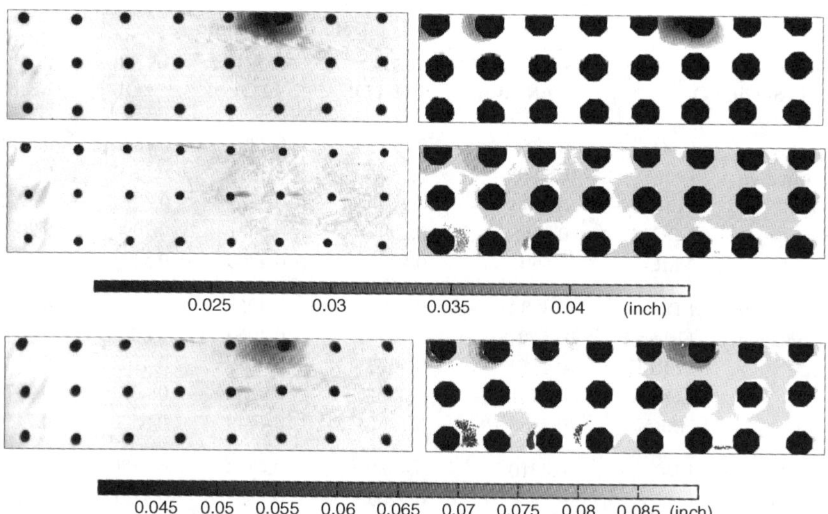

FIGURE 13.12 Right column shows the results for Section F by the classification approach (from top to bottom): first layer, second layer, and two layers. The corresponding x-ray thickness maps are given on the left column.

TABLE 13.7
Classification of Multiple NDI Inputs from Section C and Section F with the Classifiers Trained with Data from Section D

Evaluation Metric		Error	RMSE (E-3)	CORR2	PSNR (dB)	$\mu \times 10^{-3}$	σ
Section C	First layer	0.0709	1.1909	0.9989	15.02	0.795	0.0012
	Second layer	0.5196	1.6636	0.9983	28.67	1.400	0.0015
	Two layers	0.5951	3.4750	0.9976	20.18	2.400	0.0035
Section F	First layer	0.0754	1.3092	0.9982	26.50	0.527	0.0015
	Second layer	0.5400	1.2886	0.9984	31.22	-0.721	0.0014
	Two layers	0.2975	4.8269	0.9927	25.18	-0.418	0.0057

a real-value function instead of discrete class labels as in Equation 13.17.

$$\hat{y} = f(\mathbf{x}; \boldsymbol{\beta}) \tag{13.17}$$

where $\mathbf{x}$ is a measurement feature vector and $\boldsymbol{\beta}$ is the parameter vector. Given a set of data $\{(\mathbf{x}_i, y_i), i = 1, \dots n\}$, the function $f(\mathbf{x}; \boldsymbol{\beta})$ should predict a real value, that is, the thickness in inches. As shown in Figure 13.2, the target of the prediction for our case can be the first layer thickness, second layer thickness, or two-layer thickness. Therefore, it will clarify which measurement is most suitable to characterize the first layer thickness, second layer thickness, or two-layer thickness. There are two ways to estimate the regression curve. The first one is to model the relationship between an input value and a response with a linear function. The second one assumes that the function is smooth and continuous.

For linear regression, the data is assumed to be generated by a linear model

$$y = \beta_0 + \beta_1 x + \varepsilon \tag{13.18}$$

and the estimation of the regression curve is given by

$$\hat{y} = \beta_0 + \beta_1 x \tag{13.19}$$

The least square procedure, which minimizes the sum-of-squares deviation of the estimated regression curve form the observed data, is applied to obtain estimates of the parameters in the regression curve. We write

$$\hat{\boldsymbol{\beta}} = (\mathbf{X}^\mathrm{T}\mathbf{X})^{-1}\mathbf{X}^\mathrm{T}\mathbf{y} \tag{13.20}$$

where there are $\hat{\boldsymbol{\beta}} = [\beta_1, \beta_0]^T$ and $y = [y_1, \dots, y_n]^T$. The ith row of $\mathbf{X}$ is $[x_i, 1]$.

The nonlinear relationship is implemented by kernel smoothing and local weighted regression. A set of grid objects $\mathbf{q} = q(i)$ $(i = 1, \dots, n)$ across the domain of the input is generated. To obtain the regression applicable to the query point, $\mathbf{q}$, the following cost function is minimized

$$J = \sum_{i=1}^{n} \left(\beta^T \mathbf{x}_{i,1} - y_i \right)^2 K\left(\frac{|\mathbf{q} - \mathbf{x}_i|}{h} \right) \tag{13.21}$$

where K is a Gaussian kernel function with a width parameter h: $K = \exp(-(x - q(i))^2/h^2)$. The solution to the above cost function is

$$\hat{\beta} = (\mathbf{Z}^T\mathbf{Z})^{-1}\mathbf{Z}^T\mathbf{v} \tag{13.22}$$

where $\mathbf{Z} = \mathbf{W}\mathbf{X}$ and $\mathbf{v} = \mathbf{W}\mathbf{y}$. $\mathbf{W}$ is a diagonal matrix with the ith diagonal element w_{ii}. $\mathbf{X}$ is a matrix of data samples with $[x_i, 1]^T$ in the ith row and $y = [y_1, y_2, ..., y_n]^T$.

The estimate at $\mathbf{q}$ is then given by

$$\hat{y}(\mathbf{q}) = \hat{\beta}^T\mathbf{q} \tag{13.23}$$

In the following experiments, a width parameter of 0.6 was used for the Gaussian kernel.

When multiple inputs are considered, the vector $\mathbf{x}$ will contain p values. To optimize the parameter vector β of the function given in Equation 13.17, an additive model is introduced to assume that the function $f(\mathbf{x}; \beta)$ consists of a sum of terms, that is,

$$f(\mathbf{x}; \beta) = E(\mathbf{y}|\mathbf{x}_1, \mathbf{x}_2, ..., \mathbf{x}_p) = s_0 + \sum_{i=1}^{p} s_j(\mathbf{x}_j) \tag{13.24}$$

where $s_0, s_1(\cdot), ... s_p(\cdot)$ are the smoothing terms in the additive model.

As described by Hastie and Tibshirani,[46] a so-called backfitting algorithm is implemented to search for the parameter vector. The backfitting algorithm is a general algorithm that can fit an additive model using any regression-type fitting mechanism. The implementation steps are shown below:

1. Initialization

$$s_0 = E(y), \; s_1^1 = s_2^1 = \cdots = s_p^1 = 0, \; m = 0$$

2. Iteration

$$m = m + 1$$

for $j = 1$ to p do:

$$R_j = \mathbf{y} - s_0 - \sum_{k=1}^{j-1} s_k^m(\mathbf{x}_k) - \sum_{k=j+1}^{p} s_k^{m-1}(\mathbf{x}_k)$$

$$s_j^m = E(R_j|\mathbf{x}_j)$$

3. Until

$$RSS = Avg\left(\mathbf{y} - s_0 - \sum_{j=1}^{p} s_j^m(\mathbf{x}_j)\right)^2$$

fails to decrease, or satisfies the convergence criterion.

In the above algorithm, $s_j^m(\cdot)$ denotes the estimate of $s_j(\cdot)$ at the mth iteration. As shown in Figure 13.13, the ET data and P-ET LOI scan can be put into the GAM module and fused to predict the variance in thickness. RSS is the error between the actual thickness map and the estimated result. A typical curve for the iteration is given in Figure 13.14.

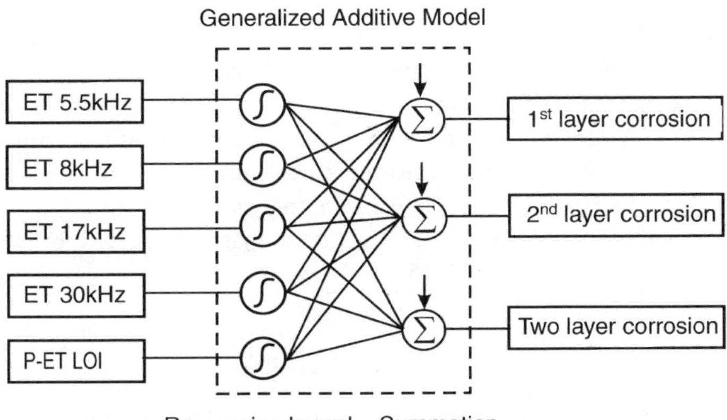

FIGURE 13.13 Framework of the generalized additive model for corrosion estimation.

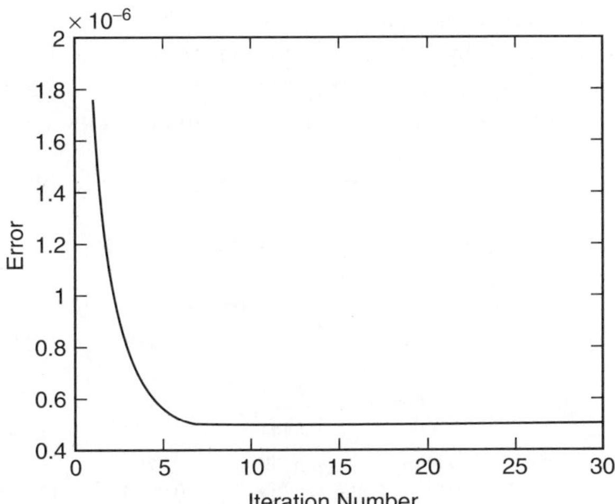

FIGURE 13.14 The iteration error of the backfitting algorithm.

TABLE 13.8
The Results of Linear Regression (Section D)

Section D	Top Layer		Bottom Layer		Two Layers	
	RMSE	CORR	RMSE	CORR	RMSE	CORR
ET 5.5	3.2298	0.9941	0.8591	0.9993	3.7750	0.9980
ET 8	2.4688	0.9974	0.8612	0.9993	3.0258	0.9990
ET 17	2.7881	0.9972	0.8422	0.9993	3.2795	0.9990
ET 30	2.7884	0.9972	0.8422	0.9993	3.2797	0.9990
P-ET LOI	1.2246	0.9983	0.7823	0.9993	1.5499	0.9993

Data from Section D is used as the training set. From Table 13.8 and Table 13.9, the suitability of the linear or nonlinear regression can be found. The nonlinear regression function is used as the smoothing term in the GAM algorithm. The parameters for the regression formula are found with this data set. The prediction is then applied to the data from Section C. The results are given in Table 13.10, from which we can see the improvement achieved by the additive model algorithm. The thickness maps estimated by the GAM method are shown in Figure 13.15.

IV. DISCUSSION

As described previously, pixel-level fusion requires that the input data have the same format. One way to achieve this transformation is through the calibration process. The calibrated data in our example is for the first layer corrosion only, and the calibration does not assure a good estimation. The experimental result indicates that the pixel-level fusion does improve the estimation. However, the best result is achieved by the averaging operation. The advantage of the multiresolution-based fusion is the representation of image features like edges and boundaries in the transform domain. The fusion can efficiently integrate all

TABLE 13.9
The Results of Nonlinear Regression (Section D)

Section D	Top Layer		Bottom Layer		Two Layers	
	RMSE	CORR	RMSE	CORR	RMSE	CORR
ET 5.5	2.6016	0.9964	0.8385	0.9993	3.1105	0.9987
ET 8	2.1304	0.9979	0.8444	0.9993	2.6619	0.9992
ET 17	1.9336	0.9984	0.8380	0.9993	2.4400	0.9994
ET 30	1.9370	0.9984	0.8389	0.9993	2.4443	0.9994
P-ET LOI	1.2818	0.9981	0.7779	0.9993	1.6269	0.9992

TABLE 13.10
The Results of GAM-Based Data Fusion (Section C)

NDI Inputs		Fusion of Multiple NDI Data						
		5.5 & P-ET	8 & P-ET	17 & P-ET	30 & P-ET	5.5 & 8 & P-ET	5.5 & 30 & P-ET	8 & 30 & P-ET
Top layer	RMSE	1.0745	1.0074	1.0241	1.0244	1.1423	0.8906	—
	CORR	0.9987	0.9989	0.9989	0.9989	0.9985	0.9991	—
Bottom layer	RMSE	0.7858	0.7900	0.8006	0.8072	—	—	0.7854
	CORR	0.9992	0.9992	0.9992	0.9992	—	—	0.9992
Two layers	RMSE	1.4530	1.41802	1.4104	1.4135	1.4564	1.2849	—
	CORR	0.9993	0.9994	0.9994	0.9994	0.9993	0.9995	—

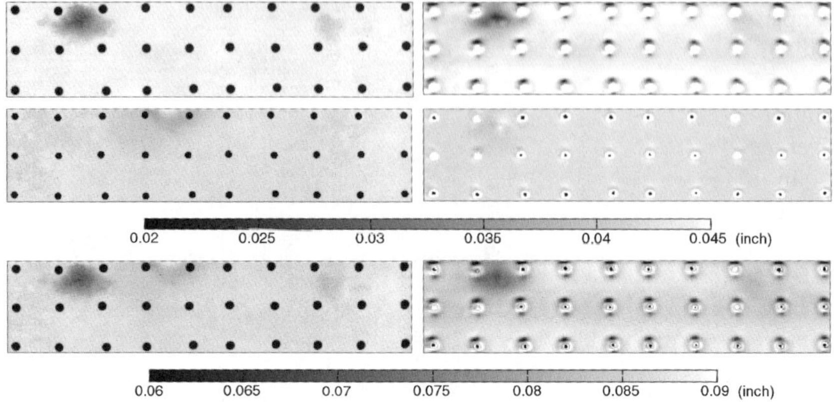

FIGURE 13.15 Right column shows the results for Section C by the GAM algorithm (from top to bottom): first layer, second layer, and two layers. The corresponding x-ray thickness maps are given on the left column.

the features in a composite image, but this might not be the optimal solution to these NDI images in this particular application.

The feature vector in the classification-based approach is created from multiple NDI inputs. In this study, the combination of the features most suitable to reveal the characteristics for each layer was examined. The disadvantage of this approach is that the direct result is not a continuous true value for thickness estimation. The methods based on Bayesian inference and Dempster–Shafer theory will generate the same discrete results. To further quantify the outputs, each class of corrosion type can be assigned the average value of material loss of this type in the definition, although such solution is not optimal. Another solution is to implement a piecewise linear regression for each type of corrosion and each NDI input. Every measurement can find its corresponding material loss value through the corresponding piecewise regression curve. The average value (material loss) can be assigned to the pixel. The combination of the results of multiple classifiers may also improve the estimation and this will be investigated in future work.

The classification result is often evaluated by the misclassification rate. This is not an appropriate measure of performance in this problem, as it does not indicate the accuracy of the corrosion damage assessment: a misclassification resulting in a 5% error is treated the same as one resulting in a 50% error. When the true value is available, the RMSE can be calculated.

The linear/nonlinear regression can set up a relationship between the measurement value and the material loss estimation. Compared with the calibration process in which only the calibration specimen is used, the linear regression implemented using training data set from actual specimen achieves a better result. The relation between the MF-ET data and the material loss is somewhat nonlinear. The efficiency of the GAM can be seen from the comparison with the linear and nonlinear regression with single NDI input, both of which are obtained using the training data set from Section C. The limitation of this approach is that the GAM does not always converge with arbitrary inputs. As shown in Table 13.10, the dashed lines indicate the divergence of the results.

In the specimen used in this study, the corrosion in the second layer is not salient as in the first layer. Thus there was little useful training or testing data for the second layer thickness estimation. It is generally more difficult to detect and quantify corrosion damage in deeper layers in lap joint structures. The problem of quantifying damage in the case where damage occurs on both layers simultaneously is also not well addressed by the specimen used in this work.

V. CONCLUSION AND FUTURE WORK

In this chapter, the fusion of MF-ET and P-ET LOI scanning data is implemented with three approaches: pixel-level fusion, classification-based approach, and a generalized additive model. The fused results give a better estimation than the individual ones. The pixel-level fusion largely depends on the calibration process and is only applied for the first layer corrosion. For the last two approaches, training data sets are needed to determine the parameters used in the estimation. Whether those parameters can be learnt from the data sets generated from calibration specimens will be investigated in the future.

In this study, only one feature from P-ET scanning is utilized, that is, the LOI point. The capability of the P-ET technique to reveal second-layer corrosion has not been fully exploited because there is no simple feature directly related to the second-layer corrosion. The method to interpret the P-ET signal is still under development.[47,48] How to involve more features from P-ET into the fusion framework is another topic for future work.

An enhanced visual inspection system named "Edge of Light" (EOL) was invented and patented by IAR/NRCC.[25,49] This system provides a fast inspection of surface deformation known as "pillowing" caused by hidden corrosion product between layers. EOL is also a solution for surface crack detection. The measure of pillowing can be further related to the material loss. Therefore, an estimation of total material loss can be derived from the EOL inspection. The EOL scanning is efficient and the result is quite intuitive. A precise quantification algorithm is under development. Future study will consider fusing the EOL image with the eddy current measurement data.

ACKNOWLEDGMENTS

This work is funded by the Defense Research and Development Canada, Air Vehicles Research Section and National Research Council Canada. Brian Lepine and Dr. Saeed Safizadeh are acknowledged for valuable discussion.

REFERENCES

1. Kinzie, R., and Peeler, D., Managing corrosion in the aging fleet: a new approach to corrosion maintenance. In *Proceedings of the 3rd Joint Conference on Aging Aircraft*, Albuquerque NM, September, 1999.

2. Brooks, C. L., and Simpson, D., Integrating Real Time Age Degradation into the Structural Integrity Process. RTO-MP-18 Fatigue in the Presence of Corrosion, RTO Report Number AC/323(AVT)TP/8, March 1999, pp. 22-1–22-13.

3. Jain, A. K., Dubuisson, M. P., and Madhukar, M. S., Multi-sensor fusion for nondestructive inspection of fiber reinforced composite materials, pp. 941–950. In *Proceedings of sixth Technology of Conference of the American Society for Composites*, 1991.

4. Gros, X. E., *NDT data fusion*, Arnold, Great Britain, 1997.

5. Gros, X. E., *Applications of NDT Data Fusion*, Kluwer Academic Publishers, Dordrecht, 2000, ISBN 0-7923-7412-6.

6. Mina, M., Udpa, S. S., and Udpa, L., A new approach for practical two dimensional data fusion utilizing a single eddy current probe, In *Review of Progress in QNDE*, Vol. 16, Thompson, D. O., and Chimenti, D. E., Eds., Plenum Press, New York, pp. 749–755, 1997.

7. Mina, M., Yim, J., Udpa, S. S., and Udpa, L., Two dimensional multi-frequency eddy current data fusion, In *Review of Progress In QNDE*, Vol. 15, Thompson, D. O., and Chimenti, D. E., Eds., Plenum Press, New York, pp. 2125–2132, 1996.

8. Matuszewski, B. J., Shark, L. K., and Varley, M. R., Region-based wavelet fusion of ultrasonic, radiographic and shearographic nondestructive testing images. In *Proceedings of 15th World Conference on Nondestructive Testing*, Roma (Italy), Oct. 15–21, 2000.

9. Gros, X. E., Liu, Z., Tsukada, K., and Hanasaki, K., Experimenting with pixel-level NDT data fusion techniques, *IEEE Trans. Instrum. Meas.*, 49(5), 1083–1090, 2000.

10. Liu, Z., Tsukada, K., Hanasaki, K., and Kurisu, M., Two-dimensional eddy current signal enhancement via multifrequecy data fusion, *Res. Nondestructive Eval.*, 11, 165–177, 1999.

11. Yim, J., Udpa, S. S., Udpa, L., Mina, M., and Lord, W., Neural network approaches to data fusion, In *Review of Progress in Quantitative NDE*, Vol. 14, Thompson, D. O., and Chimenti, D. E., Eds., Plenum Press, New York, pp. 819–8261995, .

12. Yim, J., Image fusion using multiresolution decomposition and LMMSE filter, Doctor dissertation, Iowa State University, Iowa, 1995.

13. Simone, G., and Morabito, F. C., NDT image fusion using eddy current and ultrasonic data, *Int. J. Comput. Math. Electr. Electron. Eng.*, 3(20), 857–868, 2001.

14. Ramuhalli, P., and Liu, Z., Wavelet neural network based data fusion for improved thickness characterization, In *Review of Progress in Quantitative NDE*, Vol. 23, Thompson, D. O., and Chimenti, D. E., Eds., Plenum Press, New York, 2003.

15. Gros, X. E., Bousigue, J., and Takahashi, K., NDT data fusion at pixel level, *NDT&E Int.*, 32, 183–292, 1999.

16. Dromigny, A., and Zhu, Y. M., Improving the dynamic range of real-time x-ray imaging systems via Bayesian fusion, *J. Nondestructive Eval.*, 3(16), 147–160, 1997.

17. Horn, D., and Mayo, W. R., NDE reliability gains from combining eddy current and ultrasonic testing, *NDT&E Int.*, 33, 351–362, 2000.

18. Francois, N., A new advanced multitechnique data fusion algorithm for NDT. In *Proceedings of 15th World Conference on Nondestructive Testing*, Rome (Italy), Oct. 15–21, 2000.

19. Dupuis, O., Kaftandjian, V., Babot, D., and Zhu, Y.M., Automatic detection and characterization of weld defects: determination of confidence levels for data fusion of radioscopic and ultrasonic images, *INSIGHT*, 3(41), 170–172, 1999, March.

20. Che, L .F., Zhou, X. J., and Cheng, Y. D., Defect classification based on networks integration with fuzzy integral, *Mech. Sci. Technol.*, 1(19), 111–112, 2000, (in Chinese).

21. Zhang, Z. L., Wang, Q., and Sun, S. H., A new fuzzy neural network architecture for multi-sensor data fusion in nondestructive testing, pp. 1661–1665. In *IEEE International Fuzzy Systems Conference Proceedings*, Seoul, Korea, Aug. 22–25, 1999.

22. Wallace, W., and Hoeppner D. W., AGARD Corrosion Handbook Volume 1 Aircraft Corrosion: Causes and Case Histories, AGARD-AG-278 Volume 1, NATO AGARD, 1985, July.

23. Brooks, C., Prost-Domasky, S., and Honeycutt, K., Monitoring the robustness of corrosion & fatigue prediction models. In *Proceedings of USAF ASIP 2001*, Williamsburg VA, December, 2001.

24. Bellinger, N. C., Forsyth, D. S., and Komorowski, J. P., Damage characterization of corroded 2024-T3 fuselage lap joints. In *Proceedings of The fifth Joint Conference on Aging Aircraft*, Orlando FL, September 10–13, 2001.

25. Komorowski, J. P., and Forsyth, D. S., The role of enhanced visual inspections in the new strategy for corrosion management, *Aircraft Eng. Aerospace Technol.*, 72(1), 5–13, 2000.

26. Bellinger, N. C., Krishnakumar, S., and Komorowski, J. P., Modelling of pillowing due to corrosion in fuselage lap joints, *CASI*, 40(3), 125–130, 1994.

27. Fahr, A., Forsyth, D. S., Chapman, C. E., Survey of nondestructive evaluation (NDE) techniques for corrosion in aging aircraft. National Research Council Canada, Technical Report LTR-ST-2238, Ottawa, Canada, 1999.

28. Forsyth, D. S., and Komorowski, J. P., NDT data fusion for improved corrosion detection, In *Applications of NDT Data Fusion*, Gros, X. E., Ed., Kluwer Academic Publisher, Dordrecht, pp. 205–225, 2001.

29. Canadian General Standards Board. Advanced manual for: eddy current test method, 1986, pp. 1–17.

30. Lepine, B. A., Giguere, J. S. R., Forsyth, D. S., Dubois, J. M. S., and Chahbaz, A., Interpretation of pulsed eddy current signals for locating and quantifying metal loss in skin lap splices, In *Review of Progress in Quantitative NDE*, Vol. 21, Thompson, D. O., and Chimenti, D. E., Eds., Plenum Press, New York, 2001.

31. Sun, K., Udpa, S. S., Udpa, L., Xue, T., and Lord, W., Registration issues in the fusion of eddy current and ultrasonic NDE data using Q-transforms, In *Review of Progress in QNDE*, Vol. 15, Thompson, D. O., and Chimenti, D. E., Eds., Plenum Press, New York, pp. 813–820, 1996.

32. Tian, Y., Tamburrion, A., Udpa, S. S., and Udpa, L., Time-of-flight measurements from eddy current tests, In *Review of Progress in Quantitative NDE*,

Vol. 22, Thompson, D. O., and Chimenti, D. E., Eds., Plenum Press, New York, pp. 593–600, 2002.

33. Liu, Z., and Forsyth, D. S., Registration of multi-modal NDI images for aging aircraft, *Res. Nondestructive Eval.*, 15, 1–17, 2004.

34. Qu, G., Zhang, D., and Yan, P., Information measure for performance of image fusion, *Electron. Lett.*, 7(38), 313–315, 2002, March.

35. Xydeas, C. S., and Petrovic, V., Objective image fusion performance measure, *Electron. Lett.*, 4(36), 308–309, 2000, February.

36. Piella, G., and Heijmans, H., A new quality metric for image fusion, *International Conference on Image Processing, ICIP, Barcelona*, 2003.

37. Liao, M., Forsyth, D. S., Komorowski, J. P., Safizadeh, M-S., Liu, Z., and Bellinger, N. C., Risk analysis of corrosion maintenance actions in aircraft structures. In *Proceedings of the 22nd Symposium of the International Committee on Aeronautical Fatigue ICAF2003*, Lucerne, Switzerland, May, 2003.

38. Adelson, E. H., Anderson, C. H., Bergen, J. R., Burt, P. J., and Ogden, J. M., Pyramid methods in image processing, *RCA Eng.*, 6(29), 33–41, 1984.

39. Liu, Z., Tsukada, K., Hanasaki, K., Ho, Y. K., and Dai, Y. P., Image fusion by using steerable pyramid, *Pattern Recognit. Lett.*, 22, 929–939, 2001.

40. Simoncelli, E., Freeman, W. T., Adelson, E. H., and Heeger, D. J., Shiftable multiscale transform, *IEEE Trans. Inf. Theory*, 2(38), 587–607, 1992.

41. Liu, Z., Safizadeh, M. S., Forsyth, D. S., and Lepine, B. A., Data fusion method for the optimal mixing of multi-frequency eddy current signals, In *Review of Progress in Quantitative Nondestructive Evaluation*, Vol. 22, Thompson, D. O., and Chimenti, D. E., Eds., Plenum Press, New York, pp. 577–584, 2002.

42. Liu, Z., Forsyth, D. S., Safizadeh, M. S., Lepine, B. A., and Fahr, A., Quantitative interpretation of multi-frequency eddy current data by using data fusion approaches, pp. 39–47. In *Proceedings of the SPIE*, Vol. 5046. San Diego, CA, 2003.

43. Forsyth, D. S., Liu, Z., Komorowski, J. P., and Peeler, D., An application of NDI data fusion to aging aircraft structures, *Sixth Joint FAA/DoD/NASA Conference on Aging Aircraft*, San Francisco, CA, September, 2002.

44. Webb, A., *Statistical Pattern Recognition*, Arnold, London, 1999.

45. Duin, R. P. W., *A Matlab Toolbox for Pattern Recognition*, Delft University of Technology, Netherlands, January 2000.

46. Hastie, T. J., and Tibshirani, R. J., *Generalized Additive Models*, Chapman and Hall, London, 1990.

47. Safizedeh, M. S., Lepine, B. A., Forsyth, D. S., and Fahr, A., Time-frequency analysis of pulsed eddy current signals, *J. Nondestructive Eval.*, 2(20), 73–86, 2001, June.

48. Liu, Z., Forsyth, D. S., Lepine, B. A., Hammad, I., and Farahbakhsh, B., Investigations on classifying pulsed eddy current signals with a neural network, *INSIGHT*, 9(45), 608–614, 2003.

49. Liu, Z., Forsyth, D. S., and Marincak, A., Preprocessing of edge of light images: towards a quantitative evaluation. In *SPIE's 8th Annual International Symposium on NDE for Health Monitoring and Diagnostics, SPIE Proceedings*, Vol. 5046. San Diego, March 2–6, 2003.

14 Fusion of Blurred Images

Filip Šroubek and Jan Flusser

CONTENTS

I. INTRODUCTION

In general, the term *fusion* means an approach to extraction of information spontaneously adopted in several domains. The goal of image fusion is to integrate combinations of complementary multisensor, multitemporal, and multiview information into one new image containing information of which the quality could not be achieved otherwise. The term *quality* depends on the application requirements.

Image fusion has been used in many application areas. In remote sensing and in astronomy,[1,2] multisensor fusion is used to achieve high spatial and spectral resolutions by combining images from two sensors, one of which has high spatial resolution and the other, high spectral resolution. Numerous fusion applications have appeared in medical imaging (see Ref. 3 or 4 for instance) such as simultaneous evaluation of a combination of computer tomography (CT), nuclear magnetic resonance (NMR), and positron emission tomography (PET) images. In the case of multiview fusion, a set of images of the same scene taken by the same sensor but from different viewpoints is fused to obtain an image with higher resolution than the sensor normally provides, or to recover the three-dimensional representation of the scene (shape from stereo). The multitemporal approach recognizes two different aims. Images of the same scene are acquired at different time instances either to find and evaluate changes in the scene or to obtain a less degraded image of the scene. The former aim is common in medical imaging, especially in change detection of organs and tumors, and in remote sensing for monitoring land or forest exploitation. The acquisition period is usually months or years. The latter aim requires the different measurements to be much closer to each other, typically in the scale of seconds, and possibly under different conditions. Our motivation for this work came from this area.

We assume that several images of the same scene called *channels* are available. We further assume all channels were acquired by the same sensor (or by different sensors of the same type) but under different conditions and acquisition parameters. Thus, all channels are of the same modality and represent similar physical properties of the scene.

Since imaging sensors and other devices have their physical limits and imperfections, the acquired image represents only a degraded version of the original scene. Two main categories of degradations are recognized: color (or brightness) degradations and geometric ones. The former degradations are caused by such factors as incorrect focus, motion of the scene, media turbulence, noise, and limited spatial and spectral resolution of the sensor; they usually result in a blurring of the image. The latter degradations originate from the fact that each image is a two-dimensional projection of a three-dimensional world. They cause deformations of object shapes and other spatial distortions of the image.

Individual channels are supposed to be degraded in different ways because of differences in acquisition parameters and imaging conditions (see Figure 14.1 for multichannel acquisition scheme). There are many sources of corruption and distortion that we have to cope with. Light rays (or other types of electromagnetic waves) reflected by objects on the scene travel to measuring sensors through a transport medium, for example, the atmosphere. Inevitably, each transport medium modifies the signal in some way. The imaging system is thus subject to blurring due to the medium's rapidly changing index of refraction, the finite broadcast bandwidth and the object motion. The source of corruption and its characteristics are often difficult to predict. In addition, the signal is corrupted inside a focusing set after reaching the sensor. This degradation is inherent to the system and cannot be bypassed, but it can often be measured and accounted for; typical examples are different lens imperfections. Finally, the signal must be stored on photographic material or first digitized with Charge-Coupled Devices (CCDs) and then stored. In both cases the recording exhibits a number

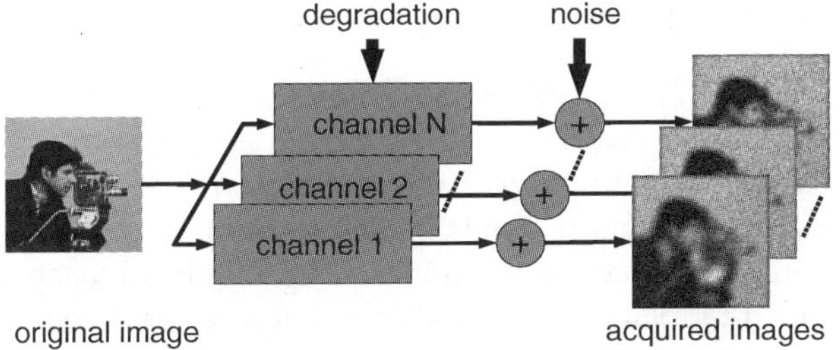

FIGURE 14.1 Multichannel acquisition model: the original scene is captured by N different channels which are subject to various degradations.

of degradations. Digital imaging systems suffer from low resolution and low sensitivity to the input signal, which are imposed by a finite number of intensity levels and a finite storage capacity. In analog systems, resolution artifacts are caused by the limited size of photographic material grain. Random noise is another crucial factor that severely affects the quality of image acquisition. In all real applications, measurements are degraded by noise. By utilizing suitable measuring techniques and appropriate devices, it can be considerably diminished, but unfortunately never cancelled.

Analysis and interpretation of degraded images is the key problem in real applications, because the degradations are, in principle, inevitable. A very promising approach to image quality enhancement is to fuse several channels with different degradations together in order to extract as much useful information as possible.

II. MULTICHANNEL IMAGE ACQUISITION MODELS

Regardless of its particular type, image degradations can be mathematically described by an operator based on an ideal representation of the scene. More formally, let $u(x, y)$ be an ideal image of the scene and let $z_1(x, y), \ldots, z_N(x, y)$ be acquired channels. The relation between each z_i and u is expressed as

$$z_i(x, y) = D_i(u(x, y)) + n_i(x, y) \tag{14.1}$$

where D_i is an unknown operator describing the image degradations of the ith channel and n_i denotes additive random noise. In the ideal situation, D_i would equal identity and n_i would be zero for each i. The major goal of the fusion is to obtain an image $\hat{u}$ as a "good estimate" of u; that means $\hat{u}$, in some sense, should be a better representation of the original scene than each individual channel z_i.

The fusion methodology depends significantly on the type of degradation operators D_i. In this work we focus on the cases where each operator D_i is a composition of image blurring and of geometric deformations caused by imaging geometry.

Under these assumptions, Equation 14.1 becomes

$$z_i(\tau_i(x,y)) = \int h_i(x,y,s,t)u(s,t)\,ds\,dt + n_i(x,y) \qquad (14.2)$$

where $h_i(x,y,s,t)$ is called the *point spread function* (PSF) of the ith imaging system at location (x,y) and τ_i stands for the co-ordinate transform, describing geometric differences between the original scene and the ith channel (in simple cases, τ_i is limited to rotation and translation, but in general complex nonlinear deformations may be present too). Having N channels, Equation 14.2 can be viewed as a system of N integral equations of the first kind. Even if all h_i's were known and neither geometric deformations nor noise were present, this system would not be generally solvable. In the sequel we simplify the model in Equation 14.2 by additional constraints and we show how to fuse the channels (that is, how to estimate the original scene) in these particular cases. The constraints are expressed as some restrictive assumptions on the PSFs and on the geometric deformations. We review five basic cases covering most situations occurring in practice.

III. PIECEWISE IDEAL IMAGING

In this simplest model, the PSF of each channel is supposed to be piecewise space-invariant and every point (x,y) of the scene is assumed to be acquired undistorted in (at least) one channel. No geometric deformations are assumed.

More precisely, let $\Omega = \bigcup_{k=1}^{K} \Omega_k$ be a support of image function $u(x,y)$, where Ω_k are its disjunct subsets. Let h_i^k be a local PSF acting on the region Ω_k in the ith channel. Since every h_i^k is supposed to be space-invariant (that is, $h_i^k(x,y,s,t) = h_i^k(x-s,y-t)$), the imaging model is defined as

$$z_i(x,y) = (u * h_i^k)(x,y) \Leftrightarrow (x,y) \in \Omega_k \qquad (14.3)$$

where $*$ stands for convolution and for each region Ω_k there exists channel j such that $h_j^k(x-s,y-t) = \delta(x-s,y-t)$.

This model is applicable in so-called *multifocus imaging*, when we photograph a static scene with a known piecewise-constant depth and focus channel-by-channel on each depth level. Image fusion then consists of comparing the channels in the image domain[5,6] or in the wavelet domain,[7,8] identifying the channel in which the pixel (or the region) is depicted undistorted and, finally, mosaicing the undistorted parts (no deconvolution is performed in this case, see Figure 14.2). To find the undistorted channel for the given pixel, a local focus measure is calculated over the pixel neighborhood and the channel which maximizes the focus measure is chosen. In most cases, the focus measures used are based on the idea of measuring the quantity of high frequencies of the image. It corresponds with an intuitive expectation that the blurring suppresses high frequencies regardless of the particular PSF. Image variance,[9] energy of a Fourier spectrum,[10] norm of image gradient,[9] norm of image Laplacian,[9] image

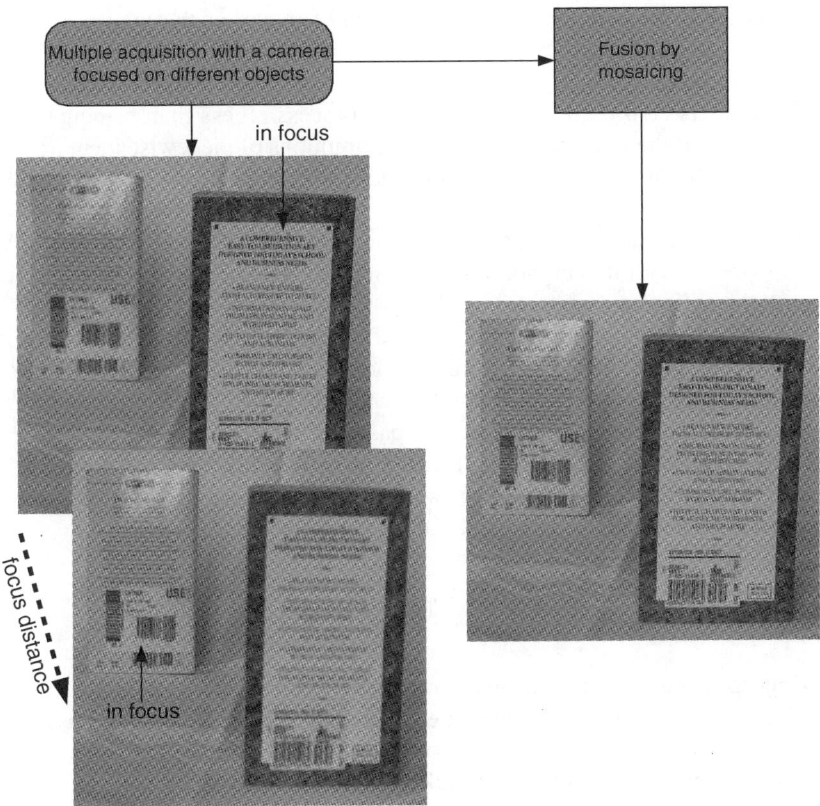

FIGURE 14.2 Two-channel piecewise ideal imaging: in each channel, one book is in focus while the other one is out of focus. Image fusion is performed by mosaicing the channel regions which are in focus.

moments,[11] and energy of high-pass bands of a wavelet transform[7,8,12] belong to the most popular focus measures. Fusion in the image domain is seriously affected by the size of the neighborhood on which the focus measure is calculated. On the other hand, fusion in the wavelet domain is very sensitive to translation changes in the source images.

A. APPLICATION IN CONFOCAL MICROSCOPY

A typical application area where piecewise ideal imaging appears is confocal microscopy of three-dimensional samples. Since the microscope has a very narrow depth of field, several images of the sample differing from each other by the focus distance are taken. Each of them shows in focus only the parts of the sample that are a certain distance from the lens, while other parts are blurred by an out-of-focus blur of various extents. These image layers form the so-called *stack image* of the sample. To obtain a focused image of the whole sample is beyond the scope of the microscope; the only possibility to get it employs a fusion of the stack image.

If the focal step used in the acquisition process is less than or equal to the depth of field of the microscope, then the assumptions of piecewise ideal imaging are fulfilled and fusion by mosaicing the undistorted parts of the individual layers can be applied.

A crucial question is how to find, for each pixel, the layer in which the given pixel (together with its neighborhood) depicted is least distorted or undistorted. Among the focus measures mentioned above, wavelet-based methods gave the best results. Their common idea is to maximize the energy of high-pass bands. Most wavelet-based focus measures ignore low-pass band(s), but Kautsky[12] pointed out that the energy of low-pass bands also reflects the degree of image blurring and, that considering both low-pass and high-pass bands increases the discrimination power of the focus measurement. We adopted and modified the idea from Ref. 12, and proposed to use a product of energies contained in low-pass and high-pass bands as a local focus measure

$$\varrho_i(x, y) = \|w_L(z_i)\|^2 \|w_H(z_i)\|^2 \tag{14.4}$$

Both energies are calculated from a certain neighborhood of the point (x, y); high-pass band energy $\|w_H(z_i)\|^2$ is calculated as the mean from three high-pass bands (in this version, we used decomposition to depth one only).

The fusion of the multifocus stack $\{z_1, \ldots, z_N\}$ is conducted in the wavelet domain as follows. First, we calculate the wavelet decomposition of each image z_i. Then a decision map $M(x, y)$ is created in accordance with a max-rule $M(x, y) = \arg \max \varrho_i(x, y)$. The decision map is the size of the subband, that is, a quarter of the original image, and it tells us from which image the wavelet coefficients should be used. The decision map is applied to all four bands and, finally, the fused image is obtained by inverse wavelet transform.

The performance of this method is shown here in fusion of microscopic images of a unicellular water organism (see Figure 14.3). The total number of the stack layers was 20; three of them are depicted. The fused image is shown on the bottom right.

In several experiments similar to this one we tested various modifications of the method. The definition 14.4 can be extended for deeper decompositions but it does not lead to any improvement. We compared the performance of various wavelets and studied the influence of the wavelet length. Short wavelets are too sensitive to noise, while long wavelets do not provide enough discrimination power. Best results on this kind of data were obtained by biorthogonal wavelets. Another possible modification is to calculate the decision map separately for each band but this also did not result in noticeable refinement. We also tested the performance of other fusion techniques. The proposed method always produced a visually sharp image, assessed by the observers as the best or one of the best.

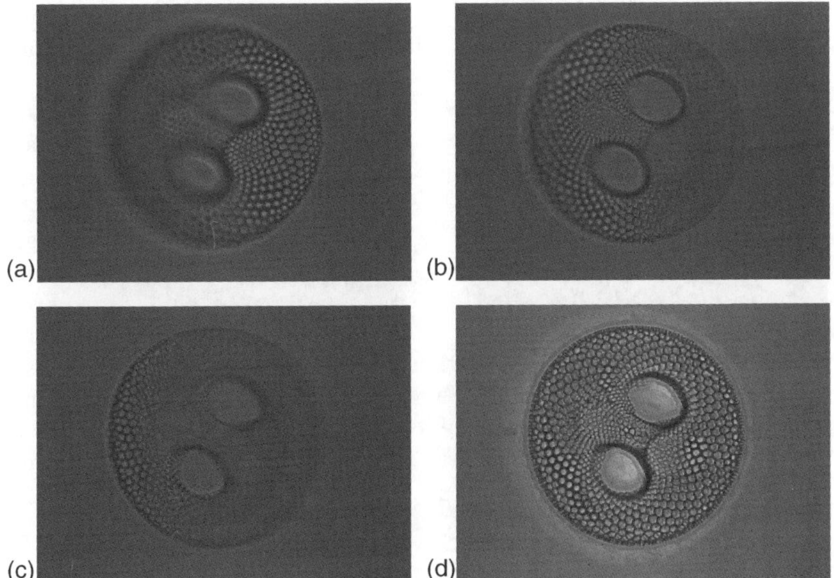

FIGURE 14.3. Fusion of a multifocus microscope image: (a)–(c) three out of 20 layers in the multifocus stack and (d) the result of the fusion in the wavelet domain.

We can thus conclude that this method is very suitable for fusion of multifocus microscope images.

IV. UNIFORMLY BLURRED CHANNELS

An acquisition model with uniformly blurred channels assumes that every PSF h_i is space-invariant within the channel, that is, $h_i(x, y, s, t) = h_i(x - s, y - t)$. Equation 14.2 then turns into the form of "traditional" convolution in each channel with no geometric deformations:

$$z_i(x, y) = (u * h_i)(x, y) + n_i(x, y) \tag{14.5}$$

This model describes, for instance, photographing a flat static scene with different (but always wrong) focuses, or repetitively photographing a scene through a turbulent medium whose optical properties change between the frames (see Figure 14.4). The image fusion is performed via multichannel blind deconvolution (MBD). It should be noted that if the PSFs were known, then this task would turn into the classical problem of *image restoration* which has been considered in numerous publications, see Ref. 13 for a survey.

Blind deconvolution in its most general form is an unsolvable problem. All methods proposed in the literature inevitably make some assumptions about the PSFs h_i and/or the original image $u(x, y)$. Different assumptions give rise to

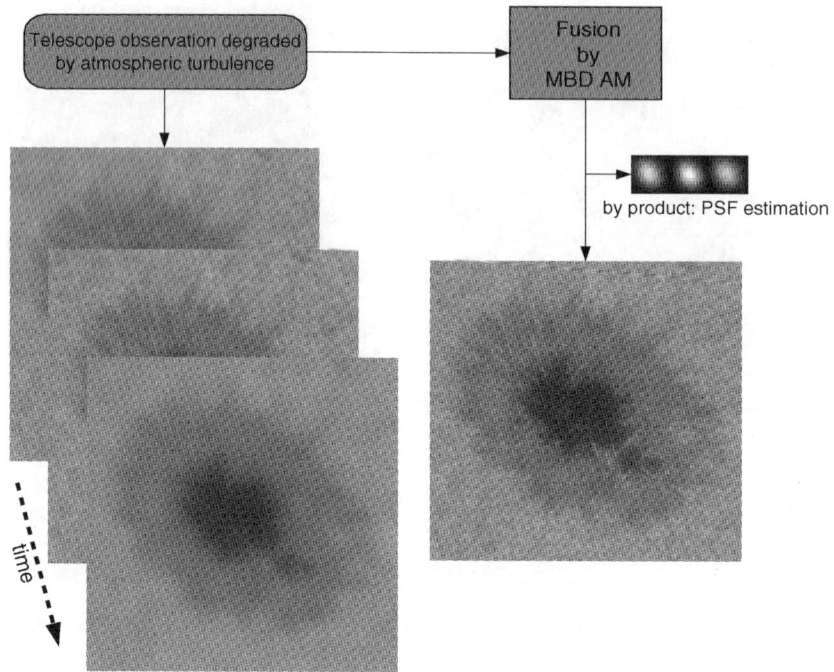

FIGURE 14.4 Images of a sunspot taken by a ground-based telescope and blurred due to perturbations of wavefronts in the Earth atmosphere. The perturbations vary in time which leads to different blurring of the individual frames. The resulting image was fused by the MBD–AM algorithm described in Section IV.A.

various deconvolution methods. There are two basic approaches to solving the MBD problem. The first one is to separately treat each channel by any single-channel deconvolution method and then to combine the results; the other is to employ deconvolution methods that are multichannel in their nature.

Numerous single-channel blind deconvolution methods have been published extensively in the literature in the last two decades (see Ref. 14 or 15 for a basic survey). However, their adaptation to the MBD problem cannot reach the power of intrinsic multichannel methods and this approach seems to be a "dead-end".

The development of intrinsic multichannel methods has begun just recently. One of the earliest methods[16] was designed particularly for images blurred by atmospheric turbulence. Harikumar and Bresler[17,18] proposed indirect algorithms (EVAM), which first estimate the PSFs and then recover the original image by standard nonblind methods. Giannakis and Heath[19,20] (and at the same time Harikumar and Bresler[21]) developed another indirect algorithm based on Bezout's identity of coprime polynomials which finds inverse filters and, by convolving the filters with the observed images, recovers the original image. Pillai and Liang[22] have proposed another intrinsically multichannel method based on the greatest common divisor which is, unfortunately, even less numerically stable than the previous methods. Pai and Bovik[23,24] came with two direct multichannel restoration algorithms that directly estimate the original image from the null space or from the range of a special matrix. To reach higher robustness, Šroubek[25] proposed an iterative deconvolution method which employs anisotropic regularization of the image and between-channel regularization of the PSFs.

A. Alternating Minimization Algorithm

In this section, we present an alternating minimization algorithm for multichannel blind deconvolution (MBD–AM) and we demonstrate that it is a powerful tool for image fusion in the case of uniformly blurred channels.

Since the blind deconvolution problem is ill posed with respect to both u and h_i, a constrained minimization technique is required to find the solution of Equation 14.5. Constraints are built on prior knowledge that we have about the system. Typical assumptions valid for the majority of real acquisition processes are the following: n_i is supposed to have zero mean and the same variance σ^2 in each channel, and the PSFs are supposed to preserve the overall brightness (mean intensity) of the image. The imposed constraints then take the forms

$$\frac{1}{|\Omega|} \int_{\Omega} (h_i * u - z_i)^2 \mathrm{d}x = \sigma^2 \tag{14.6}$$

$$\int_{\Omega} h(x)\mathrm{d}x = 1 \tag{14.7}$$

(To simplify the notation, we drop the two-dimensional co-ordinates (x, y) or, if required, we only write x.)

Let $Q(u)$ and $R(h)$ denote some regularization functionals of the estimated original image u and blurs $h \equiv \{h_1, ..., h_N\}$, respectively. The constrained minimization problem is formulated as $\min_{u,h} Q(u) + \gamma R(h)$ subject to Equation 14.6 and Equation 14.7. The unconstrained optimization problem, obtained by means of Lagrange multipliers, is to find u and h which minimize the functional

$$E(u, h) = \frac{1}{2} \sum_{i=1}^{N} \|h_i * u - z_i\|^2 + \lambda Q(u) + \gamma R(h) \tag{14.8}$$

where λ and γ are positive parameters which penalize the regularity of the solutions u and h. The crucial questions are how to construct functionals Q and R and whether the global minimum can be reached. We propose an alternating minimization algorithm that iteratively searches for a minimum of Equation 14.8. Constraint Equation 14.7 was dropped since it will be automatically satisfied in the algorithm if the initial blurs satisfy the constraint. We now proceed the discussion with possible choices of $Q(u)$ and $R(h)$.

1. Regularization of the Image $Q(u)$

Regularization of Equation 14.5 with respect to the image function can adopt various forms. The classical approach of Tichonov chooses $Q(u) = \int_\Omega |\nabla u(x)|^2 dx$, where ∇u denotes the gradient of u. Apart from easy implementation, this regularization is not suitable, since the L_2 norm of the image gradient penalizes too much the gradients corresponding to edges and an oversmoothing effect is observed. In real images, object edges create sharp steps that appear as discontinuities in the intensity function. It is the space of bounded variation (BV) functions that is widely accepted as a proper setting for real images. Rudin[26] first demonstrated very good anisotropic denoising properties of the total variation (TV) $Q_{TV}(u) = \int |\nabla u(x)| dx$. Existence and uniqueness of the minimum of TV is possible only in the BV-space, in which case ∇u denotes the gradient of u in the distributional sense. The same holds true for a more general case of convex functions of measures

$$Q_\phi(u) = \int \phi(|\nabla u(x)|) dx$$

where ϕ is a strictly convex, nondecreasing function that grows at most linearly. Examples of $\phi(s)$ are s (TV), $\sqrt{1 + s^2} - 1$ (hyper-surface minimal function) or $\log(\cosh(s))$. For nonconvex functions nothing can be said about the existence of the minimum. Nevertheless, nonconvex functions, such as $\log(1 + s^2), s^2/(1 + s^2)$ or $\arctan(s^2)$ (Mumford–Shah functional[27]), are often used since they provide better results for segmentation problems.

2. Regularization of the Blurs $R(h)$

Regularization of the blurs h_i's directly follows from our model, Equation 14.5, and can be derived from the mutual relations of the channels. The blurs are assumed to have finite support S of the size (s_1, s_2) and certain channel disparity is necessary. The disparity is defined as weak coprimeness of the channel blurs, which states that the blurs have no common factor except a scalar constant. In other words, if the channel blurs can be expressed as a convolution of two subkernels then there is no subkernel that is common to all blurs. An exact definition of weakly coprime blurs can be found in Ref. 20. The channel coprimeness is satisfied for many practical cases, since the necessary channel disparity is mostly guaranteed by the nature of the acquisition scheme and random processes therein. We refer the reader to Ref. 18 for a relevant discussion.

Under the assumption of channel coprimeness, we can see that any two correct blurs h_i and h_j satisfy $\|z_i * h_j - z_j * h_i\|^2 = 0$ if the noise term in Equation 14.5 is omitted. We therefore propose to regularize the blurs by

$$R(h) = \frac{1}{2} \sum_{1 \le i < j \le N} \|z_i * h_j - z_j * h_i\|^2 \tag{14.9}$$

This regularization term does not penalize spurious factors, that is, $f * h_i$ for any factor f are all equivalent. We see that the functional $R(h)$ is convex but far from strictly convex. The dimensionality of the null space of $R(\hat{h})$ is proportional to the degree of size overestimation of $\hat{h}_i$ with respect to the size of the original blurs h_i's. Therefore to use the above regularization, we have to first estimate S of the original blurs and impose this support constraint in R. The size constraint is imposed automatically in the discretization of R, which is perfectly plausible since the calculations are done in the discrete domain anyway. An exact derivation of the size of the null space is given in Ref. 18.

3. Iterative Minimization Algorithm

We consider the following minimization problem

$$E(u,h) = \frac{1}{2} \sum_{i=1}^{N} \|h_i * u - z_i\|^2 + \frac{\gamma}{2} \sum_{1 \le i < j \le N} \|z_i * h_j - z_j * h_i\|^2$$

$$+ \lambda \int_{\Omega} \phi(|\nabla u(x)|)\mathrm{d}x \tag{14.10}$$

$E(u,h)$, as a function of variables u and h, is not convex due to the convolution in the first term. On the other hand, the energy function is convex with respect to u if h is fixed and it is convex with respect to h if u is fixed. The minimization sequence (u^n, h^n) can be thus built by alternating between two minimization subproblems

$$u^n = \arg \min_u E(u, h^{n-1}) \quad \text{and} \quad h^n = \arg \min_h E(u^n, h) \tag{14.11}$$

for some initial h^0 with the rectangular support S. The advantage of this scheme lies in its simplicity, since for each subproblem a unique minimum exists that can be easily calculated. However, we cannot guarantee that the global minimum is reached this way, but thorough testing indicates good convergence properties of the algorithm for many real problems.

The solution of the first subproblem in Equation 14.10 formally satisfies the Euler–Lagrange equation

$$\frac{\partial E}{\partial u} = \sum_{i=1}^{N} h_i' * (h_i * u - z_i) - \lambda \operatorname{div}\left(\frac{\mathrm{d}\phi(|\nabla u|)}{\mathrm{d}|\nabla u|} \frac{\nabla u}{|\nabla u|}\right) = 0 \tag{14.12}$$

where the prime means mirror reflection of the function, that is, $h_i'(x,y) = h_i(-x,-y)$. One can prove (see for example Ref. 28) that a unique solution exists in the BV-space, where the image gradient is a measure. To circumvent the difficulty connected with implementing the measure and with the nonlinearity of the divergence term in Equation 14.12, the solution can be found by relaxing ϕ and following a half-quadratic algorithm originally proposed in Ref. 29 and generalized for convex functions of measures in Ref. 28.

The solution of the second subproblem in Equation 14.11 formally satisfies the Euler–Lagrange equations

$$\frac{\partial E}{\partial h_k}=u'*(u*h_k-z_k)-\gamma\sum_{\substack{i=1\\i\neq k}}^{N}z_i'*(z_i*h_k-z_k*h_i)=0,\quad k=1,\dots,N \qquad (14.13)$$

This is a set of linear equations and thus finding h is a straightforward task.

It is important to note that the algorithm runs in the discrete domain and that a correct estimation of the weighting constants, λ, and γ, and mainly of the blur support S is crucial. In addition, the algorithm is iterative and the energy (Equation 14.10) as a function of the image and blurs does not have one minimum, so the initial guess g^0 plays an important role as well. The positive weighting constants λ and γ are proportional to the noise levels σ and can be calculated in theory from the set of Equation 14.6, Equation 14.12 and Equation 14.13 if the noise variance is known. This is, however, impossible to carry out directly and techniques such as generalized cross validation must be used instead. Such techniques are computationally very expensive and we suggested an alternative approach in Ref. 25 which uses bottom limits of λ and γ. Estimation of the size of the blur support S is even more vexatious. Methods proposed in Refs. 18,20 provide a reliable estimate of the blur size only under ideal noise-free conditions. In the noisy case they suggest a full search, that is, for each discrete rectangular support S estimate the blurs and compare the results.

B. Experiment with Artificial Data

First, we demonstrate the performance of the MBD–AM algorithm on images degraded by computer-generated blurring and noise and we compare the results with two recent methods — Harrikumar's EVAM and Pai's method.

For the evaluation, we use the percentage mean-square error of the fused image $\hat{u}$, defined as

$$\text{PMSE}(u) \equiv 100\frac{\|\hat{u}-u\|}{\|u\|} \qquad (14.14)$$

Although the mean-square error does not always correspond to visual evaluation of the image quality, it has been commonly used for quantitative evaluation and comparison.

A test image of size 250×250 in Figure 14.5(a) was first convolved with four 7×7 PSFs in Figure 14.5(b) and then white Gaussian noise at five different levels (SNR = 50, 40, 30, 20, and 10 dB, respectively) was added. This way, we simulated four acquisition channels ($N = 4$) with a variable noise level. The size of the blurs and the noise level were assumed to be known. All three algorithms were therefore started with the correct blur size $S = (7,7)$. In the case of MBD–AM, λ and γ were estimated as described in Ref. 25 and the starting position h^0 was set to the delta functions. The reconstructed images and blurs are

FIGURE 14.5 Synthetic data: (a) original 250×250 image of the Tyn church, (b) four 7×7 PSFs, and (c) four blurred channels.

shown in Figure 14.6 with the percentage mean-square errors summarized in Table 14.1. Note that Pai's method reconstructed only the original image and not the blurs. The performance of the EVAM method quickly decreases as SNR decreases, since noise is not utilized in the derivation of this method. Pai's method shows superior stability but for lower SNR the reconstructed images are still considerably blurred. Contrary to the previous two methods, the MBD–AM algorithm is stable and performs well even for lower SNRs (20 dB, 10 dB). One slight drawback is that the output increasingly resembles a piecewise constant function which is due to the variational regularization $Q(u)$.

C. Experiment with Real Data

The following experiment was conducted to test the applicability of MBD–AM on real data. Four images of a bookcase were acquired with a standard digital camera focused to 80 cm (bookcase in focus), 40, 39, and 38 cm distance, respectively. The acquired data were stored as low resolution 640×480 24 bit color images and only the central rectangular part of the green band of size 250×200 was used for the fusion. The central part of the first image, which captures the scene in focus, is shown in Figure 14.7(a). Three remaining images, Figure 14.7(c), were used as the input for the MBD–AM algorithm. The parameter $\lambda = 1.6 \times 10^{-4}$ was estimated experimentally by running the

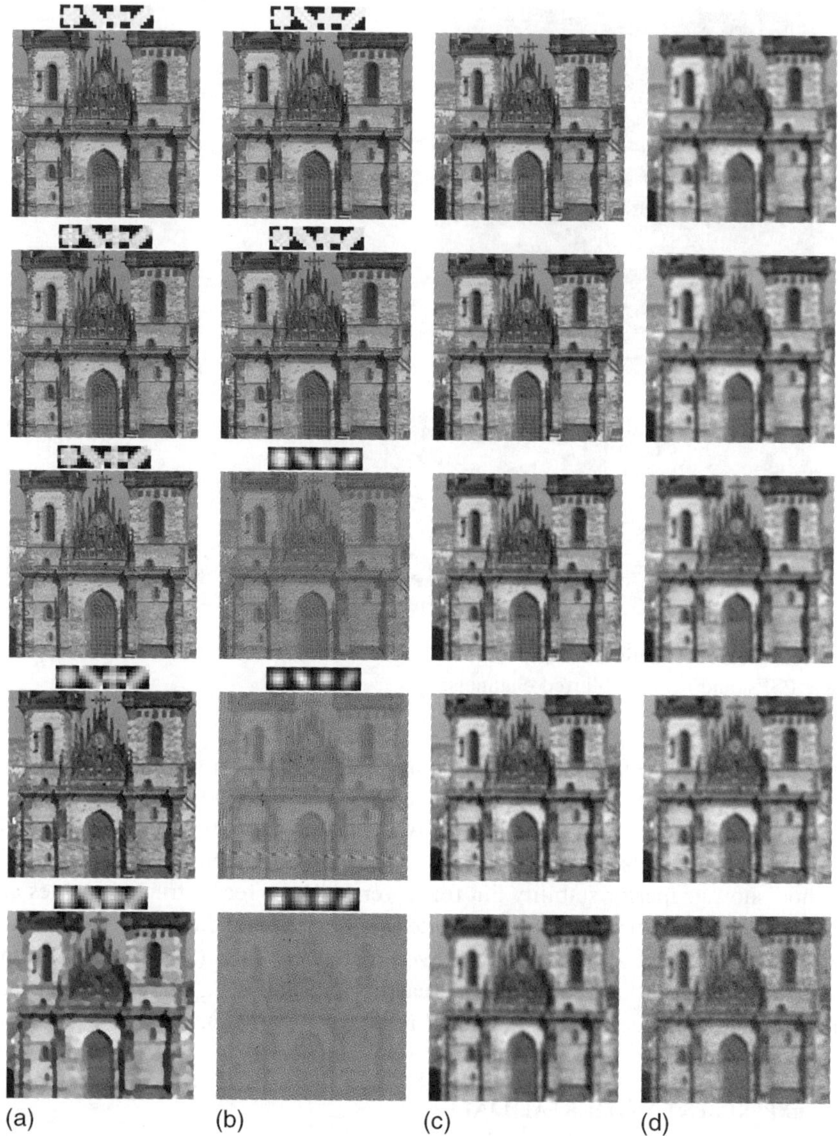

(a)   (b)   (c)   (d)

FIGURE 14.6 Reconstruction of the test image and blurs from four degraded images using (a) MBD–AM, (b) EVAM, and (c) Pai's method. The first of the four degraded channels is in column (d) for comparison. From top to bottom SNR = 50, 40, 30, 20, and 10 dB, respectively.

algorithm with different λ's and selecting the most visually acceptable results. The parameter γ was calculated as described in Ref. 25. A defocused camera causes image degradation approximately modeled by cylindrical blurs. A cepstrum analysis in Ref. 30 was used to estimate diameters of these blurs, which

TABLE 14.1
Performance of the MBD–AM, the EVAM and the Pai's Algorithms on the Data in Figure 14.5

SNR (dB)	MBD–AM	EVAM	Pai
50	0.93	0.99	5.09
40	2.61	2.99	7.87
30	5.17	24.1	10.9
20	10.2	35.7	13.9
10	15.3	38.3	16.4

The table shows percentage mean-square error of the fused image.

were determined to be around eight pixels. Obtained results after ten iterations are shown in Figure 14.7(b). Further iterations did not produce any visual enhancement. Simple visual comparison reveals that the letters printed on shelf backs are more legible in the restored image but still lack the clarity of the focused image, and that the reconstructed blurs resemble the cylindrical blurs as was expected. It is remarkable how successful the restoration was, since one

FIGURE 14.7 Real bookcase images: (a) 250×250 image acquired with the digital camera set to the correct focus distance of 80 cm, (b) MBD–AM fused image and estimated 10×10 blurs after 10 iterations ($\lambda = 1.6 \times 10^{-4}$) obtained from three blurred images in (c) that have false focus distance 40 cm, 39 cm and 38 cm. Reprinted from [25] with permission of the IEEE.

would expect that the similarity of blurs would violate the coprimeness assumption. It is believed that the algorithm would perform even better if a wider disparity between blurs was assured.

V. SLIGHTLY MISREGISTERED BLURRED CHANNELS

This is a generalization of the previous model which allows between-channel shifts (misregistrations) of extent up to a few pixels.

$$z_i(x + a_i, y + b_i) = (u * h_i)(x, y) + n_i(x, y) \tag{14.15}$$

where a_i, b_i are unknown translation parameters of the ith channel. This model is applicable in numerous practical tasks when the scene or the camera moves slightly between consecutive channel acquisitions (see Figure 14.8). Such a situation typically occurs when the camera is subject to vibrations or in multitemporal imaging if the scene is not perfectly still. Sometimes a subpixel between-channel shift is even introduced intentionally in order to enhance spatial resolution of the fused image (this technique is called *superresolution imaging*, see Ref. 31 for a survey and other references).

Images degraded according to this model cannot be fused by the methods mentioned in Section IV.A. If they were applied, the channel misregistrations would lead to strong artifacts in the fused image. On the other hand, the misregistrations considered in this model are too small to be fully removed by image registration techniques (in case of blurred images, registration methods usually can suppress large spatial misalignments but seldom reach subpixel accuracy).

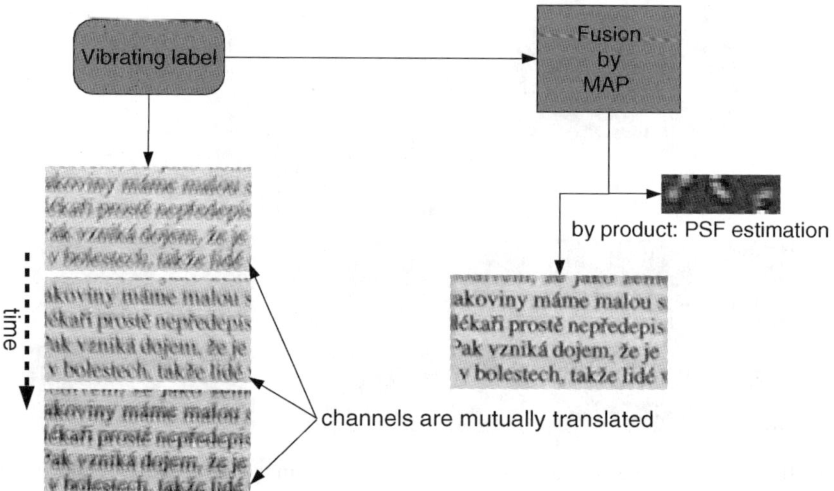

FIGURE 14.8 Multiple acquisition of a vibrating text label. Motion blur of various kind and small spatial misalignments of the individual frames can be observed. The fused image was achieved by the MAP algorithm described in Section V.A.

Fusion of images degraded according to this model requires special blind deconvolution methods, which can — in addition to the deconvolution itself — identify and compensate the between-channel misregistration. A successful method based on a stochastic approach is described below.

A. Maximum *A Posteriori* Probability Algorithm

Equation 14.15 can be expressed into equivalent form

$$z_i(x, y) = u(x, y) * h_i(x - a_i, y - b_i) + n_i(x, y) \tag{14.16}$$

which can be further rewritten as

$$z_i(x, y) = (u * g_i)(x, y) + n_i(x, y) \tag{14.17}$$

where g_i is a shifted version of the original PSF h_i : $g_i(x, y) = h_i(x - a_i, y - b_i)$. We can therefore work only with g_i and use the MBD–AM algorithm (Equation 14.11). In this case, the estimate of the blur size has to include also the maximum shift between the channels. Since this is difficult to determine, standard MBD techniques including MBD–AM in its present form cannot be applied.

To overcome the difficulties connected with the parameter estimation, we adopt in Ref. 32 a stochastic approach to the minimization problem. The restoration can be formulated then as a maximum *a posteriori* (MAP) estimation. We assume that the matrices u, $g = \{g_1, ..., g_N\}$ and $z = \{z_1, ..., z_N\}$ are random vector fields with given probability density functions (PDFs) $p(u)$, $p(g)$ and $p(z)$, respectively, and we look for such realizations of u and g which maximize the *a posteriori* probability $p(u, g|z)$. The MAP estimation is equivalent to minimizing $-\log(p(u, g|z))$. The only two assumptions that we must make in addition to those in the energy minimization problem are: u and g are supposed to be statistically independent and n_i is white (that is, uncorrelated) Gaussian noise. Using the Bayes rule, the relation between *a priori* densities $p(u)$, $p(g)$ and the *a posteriori* density is $p(u, g|z) \propto p(z|u, g)p(u)p(g)$. The conditional PDF $p(z|u, g)$ follows from our model, Equation 14.5, and from our assumption of white noise. The blur PDF $p(g)$ can be derived from the regularization $R(g)$, which is also Gaussian noise with a covariance matrix that can be easily calculated. If the image PDF $p(u)$ is chosen in such a way that $-\log(p(u)) \propto Q(u)$ then the MAP estimation is almost identical to the minimization problem, Equation 14.10, for $\lambda = \sigma^2$ and $\gamma = |S|/2$ and we can use the alternating iterative algorithm. To improve stability of the algorithm against the overestimation of S and thus handle inaccurate registration, it suffices to add the constraint of positivity $h(x) > 0$ to Equation 14.7 and perform in Equation 14.11 the minimization subject to the new constraints.

Setting appropriately initial blurs can help our iterative algorithm to converge to the global minimum. This issue is especially critical for the case of overestimated blur size. One can readily see that translated versions of the correct blurs are all equivalent as long as they fit into our estimated blur size. We have seen that the regularization of the blurs R is unable to distinguish between the correct blurs and the correct blurs convolved with an arbitrary spurious factor. This has a negative impact on the convergence mainly if channel misalignment occurs, since new local minima appear for blurs that cope with the misalignment by convolving the correct blurs with an interpolating kernel. To get closer to the correct solution, we thus propose to set the initial blurs g^0 to delta functions positioned at the centers of gravity of blurs $\hat{g} = \arg\min R(g)$. This technique enables us to compensate for the channel shifts right from the start of the algorithm and get away from the incorrect interpolated solutions.

B. EXPERIMENT WITH MISREGISTERED IMAGES

Although the MAP fusion method can also be applied to registered channels, its main advantageous property, that discriminates it from other methods, is the ability to fuse channels which are not accurately registered. This property is illustrated by the following experiment.

The 230×260 test image in Figure 14.9(a) was degraded with two different 5×5 blurs and noise of SNR $= 50$ dB. One blurred image was shifted by 5×5 pixels and then both images were cropped to the same size; see Figure 14.9(c). The MAP algorithm was initialized with the overestimated blur size 12×12. The fused image and the estimated blur masks are shown in Figure 14.10. Recovered blurs contain negligible spurious factors and are properly shifted to compensate for the misregistration. The fused image is, by visual comparison, much sharper than the input channels and very similar to the original, which demonstrates excellent performance. This conclusion is supported also by the real experiment shown in Figure 14.8, where both blurring and shift were introduced by object vibrations. Unlike the input channels, the text on the fused image is clearly legible.

VI. HEAVILY MISREGISTERED BLURRED CHANNELS

This model is a further generalization of the previous model. Blurring of each channel is still uniform and is modeled by a convolution, but significant misregistrations between the channels are allowed.

$$z_i(\tau_i(x, y)) = (u * h_i)(x, y) + n_i(x, y) \qquad (14.18)$$

In this model there are almost no restrictions on the extent and the type of τ_i; it may have a complex nonlinear form (the only constraint is that the individual frames must have sufficient overlap in the region of interest). This is a very realistic model of photographing a flat scene, where the camera moves in three-dimensional space in an arbitrary manner (see Figure 14.11).

Because of the complex nature of τ_i, it cannot be compensated for during the deconvolution step. Thus, fusion of images degraded according to this model is a two-stage process — it consists of image registration (spatial alignment) followed by MBD discussed in the previous section. Since all

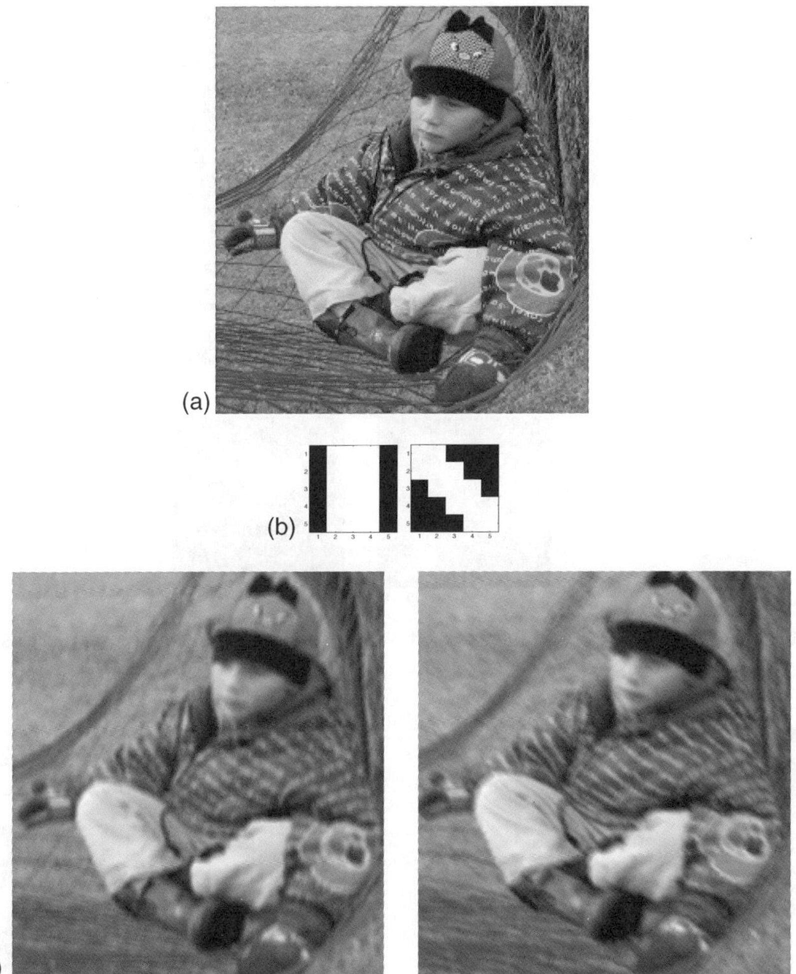

FIGURE 14.9 (a) original test image 230×260 pixels, (b) two 5×5 PSFs, and (c) blurred and shifted images.

deconvolution methods require either perfectly aligned channels (which is not realistic) or allow, at most, small shift differences, the registration is a crucial step of the fusion.

Image registration in general is a process of transforming two or more images into a geometrically equivalent form. It eliminates the degradation effects caused by geometric distortion. From a mathematical point of view, it consists of approximating τ_i^{-1} and of resampling the image. For images which are not blurred, the registration has been extensively studied in the recent literature (see Ref. 33 for a survey). However, blurred images require special registration techniques. They can, as well as the general-purpose registration methods, be

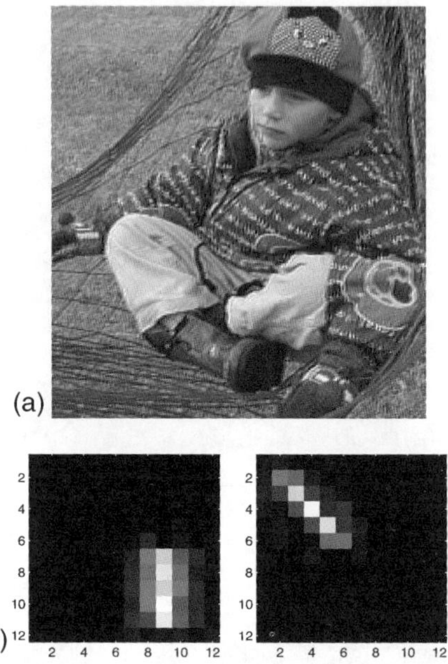

FIGURE 14.10 MAP image fusion: (a) fused image and (b) estimated blur masks with between-channel shift.

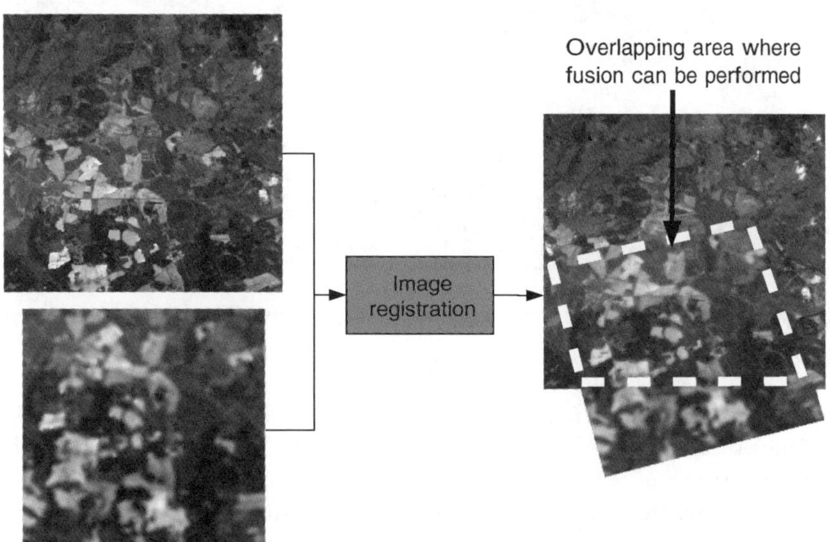

FIGURE 14.11. Two satellite images differing from one another by amount of blur due to different spatial resolution and by shift, rotation, and scaling (left). After the registration (right), the MAP fusion algorithm from the previous section can be applied on the overlapping area. The registration of these images was performed in IMARE Toolbox[50] by means of invariant-based method,[42] courtesy of Barbara Zitová.

divided in two groups — global and landmark-based ones. Regardless of the particular technique, all feature extraction methods, similarity measures, and matching algorithms used in the registration process must be insensitive to image blurring.

Global methods do not search for particular landmarks in the images. They try to estimate directly the between-channel translation and rotation. Myles and Lobo[34] proposed an iterative method working well if a good initial estimate of the transformation parameters is available. Zhang et al.,[35,36] proposed to estimate the registration parameters by bringing the channels into canonical form. Since blur-invariant moments were used to define the normalization constraints, neither the type nor the level of the blur influences the parameter estimation. Kubota et al.[37] proposed a two-stage registration method based on hierarchical matching, where the amount of blur is considered as another parameter of the search space. Zhang and Blum[38] proposed an iterative multiscale registration based on optical flow estimation in each scale, claiming that optical flow estimation is robust to image blurring. All global methods require considerable (or even complete) spatial overlap of the channels to yield reliable results, which is their major drawback.

Landmark-based blur-invariant registration methods have appeared very recently, just after the first paper on the moment-based blur-invariant features.[39] Originally, these features could only be used for registration of mutually shifted images.[40,41] The proposal of their rotational-invariant version[42] in combination with a robust detector of salient points[43] led to registration methods that are able to handle blurred, shifted and rotated images.[44,45]

Although the above-cited registration methods are very sophisticated and can be applied to almost all types of images, the result rarely tends to be perfect. The registration error usually varies from subpixel values to a few pixels, so only fusion methods sufficiently robust to between-channel misregistration can be applied to channel fusion.

VII. CHANNELS WITH SPACE-VARIANT BLURRING

This model comprises space-variant blurring of the channels as well as nonrigid geometric differences between the channels.

$$z_i(\tau_i(x, y)) = \int h_i(x, y, s, t)u(s, t)\mathrm{d}s\,\mathrm{d}t + n_i(x, y) \qquad (14.19)$$

The substantial difference from the previous models is that image blurring is no longer uniform in each frame and thus it cannot be modeled as a convolution. Here, the PSF is a function of spatial co-ordinates (x, y) which makes the channel degradation variable depending on the location. This situation typically arises when photographing a three-dimensional scene by a camera with a narrow depth of field. Differently blurred channels are obtained by changing the focus distance of the camera (see Figure 14.12). Unlike piecewise ideal imaging, the depth of the scene can vary in a continuous manner and the existence of at least one "ideal"

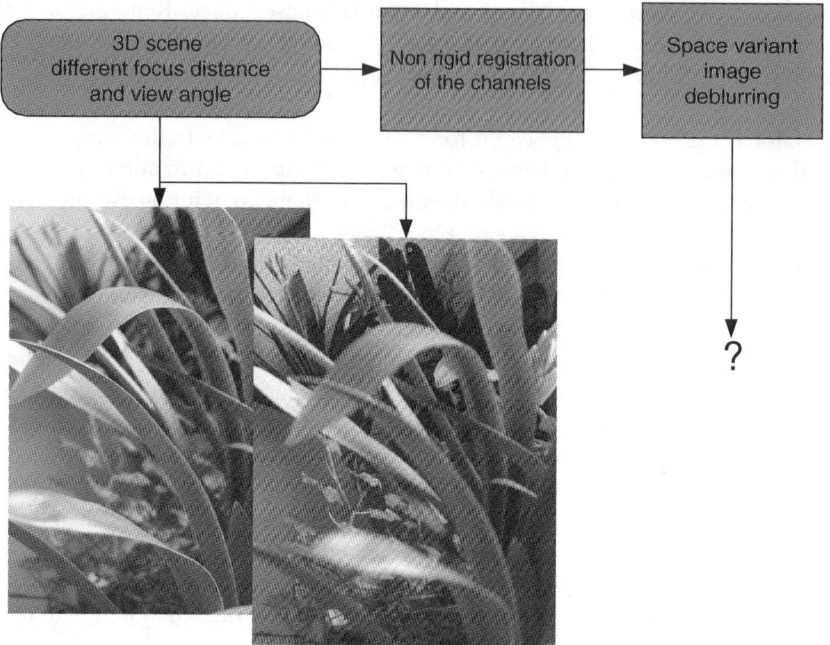

FIGURE 14.12 Space-variant blurring of the channels. Two pictures of a complex three-dimensional scene taken with a variably focused camera. The camera also changed its position and viewing angle between the acquisitions which lead to projective geometric deformation between the channels.

picture for each location is not guaranteed. Another example is photographing a dynamic scene where different parts move by different velocity and/or in different directions (see Figure 14.13).

Space-variant blurring is not a simple extension of the previous models. It requires qualitatively new approaches and methods. As for the previous model, the image fusion consists of image registration and multichannel blind deblurring but there is a significant difference. While the registration methods can be in principle the same, the techniques used here in the second step must be able to handle space-variant blurring.

Up until now, no papers have been published on multichannel space-variant deblurring. There are, however, a few papers on single-channel space-variant image deblurring, usually originating from space-invariant deconvolution methods. Guo et al.[46] proposed to divide the image into uniformly blurred regions (if possible) and then to apply a modified expectation-maximization algorithm in each region. You and Kaveh[47] considered parameterized PSF and used anisotropic regularization for image deblurring. Cristobal and Navarro[48] applied multiscale Gabor filters to the restoration. However, the extension of the above methods to the multichannel framework is questionable.

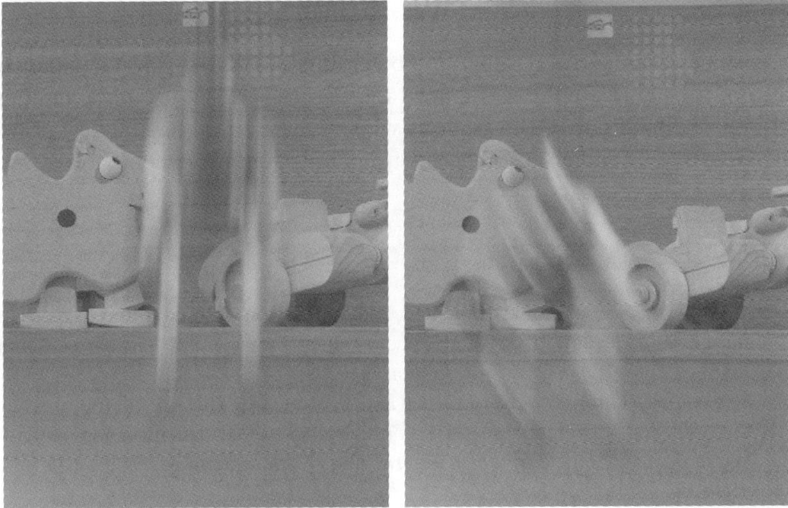

FIGURE 14.13 Space-variant degradation of the channels due to motion blur.

No doubt the solution to this problem is a big challenge for image fusion for the near future. Prospective methods should employ all available *a priori* information, such as a depth map or relief model. They may comprise depth-based, defocus-based or depth- and defocus-based segmentation of the input channels in order to find regions of the same type of blur. Nevertheless, a general solution probably does not exist.

VIII. CONCLUSION

In this chapter, we presented an overview of image fusion methods for the case where the input channels are blurred, noisy, and geometrically different. One has to face this problem in various application areas where the picture of the scene is taken under nonideal conditions. Mathematically, this task is ill posed and cannot be resolved by inverting all degradation factors. The only solution is multiple acquisition of the scene and consequent fusion of all acquired channels. It is believed that if the channel degradations are different, the channels can be fused together in such a way that the information missing in one channel can be supplemented by the others. The fusion approaches and methods differ from each other according to the type of assumed degradations of the channels. Here we classified the possible degradations into five major groups: piecewise ideal imaging, uniform blurring, slight and heavy channel misregistration, and space-variant blurring of the channels. For each category, except the last one, we presented reliable fusion methods whose performance was experimentally verified.

ACKNOWLEDGMENTS

This work has been partially supported by grant No. 102/04/0155 from the Grant Agency of the Czech Republic. Figure 14.4, Figure 14.7, Figure 14.8, and Figure 14.11 are modifications of illustrations which appeared in the authors' previous publications[25,32,49] and are used here with permission of the coauthors.

REFERENCES

1. Chavez, P., Sides, S., and Anderson, J., Comparison of three different methods to merge multiresolution and multispectral data: Landsat TM and SPOT panchromatic, *Photogrammetric Eng. Remote Sensing*, 57, 295–303, 1991.
2. Duport, B., Girel, J., Chassery, J., and Pautou, G., The use of multiresolution analysis and wavelets transform for merging SPOT panchromatic and multispectral image data, *Photogrammetric Eng. Remote Sensing*, 69(9), 1057–1066, 1996.
3. Oldmixon, E., and Carlsson, K., Methods for large data volumes from confocal scanning laser microscopy of lung, *J. Microsc. — Oxford*, 170, 221–228, 1993.
4. Li, H., Deklerck, R., Decuyper, B., Hermanus, A., Nyssen, E., and Cornelis, J., Object recognition in brain CT-scans: knowledge-based fusion of data from multiple feature extractors, *IEEE Trans. Med. Imaging*, 14, 212–229, 1995.
5. Li, S., Kwok, J., and Wang, Y., Combination of images with diverse focuses using the spatial frequency, *Inf. Fusion*, 2(3), 169–176, 2001.
6. Li, S., Kwok, J., and Wang, Y., Multifocus image fusion using artificial neural networks, *Pattern Recognit. Lett.*, 23(8), 985–997, 2002.
7. Li, H., Manjunath, B., and Mitra, S., Multisensor image fusion using the wavelet transform, *Graphical Models Image Process.*, 57(3), 235–245, 1995.
8. Zhang, Z., and Blum, R., A categorization of multiscale-decomposition-based image fusion schemes with a performance study for a digital camera application. In *Proceedings of the IEEE*, Vol. 87, 1999, August.
9. Subbarao, M., Choi, T., and Nikzad, A., Focusing techniques, *J. Opt. Eng.*, 32, 2824–2836, 1993.
10. Subbarao, M., and Tyan, J. K., Selecting the optimal focus measure for autofocusing and depth-from-focus, *IEEE Trans. Pattern Anal. Machine Intell.*, 20, 864–870, 1998.
11. Zhang, Y., Zhang, Y., and Wen, C., A new focus measure method using moments, *Image Vis. Comput.*, 18, 959–965, 2000.
12. Kautsky, J., Flusser, J., Zitová, B., and Šimberová, S., A new wavelet-based measure of image focus, *Pattern Recognit. Lett.*, 23, 1785–1794, 2002.
13. Banham, M., and Katsaggelos, A., Digital image restoration, *IEEE Signal Process. Mag.*, 14(2), 24–41, 1997.
14. Sezan, M. I., and Tekalp, A. M., Survey of recent developments in digital image restoration, *Opt. Eng.*, 29, 393–404, 1990.
15. Kundur, D., and Hatzinakos, D., Blind image deconvolution, *IEEE Signal Process. Mag.*, 13(3), 43–64, 1996.
16. Schulz, T. J., Multiframe blind deconvolution of astronomical images, *J. Opt. Soc. Am. A*, 10, 1064–1073, 1993.

17. Harikumar, G., and Bresler, Y., Efficient algorithms for the blind recovery of images blurred by multiple filters. In *Proceedings IEEE International Conference on Image Processing*, Vol. 3, Lausanne, Switzerland, 1996.

18. Harikumar, G., and Bresler, Y., Perfect blind restoration of images blurred by multiple filters: theory and efficient algorithms, *IEEE Trans. Image Process.*, 8(2), 202–219, 1999.

19. Giannakis, G. B., and Heath, R. W., Blind identification of multichannel FIR blurs and perfect image restoration. In *Proceedings 13th International Conference on Pattern Recognition*, 1996.

20. Giannakis, G., and Heath, R., Blind identification of multichannel FIR blurs and perfect image restoration, *IEEE Trans. Image Process.*, 9(11), 1877–1896, 2000.

21. Harikumar, G., and Bresler, Y., Exact image deconvolution from multiple FIR blurs, *IEEE Trans. Image Process.*, 8, 846–862, 1999.

22. Pillai, S., and Liang, B., Blind image deconvolution using a robust GCD approach, *IEEE Trans. Image Process.*, 8(2), 295–301, 1999.

23. Pai, H.-T., and Bovik, A., Exact multichannel blind image restoration, *IEEE Signal Process. Lett.*, 4(8), 217–220, 1997.

24. Pai, H.-T., and Bovik, A., On eigenstructure-based direct multichannel blind image restoration, *IEEE Trans. Image Process.*, 10(10), 1434–1446, 2001.

25. Šroubek, F., and Flusser, J., Multichannel blind iterative image restoration, *IEEE Trans. Image Process.*, 12(9), 1094–1106, 2003.

26. Rudin, L., Osher, S., and Fatemi, E., Nonlinear total variation based noise removal algorithms, *Physica D*, 60, 259–268, 1992.

27. Mumford, D., and Shah, J., Optimal approximation by piecewise smooth functions and associated variational problems, *Comm. Pure Appl. Math.*, 42, 577–685, 1989.

28. Aubert, G., and Kornprobst, P., *Mathematical Problems in Image Processing*, Springer, New York, 2002.

29. Chambolle, A., and Lions, P., Image recovery via total variation minimization and related problems, *Numer. Math.*, 76(2), 167–188, 1997.

30. Chang, M., Tekalp, A., and Erdem, A., Blur identification using the bispectrum, *IEEE Trans. Signal Process.*, 39(10), 2323–2325, 1991.

31. Park, S., Park, M., and Kang, M., Super-resolution image reconstruction: a technical overview, *IEEE Signal Proc. Mag.*, 20(3), 21–36, 2003.

32. Šroubek, F., and Flusser, J., Shift-invariant multichannel blind restoration. In *Proceedings of the third International Symposium on Image and Signal Processing and Analysis, ISPA'03*, Rome, 2003, September.

33. Zitová, B., and Flusser, J., Image registration methods: a survey, *Image Vis. Comput.*, 21, 977–1000, 2003.

34. Myles, Z., and Lobo, N. V., Recovering affine motion and defocus blur simultaneously, *IEEE Trans. Pattern Anal. Machine Intell.*, 20, 652–658, 1998.

35. Zhang, Y., Wen, C., and Zhang, Y., Estimation of motion parameters from blurred images, *Pattern Recognit. Lett.*, 21, 425–433, 2000.

36. Zhang, Y., Wen, C., Zhang, Y., and Soh, Y. C., Determination of blur and affine combined invariants by normalization, *Pattern Recognit.*, 35, 211–221, 2002.

37. Kubota, A. Kodama, K., and Aizawa, K., Registration and blur estimation methods for multiple differently focused images. In *Proceedings International Conference on Image Processing*, Vol. II, 1999.

38. Zhang, Z., and Blum, R., A hybrid image registration technique for a digital camera image fusion application, *Inf. Fusion*, 2, 135–149, 2001.
39. Flusser, J., Suk, T., and Saic, S., Recognition of blurred images by the method of moments, *IEEE Trans. Image Process.*, 5, 533–538, 1996.
40. Flusser, J., and Suk, T., Degraded image analysis: an invariant approach, *IEEE Trans. Pattern Anal. Machine Intell.*, 20(6), 590–603, 1998.
41. Bentoutou, Y., Taleb, N., Mezouar, M., Taleb, M., and Jetto, L., An invariant approach for image registration in digital subtraction angiography, *Pattern Recognit.*, 35, 2853–2865, 2002.
42. Flusser, J., and Zitová, B., Combined invariants to linear filtering and rotation, *Int. J. Pattern Recognit. Art. Intell.*, 13(8), 1123–1136, 1999.
43. Zitová, B., Kautsky, J., Peters, G., and Flusser, J., Robust detection of significant points in multiframe images, *Pattern Recognit. Lett.*, 20, 199–206, 1999.
44. Flusser, J., Zitová, B., and Suk, T., Invariant-based registration of rotated and blurred images. In *Proceedings IEEE 1999 International Geoscience and Remote Sensing Symposium*. Los Alamitos, 1999, June.
45. Flusser, J., Boldyš, J., and Zitová, B., Moment forms invariant to rotation and blur in arbitrary number of dimensions, *IEEE Trans. Pattern Anal. Machine Intell.*, 25(2), 234–246, 2003.
46. Guo, Y., Lee, H., and Teo, C., Blind restoration of images degraded by space-variant blurs using iterative algorithms for both blur identification and image restoration, *Image Vis. Comput.*, 15, 399–410, 1997.
47. You, Y.-L., and Kaveh, M., Blind image restoration by anisotropic regularization, *IEEE Trans. Image Process.*, 8(3), 396–407, 1999.
48. Cristobal, G., and Navarro, R., Blind and adaptive image restoration in the framework of a multiscale Gabor representation. In *Proceedings IEEE Time-Frequency Time-scale analysis*, 1994.
49. Flusser, J., Boldyš, J., and Zitová, B., Invariants to convolution in arbitrary dimensions, *J. Math. Imaging Vis.*, 13, 101–113, 2000.
50. Zitová, B., and Flusser, J. IMARE — image registration toolbox for MATLAB. Freeware, http://www.utia.cas.cz/user_data/zitova/IMARE.htm.

15 Gaze-Contingent Multimodality Displays for Visual Information Fusion: Systems and Applications

Stavri Nikolov, Michael Jones, Iain Gilchrist, David Bull, and Nishan Canagarajah

CONTENTS

I. INTRODUCTION

In recent decades, various display techniques have been developed and used to present an observer or interpreter with multiple images of the same object or scene, coming from different instruments, sensors, or modalities. Choosing an appropriate display technique determines, to a large extent, whether and how the human observer will be able to combine or fuse this incoming visual information. Traditional image fusion systems align or register the input images, and then, using some mathematical algorithm, combine the image information to produce a single fused image. This fused image is usually used for human viewing or interpretation, or in rarer cases is subjected to further computer processing.

An alternative approach to multisensor or multimodality image fusion is to use the input images, or some of their features, to construct an integrated display, and then leave the human visual system to fuse this information. Thus, the visualization and perception of the images become important parts of the fusion process. By optimizing the display and tailoring the visual information that is presented, it is often possible to improve the perception of this information and to enhance the understanding and performance of the observer. It is exactly this kind of approach that has been used by the authors to construct a special kind of integrated display, called a gaze-contingent multimodality display (GCMMD). The way such display systems can be built and their application in a number of areas, such as medical image fusion, remote sensing fusion, and perception of multilayered maps, are all discussed in this chapter.

The chapter is structured as follows. In Sections II and Section III, a short survey of multimodality and focus + context displays (FCDs) is given. The next two sections discuss how eye-trackers in general (Section IV), and gaze-contingent displays (GCDs) in particular (Section V), have been used by psychologists in the past in different applications. Section VI presents a brief overview of gaze-contingent multiresolution displays (GCMRDs) and how they can be used for image and video compression, and for foveated imaging. The following Section VII presents a detailed overview of our work on GCMMDs including a number of different applications such as two-dimensional image fusion, three-dimensional volumetric image fusion and multilayered map perception. Finally, in Section VIII, we discuss implementation and performance issues for several different GCMMDs, and also how various eye-trackers can be integrated in such systems.

II. MULTIMODALITY DISPLAYS

"Multimodality" is a term used, within the literature, with slightly different meanings. Within psychology and other human-centered sciences, it is often taken to mean the use of multiple [human] senses (such as vision, hearing or touch) to perceive, or interact with, data. Thus, in this context, "multimodality display" means presentation of data for perception via multiple senses. Within other fields such as medical imaging, remote sensing and surveillance, "multimodality" usually means the acquisition of information, usually from the same object, region or scene, using multiple imaging devices [modalities], for example, ultrasound and magnetic resonance images in a medical context, or visible and infrared images in a surveillance application. Here, we use this second meaning of multimodality.

Amongst the application areas in which multimodality imaging is common, medical imaging is one of the most important. Some natural pairings of modalities exist. For example, soft tissue is well imaged, owing to its high water content, using magnetic resonance imaging (MRI), whereas the high radiographic density of bone, tooth enamel and other hard materials may be imaged better using x-rays (the three-dimensional form being imaged via computed tomography [CT]). Another useful pairing is MRI and ultrasound imaging. The use of MRI intraoperatively requires specialized and costly surgical tools that are compatible with high magnetic fields. Ultrasound can be used without such adaptations. By exploiting both, high-detail pre-operative MRI scans can be registered with real-time, intraoperative ultrasound; even though soft tissue may change shape during surgery, multimodality imaging allows the utilization of MRI data to take into account any change in soft tissue shape throughout the operation.

Some of the simplest and most commonly used multimodality image displays include an adjacent display, with or without a linked cursor, a "chessboard" display and a transparency/opacity weighted display. Adjacent displays present images in multiple windows or on multiple monitors. This is a simple method — but one of the most effective — for integrated two-dimensional display (see example in Figure 15.1, top row), especially when a linked cursor is added to the display to assist the observer in relating corresponding features in the images.[1,2] In chessboard displays, also known as alternate pixel, interleaved pixel, or split-screen displays,[1,3,4] the two input images are displayed as a "chessboard" (Figure 15.1, bottom left), where pixels from the first image occupy the "white" squares and pixels from the second one occupy the "black" squares. The observer's visual system fuses the adjacent pixels into a single representation. Some of the main problems associated with this technique are: (1) the spatial resolution of the display is effectively halved, and (2) when color is used and the

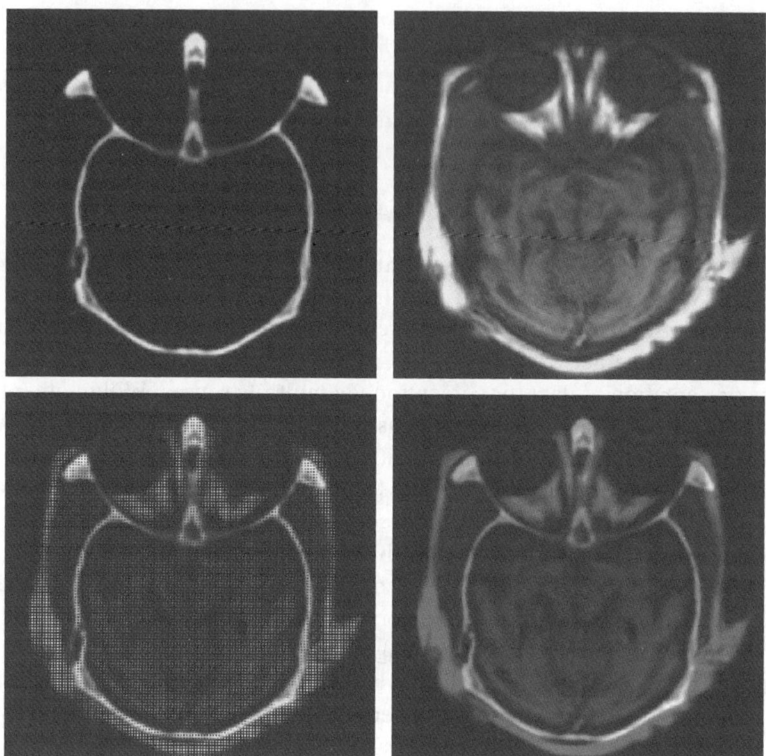

FIGURE 15.1 Traditional multimodality displays of CT (top left) and MRI (top right) images, adjacent display (top row), "chessboard" display (bottom left), and transparency weighted display: 60% CT and 40% MRI (bottom right). (Images courtesy of Oliver Rockinger, Metapix. With permission.)

square size of the chessboard is small, the relative difference in visual acuity and chromatic acuity can cause perceived colors to "bleed" beyond their true bounds. The transparency/opacity weighted display was initially proposed in Ref. 5 and has been used in many software systems in recent years. Transparency values are assigned to each pixel, which constitute the so-called "α-channel" of the image. This makes it possible to fade one image in and the other out (Figure 15.1, bottom right). As with chessboard displays, transparency/opacity weighted displays are generally used only in cases of two input images. Color weighted displays have also been described by some authors.[6]

III. FOCUS + CONTEXT DISPLAYS

FCDs are those which may be considered as including two distinct components: the "focus", which provides detail around a compact local region, and the "context", which provides a less detailed, global view. There may or may not exist a clear boundary between these two components. Focus + context visualization techniques start from three premises:[7] (1) the user needs both overview (context) and detail information (focus) simultaneously, (2) information needed in the overview can be different than that needed in the detail, and (3) focus + context information can be combined within a single (dynamic) display. Much of the display methodology may be considered analogous to the human visual system, which combines high-acuity foveal information with the less detailed information from the rest of the retina. A number of FCD techniques have been proposed in the literature, for example, Spence and Apperley's bifocal displays,[8] Furnas's fisheye views,[9] Lamping, Rao, and Pirolli's hyperbolic tress[10] (see Figure 15.2, left, for an example of how hyperbolic trees have been used in Inxight Tree Studio), Keahey's nonlinear magnification displays[11] (see Figure 15.2, right, for an example of nonlinear magnification of a map image), and Blackwell, Jansen and Marriott's restricted focus views.[12] A very good review of different visual transfer functions with a classification is given in Ref. 13. In Ref. 14 visual transfer functions are extended from two-dimensional to three-dimensional. In order to overcome occlusions in three-dimensional space, the functions are applied to the line of sight between the observer's viewpoint and the object of interest. This three-dimensional distortion viewing method has been applied only for visualization of graph structures. Other detail-in-context approaches to three-dimensional data visualization include Fairchild's SemNet[15] and Mitra's aircraft maintenance approach.[16] The main hypothesis behind all focus + context visualization methods is that it may be possible to create better cost structures of information by displaying more peripheral information at reduced detail in combination with the information in focus, dynamically varying the detail in parts of the display as the user's attention changes.

Whilst the focus + context paradigm is a common one, more powerful approaches, such as allowing a general spatial distribution of detail around the fixation point, may also be developed. However, the focus + context paradigm does match well with the idea of foveated imaging; the focus may be seen as the region whose image appears within the fovea, and the context as the region whose image appears on the surrounding regions of the retina.

In the case where the context information comes from one image modality and the context comes from another one, we can talk about multimodality focus + context displays (MMFCDs). An increasing number of visualization systems make use of such displays. An example is shown in Figure 15.3. Multimap[17] offers its users the ability to view online (on their web site) such displays where a focal region from a street map can be overlaid on top of an aerial photograph. Figure 15.4 shows three MMFCDs at different scales. How scale in such displays affects the ability of the user to combine the focus + context information is a very interesting question and one deserving investigation in the future. An initial observation, though, is that having image features (e.g., roads, junctions) in the two modalities with comparable size (Figure 15.4 bottom, and to an extent middle) aids the fusion process.

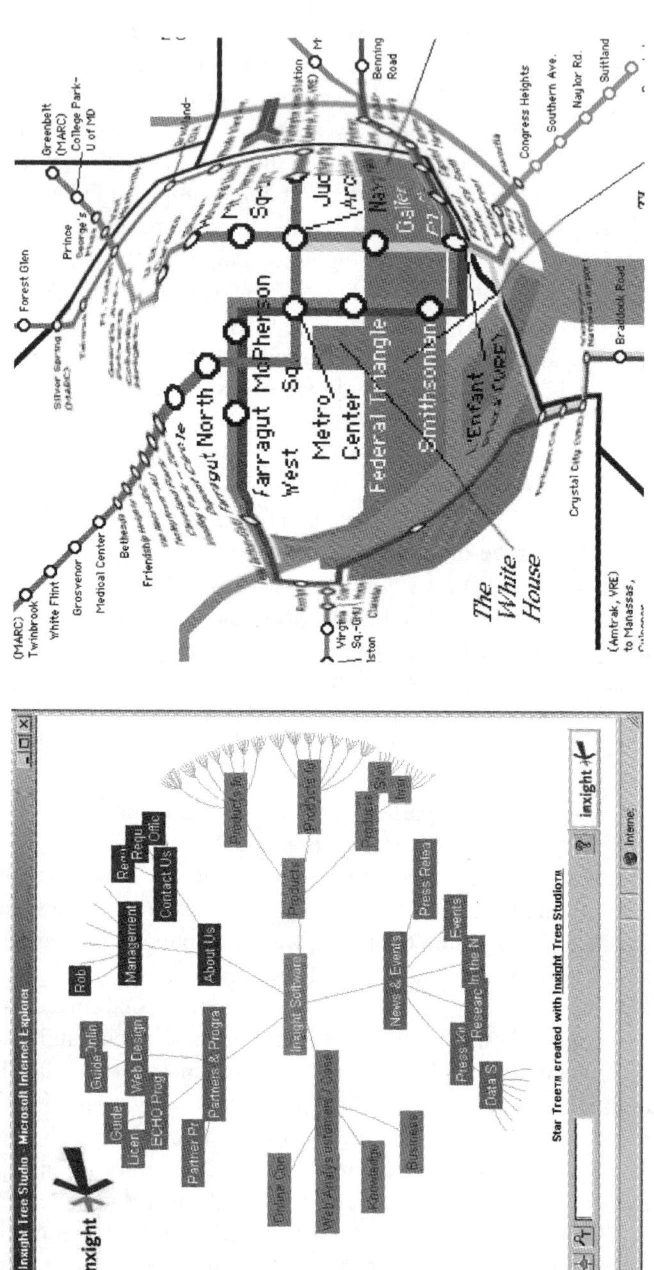

FIGURE 15.2 (See color insert following page 236) Focus + context displays: Inxight Software using hyperbolic trees (left) and a Washington subway map using nonlinear magnification (right). (Images courtesy of Inxight [left] and Alan Keahey, Visintuit [right].)

FIGURE 15.3 (See color insert) An example of a multimodality display in everyday life: a map overlaid on top of an aerial photograph. (Maps and Images courtesy of Multimap and Getmapping Plc.)

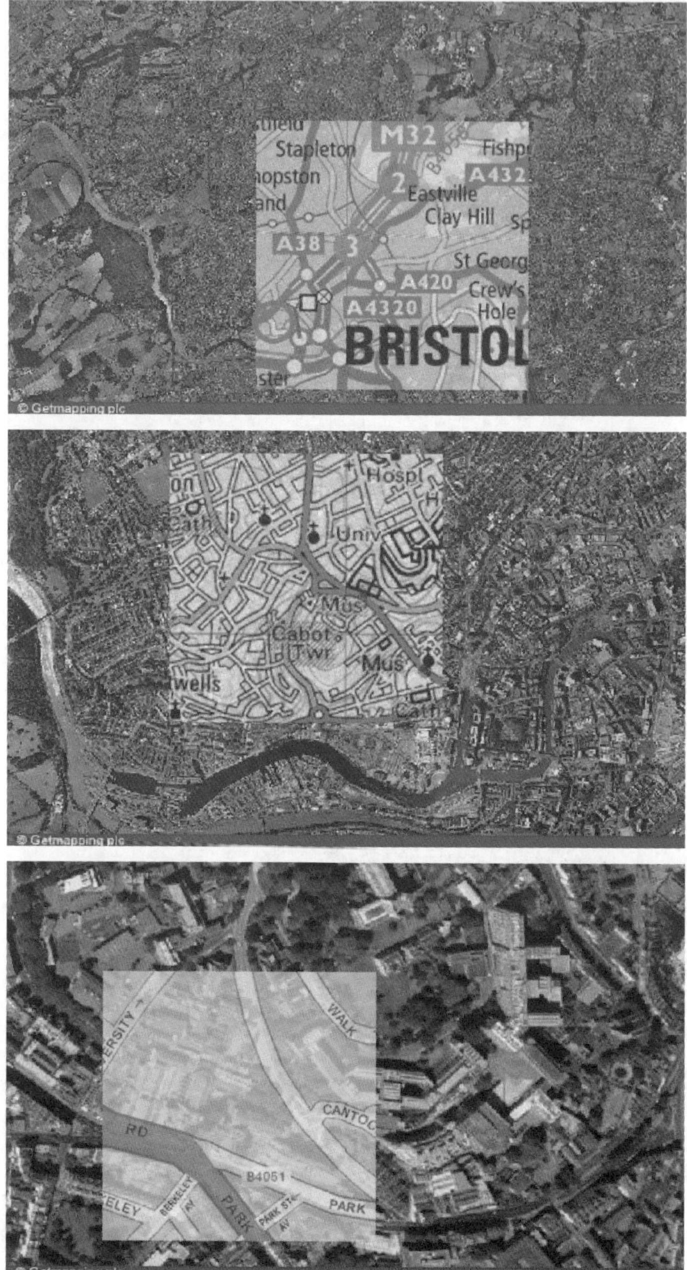

FIGURE 15.4 (See color insert) Multimodality displays of different scales (different scale maps overlaid on top of aerial photographs): 1:100000 (top), 1:25000 (middle), 1:5000 (bottom). (Maps and Images courtesy of Multimap and Getmapping Plc.)

IV. EYE-TRACKING APPLICATIONS

There are a growing number of systems and applications which use eye-tracking technology for various purposes. Here, we use the term eye-tracking in its broad sense of tracking both where the eyes are and of what they are looking at (in the literature, the term gaze-tracking is sometimes used in the second case). Eye-tracking applications according to Duchowski[18] can broadly be dichotomized from a system analysis point of view as *diagnostic* or *interactive*. In a diagnostic role, the eye-tracker provides information about the observer's visual and attentional processes. In an interactive role, the eye-tracker serves as an input device. An interactive system responds to the observer's actions and interacts with him. In this instance, this is done based on the observer's eye movements. Duchowski[18] also distinguishes between two types of interactive systems: (1) *selective*, where the point of gaze is used as a pointing device, and (2) *gaze-contingent*, where the observer's gaze is used to change the rendering of complex information displays. For a survey of eye-tracking applications see Ref. 19.

V. GAZE-CONTINGENT DISPLAYS

The human visual system can only resolve detailed information within a very small area at the center of vision. Resolution rapidly drops in the visual periphery. Effectively, at any one time our visual system processes information only from a relatively small region centered around the current fixation point. Real-time monitoring of gaze position, using various kinds of eye-tracking devices, permits the introduction of display changes that are contingent upon the spatial or temporal characteristics of eye movements. Such displays, called GCDs, have been described in numerous publications (e.g., Refs. 20–23) and have been used in various applications, for example, reading, image and scene perception, virtual reality and computer graphics, and visual search studies. An excellent survey of GCDs can be found in Ref. 24.

In its classical form, a GCD has a window, centered around the observer's fixation point, which is modified in real-time while the observer moves their eyes around the display. Here, by "modified" we mean that it is rendered differently from the rest of the display (see Figure 15.5). In fact, a GCD is a dynamic focus + context display which is synchronized with the eye movements. In Refs. 18,19 two main types of gaze-contingent applications are considered: *screen-based* and *model-based*; the main difference between the two being what is rendered on the display. In the first case, this is an image (a matrix of pixels) while in the second this is a model, for example, a polygonal model of graphics objects. Both screen-based and model-based GCDs have been studied extensively over the last 20 years. For a detailed study of model-based GCDs and level-of-detail (LOD) rendering techniques in computer graphics see Ref. 25.

gaze-contingent display (GCD)

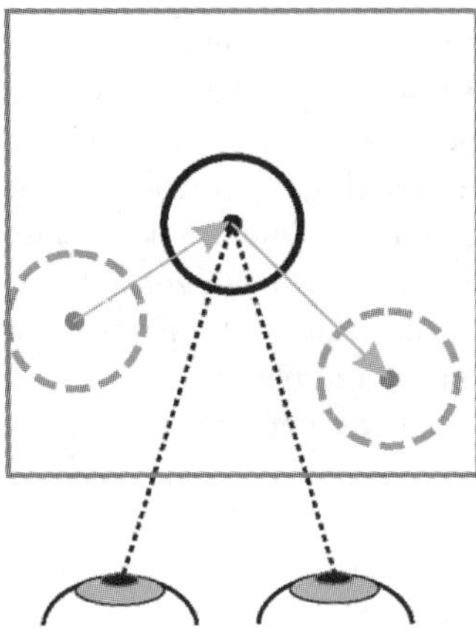

FIGURE 15.5 Gaze-contingent display: in most GCDs the window information (dark gray) is rendered differently from the background information (light gray).

VI. GAZE-CONTINGENT MULTIRESOLUTION DISPLAYS

Gaze-Contingent Multiresolution Displays (GCMRDs) are GCDs in which image resolution varies, with high-resolution information being presented at the center of vision and low-resolution information in the periphery. Alternative names under which the same idea is described are variable-resolution displays[24] or foveated displays.[26,27] The earliest and best known application of GCMRDs is in flight simulators. Some other areas where two-dimensional GCMRDs have been successfully used include virtual reality, large immersive displays, videoconferencing and tele-operation. A number of studies[28,29] have been carried out to determine display parameters (window size and border, and peripheral degradation) of both imperceptible and perceptible GCMRDs. A theoretical evaluation of variable-resolution displays and the gain in using them can be found in Ref. 24. In the same study the behavioral consequences associated with using variable-resolution displays are also reviewed in both theoretical and applied contexts. Generally, GCMRDs can greatly conserve processing time and bandwidths. Foveated displays have been successfully used for video compression by a number of researchers; see for example.[30–32] An example frame from a foveated video sequence is shown in Figure 15.6 (for more details of this technique see Ref. 32).

GCMRDs can improve the quality of the displays, especially in the case of virtual reality and immersive displays, and reduce the likelihood of simulator sickness. They can also serve as low-vision enhancement tools.[33,34]

VII. GAZE-CONTINGENT MULTIMODALITY DISPLAYS

Gaze-Contingent Multimodality Displays (GCMMDs)[35] are GCDs in which information from one image modality is presented at the center of vision while

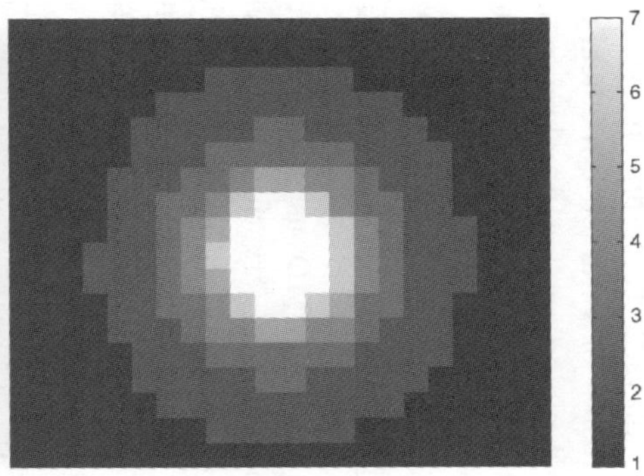

FIGURE 15.6 (See color insert) GCMRD: one frame of the Akiyo sequence (left), foveated frame (middle) with a foveation map (right). The fixation point lies on the drawn circle (middle). The 16×16 pixel blocks are grouped together in regions, with each region being filtered with a different cutoff frequency.

information from another (different) modality is presented in the periphery. Some applications of two-dimensional GCMMDs include image fusion[35] and multi-layered geographical map displays.[36] Three-dimensional GCMMDs have also been studied in Ref. 37 for fusion of volumetric medical images using a three-dimensional gaze-tracker and region-enhanced volume rendering.[38] Below we will summarize the main findings of our studies involving GCMMD in these three areas.

A. TWO-DIMENSIONAL GCMMD FOR IMAGE FUSION

In Ref. 35 we proposed the use of GCMMDs as an alternative to traditional image fusion techniques. Some of our initial observations[35] made when using GCMMDs of several different types of images, for example, medical, surveillance or remote sensing images (see the rest of this section for examples) are:

1. In most cases, when using GCMMDs we were able to fuse the information from the two input modalities but only if the display parameters were carefully set.
2. A circular gaze-contingent window provided best results.
3. When the window region was from a color image, (or pseudo-coloring was applied to the window), while the peripheral region was taken from a gray-scale image, it was found that the resulting display does not aid the fusion process but on the contrary is very distracting. Since color discrimination is poor in peripheral vision, having only gray-scale information displayed there should not be a problem. With such displays, however, the border between the (color) window and the (gray-scale) background is very noticeable and this is probably the reason why the two cannot be merged easily in a single percept.
4. Any window update delays, flicker or gaze-position estimation inaccuracies, (for example, due to poor eye-tracker resolution [temporal or special] or calibration), were found to be very distracting, even if minimal, and completely "block" the fusion process. Thus eye-trackers providing high temporal (and spatial) resolution are preferable when building GCMMDs.

A brief comparison of GCMMD with other commonly used multimodality displays, provided in Ref. 35, will be summarized here. The GCMMD is a dynamic gaze-contingent MMFCD. Unlike the adjacent display, and much like the "chessboard" and transparency weighted displays, the GCMMD display is an integrated display. When compared to the "chessboard" display, it does not have the problem of reducing the spatial resolution. In a way it is similar to "cut-and-paste" displays, which have been widely used in medical imaging, but with the main advantage of the window being continuously updated in real-time to be always around the fixation point. When compared to transparency weighted displays, GCMMD presents a localized gaze-contingent combination, instead of showing a "global" weighted combination of the two images. They also avoid the reduction in dynamic range typical of transparency weighted displays.

The main difficulty associated with GCMMDs, as with other kinds of GCDs, is the optimization of the display parameters to achieve optimal image fusion. Little is known about how exactly our visual system combines and jointly processes the foveal and peripheral information. Human visual resolution ability drops off dramatically away from the axis of fixation.[39,40] As a result, the extent to which the proposed fusion method will be successful will depend on a number of factors. First, if the window is too large, then no information can be picked up from the context. Second, if the visual content that is in the context is finely scaled then this information will not be easily processed by peripheral vision and may give no additional benefit when compared to a blank background. Importantly, however, visual acuity is dramatically affected in peripheral vision by the presence of additional information.[40] If the amount of peripheral information is reduced and, only that information that is task relevant is presented, this should give a real performance advantage. We are in the process of carrying out several experiments to try to measure quantitatively the visual span in image fusion for various tasks, to analyze observers' scanpaths and to visually search for different multimodality displays. Three different applications of GCMMD for visual information fusion in the areas of medical imaging, surveillance and remote sensing are described below.

1. Medical Images

In management of patients undergoing radiotherapy or skull based surgery, MRI and CT provide complementary information, that is, normal and pathological soft tissues are better visualized by MRI, while bone structures are better depicted by CT. In Figure 15.1 (top row), a registered pair of MRI and CT images are displayed. A GCMMD, with the circular window at two different gaze positions, is shown in Figure 15.7. This new approach allows the observer to continuously investigate the soft tissue from the MRI image in the central part of vision, while being presented with the skull structure as context in peripheral vision. Other medical imaging modalities may also benefit from our approach; for example, the low-resolution functional information provided by Positron Emission Tomography (PET) or Single Photon Emission Computed Tomography (SPECT) scanning is best interpreted with reference to spatial information provided by a modality such as MRI. The GCMMD setup can be compared with the "chessboard" display and transparency weighted display in Figure 15.1 (bottom row). A three-dimensional GCMMD of volumetric medical data is described in Section VII.C.

2. Surveillance Images

Example surveillance infrared (IR) and visible images from the "UN camp" sequence[41] are shown in Figure 15.8 (top row). Two GCMMDs of these

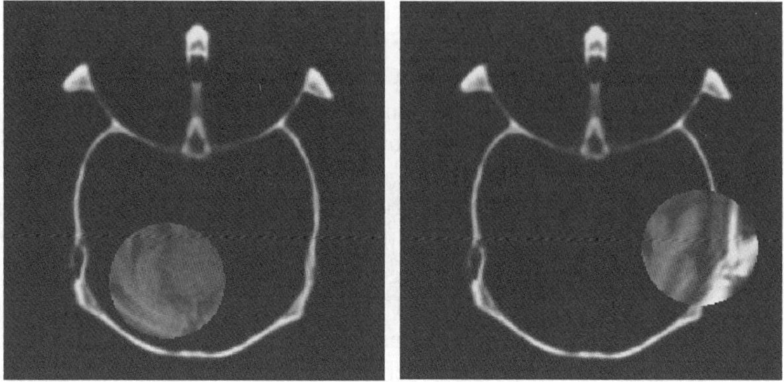

FIGURE 15.7 GCMMD (two different gaze positions) of the CT and MRI images in Figure 15.1. Image size: 256 × 256 pixels; circular window (diameter = 80 pixels).

images, at different gaze positions, are also given in Figure 15.8 (bottom row). Here, a GCMMD provides the observer with the ability to view the thermal intensity information in the IR image in the context of the visible image. While the GCMMD of the gray-scale images allows easy fusion of the window and context (Figure 15.8, top right), the GCMMD of their pseudo-colored versions does not lead to successful fusion, possibly because of the bright and different (red and green) colors used. Another set of surveillance IR and visible images from Octec, demonstrate the effect of using different window sizes and borders (see Figure 15.9). Here, in both GCMMDs, a centered Gaussian function is used in both GCMMDs for the GCD mask, but the window has different sizes. If a smaller window is used (Figure 15.9, right), the user can generally fuse the information better than if a larger window is used (Figure 15.9, left).

3. Remote Sensing Images

Image fusion of multisensor data provides a way to present, in a single image, information from multiple spatial resolutions and radiometric bands. The different sensors usually provide complementary information about the area under investigation, which may be fused for the purpose of better visual perception or further image processing, e.g. classification. Examples of remote sensing image fusion include fusion of: (1) Landsat Thematic Mapper (TM) and Advanced Very High Resolution Radiometer (AVHRR) data, (2) Landsat TM and SPOT data, and (3) Landsat TM and Synthetic Aperture Radar (SAR) data. In most cases, fused results combine the spectral information of a coarse-resolution image with the spatial resolution of a finer image. A GCMMD of Landsat and SPOT images was shown in Ref. 35, where, similar to the GCMMD of medical images, the high-resolution SPOT

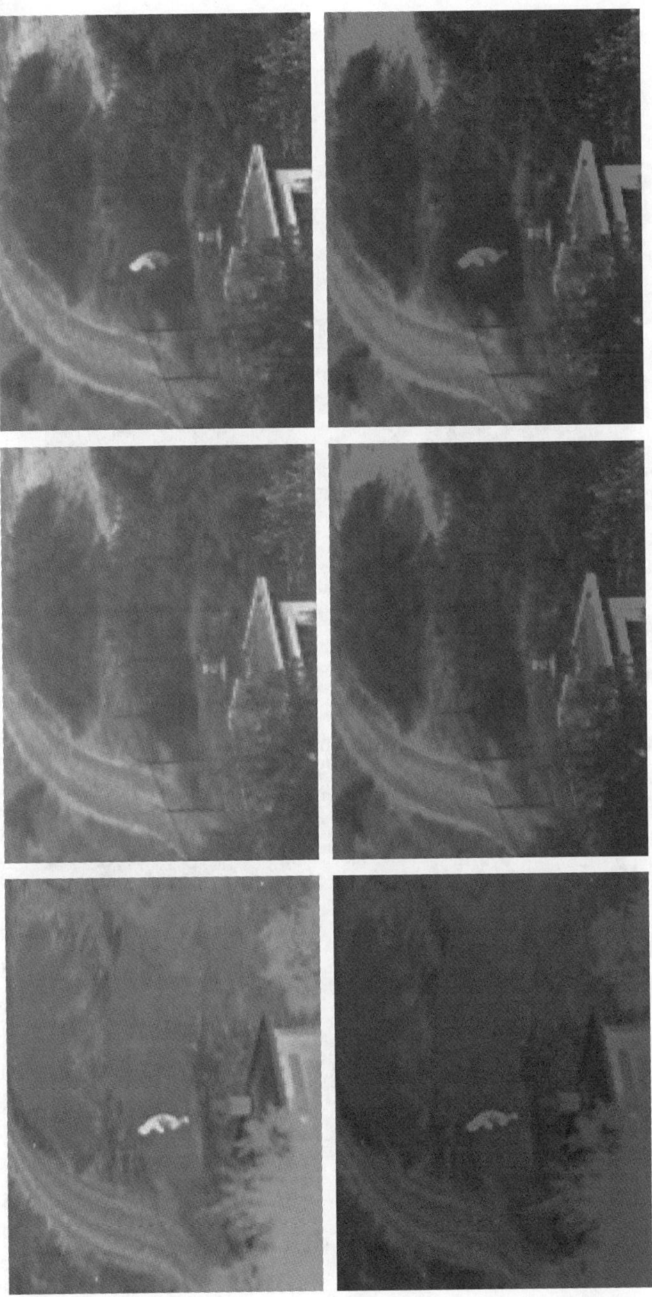

FIGURE 15.8 **(See color insert)** Original surveillance infrared (top left) and visible (top middle) images. The same images pseudo-colored in red (infrared image, bottom left) and green (visible image, bottom middle). Two GCMMDs of these respective images (right column). (Images courtesy of Alexander Toet, TNO Human Factors Institute, The Netherlands.)

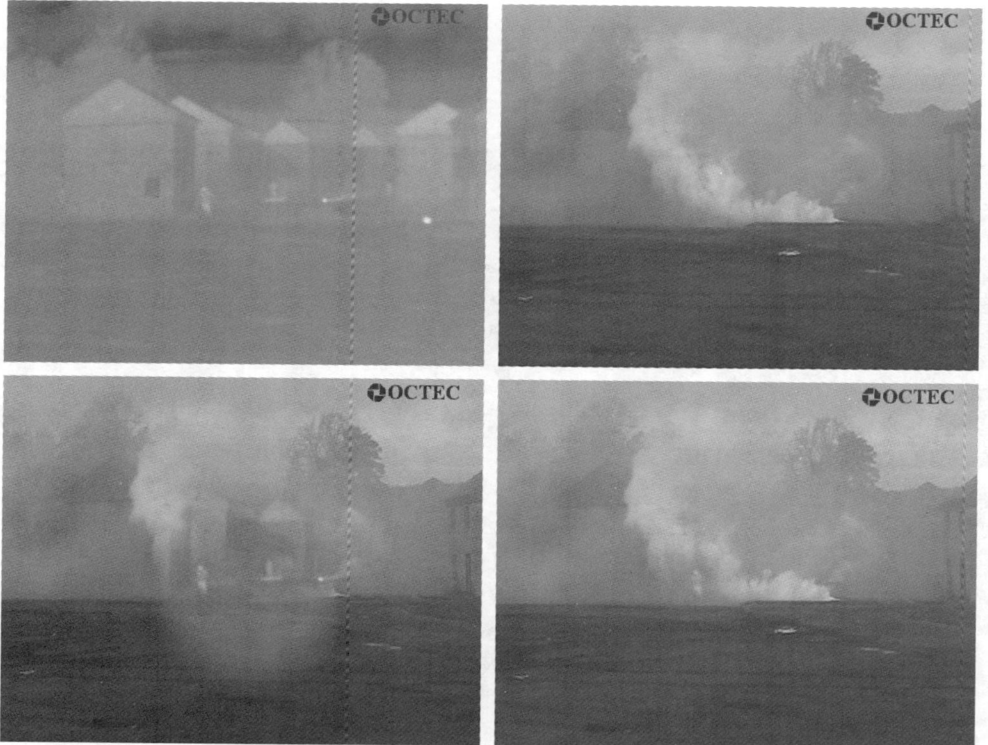

FIGURE 15.9 **(See color insert)** Original surveillance infrared (top left) and visible (top right) images. Two GCMMDs (different window sizes) of these images (bottom row). (Images courtesy of Octec Ltd., UK.)

panchromatic image provides the gaze-contingent window information, while the coarse-resolution natural color Landsat image is seen in peripheral vision. Thus, the viewer can visually examine the SPOT image in the context of the Landsat image.

B.　GCMMD of Multilayered Maps

In Ref. 36 we extended the idea of using GCMMD for visual information fusion to multilayered geographical maps. An example below (Figure 15.10) shows a GCMMD based on an aerial photograph of Bristol and the corresponding street map. This is a multimodality display of a remote sensing image and a geographical map. Here the observer can view the street map in the context of the aerial photograph. Such GCMMDs can be easily constructed using, for example, web services such as Multimap[17] (see also Figure 15.3) or the online atlas provided by Environment Australia.[42] Their web site contains multilayered geographical maps and allows users to construct, view and print multilayered maps of Australia's coastal zones. Another example of a GCMMD is a multilayered road map of Benton County (Washington) shown in Figure 15.11.

Multilayered information representations are increasingly favored in digital map production. Having such representations, different map layers can be "switched on" or "off" depending on the task at hand, and thus determining the visual complexity of the map display. In a more general sense, any kind of multilayered visual information map can be used to construct a GCMMD. In Figure 15.12 the scheme of a multi-dimensional geographical map is shown. One could construct a GCMMD by creating "foreground" and "background" bitmap images (FI and BI). By careful selection of the information to be presented in them, various map viewing and reading experiments can be designed. More information about the implementation of such displays is given in Section VIII.A.1.

C.　Three-Dimensional Volumetric GCMMDs

Two components must be integrated in order to produce a three-dimensional volumetric GCMMD: a three-dimensional gaze tracking subsystem, and a three-dimensional rendering algorithm that can implement either a focus + context paradigm, or some other form of spatially variant rendering.

1.　Challenges of Three-Dimensional Gaze-Tracking

Three-dimensional gaze-tracking warrants some description. Three-dimensional data may be represented on flat display devices, such as computer monitors. For the static case, and remembering that we here consider volumetric data within

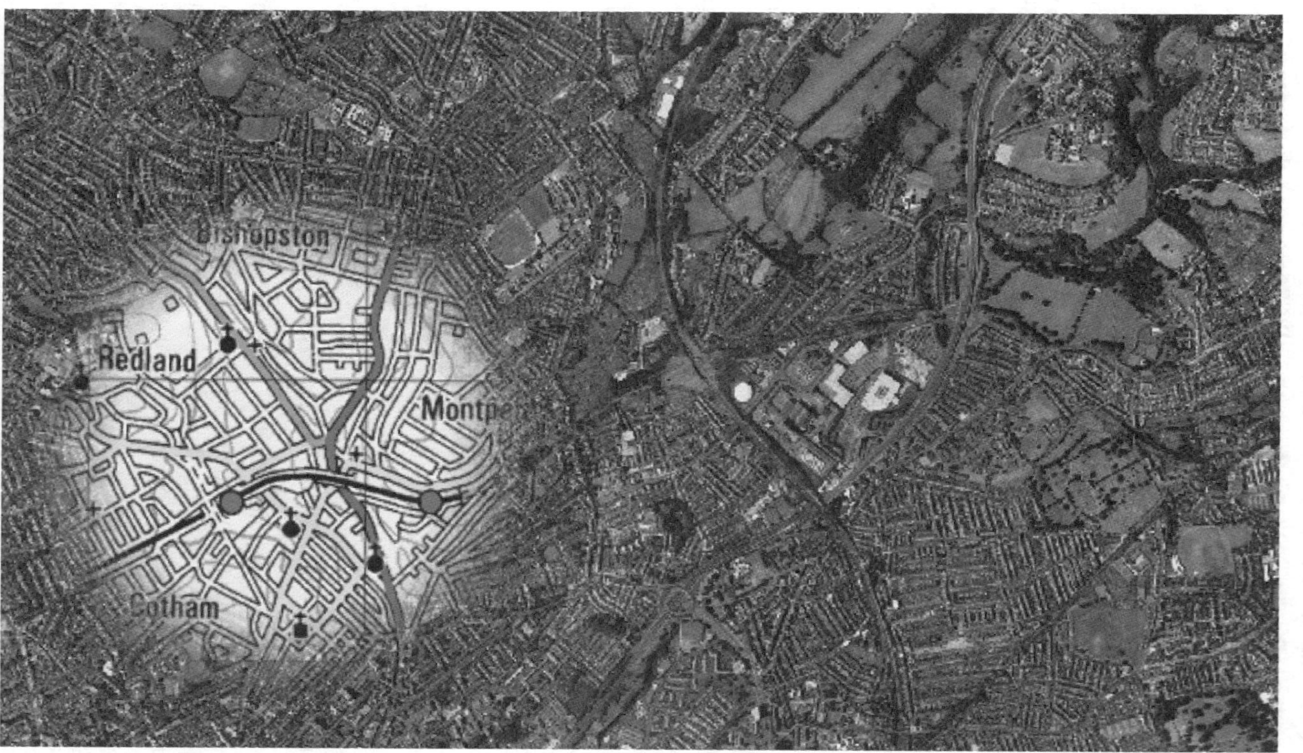

FIGURE 15.10 (See color insert) A GCMMD using the new system: an aerial photograph of Bristol, UK, fused with an ordnance survey map.

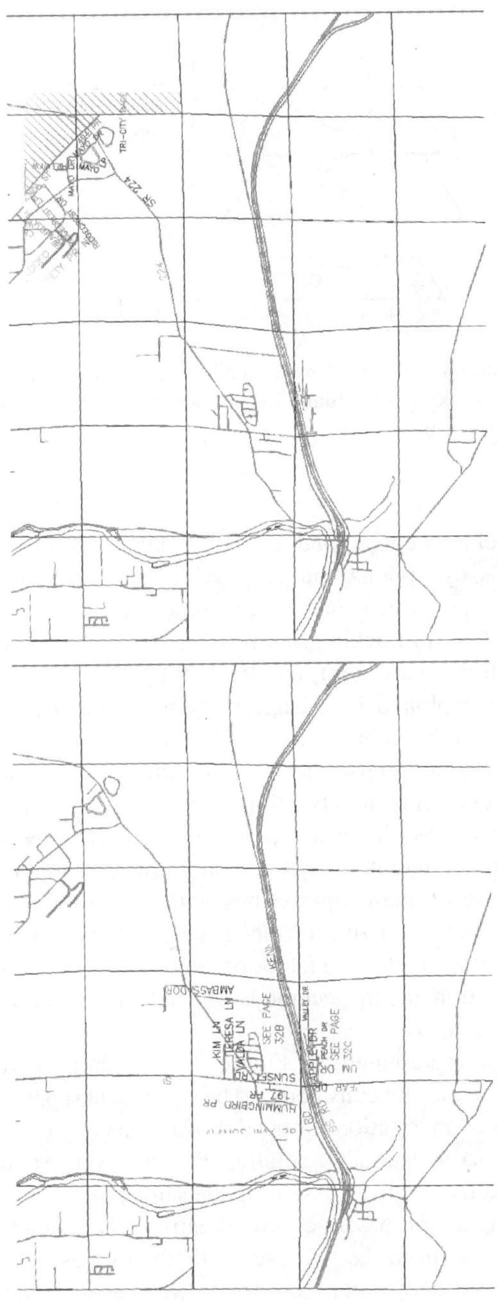

FIGURE 15.11 (See color insert) GCMMDs of a multilayered road map of Benton County (Washington).

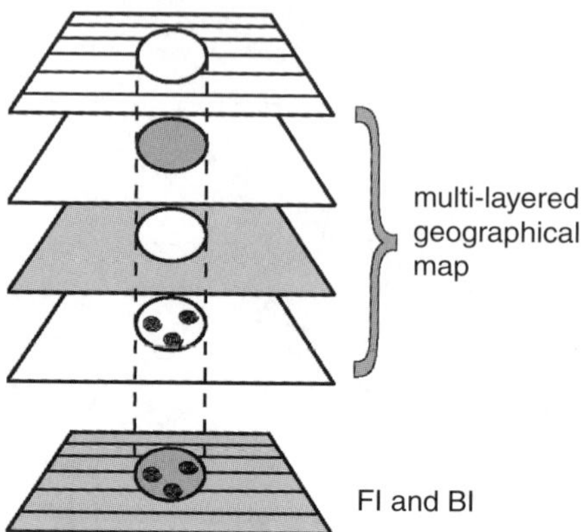

FIGURE 15.12 Construction of the "foreground" image (to provide information inside the window) and the "background" image (to provide information outside the window) from the multiple layers of the map.

which a display pixel may correspond to a continuum of depths, the direction in which an observer's eye points cannot provide a three-dimensional fixation position. Two solutions, which may be combined, exist. Firstly, stereoscopic display may be used. With tracking of two eyes, each of which sees an image rendered from a different viewpoint, the depth-dependent disparity between the two images may be exploited by triangulating three-dimensions gaze vectors from each eye. Secondly, motion may be used. By direct analogy to the perception of depth via motion parallax[43] and kinetic depth,[44] fixation depth may be inferred from eye movements. Since this may be achieved without stereoscopic display, a three-dimensional volumetric GCD may be implemented on a standard computer monitor (with the addition of suitable gaze-tracking technology). The first of these approaches suffers from the relatively short distance between the eyes relative to the distance of the point being fixated. Within typical display geometries, a factor of ten or greater may be involved and this causes poor fixation depth accuracy. The advantage of motion may be illustrated *via* two scenarios.

In one scenario, represented in Figure 15.13, an observer watching a smoothly rotating cube may be considered to be equivalent (approximately — the ocular motor system compensation for head movements means that these are, in practice, not identical) to smoothly rotating the observer around a stationary cube. Over the period from t_1 to t_2, assuming the same point in the cube is fixated throughout, the triangulation baseline may effectively be extended by selecting appropriate gaze vectors from the two eyes at distinct times.

In a second scenario, shown in Figure 15.14, when an observer fixates a point in the rotating cube, the rate of eye movement is (approximately, under perspective projection) proportional to the distance of the point from the axis of

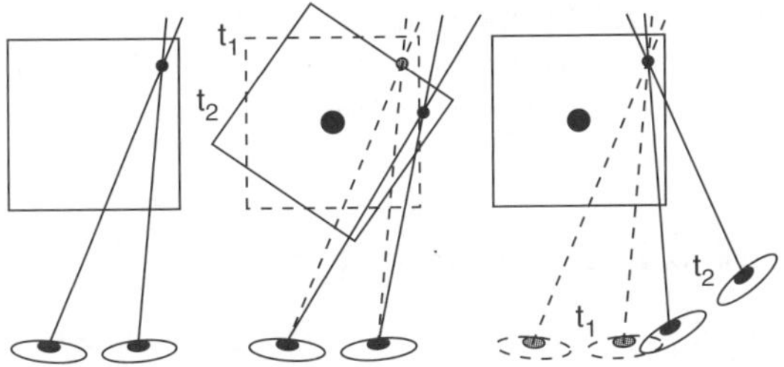

FIGURE 15.13 Triangulation based on the observer fixating the same point within an object as it rotates over time.

rotation and changes sign, dependent upon whether the point is in front of, or behind, the axis of rotation. The rate of eye movement may thus be used to infer relative depth.

Three-dimensional gaze-tracking however, remains a challenging task. Some past work has been performed looking at three-dimensional gaze-tracking interfaces,[45,46] and for utilizing the fixation depth to compensate for problems caused when the eyes try to accommodate vergence/focus discrepancies in stereoscopic displays.[47] Gaze-tracking and volume rendering have previously been combined, but this has used two-dimensional, not three-dimensional, gaze-tracking.[48] When triangulation is used, small errors in estimated gaze angles commonly manifest as large errors in axial fixation position estimates. We do not detail here the calibration process that we have employed, but note that this is important if accurate fixation estimates are to be obtained.

As well as the technical difficulties, there are also physiological issues related to how the eyes move. When changing fixation, from point A to point B in three-dimensional space, the usual pattern is a saccade followed by a vergence movement.[49] The saccade results in the eyes triangulating upon a point the same distance away as A, but along the line of sight of B. The vergence movement then

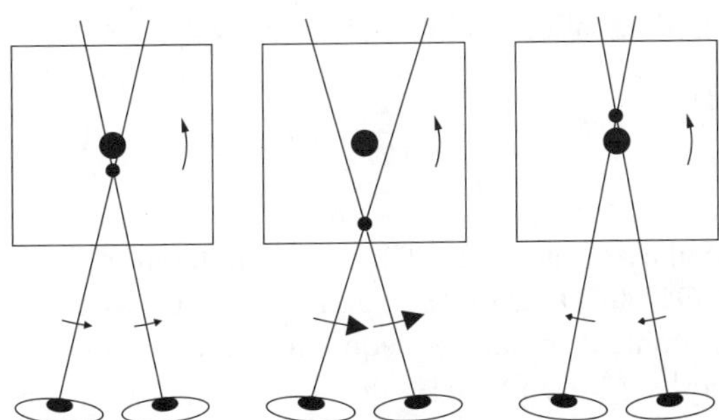

FIGURE 15.14 Fixation depth relative to the axis of rotation estimated from the rate and direction of eye movements.

brings the eyes to triangulate upon B, taking perhaps a substantial part of a second. The result is that whilst lateral target position estimation may be quickly established, the axial component takes longer.

2.   Three-Dimensional Spatially Variant Rendering

There is a wide variety of methods for rendering volumetric data, including multiplanar slicing, surface fitting and direct volume rendering (DVR).[50–52] Each produces opaque or, often, partially opaque representations of structures within a volume. For each, a multimodality form can be developed in which spatially variant rendering is applied, allowing one modality to be examined in the spatial context of another.

In DVR, raw volumetric data is fed directly to a rendering engine without use of an intermediate representation (such as polygonal models of surfaces within the volume). Rays are projected from the viewpoint, through each pixel and through the data volume. Samples are taken within the data at [usually] regular positions along each ray. A method is needed to produce a single pixel color and intensity from the samples on a single ray. One of the simplest methods is to take the maximum value, which results in a maximum intensity projection (MIP); integrated and average intensity projections also exist (IIP and AIP, respectively). More realistic representations may be produced, however, by using varying opacity, distributed throughout the data volume.

Opacity-based rendering applies a transfer function (TF) to the data set. Essentially, this TF specifies which regions of the data will be visible; and the degree to which those that are visible, are opaque. A similar function may map the raw data to an artificial color representation; conversely, if the original data are already coloring, they may be used directly. Several methods exist to take these voxel opacity and color sets and render an image from them. We consider here the front-to-back compositing method.

In front-to-back compositing, the opacity and color data sets are sampled along each ray to give sample sets $\{\alpha_i\}$ and $\{c_i\}$; successive samples from these are combined to give the final pixel color. Beginning from the front-most sample, the following paired equations are applied iteratively

$$C' = C + c_i \alpha_i T \qquad T' = T(1 - \alpha_i) \qquad\qquad (15.1)$$

where C is replaced by C' (and T by T') after each iteration. To develop a form of this suitable for GCMMD, with $\{\alpha_i\}$ and $\{c_i\}$ taken to be opacity and color samples for the context modality, equivalents are produced for the focus modality, denoted by $\{\beta_i\}$ and $\{d_i\}$.

A spatially variant function ω is then introduced which acts to blend between the two modalities: in regions where $\omega = 1$, only the focus modality contributes to the rendered image; in regions where $\omega = 0$, only the context modality contributes. This may be achieved via the following iterative paired equations

$$C' = C + [c_i \bar{\omega}_i \alpha_i + d_i \omega_i \beta_i]T \quad T' = T(1 - [\bar{\omega}_i \alpha_i + \omega_i \beta_i]) \qquad (15.2)$$

where the shorthand notation, $\bar{\omega}_i = 1 - \omega_i$, is adopted.

A method of calculating ω is needed. Many spatially variant forms could be applied. For example, if $\vec{P}_i$ is taken to be vector holding the position of sample i, then the distance of this from the center of the focus region, $\vec{C}$ can be calculated as $z_i = |\vec{P}_i - \vec{C}|$; and this compared with the radius of the focus region, r_0, to calculate ω_i, the value of ω at this sample position

$$\omega_i = \begin{cases} 1 & z_i < r_0 \\ 0 & \text{otherwise} \end{cases} \qquad (15.3)$$

In order to implement a transparency or opacity weighted display which allows a portion of the focus + context modalities to contribute (and thus be seen) throughout the volume, it would be necessary to modify this equation, so that ω is constrained to a range $[a, b]$, $0 < a < b < 1$, rather than the $[0,1]$ range of Equation 15.3.

Alternative forms for ω may be developed. In Ref. 53, ω is a function of two variables, r and a. We define $\overrightarrow{VC}$, to be the vector from the viewpoint to the center of the focus region. r, the radial distance, is then the distance from the sample to the closest point on $\overrightarrow{VC}$; taking $\vec{R}$ to be this closest point, a, the axial distance, is the distance from $\vec{R}$ to $\vec{C}$. This function for ω allows a to be considered analogous to focal depth, with r analogous to angular deviation from the fixation point.

3. Three-Dimensional GCMMDs of Medical Images

Figure 15.15 shows a typical multimodality image rendered within the volumetric GCMMD. It is seen that the bone structure is well represented (*via* the CT data) and can provide, in both the anatomical and visual senses, a skeleton within which the color histological soft tissue structure may be interpreted.

Figure 15.16 shows the effect of swapping the two modalities. Additionally, the context modality (here the color histology data) has been rendered in a sparse fashion. A threshold has been introduced and if, at any point, the contribution that a sample would make to the final pixel intensity is below this threshold, then it is ignored. This is implemented by comparing the $\omega_i \beta_i$ term in the calculation of C' in Equation 15.2 with the threshold, s. If it is less than this threshold, then $\omega_i \beta_i$ is replaced with zero (the calculation of T' is unaffected). Essentially, this allows the histological data to be simplified; only the strongest features are preserved. Thus, the color histological data

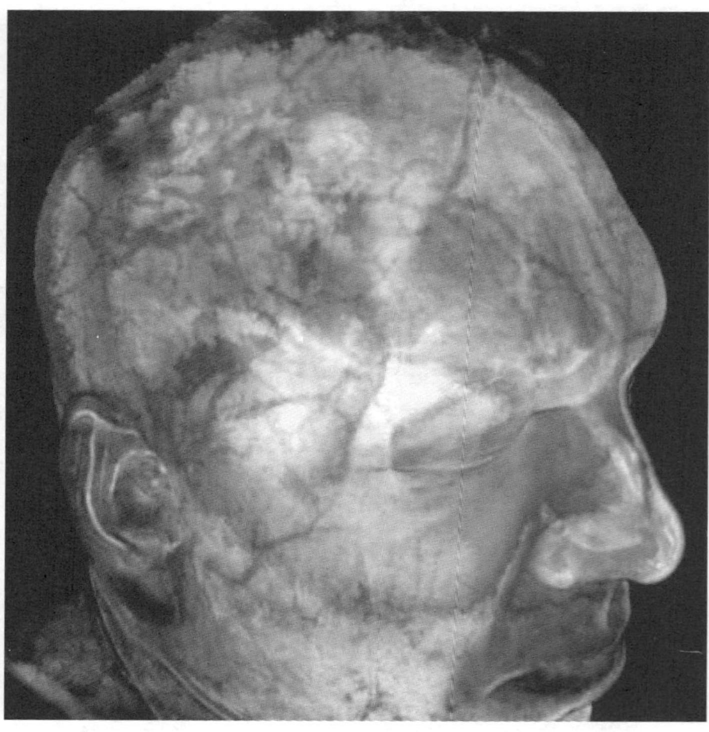

FIGURE 15.15 (See color insert) Three-dimensional GCMMD of CT (context) and color histological (window) images from the Visible Human Project. (Data courtesy of the National Library of Medicine, USA.)

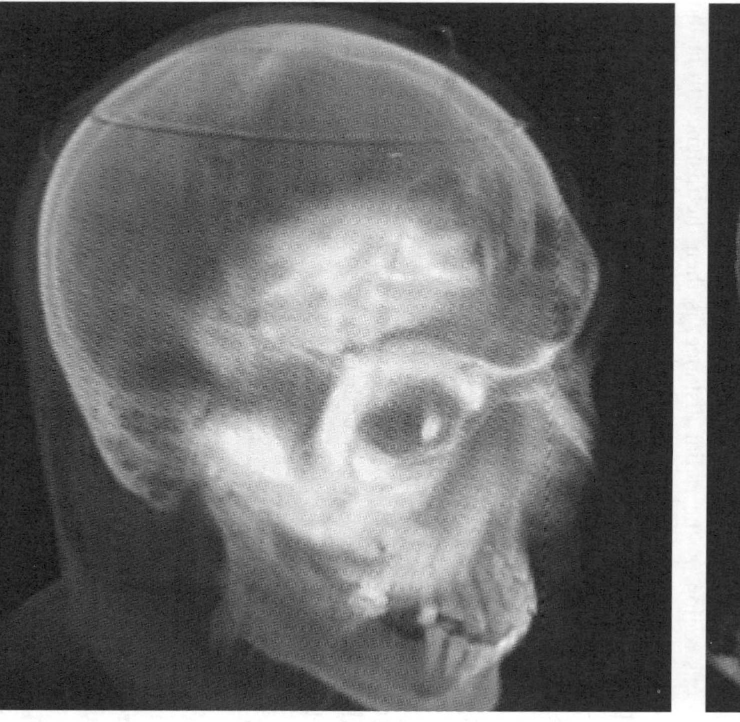

FIGURE 15.16 (See color insert) Three-dimensional GCMMD of color histological (context) and CT (window) images from the Visible Human Project. A sparse rendering method has been applied to the context data. (Data courtesy of the National Library of Medicine, USA.)

fulfils the same role as the CT data did in Figure 15.15: it provides a spatial context, but does not overwhelm the image.

4. Focus–Context Boundary in Three-Dimensional GCMMDs

Within two-dimensional focus + context GCDs the form of the boundary between the focus and context may not be critical.[28,29] Within three-dimensional GCDs, however, the boundary itself has a three-dimensional form. Whilst it may be inferred from the two-dimensional case that the form of the projected boundary silhouette should not be critical, no such inference may be drawn for that part of the boundary which is seen face-on, through which the observer sees the focus region. If a hard boundary is used, then small fluctuations in the axial component of the fixation position estimate may cause features from both modalities to flicker abruptly in and out of view. For this reason, a blended boundary is preferred.

VIII. GCMMD IMPLEMENTATION AND PERFORMANCE

A. Two-Dimensional GCMMD

1. Implementation

Several hardware and software GCD implementations have been proposed in the past. For a more thorough review see Refs. 24,25. In Ref. 55 an ATVista graphics card was used; this has the ability to mix an internally stored and an externally generated image into a composite image. By programming the internal image to be moveable and partly transparent, a GCD can be created. In a later implementation van Diepen[56] used a custom-built video switcher and three synchronized video boards. Perry and Geisler used Laplacian pyramids to create variable-resolution displays in Refs. 30,57. A number of researchers[23,58,59] describing experiments with the EyeLink I and II systems,[60] have used the GCD software implementation that was developed by SR Research. It uses windows with different sizes and shapes and performs differential updates. No control over the border of the window is available. Our initial software GCD implementation, using an EyeLink I system which extends this technique to various window shapes and borders by using masking, is described in detail in Ref. 35.

Recently we designed a new GCD implementation using texture mapping and OpenGL.[54] Perhaps the closest implementation to the one described in this paper is the system designed by Watson,[61] which uses texture mapping but no eye-tracking. In Ref. 61 texturing was accomplished in real time with the *fbsubtexload* command and *FAST_DEFINE* on a Reality Engine SGI computer. Our new OpenGL GCD implementation is described in detail in Refs. 54,62. The main components of this system are presented briefly below.

The system we have developed is a new type of screen-based GCD using texture mapping and OpenGL. We had a number of key objectives when designing this system.[54] We wanted this system: (1) to be platform independent, that is, to run on different computers and under different operating systems, (2) to be eye-tracker independent, that is, not to be part of the eye-tracker system but rather to be more like a real-time focus + context display that can be easily integrated with any eye-tracker, (3) to be flexible, that is, to provide for straightforward modification of the main GCD parameters, including size and shape of the window and its border, (4) to provide also the ability to perform local real-time image analysis of the GCD window, and (5) to implement a GCMMD. Specific information about how our new two-dimensional GCD system has been integrated with two commercial eye-trackers can be found in Section VIII.C.

The new GCD system is based on a simple two-polygon OpenGL display; one fixed full screen quadrangle for the background and a smaller overlaid quadrangle for the moving window. The moving window polygon is bound to a set of input coordinates, which come from the eye-tracker. From the input coordinates, the vertex coordinates and texture coordinates of the moving window are calculated. The geometry of the GCD is shown in Figure 15.17 (left), where O_x and O_y are the coordinates of the window center, W_m is the maximum dimension of the image, W and H are the image width and height, and b_x and b_y are the horizontal and vertical distances between the centered image border and the edge of the blank texture.

Three textures are used: a foreground texture, a background texture, and an alpha mask texture. Each dimension of an OpenGL texture must be a power of two (this is standard OpenGL; an extension exists which does not have this restriction), though not necessarily square. This means, for example, that to fit in a full 1024×768 graphic as an input modality, a 1024×1024 texture must be used. The alpha texture is a *RGBA*8 texture created by repacking an 8 bit grayscale image as 32 bit bitmap. This format was chosen because it is supported in almost

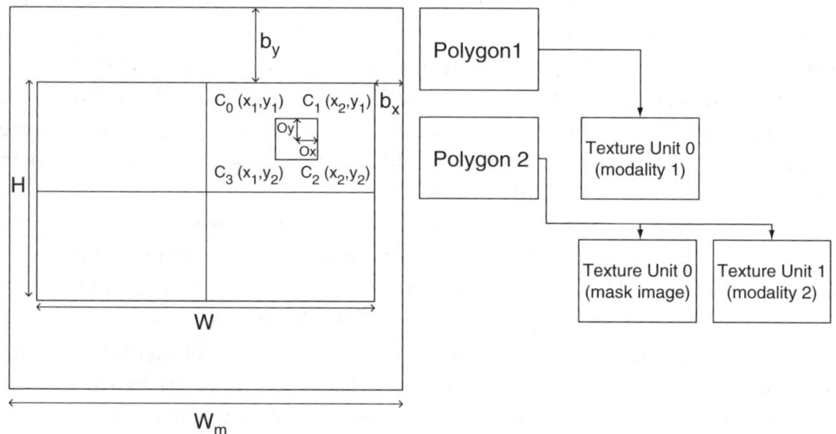

FIGURE 15.17 The new GCD system geometry (left) and program flow (right).

all OpenGL implementations. However, the foreground and background images can be set to any of the *RGB */RGBA ** family of textures. Thus the system is not restricted to grayscale images; color images can be used in the same manner. The background texture was mapped on to the fixed polygon then the moving window polygon was texture mapped, first with the alpha mask then the foreground texture (see Figure 15.17, right). Since Version 1.3, OpenGL has supported multi-texturing, in which multiple texture units may be utilized during a single pass, allowing more efficient rendering. Two different blending phases are used: (1) blending between texture units to produce a texture fragment, and (2) blending between this texture fragment and the current contents of the frame-buffer. The blending between texture units is governed by the *glTexEnvi* command and the method of texture application chosen. The alpha texture was painted on using *GL_ADD*. The use of a blank texture results in the alpha channel being added to the polygon. The foreground texture was modulated using the alpha channel information. For blending between polygons OpenGL requires the *glBlendFunc* source and destination factors to be defined. In our GCD the factors were set to *GL_ONE_MINUS_SRC_ALPHA* for the source and *GL_SRC_ALPHA* for the destination. The size of the GCD window is dependant on the offsets O_x and O_y in Figure 15.17 and can be changed at every frame by passing a new value of *BoundCalc to DrawGLScene*. The boundary and the shape of the window are dependant solely on the alpha mask. Since the mask is just an array of pixels it can be generated in any way the user chooses, for example, the further the pixels are from the center the darker they become (in other words, the more transparent the window is). It is possible to repack the images into textures on the fly while the GCD is operating and on a sufficiently fast computer there is little noticeable lag.

For our experiments with multilayered geographical maps (see Section VII.B), we used maps stored in the drawing web format (DWF). DWF is an open vector file format developed by Autodesk for the transfer of drawings over networks. In order to view DWF files one can use a standard web browser plug-in, such as Autodesk's WHIP[63] (this is the viewer we have used) or Volo View.[64] Then, inside the browser window it is possible to pan and zoom in or out, and most importantly for our studies, toggle different map layers "on" or "off". In order to use the GCMMD system described above, "foreground" and "background" images are generated [shown] in WHIP and stored as bitmap files, each one containing information from a number of layers of the DWF map (see Figure 15.12). These bitmap images are then used to construct a GCMMD using texture mapping. There is no need for spatial registration of the input images, since identical WHIP viewing parameters (position and magnification) are used for each.

2. Performance

The total time to make a GCD change is a function of: (1) the eye-tracker sampling rate, (2) the saccade detection delay, (3) the software drawing routine speed, and (4) the refresh rate of the monitor. The OpenGL two-dimensional GCD system we have designed tries to optimize the performance of (3) while also providing substantial flexibility in changing the GCD window parameters depending on the particular application. All other variables are very much hardware specific, that is, dependent on the eye-tracker (1) and (2), and the monitor (4), being used. Some performance issues related to particular eye-trackers are discussed in Section VIII.C.

Using a frame rate counter taken from Ref. 65 the new system has been tested on a number of different computers. It performs at or above the critical flicker frequency (CFF) on all systems tested. The performance varied between 60 and 92 frames per second (FPS). On Nvidia Geforce 2 and Nvidia Geforce 4 cards[66] the system runs at 60 FPS irrespective of the central processor unit (CPU). It was tested on a 650 MHz AMD Athlon classic with a Geforce 2, on a 2.0 GHz AMD Athlon XP with a Geforce 4 and on a 2.0 GHz Intel Pentium 4 systems with Geforce 2 and Geforce 4 cards. The peak performance of 92 FPS was attained on a 2.0 GHz Intel Pentium 4 system with an onboard Intel 845GL graphics card. The performance differences may be due to the graphics processing unit (GPU), the processor chipset, the interface between the graphics subsystem and the CPU, or the CPU load itself. It may be that when other intensive tasks are running in parallel and consuming processor power, the relative performance of systems using the Nvidia cards would improve.

When evaluating GCDs, and also GCMMDs, it is also necessary to distinguish between two main types of fixation-dependent changes;[19] those affecting perception and those affecting [task] performance. Generally, perception is more sensitive than performance; it is possible to degrade a display to a substantial degree without necessarily degrading performance.

B. THREE-DIMENSIONAL VOLUMETRIC GCMMD

1. Implementation

Two prototype three-dimensional GCD demonstrators have been implemented by the authors, both of which use mirrors to provide stereoscopic display. A pair of "cold mirrors" (these reflect visible light but transmit infrared light) separate the optical paths used for display and for gaze-tracking. The first demonstrator makes use of mirrors mounted behind two eyepieces and is designed to be used with a single conventional computer monitor, the left and right eye images being displayed on the left and right sides of the screen. Figure 15.18 shows a schematic (left) and photographs (middle and right) of the first demonstrator. Around the perimeter of each of the eyepieces seen in the bottom photograph, are the ends of optical fibers that are used to provide four, infrared illumination spots. The second, newer, demonstrator is a self-contained unit with two built-in LCD displays and provides considerably greater fidelity and field-of-view. Figure 15.19 shows a photograph of the second demonstrator. This uses large cold mirrors;

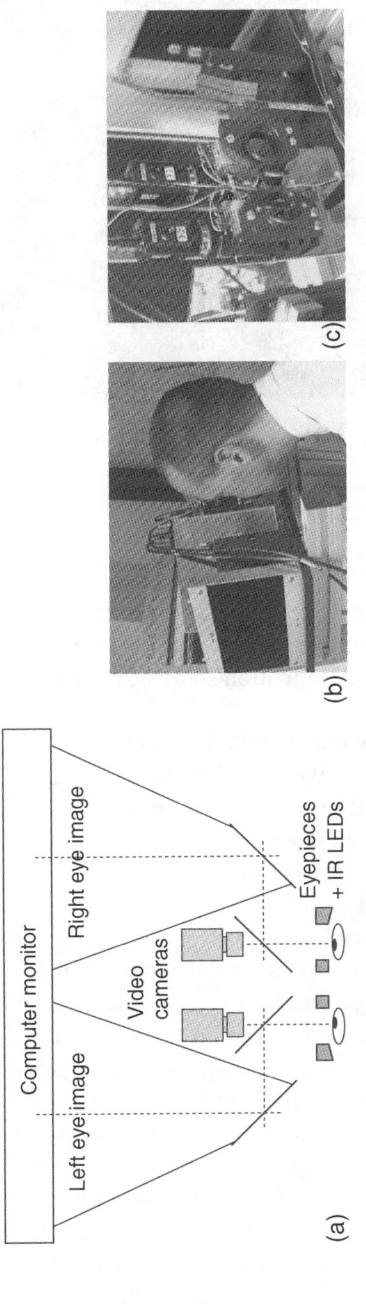

FIGURE 15.18 The old three-dimensional gaze-tracker system, UBHT and University of Bristol, UK.

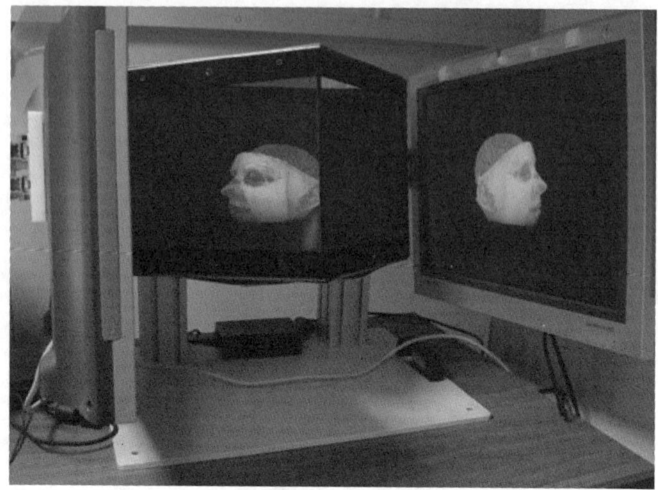

FIGURE 15.19 (See color insert) The new three-dimensional gaze-tracker and display system, UBHT and University of Bristol, UK, and Griffith University, Australia.

observers move forward to place the bridge of their nose close to where these join and, looking forward, see the three-dimensional image. Illumination is provided through the cold mirrors. Whilst preliminary experiments have been performed with different illumination sources, no final arrangement has yet been selected.

The new prototype is implemented on a PC with three monitor outputs. One dual-head graphics card drives the left and right LCD monitors. A separate single-head graphics card drives a utility monitor which is used for normal PC operations, software development and controlling the GCMMD software. The use of this separate monitor ensures that no extraneous visual information is presented to the observer using the GCD.

Within the current demonstrators, region-enhanced rendering algorithms have been implemented in software using a combination of C++ and assembler (to utilize the fast single-instruction multiple-data (SIMD) instruction sets present on many PC processors). Within the near future, however, this approach will be supplemented by one that exploits the increasing programmability of modern day graphics cards. Supported by suitable software tools, such cards can already implement a wide variety of custom rendering algorithms and this trend appears set to continue. Graphics cards support very high bandwidth paths between graphics processor and graphics memory and this bandwidth enables effective high-speed rendering. Where graphics card implementation may fall down, is on the amount of memory held on the card. This is usually substantially less than that supported on the main system memory; since volumetric data sets are often large, this can be problematic.

Since volumetric rendering is so computationally demanding, we use a significant amount of previsualization computation. During this phase, images are generated containing only context information, with no focus region added. These images are generated using a succession of regularly spaced data volume rotations. During gaze-contingent visualization, this sequence is played backwards and forwards: rotation and other forms of movement are hugely beneficial in improving depth perception via motion parallax and kinetic depth effects.[43] The precalculated context-only images are then combined (we skirt some issues here; see our other publications for details) with focus information, so that a GCD results, with the data volume rocking back and forth. This is somewhat restrictive, requiring a lengthy computation phase to be performed for each viewpoint change, but does allow implementation without excessive computing power. The use of PC clusters and multiresolution rendering (and graphics card processing, as mentioned previously) are other approaches which are being researched.

Within these demonstrators a separate software program performs real-time gaze tracking. This communicates via standard networking links with the rendering software and, as such, may be run either on the same machine or on a different one. Processing requirements and implementation difficulties associated with supporting multiple graphics cards, frame-grabbers, and other standard cards within a single PC have led to implementation on separate PCs being the most successful to date. The gaze-tracking program receives transformation matrix information from the rendering program; this encodes the instantaneous position and orientation of the rocking data volume. The gaze-tracking program then processes incoming video images from the cameras shown in Figure 15.18. Synchronization between the two programs is an issue. If these programs are running on different machines, then the time-stamps sent between them, and used to achieve this synchronization, must be adjusted to compensate for the different clocks within each machine.

2. Performance

A number of measures of performance may be monitored for the three-dimensional GCMMD. These include assessment of: (1) the rendering engine speed and fidelity, (2) the gaze-tracking accuracy and robustness and synchronization errors between these two components, and (3) observer behavior when engaged in higher level tasks using the GCMMD. As of yet, we have only begun to measure performance. The following few paragraphs outline our achievements to date.

Rendering engine speed is dependent upon a large number of parameters including: the data set and rendered image resolutions; the forms of projection and antialiasing employed during projection; whether color rendering is used; the size and shape of the focus region; and, how the focus region is blended with the context region. Our stereoscopic region-enhanced volume rendering implementation runs multithreaded codes on a dual Athlon MP2000+ Linux system. The inner loops of the weighted projection summations have been hand optimized using the MMX (rapid, parallel integer arithmetic) and SSE (rapid, parallel single-precision floating point) instruction sets. Data set sampling is performed using either nearest neighbor or tri-linear interpolation. Large variations in rendering rates result from adjustment of the many rendering parameters and sufficient speed can be achieved for real-time rendering only for low-resolution rendering of small data volumes. For example, using a volume data resolution of 256^3 voxels at a display resolution of 256^2 pixels, 12 stereo FPS can be generated. This figure indicates that software implementation is becoming feasible on standard PCs.

With regards to gaze-tracking accuracy, Figure 15.20 shows reconstructed fixation positions generated using the three-dimensional gaze-tracker. This figure contains the data from one run of the calibration process, in which the observer fixates in turn upon a succession of brightly lit, small wire-cross targets. Here, each target has been given a color and the reconstructed fixation positions given matching colors, so that tracker performance can be verified. The eye video images used to perform gaze-tracking have also been included in the figure and

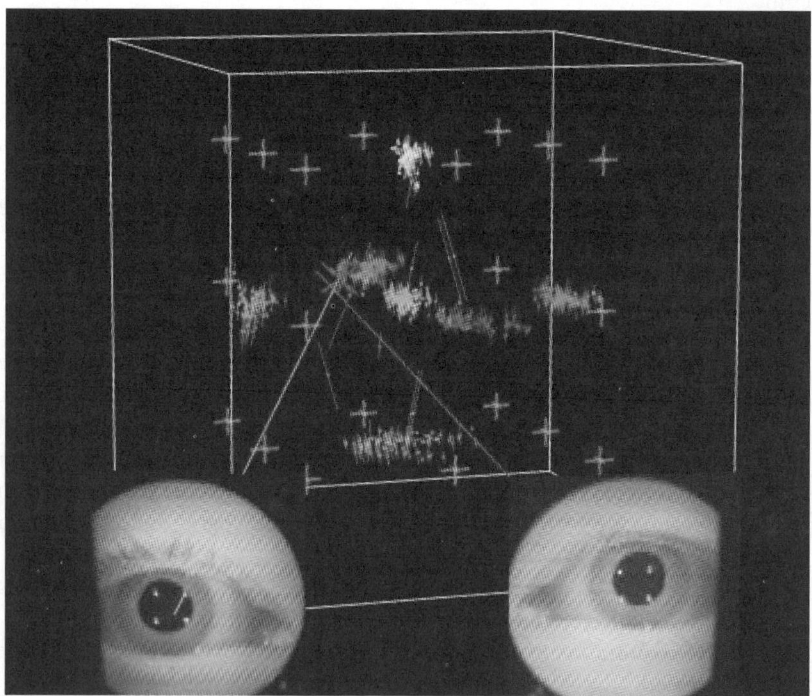

FIGURE 15.20 (See color insert) Reconstructed fixation positions from the three-dimensional gaze-tracker.

within these can be seen the four specular reflections from the optical fibers used to illuminate each eye.

The wire-frame cube shown in Figure 15.20 was displayed, $120 \times 120 \times 120$ mm in size, with its center positioned 300 mm from the subject. A set of seven small-cross targets, a subset of the 27 positions on a regular $3 \times 3 \times 3$ grid, were positioned inside the cube boundaries; intertarget distance was 37 mm. Each target in turn was displayed brightly lit for a period of 2 sec, during which the subject fixated upon it. Calibration was performed with the cube rocking back and forth through an angular range of 90 over a period of 6 sec.

The initial 0.3 sec of data following a change in highlighted target were ignored: these represented the observer reaction time and saccade duration. An optimization algorithm adjusted the model parameters to obtain a best fit between the actual target positions and the reconstructed estimates. Error terms were calculated for each instantaneous estimate by comparing its position to the actual target position. For performance analysis, error components measured radially and axially relative to the observation axis were recorded. For model parameter optimization, Cartesian error distances for each instantaneous estimate across the full calibration sequence were summed, and this error metric minimized. Average radial and axial errors, and the standard deviation of these errors, were calculated. This step produced axial and lateral error estimates for each of the targets for each of the filtering schemes. Error estimates were then averaged across all the targets to produce mean positional axial and lateral error estimates of 3.0 ± 1.3 mm radially and 7.5 ± 5.2 mm axially. It should be noted that these figures apply to errors within a single experimental run, with parameters optimized to cluster the reconstructed fixations around the targets. Tracking performance within a real GCD is likely to be significantly worse. Nonetheless, we feel that sufficient success has been achieved at this stage to warrant further - development.

C. INTEGRATION WITH EYE-TRACKERS

So far we have designed a number of GCMMDs making use of several different eye-trackers. The new OpenGL implementation of our two-dimensional GCD has been integrated and tested with two very different commercial video-based eye-trackers. The first one, the EyeLink II eye-tracker[60] from SR Research, Canada, is one of the most accurate video-based eye-trackers in the world, providing spatial resolutions of $0.01°$ and data rate of 500 Hz. More details about it can be found in Section VIII.C.1. The second one, the faceLAB system[67] from Seeing Machines, Australia, is a portable video-based eye-tracker with a lower spatial resolution and a typical data rate of 60 Hz. This system is described briefly in Section VIII.C.2. For our three-dimensional GCMMD we have used two prototype, in-house built, stereo gaze-trackers, about which more information is given in Section VIII.C.3.

1. Eye Link II from SR Research

The EyeLink II system[60] shown in Figure 15.21 uses corneal reflections in combination with pupil tracking which leads to stable tracking of eye position. EyeLink II has one of the highest spatial resolutions (noise-limited at $<0.01°$) and fastest data rates (500 samples per second) of any video-based eye-tracker system. It consists of three miniature cameras (see Figure 15.21) mounted on a leather-padded headband. Two eye cameras allow binocular eye-tracking. An optical head-tracking camera integrated into the headband allows accurate tracking of the observer's point of gaze while allowing natural head motion and speech. The EyeLink II system is very suitable for saccade analysis and smooth pursuit studies. In the context of this research, however, the most important feature of the EyeLink system is its ability to provide on-line gaze position data available with delays as low as 3 msec, which makes the system ideal for GCD applications. Indeed, when we integrated the new GCD with the EyeLink system there was no visible lag for a wide range of applications. The GCD code was integrated with the EyeLink *GCWindow* experiment code,[60] by replacing the "rendering" parts of the code with our new OpenGL implementation.

2. FaceLAB from SeeingMachines

FaceLAB,[67] from Seeing Machines (Canberra, Australia), includes a stereoscopic vision system and provides head and gaze-tracking within a self-contained PC that can readily be networked. Video input in faceLAB is obtained by a stereo-camera head (Figure 15.22). The system does not require any external devices to be worn. All measurements are calculated at 60 FPS (60 Hz) and can be logged to disk or streamed over a TCP/IP network or an RS-232 serial port. Via networking, any subset of the many gaze-tracking parameters may easily be linked into a user's program. This modular approach makes implementation of a rapid and straightforward GCD. The emphasis within the design of faceLAB has been upon allowing free-head movement and tracking under a variety of real-world lighting conditions. The spatial accuracy of gaze-tracking is in the range of $2°-10°$, so if gaze-tracking accuracy is of the utmost importance, then other tracking systems may be preferred. The performance of the system was rather slow with a significant lag in some cases because of the slow data rate (60 Hz) of the faceLAB system. Large window sizes lead to slightly improved perception in the case of GCMRDs or GCMMDs because they partly mitigated the effects of the time lag.

3. Three-Dimensional Gaze-Tracker

Since the prototype three-dimensional gaze-tracker and gaze-contingent rendering software have both been developed fully in-house, with custom software written in each case, integration has been straightforward, with tracking or rendering parameters passed between the two programs as required. The communication protocol between these software components now adopts a modular approach and

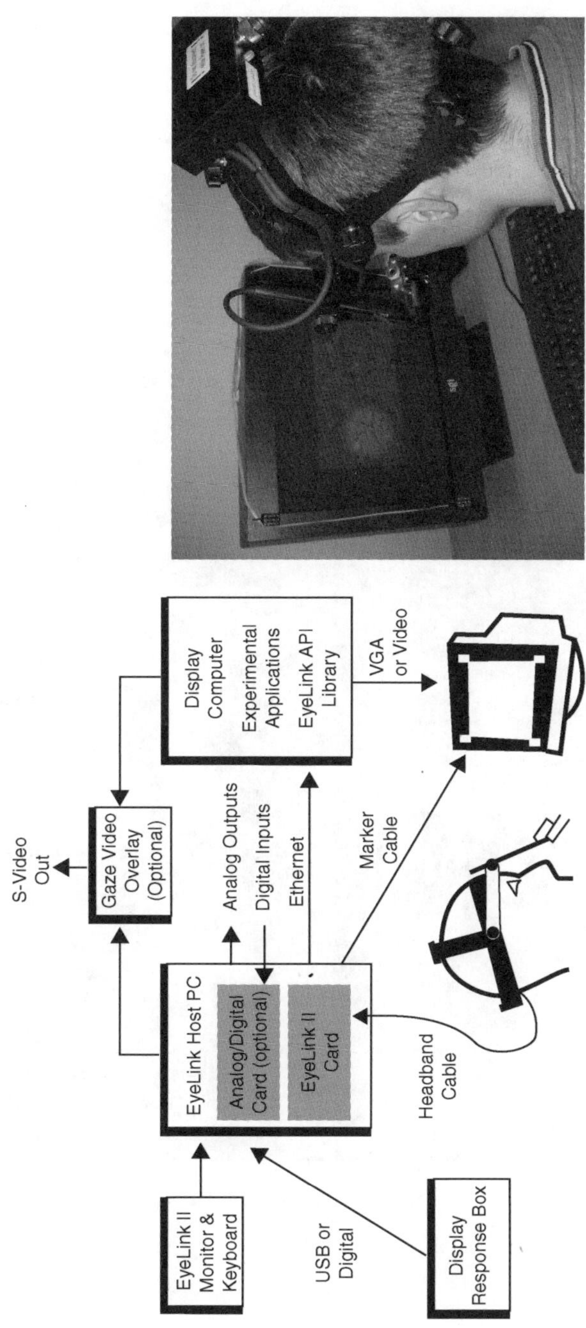

FIGURE 15.21 The Eyelink II system (left) from SR Research, Canada. A GCMMD using the EyeLink system (right).

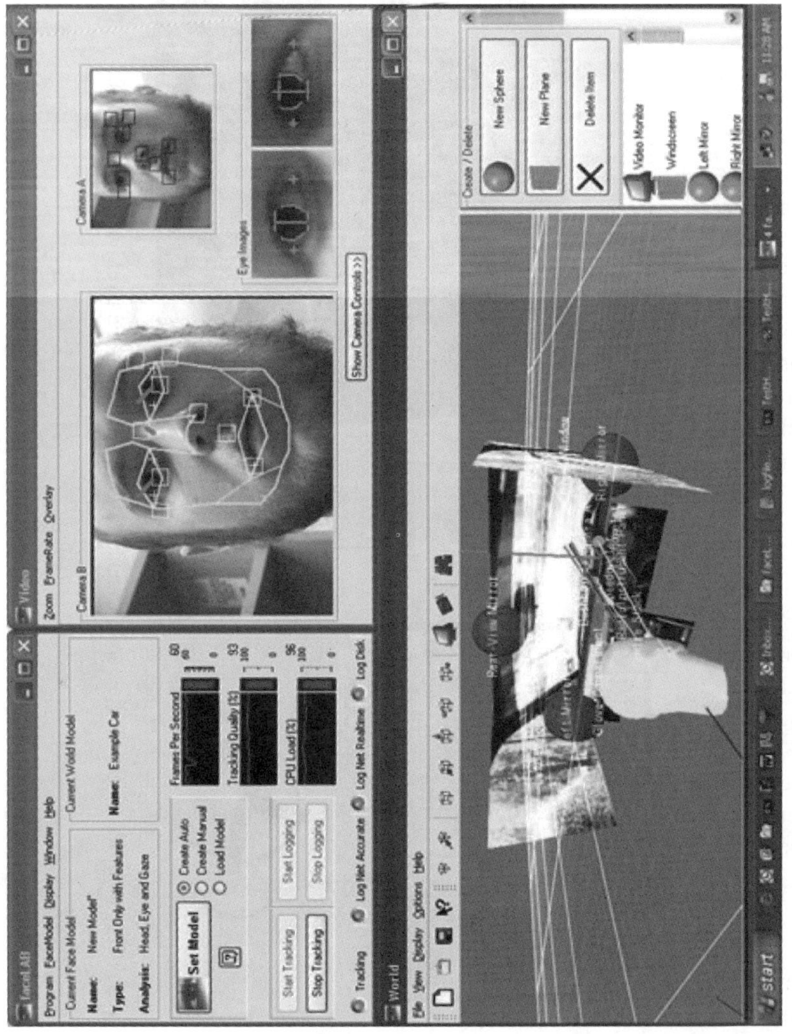

FIGURE 15.22 (See color insert) The faceLAB system from Seeing Machines, Australia.

arbitrary display geometries are supported, allowing for coordinate transformation between data set reference frames, observer reference frames and gaze-tracker reference frames. One of the active research questions now is whether information within the rendered image — or sequence of rendered images — can be usefully applied to improve the gaze-tracking accuracy. In a GCD, in contrast to wholly diagnostic gaze-tracking applications, what is being presented to the observer is generated within the computer. Since eye motion is tightly coupled to the scene viewed it should be possible to exploit this link.

IX. CONCLUSIONS

In this chapter we have summarized the results of our recent work on the design and use of GCMMDs. Such displays can provide a viable alternative to classical image fusion approaches, in that they allow us to adapt the multimodality visual information that is presented to the observer and, hence, to optimize the display for different image fusion tasks. A number of applications that can benefit from the use of GCMMDs, including fusion of two-dimensional medical, surveillance and remote sensing images, GCMMDs of multilayered information maps, and volumetric image fusion, have been described in this work. They all show great potential and flexibility to tailor large amounts of multisource, multiview, or multilayered visual information so that it can be better perceived and interpreted by the human visual system. If the amount of peripheral information is reduced, and only task relevant information is presented, this should give a real performance advantage. More experiments are needed to determine the visual span or optimal GCD parameters for various scenarios and images, and for different tasks. The design principles behind GCMMDs have been discussed in detail in this chapter, as well as the implementation and performance of such displays. Several GCMMDs using commercial and in-house built eye-trackers have been constructed and tested. Our recent eye-tracker independent implementation of a GCD using OpenGL opens the way for many more such systems to be developed and used in the above mentioned and future research areas.

ACKNOWLEDGMENTS

Financial support from the Royal Society (UK) is gratefully acknowledged. Mike Jones expresses thanks for the support given to him in his previous position within the United Bristol Healthcare NHS Trust (UBHT) where a portion of this work was carried out. The authors would like to thank Tim Newman, for developing the initial prototype of the OpenGL two-dimensional GCD, Dimitris Agrafiotis, for generating the images shown in Figure 15.6, and Ryan Krumins for his contribution towards implementation of the faceLAB-based GCD.

REFERENCES

1. Hawkes, D. J., Hill, D. L. G., Lehmann, E. D., Robinson, G. P., Maisey, M. N., and Colchester, A. C. F., Preliminary work on the integration of SPECT images with the aid of registered MRI images and an MR derived neuro-anatomical atlas, In *3D Imaging in Medicine: Algorithms, Systems, Applications*, Hoehne, K. H., Fuchs, H., and Pizer, S. M., Eds., Springer, Berlin, pp. 241–252, 1990.
2. Stokking, R., Integrated Visualization of Functional and Anatomical Brain Images, Ph.D. thesis, University of Utrecht, Utrecht, 1998.
3. Rehm, K., Strother, S. C., Anderson, J. R., Schaper, K. A., and Rottenberg, D. A., Display of merged multimodality brain images using interleaved pixels with independent color scales, *J. Nucl. Med.*, 35(11), 1815–1821, 1994.
4. Robb, R. A., and Hanson, D. P., The ANALYZE software system for visualization and analysis in surgery simulations, In *Computer-Integrated Surgery*, Taylor, R. H., Lavallee, S., Burdea, G. C., and Moesges, R., Eds., MIT Press, Cambridge, MA, pp. 175–189, 1996.
5. Porter, T., and Duff, T., Compositing digital images, *Comput. Graph.*, 18, 253–259, 1984.
6. Hill, D. L. G., Combination of 3D Medical Images from Multiple Modalities, Ph.D. thesis, University of London, Image Processing Group, UMDS, Guy's Campus, St Thomas's Street, London SE1 9RT, UK, December, 1993.
7. Card, S. K., Mackinley, J. D., and Shneiderman, B., *Readings in Information Visualization: Using Vision to Think*, Morgan Kaufmann Publishers, San Francisco, CA, 1999.
8. Spence, R., and Apperley, M., Database navigation: an office environment for professionals, *Behav. Inf. Technol.*, 1(1), 43–54, 1982.
9. Furnas, G. W., Generalized fisheye views, pp.16–23. In *Proceedings of CHI' 86, ACM Conference on Human Factors in Computing Systems, New York*, 1986.
10. Lamping, J., Rao, R., and Pirolli, P., A focus + context technique based on hyperbolic geometry for visualizing large hierarchies, pp. 401–408. In *Proceedings ACM Conference Human Factors in Computing Systems, CHI*, ACM, 1995.
11. Keahey, T. A., The generalized detail-in-context problem. In *Proceedings of the IEEE Symposium on Information Visualization*, IEEE, 1998.
12. Blackwell, A. F., Jansen, A. R., and Marriott, K., Restricted focus viewer: a tool for tracking visual attention. Lecture notes in artificial intelligence, In *Theory and Application of Diagrams*, Anderson, M., Cheng, P., and Haarslev, V., Eds., Vol. 1889, Springer, Berlin, pp. 162–177, 2000.
13. Leung, Y. K., and Apperley, M. D., A review and taxonomy of distortion-orientation presentation techniques, *Trans. Comput.–Hum. Interact.*, 1, 126–160, 1994.
14. Carpendale, M. S. T., Cowperthwaite, D. J., and Fracchia, F. D., Extending distortion viewing from 2D to 3D, *IEEE Comput. Graph. Appl.*, 42–51, 1997.
15. Fairchild, K. M., Poltrock, S. T., and Furnas, F. W., *SemNet: Three-Dimensional Graphic Representations of Large Knowledge Bases*, Lawrence Erlbaum Associates, Hillsdale, NJ, pp. 201–234, 1988.
16. Mitra, D. A., *A Fisheye Presentation Strategy: Aircraft Maintenance Data*, Lawrence Erlbaum Associates, Hillsdale, NJ, pp. 875–880, 1990.
17. MultiMap, UK aerial photo coverage; http://www.multimap.com.

18. Duchowski, A. T., *Gaze Tracking Methodology: Theory and Practice*, Springer, London, 2003.

19. Duchowski, A. T., A breadth-first survey of eye tracking applications, *Behav. Res. Methods, Instrum., Comput. (BRMIC)*, 34(4), 455–470, 2002.

20. McConkie, G. W., and Rayner, K., The span of the effective stimulus during a fixation in reading, *Percept. Psychophys.*, 17, 578–586, 1975.

21. Saida, S., and Ikeda, M., Useful visual field size for pattern perception, *Percept. Psychophys.*, 25, 119–125, 1979.

22. Rayner, K., Eye movements in reading and information processing: 20 years of research, *Psychol. Bull.*, 124, 372–422, 1998.

23. Pomplun, M., Reingold, E. M., and Shen, J., Investigating the visual span in comparative search: the effects of task difficulty and divided attention, *Cognition*, 81, B57–B67, 2001.

24. Parkhurst, D. J., and Niebur, E., Variable resolution displays: a theoretical, practical, and behavioral evaluation, *Hum. Factors*, 44(4), 611–629, 2002.

25. Luebke, D., Reddy, M., Cohen, J., Varshney, A., Watson, B., and Huebner, R., *Level of Detail for 3D Graphics*, Morgan Kaufmann Publishers, San Francisco, CA, 2002.

26. Weiman, C. F. R., Video compression via log polar mapping, pp. 266–277. In *Real Time Image Processing II*, SPIE 1295, 1990.

27. Kortum, P. T., and Geisler, W. S., Implementation of a foveated image-coding system for bandwidth reduction of video images, In *Human Vision and Electronic Imaging*, Rogowitz, B., and Allebach, J., Eds., SPIE 2657, pp. 350–360, 1996.

28. Loschky, L. C., and McConkie, G. W., User performance with gaze contingent displays, pp. 97–103. In *Proceedings of the Eye Tracking Research and Applications Symposium 2000*, ACM SIGGRAPH, 2000.

29. Reingold, E. M., and Loschky, L. C., Reduced saliency of peripheral targets in gaze-contingent multi-resolution displays: blended versus sharp boundary windows, pp. 89–94. In *Proceedings of the Eye Tracking Research and Applications Symposium 2002*, ACM SIGGRAPH, 2002.

30. Geisler, W. S., and Perry, J. S., A real-time foveated multi-resolution system for low-bandwidth video communication, In *Human Vision and Electronic Imaging*, Rogowitz, B., and Pappas, T., Eds., SPIE 3299, pp. 294–305, 1998.

31. Sheikh, H. R., Liu, S., Evans, B. L., and Bovic, A. C., Real time foveation techniques for H.263 video encoding in software, pp. 1781–1784. In *IEEE International Conference on Acoustics, Speech, and Signal Processing*, IEEE, 3, 2001.

32. Agrafiotis, D., Canagarajah, C. N., and Bull, D. R., Perceptually optimised sign language video coding based on eye tracking analysis, *Electron. Lett.*, 39(24), 1703–1705, 2003.

33. Dagnelie, G., and Massof, R. W., Toward an artificial eye, *IEEE Spectr.*, 33(5), 20–29, 1996.

34. Geisler, W. S., and Perry, J. S., Real-time simulation of arbitrary visual fields, pp. 83–87. In *Proceedings of the Eye Tracking Research and Applications Symposium 2002*, ACM SIGGRAPH, 2002.

35. Nikolov, S. G., Jones, M. G., Gilchrist, I. D., Bull, D. R., and Canagarajah, C. N., Multi-modality gaze-contingent displays for image fusion, pp. 1213–1220. In *Proceedings of the Fifth International Conference on Information Fusion (Fusion*

2002), Annapolis, MD, International Society of Information Fusion (ISIF), July 8–11, 2002.

36. Nikolov, S. G., Gilchrist, I. D., and Bull, D. R., Gaze-contingent multi-modality displays of multi-layered geographical maps, pp. 325–332. In *Proceedings of the Fifth International Conference on Numerical Methods and Applications (NM&A02), Symposium on Numerical Methods for Sensor Data Processing, Borovetz, Bulgaria*. LNCS 2542, Springer, Berlin, 2003.

37. Nikolov, S. G., Jones, M. G., Bull, D. R., Canagarajah, C. N., Halliwell, M., and Wells, P. N. T., Focus + context visualization for image fusion, pp. WeC3-3–WeC3-9. In *the Fourth International Conference on Information Fusion (Fusion 2001), Montreal, Canada*, Vol. I. International Society of Information Fusion (ISIF), August 7–10, 2001.

38. Jones, M. G., and Nikolov, S. G., Volume visualization via region-enhancement around an observer's fixation point, pp. 305–312. In *MEDSIP (International Conference on Advances in Medical Signal and Information Processing), Bristol, UK*, Number 476. IEEE, September 4–6, 2000.

39. Anstis, S. M., A chart demonstrating variation in acuity with retinal position, *Vis. Res.*, 14, 589–592, 1974.

40. Bouma, H., Interaction effects in parafoveal letter recognition, *Nature*, 226, 177–178, 1970.

41. Toet, A., IJspeert, J. K., Waxman, A. M., and Aguilar, M., Fusion of visible and thermal imagery improves situational awareness, *Displays*, 24, 85–95, 1997.

42. Environment Australia, Australian Coastal Atlas (part of the Australian Natural Resources Atlas); http://www.ea.gov.au/coasts/atlas/index.html.

43. Ware, C., *Information Visualization Perception for Design*, Morgan Kaufmann, San Francisco, CA, 2000.

44. Goldstein, E. B., *Sensation and Perception*, 4th ed., Brooks/Cole, Pacific Grove, 1996.

45. Istance, H., and Howarth, P., Using eye vergence to interact with objects in the three dimensional space created by stereoscopic display, *IEE Colloquium (digest)*, 126, 2/1–2/5, 1996.

46. Talmi, K., and Liu, S., Eye and gaze-tracking for visually controlled interactive stereoscopic displays, *Signal Process. Image Commun.*, 14, 799–810, 1999.

47. Shiwa, S., Omura, K., and Kishino, F., Proposal for a three dimensional display with accommodative compensation: 3DDAC, *J. Soc. Inf. Display*, 4, 255–261, 1996.

48. Levoy, M., and Whitaker, R., Gaze-directed volume rendering, *Comput. Graph.*, 24, 217–223, 1990.

49. Westheimer, G., Oculomotor control: the vergence system, In *Eye Movements and Psychological Processes*, Monty, A. R., and Senders, J. W., Eds., Lawrence Erlbaum Associates, Hillsdale, NJ, pp. 55–64, 1976.

50. Drebin, R. A., Carpenter, L., and Hanrahan, P., Volume rendering, *Comput. Graph.*, 22, 65–74, 1988.

51. Levoy, M., Display of surfaces from volume data, *IEEE Comput. Graph. Appl.*, 8, 29–37, 1988.

52. Kaufman, A., Cohen, D., and Yagel, R., Volume graphics, *Computer*, 26, 51–64, 1993.

53. Jones, M. G., and Nikolov, S. G., Dual-modality two-stage direct volume rendering, pp. 845–851. In *Proceedings of the Sixth International Conference on*

Information Fusion (Fusion 2003), Cairns, Qld., Australia. International Society of Information Fusion (ISIF), July 8–11, 2003.

54. Nikolov, S. G., Newman, T. D., Jones, M. G., Gilchrist, I. D., Bull, D. R., and Canagarajah, C. N., Gaze-contingent display using texture mapping and OpenGL: system and applications. In *Proceedings of the Eye Tracking Research and Applications (ETRA) Symposium 2004, San Antonio, TX,* ACM SIGGRAPH, March 22–24, 2004, pp. 11–18.

55. van Diepen, P. M. J., De Graef, P., and Van Rensbergen, J., On-line control of moving masks and windows on a complex background using the ATVista Videographics Adapter, *Behav. Res. Methods, Instrum. Comput.,* 26, 454–460, 1994.

56. van Diepen, P. M. J., De Graef, P., and Van Rensbergen, J., A pixel-resolution video switcher for eye-contingent display changes, *Spat. Vis.,* 10, 335–344, 1997.

57. Geisler, W. S., and Perry, J. S., Variable-resolution displays for visual communication and simulation, *Soc. Inf. Display,* 30, 420–423, 1999.

58. Pomplun, M., Reingold, E. M., and Shen, J., Peripheral and parafoveal cueing and masking effects on saccadic selectivity in a gaze-contingent window paradigm, *Vis. Res.,* 41, 2757–2769, 2001.

59. Reingold, E. M., Charness, N., Pomplun, M., and Stampe, D. M., Visual span in expert chess players: evidence from eye movements, *Psychol. Sci.,* 12, 48–55, 2001.

60. SR Research, EyeLink; http://www.eyelinkinfo.com.

61. Watson, B., Walker, N., Hodges, L. F., and Worden, A., Managing level of detail through peripheral degradation: Effects on search performance with a head-mounted display, *ACM Trans. Comp.–Hum. Interact.,* 4(4), 323–346, 1997.

62. Newman, T., Gaze-Contingent Displays Using Texture Mapping, M.Eng. Thesis, University of Bristol, Bristol, 2003.

63. Autodesk, WHIP! Viewer; http://www.autodesk.com/cgi-bin/whipreg.pl.

64. Autodesk, Volo View; http://www.autodesk.com/voloview.

65. Vahonen, S., Samu's OpenGL programs; http://www.students.tut.fi/vahonen/computer/OpenGL/.

66. NVIDIA, GeForce graphics cards; http://www.nvidia.com.

67. Seeing Machines, faceLAB; http://www.seeingmachines.com.au.

16 Structural and Information Theoretic Approaches to Image Quality Assessment

Kalpana Seshadrinathan, Hamid R. Sheikh, Alan C. Bovik, and Zhou Wang

CONTENTS

I. INTRODUCTION

Digital images and video are prolific in the world today owing to the ease of acquisition, processing, storage and transmission. Many common image processing operations such as compression, dithering and printing affect the quality of the image. Advances in sensor and networking technologies, from the internet to wireless networks, has led to a surge of interest in image communication systems. Again, the communication channel tends to distort the image signal passing through it and the quality of the image needs to be closely monitored to meet the requirements at the end receiver. The integrity of the image also needs to be preserved, irrespective of the specific display system that is used by the viewer. In this chapter, we describe state-of-the-art objective quality metrics to assess the quality of digital images.

In all the applications mentioned above, the targeted receiver is the human eye. Compression and half-toning algorithms are generally designed to generate images that closely approximate the original reference image, as seen by the human eye. In fact, most of the images encountered in day-to-day life are meant for human consumption. The measure of quality depends on the intended receiver and we focus on applications where the ultimate receiver is the Human Visual System (HVS).

Subjective assessment of image quality involves studies where participants are asked to assign a score to an image after viewing it under certain fixed environment conditions such as viewing distance, display characteristics and so on. Typically, the same image is shown to a number of subjects and final scores are assigned after accounting for any variability in the assessment. However, subjective assessment studies are tedious and impossible to do for every possible image in the world. The goal of quality assessment research is to objectively predict the quality of an image to approximate the score a human might assign to it.

The HVS is very good at evaluating the quality of an image blindly, that is, without a reference "perfect" image to compare it against. It is, however, rather difficult to perform this task automatically using a computer. Although the term *image quality* is used, what we are actually referring to is *image fidelity*. We assume that a reference image is available and the quality of the test image is determined by how close it is to the reference image perceptually.

Objective measures for image quality play an important role in evaluating the performance of image fusion algorithms.[1] In many applications, the end receivers of the fused images are humans. Several fusion algorithms presented in the literature have been evaluated using subjective criteria,[2] as well as measures such as the mean square error (MSE) described below.[3] Measures to evaluate objective quality are hence useful to evaluate the effectiveness of a fusion algorithm, as well as to optimize various parameters of the algorithm to improve performance.

MSE between the reference and test images is a commonly used metric for image quality. Let $\vec{x} = \{x_i, i = 1, 2, \ldots, N\}$ denote a vector containing the reference image pixel values, where i denotes a spatial index. Similarly, let $\vec{y} = \{y_i, i = 1, 2, \ldots, N\}$ denote the test image. Then the MSE between $\vec{x}$ and $\vec{y}$ is defined as

$$\text{MSE}(\vec{x}, \vec{y}) = \frac{1}{N} \sum_{i=1}^{N} (y_i - x_i)^2 \qquad (16.1)$$

This metric is known to correlate very poorly with visual quality, but is still widely used due to its simplicity. The failure of MSE as a metric for quality is illustrated here using a number of examples where all distortions have been adjusted in strength such that the MSE between the reference and distorted image is 50. The visual quality of the images are, however, drastically different. The MSE is only a function of the difference between corresponding pixel values and implicitly assumes that the perceived distortion is independent of both the actual value of the pixel in the reference image and neighboring values. This directly contradicts the luminance masking and contrast masking properties of the HVS.[4] In reality, the perception of distortion varies with the actual image at hand, as illustrated in Figure 16.1. Figure 16.1(a) shows the original "Buildings" image. Figure 16.1(c) shows a blurred version of the same image obtained by convolving the image with a Gaussian window. Figure 16.1(e) shows a JPEG compressed version of the image. Figure 16.1(b), (d) and (f) show the original, Gaussian blurred and JPEG compressed "Parrots" image, respectively. The "Parrots" image has relatively large smooth areas, corresponding to low frequencies, and the blurring distortion is very pronounced. The blocking artifacts of JPEG compression are also visibly annoying in this image. This is, however, not the case in the "Buildings" image. Additionally, just adding a constant to every pixel in the image leads to a large MSE, but almost insignificant loss in visual quality. This is illustrated in

FIGURE 16.1 Examples of additive distortions (MSE = 50).

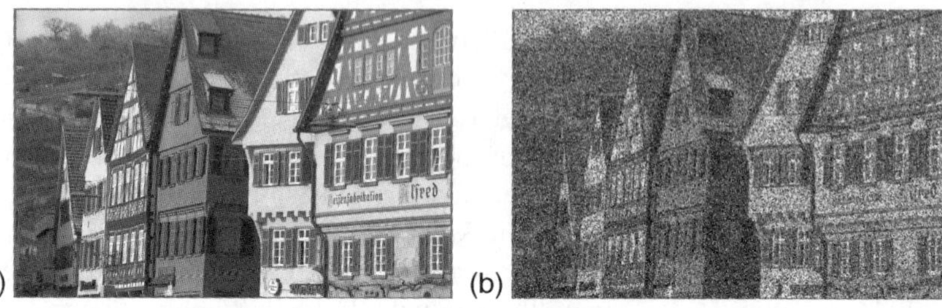

FIGURE 16.2 Examples of additive distortions (MSE = 50).

Figure 16.2(a). Salt and pepper noise is a form of extremely impulsive noise where randomly selected pixels in the image are turned black or white. This kind of noise is usually visible to the eye even at very low strengths and the image has extremely poor quality at a MSE of 50, as illustrated in Figure 16.2(b). The reference images shown here are available from the database in Ref. 5. These examples clearly illustrate the failure of MSE as a good metric for visual quality.

Traditional approaches to image quality assessment use a bottom–up approach, where models of the HVS are used to derive quality metrics. Bottom–up HVS-based approaches are those that combine models for different properties of the HVS in defining a quality metric. The response of the HVS to variations in luminance is a nonlinear function of the local mean luminance and this is commonly referred to as luminance masking. It is called masking because the variations in the distorted signal are masked by the base luminance of the reference image. The HVS has a bandpass characteristic and its frequency response can be characterized by the Contrast Sensitivity Function (CSF). Experiments are conducted to detect the threshold of visibility of sine waves of different frequencies to determine the CSF. Contrast masking refers to the masking of certain frequency and orientation information due to the presence of other components that have similar frequency and orientation. Bottom-up HVS-based quality metrics use different models to account for the luminance masking, contrast sensitivity, and contrast masking features of the HVS, and normalize the error terms by corresponding thresholds. The final step involves combining these normalized error terms to obtain either a quality map for the image at every pixel or a single number representing the overall visual quality of an image. A commonly used metric for error pooling is the Minkowski error. A detailed description of HVS-based quality measures can be found in Ref. 6.

Recent approaches to quality assessment, however, follow a top-down approach where the hypothesized functionality of the HVS is modeled. In this chapter, we describe two such approaches that have been shown to be competitive with bottom-up HVS-based approaches in predicting image quality. These methods additionally demonstrate advantages over bottom-up HVS-based measures in several aspects.[7,8]

Structural SIMilarity (SSIM) approaches to image quality assume that the HVS has evolved to extract structural information from an image.[8] The quality of the image, as perceived by the human eye, is hence related to the loss of structural information in the image. The error metrics used here correspond to measures used to quantify structural distortions and are more meaningful than simple signal similarity criteria like MSE. A detailed description, and the intuition behind this approach, is presented in Section II.

A related recent top-down approach hypothesizes that the test image is the output of a communication channel through which the reference image passes and that image quality is related to the *mutual information* between these images.[7] Statistical models that accurately characterize the source and the channel are the key to the success of this approach in relating statistical information measures to perceived distortion. Natural images are images obtained from the real world and form a small subspace of the space of all possible signals.[9] A computer that generates images randomly is unlikely to produce anything that even contains objects resembling those in natural images. The statistical properties of the class of natural images have been studied by various researchers and these natural scene models are used in the information theoretic development here. The details of this approach and the derivation of the quality metric are presented in Section III.

We briefly introduce the notation used throughout this chapter. $\vec{x}$ represents a vector x and a bold face character $\mathbf{X}$ represents a matrix X. Capital letters are used to denote random variables and the symbol $\hat{X}$ is used to represent the estimated value of the random variable X. Greek characters are used to denote constants.

Section IV presents experimental results that demonstrate the success of structural and information-theoretic approaches in image quality assessment. Finally, we conclude this chapter in Section V with a brief summary of the two paradigms of quality assessment presented here.

II. THE STRUCTURAL SIMILARITY PARADIGM

Traditional bottom-up HVS-based measures of image quality have several limitations.[8] The working of the HVS is not yet fully understood and the accuracy of the models for the HVS used in quality assessment is unclear. Models of the frequency response of the HVS, for example, are typically obtained by showing human subjects relatively simple patterns like global sinusoidal gratings. Masking behavior of the HVS is modeled using data obtained by showing human subjects superposition of two or three of these sinusoidal patterns. Images of the real world, however, are quite complex and contain several structures that are the superposition of hundreds of simple sinusoidal patterns. It is difficult to justify the generalization of the models obtained from these simple experiments to characterize the HVS. Also, typical experiments performed to understand the properties of the HVS operate at the threshold of visual perception. However, quality assessment deals with images that are perceptibly distorted, known as suprathreshold image distortion and it is not clear how models developed for near visibility generalize to models that quantify perceived distortion. Finally, an implementation involving accurate models for the HVS might be too complex for most practical applications. A top-down approach could lead to a simplified algorithm that works adequately, as long as the underlying hypothesis characterizes the *primary features* of the distortion that the HVS perceives as loss of quality.

To overcome these limitations, a new framework for image quality assessment has been proposed that assumes that the HVS has evolved to extract *structural information* from images. Hence, a measure of the structural information change can be used to quantify perceived distortions.[8] This is illustrated in Figure 16.3. All images shown here have approximately the same MSE with respect to the reference image. Clearly, the mean shifted and contrast enhanced image have very high perceptual quality despite the large MSE. However, this is not the case in the blurred, JPEG2000 compressed and additive white Gaussian noise (AWGN) corrupted images. This can be attributed to the fact that there is no loss of structural information in the former case, but this is not true in the latter case. The mean shifting and contrast stretching operations are invertible (except at the points where the luminance saturates) and the original image can be fully recovered. However, the blurring and compression are not easily invertible transformations. Blurring can be inverted, in some cases, by deconvolution when none of the frequency components are zeroed out. The HVS, however, is not able to easily invert the transformation and extract the structural information in the image. In this sense, there is no loss of structural information in the mean-shifted and contrast stretched images. Furthermore, the mean shifted and contrast stretched images have only luminance and contrast changes, as opposed to the blurred images that have severe structural distortions. The luminance and contrast of an image depend on the illumination, which does not affect the structural information in the image. The good visual quality of the mean shifted and contrast stretched images, despite the large MSE, can be attributed in the structural framework to the fact that there is almost no loss of structural information in these images.

The mathematical formulation of the SSIM index is given in Section II.A. Use of this index to predict image quality and illustrative examples of the performance of this algorithm are presented in Section II.B.

A. The Structural Similarity Index

Figure 16.4 shows a block diagram of the SSIM quality assessment algorithm. The luminance of an object is the product of the reflectance and illumination of the object and is independent of the structure of the object. The structural information in an image is defined as those attributes that are independent of the illumination. Hence, to quantify the loss of structural information in an image, the effects of luminance and contrast are first cancelled out. The structure comparison is then carried out between the luminance and contrast normalized signals. The final quality score is a function of the luminance, contrast and structure comparisons, as shown in Figure 16.4.

FIGURE 16.3 Illustrative examples of structural distortions: (a) original 'Boats' image; (b) contrast enhanced image; (c) mean shifted image; (d) blurred image; (e) JPEG2000 compressed image; (f) AWGN corrupted image.

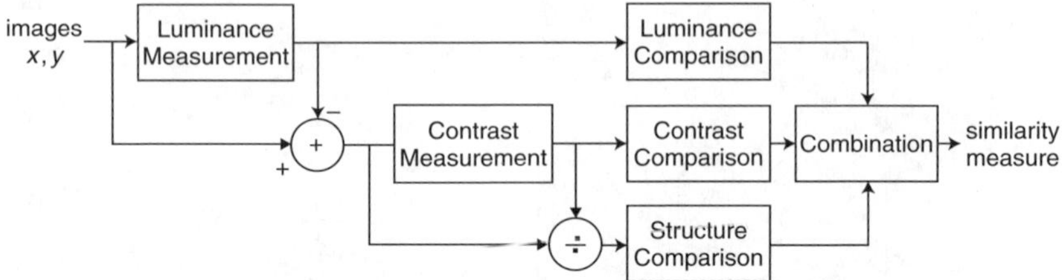

FIGURE 16.4 Block diagram of the SSIM quality assessment system. (From Wang, Z., and Bovik, A. C., Structural approaches to image quality assessment, In *Handbook of Image and Video Processing*, Bovik, A. C., Ed., Academic Press, submitted for publication. With permission.)

Let $\vec{x}$ and $\vec{y}$ represent vectors containing pixels from the reference and distorted images, respectively. The reference image is assumed to have "perfect" quality. The SSIM algorithm operates in three stages: luminance, contrast, and structure comparison.

First, the luminance of the two signals is compared. The luminance of an image is estimated using its mean intensity and is given by

$$\mu_x = \frac{1}{N} \sum_{i=1}^{N} x_i \tag{16.2}$$

where N represents the number of pixels in $\vec{x}$. The luminance comparison function $l(\vec{x}, \vec{y})$ is then a function of the luminance of the reference and test images, μ_x and μ_y, respectively. Then, the luminance of the images is normalized by subtracting out the mean luminance. The resulting signal, given by $\vec{x} - \mu_x$, can then be thought of as the projection of the image $\vec{x}$ onto an N-dimensional hyperplane defined by

$$\sum_{i=1}^{N} x_i = 0 \tag{16.3}$$

The second stage is to compare and normalize the contrasts of the two images. The contrast is defined as the estimate of the standard deviation of the image intensities and is given by

$$\sigma_x = \left(\frac{1}{N-1} \sum_{i=1}^{N} (x_i - \mu_x)^2 \right)^{1/2} \tag{16.4}$$

The factor of $N - 1$ is used in the denominator to obtain an unbiased estimate of the standard deviation. The contrast comparison function $c(\vec{x}, \vec{y})$ is then a function of the contrasts of the reference and test signals, σ_x and σ_y, respectively. The contrasts of $\vec{x}$ and $\vec{y}$ are then normalized by dividing them by their own standard deviations. The structure comparison is then performed on these normalized signals, given by $\vec{x} - \mu_x/\sigma_x$ and $\vec{y} - \mu_y/\sigma_y$, to obtain the index $s(\vec{x}, \vec{y})$.

Finally, the three components are combined to obtain the overall SSIM given by

$$\mathrm{SSIM}(\vec{x}, \vec{y}) = f(l(\vec{x}, \vec{y}), c(\vec{x}, \vec{y}), s(\vec{x}, \vec{y})) \tag{16.5}$$

The three components used to predict image quality are relatively independent, as we cancel out the effect of each one of them by normalization before computing the next component.

We now define the three functions for luminance, contrast and structure comparisons. These functions are designed to satisfy the following properties:

1. Symmetry: $S(\vec{x}, \vec{y}) = S(\vec{y}, \vec{x})$
2. Bounded: $S(\vec{x}, \vec{y}) \leq 1$
3. Unique maximum: $S(\vec{x}, \vec{y}) = 1$ if and only if $\vec{x} = \vec{y}$

For luminance, $l(\vec{x}, \vec{y})$ is defined by

$$l(\vec{x}, \vec{y}) = \frac{2\mu_x\mu_y + C_1}{\mu_x^2 + \mu_y^2 + C_1} \tag{16.6}$$

where the constant C_1 is added to increase stability when the denominator becomes very small. One choice for C_1 given in Ref. 8 is $C_1 = (K_1L)^2$, where L is the range of the pixel values and $K_1 \ll 1$ is a small constant. Equation 16.6 is qualitatively consistent with Weber's law which is widely used to model luminance masking in the HVS. The Weber's law states that the just noticeable difference in luminance is directly proportional to the background luminance. Hence, the HVS is sensitive to relative luminance changes, not the absolute change. Letting R represent the fractional change in luminance, the luminance of the distorted signal can be written as $\mu_y = (1 + R)\mu_x$. Then, we have

$$l(\vec{x}, \vec{y}) = \frac{2(1 + R)}{1 + (1 + R)^2 + \dfrac{C_1}{\mu_x^2}} \tag{16.7}$$

If C_1 is small enough compared to μ_x^2, then $l(\vec{x}, \vec{y})$ is a function of just R and this is consistent with Weber's law.

The contrast comparison function is given by

$$c(\vec{x}, \vec{y}) = \frac{2\sigma_x\sigma_y + C_2}{\sigma_x^2 + \sigma_y^2 + C_2} \tag{16.8}$$

where $C_2 = (K_2L)^2$ is chosen as earlier and again, $K_2 \ll 1$. For the same absolute change in contrast, this measure is less sensitive to high base contrast than low base contrast. This is also qualitatively consistent with the contrast masking feature of the HVS.

The structure comparison is then performed between the luminance and contrast normalized images, $(\vec{x} - \mu_x)/\sigma_x$ and $(\vec{y} - \mu_y)/\sigma_y$. These lie in the hyperplane defined by $\sum_{i=1}^{N} x_i = 0$. The correlation between these vectors is used as the measure to quantify SSIM between the images. The correlation is defined by

$$s(x, y) = \frac{\sigma_{xy} + C_3}{\sigma_x \sigma_y + C_3} \tag{16.9}$$

where σ_{xy} is given by

$$\sigma_{xy} = \frac{1}{N-1} \sum_{i=1}^{N} (x_i - \mu_x)(y_i - \mu_y) \tag{16.10}$$

Geometrically, $s(\vec{x}, \vec{y})$ corresponds to the cosine of the angle between these vectors in the hyperplane.

Finally, these three components are combined to obtain the SSIM index between $\vec{x}$ and $\vec{y}$ using

$$\mathrm{SSIM}(\vec{x}, \vec{y}) = l(\vec{x}, \vec{y})^\alpha c(\vec{x}, \vec{y})^\beta s(\vec{x}, \vec{y})^\gamma \tag{16.11}$$

where $\alpha, \beta, \gamma > 0$ are parameters to adjust the relative importance of these parameters. Specific values for these constants given by $\alpha = \beta = \gamma = 1$ and $C_3 = C_2/2$ have been shown to be effective in Ref. 8.

B. SSIM INDEX IN IMAGE QUALITY ASSESSMENT

For image quality assessment, the SSIM index is applied locally rather than globally. This is because image features are highly nonstationary. Additionally, using local windows provides a quality map of an image, as opposed to a single index for the entire image, and can provide valuable information about local quality.

The quantities $\mu_x, \sigma_x, \mu_y, \sigma_y$ and σ_{xy} are computed in a local sliding window, that is, moved pixel by pixel over the entire image. To avoid blocking artifacts, the resulting values are weighted using a circularly symmetric 11×11 Gaussian function. The weighting function, $\vec{w} = \{w_i, i = 1, 2, \ldots, N\}$, has a standard deviation of 1.5 samples and is normalized to have a unit sum.

$$\sum_{i=1}^{N} w_i = 1$$

The estimates of μ_x, σ_x and σ_{xy} are then modified accordingly as

$$\mu_x = \sum_{i=1}^{N} w_i x_i \tag{16.12}$$

$$\sigma_x = \left(\sum_{i=1}^{N} w_i (x_i - \mu_x)^2 \right)^{1/2} \qquad (16.13)$$

$$\sigma_{xy} = \sum_{i=1}^{N} w_i (x_i - \mu_x)(y_i - \mu_y) \qquad (16.14)$$

The constants K_1 and K_2, used in the definition of $l(\vec{x}, \vec{y})$ and $c(\vec{x}, \vec{y})$, are chosen to be 0.01 and 0.03 experimentally. The overall quality of the entire image is defined to be the Mean SSIM (MSSIM) index and is given by

$$\text{MSSIM}(\vec{X}, \vec{Y}) = \frac{1}{M} \sum_{i=1}^{M} \text{SSIM}(\vec{x}_i, \vec{y}_i) \qquad (16.15)$$

where $\vec{x}$ and $\vec{y}$ are the reference and test images, x_i and y_i are the pixels in the ith local window and M is the total number of windows in the image. A MATLAB™ implementation of the SSIM algorithm is available at Ref. 11.

Figure 16.5 shows the performance of the SSIM index on an image. Figure 16.5(a) and (b) show the original and JPEG compressed "Church and Capitol" images. The characteristic blocking artifacts of JPEG compression are clearly visible in the background of the image, on the roof of the church, in the trees and so on. Also, compression causes loss of high frequency information and the ringing artifacts are clearly visible along the edges of the Capitol dome. Figure 16.5(c) clearly illustrates the effectiveness of SSIM in capturing the loss of quality in these regions. Brighter regions correspond to better visual quality and the map has been scaled for better visibility. The SSIM index clearly captures the loss of quality in the trees and the roof, and also captures the ringing artifacts along the edge of the Capitol. Figure 16.5(d) shows the absolute error map between the images. This clearly fails to capture the distortion present in different regions of the image adequately.

III. THE INFORMATION THEORETIC PARADIGM

The information theoretic paradigm approaches the quality assessment problem as an information fidelity problem, as opposed to a signal fidelity problem. MSE and SSIM are examples of signal fidelity criteria; MSE is a simple mathematical criterion, while SSIM attempts to measure closeness between signals in the perceptual domain. However, information fidelity criteria attempt to relate visual quality to the amount of information shared between the reference and test images.[7] This shared information can be quantified by the commonly used statistical measure, namely mutual information.

Here, the test image is assumed to be the output of a communication channel whose source is the reference image. The communication channel consists of a distortion channel as well as the HVS. The distortion channel models various operations like compression, blurring, additive noise, contrast enhancement, and

FIGURE 16.5 Illustrative example of SSIM. (a) Original "Church and Capitol" image, (b) JPEG compressed image, (c) SSIM Quality map, and (d) Absolute Error Map.

FIGURE 16.5 (Continued).

so on, that lead to loss or enhancement of visual quality of the image. The HVS itself acts as a distortion channel as it limits the amount of information that is extracted from an image that passes through it.[12] This is the consequence of various properties of the HVS like luminance, contrast, and texture masking, which make certain distortions imperceptible. In fact, image compression algorithms rely on these properties of the HVS to successfully reduce the number of bits used to represent an image without affecting the visual quality.

Information theoretic analysis requires accurate modeling of the source and the communication channel to quantify the information shared between the source and the output of the communication channel. Source modeling is accomplished using statistical models for *natural images*. Natural images are those that represent images of the real world, not necessarily images of nature. The statistical properties of such images have been studied by numerous researchers in the context of several applications such as compression or de-noising.[13,14] These natural scene models attempt to characterize the distributions of natural images that distinguish them from images generated randomly by a computer.

In Section III.A, we present the natural scene model that is used in the quality assessment algorithm. Section III.B discusses the distortion model and Section III.C presents the HVS model used here. The algorithm for quality assessment is presented in Section III.D. We present certain illustrative examples that describe the properties of this novel quality measure in Section III.E.

A. Natural Scene Model

The semiparametric class of Gaussian scale mixtures (GSM) has been used to model the statistics of the wavelet coefficients of natural images.[15] A random vector $\vec{Y}$ is a GSM if $\vec{Y} \sim Z\vec{U}$ where Z is a scalar random variable, $\vec{U}$ is a zero mean Gaussian random vector and Z and $\vec{U}$ are independent. Z is called the *mixing* density. The GSM density can be represented as the integral of Gaussian density functions weighted by the mixing density; hence the term "mixtures." This class of distributions has heavy-tailed marginal distributions and the joint distributions exhibit certain nonlinear dependencies that are characteristic of the wavelet coefficients of natural images.[16]

Here, we model a coefficient and a collection of its neighbors in each subband of the wavelet decomposition of an image as a GSM. Specifically, we use the steerable pyramid which is a tight frame representation and splits the image into a set of subbands at different scales and orientations.[17] We model each subband of the wavelet decomposition by a random field; $C = \{\vec{C}_i, i \in I\}$, given by

$$C = ZU = \{Z_i\vec{U}_i, \quad i \in I\} \tag{16.16}$$

where $Z = \{Z_i, i \in I\}$ is the mixing field, $U = \{\vec{U}_i, i \in I\}$ is an M-dimensional zero-mean Gaussian vector random field with covariance matrix $\mathbf{C_U}$ and I denotes a set of spatial indices. Also, U is assumed to be white, that is, $\vec{u}_i$ is uncorrelated with $\vec{u}_j$ if $i \neq j$. Each subband of the wavelet decomposition is partitioned into nonoverlapping blocks of M coefficients each and each block is modeled as the vector $\vec{C}_i$.

This model has certain properties that make it analytically tractable. Each $\vec{C}_i$ is normally distributed given Z_i. Also, given Z, $\vec{C}_i$ is independent of $\vec{C}_j$ if $i \neq j$. Methods to estimate the multiplier Z and the covariance matrix $\mathbf{C_U}$ have been described in detail in Refs. 14,15. This GSM model is used as the source model for natural images in the following discussion.

B. Distortion Model

The distortion model that is used here is a signal attenuation and additive noise model in the wavelet domain given by

$$D = GC + \nu = \{g_i \vec{C}_i + \vec{v}_i, i \in I\} \tag{16.17}$$

where C denotes the random field that represents one subband of the reference image, $D = \{\vec{D}_i, i \in I\}$ denotes the random field representing the corresponding subband of the test image, G represents a deterministic scalar attenuation field and ν is a stationary, additive, zero-mean Gaussian noise field with covariance $\mathbf{C}_\nu = \sigma_\nu^2 \mathbf{I}$, where $\mathbf{I}$ is the identity matrix.

This model is both analytically tractable and computationally simple. This model can be used to describe most commonly occurring distortion types *locally*. The deterministic gain field G captures the loss of signal energy in subbands due to various operations like compression and blurring. The additive noise field accounts for local variations in the attenuated signal. Additionally, changes in image contrast can also be described locally by a combination of these two factors. For most practical distortions, g_i would be less than unity, but it could take larger values when the image is contrast enhanced, for instance.

C. HVS Model

The HVS model that is used here is also described in the wavelet domain. Natural scene models are in some sense the dual of HVS models, as the HVS has evolved by observing natural images.[18] Hence, many aspects of the HVS have already been incorporated in the natural scene model. It was experimentally determined that just an additive noise model for the HVS gives a marked improvement in the performance of the quality assessment algorithm.[7]

The noise added by the HVS is modeled as a stationary, additive noise field $N = \{\vec{N}_i, i \in I\}$, where the $\vec{N}_i$ are zero-mean, uncorrelated Gaussian random vectors. We then have

$$E = C + N \tag{16.18}$$

$$F = D + N' \tag{16.19}$$

where E and F denote the output of the communication channel (that is the HVS in this case) when the inputs are the reference and test images, respectively. The noise field N is assumed to be independent of C and the covariance of N, given by $\mathbf{C_N}$, is modeled using $\mathbf{C_N} = \sigma_N^2 \mathbf{I}$. N' is modeled similarly. σ_N^2 is the variance of the HVS noise and is a parameter of the model that is derived empirically to optimize the performance of the algorithm. Although the performance of the quality assessment algorithm is affected by the choice of σ_N^2, it is quite robust to small changes in the value.

D. The Visual Information Fidelity Measure

Let $\vec{C}^N = \{C_1, C_2, \ldots C_N\}$ denote N elements from C. Let $\vec{E}^N, \vec{F}^N, \vec{D}^N, Z^N$ and $\vec{U}^N$ be defined similarly. Also, let $\hat{Z}_i$ and $\hat{g}_i$ denote the estimated value of Z_i and g_i at coefficient i, respectively. Similarly, $\hat{\sigma}_N$ and $\hat{\sigma}_v$ represent the estimated variances of the HVS noise and the noise in the distortion model, respectively. Let the eigen decomposition of the covariance matrix $\mathbf{C_U}$ be given by

$$\mathbf{C_U} = \mathbf{Q}\Lambda\mathbf{Q}^{\mathrm{T}} \tag{16.20}$$

Then, it can be shown[7] using the models described above that

$$I(\vec{C}^N, \vec{E}^N | Z^N) = \frac{1}{2} \sum_{i=1}^{N} \sum_{j=1}^{M} \log_2 \left(1 + \frac{\hat{Z}_i^2 \lambda_j}{\hat{\sigma}_N^2} \right) \tag{16.21}$$

$$I(\vec{D}^N, \vec{F}^N | Z^N) = \frac{1}{2} \sum_{i=1}^{N} \sum_{j=1}^{M} \log_2 \left(1 + \frac{\hat{g}_i^2 \hat{Z}_i^2 \lambda_j}{\hat{\sigma}_N^2 + \hat{\sigma}_v^2} \right) \tag{16.22}$$

where $I(\vec{C}^N, \vec{E}^N | Z^N)$ represents the mutual information between the random field representing the reference image coefficients and the output of the HVS channel, conditioned on the mixing field.[7] Here, λ_i denote the eigen values of the covariance matrix $\mathbf{C_U}$.

Notice that the form of this equation is very similar to the Shannon capacity of a communication channel. This is not surprising as the capacity of a communication channel is in fact defined by the mutual information between the source and the output of the channel. The quantity in the LHS of Equation 16.21 can be interpreted as the reference image information, that is, the amount of information that can be extracted by the HVS from an image that passes through it. Similarly, the quantity in the LHS of Equation 16.22 can be thought of as the amount of information that can be extracted by the HVS from the reference image

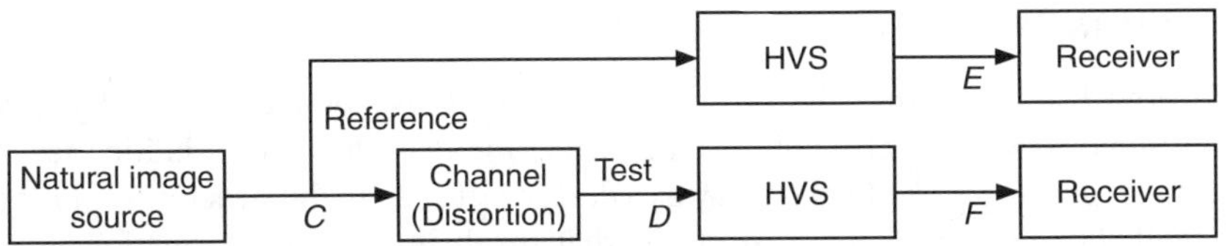

FIGURE 16.6 Block diagram of the VIF quality assessment system. (From Sheikh, H. R., and Bovik, A. C., Information theoretic approaches to image quality assessment, In *Handbook of Image and Video Processing*, A. C. Bovik, Ed., Academic Press, submitted for publication. With permission.)

after it has passed through the distortion channel. The visual quality of the distorted image should relate to the amount of information that can be extracted by the HVS from the test image relative to the reference image information. If the amount of information that is extracted is very close to the reference image information, then the visual quality of the distorted image is very high since no loss of information occurs in the distortion channel.

The ratio of the two information measures has been shown to relate very well with visual quality.[7] Figure 16.6 illustrates the block diagram to compute the Visual Information Fidelity (VIF) measure. Thus, the VIF criterion is given by

$$\text{VIF} = \frac{\sum\limits_{j \in \text{subbands}} I(\vec{D}^{N,j}, \vec{F}^{N,j} | Z^{N,j})}{\sum\limits_{j \in \text{subbands}} I(\vec{C}^{N,j}, \vec{E}^{N,j} | Z^{N,j})} \tag{16.23}$$

where $\vec{C}^{N,j}$ represents a set of N vectors from the jth subbands. The VIF index for the entire image is hence calculated as the sum of this ratio of information measures over all subbands of interest, assuming that the random fields representing the subbands are all independent of each other. Although this assumption is not strictly true, it considerably simplifies the analysis without adversely affecting prediction accuracy.

Notice that the calculation of the VIF criterion involves the estimation of several parameters in the model. Z^N and $\mathbf{C_U}$ are parameters of the GSM model and ways to obtain the maximum likelihood estimates are discussed in Ref. 14. Since we calculate the parameters of the model from the reference image, we are implicitly assuming that the random field C is ergodic. The parameters of the distortion model, namely $\hat{g}_i$ and $\hat{\sigma}_v^2$ can also be obtained easily using linear regression as the reference image is available.[7] The gain field G is assumed to be constant over small blocks and is estimated using the reference and test image coefficients in these blocks. Finally, as mentioned earlier, the variance of the HVS noise modeled by $\hat{\sigma}_N^2$ was obtained experimentally.

E. VIF in Image Quality Assessment

We now briefly discuss the properties of VIF. It is bounded below by zero. Additionally, VIF exactly equals unity when the distorted image is identical to the reference image. Note that this was a design criterion in the structural approach as well. For most practical distortions that result in loss of information in the distortion channel, VIF takes values between 0 and 1. Finally, VIF can capture improvements in the quality of the image caused by, for instance, operations like contrast enhancement. In these cases, VIF takes values larger than unity. This is a remarkable property of VIF that distinguishes it from other metrics for image quality. Most other metrics assume that the reference image is of "perfect" quality and quantify only the *loss* in quality of the test image.

Figure 16.7 presents an illustrative example of the power of VIF in predicting image quality. Figure 16.7(a) shows the reference "Church and Capitol" image and Figure 16.7(b) shows the JPEG compressed version of the image. These are the same images on which the performance of the SSIM algorithm was illustrated earlier in Section II.B. Figure 16.7(c) shows the information map of the reference image. This corresponds to the denominator of the VIF measure and shows the spread of statistical information in the reference image. It is seen that the information is high in regions of high frequency, but is relatively low in the smooth regions of the image. Figure 16.7(d) shows the VIF quality map of the image and illustrates the loss of information due to the distortion. Brighter regions correspond to better quality and the map has been contrast stretched for better visibility. The VIF measure is also successful in predicting the loss of quality in specific regions of the image that we visually noted. This includes the blocking artifacts in the background and roof of the church and the ringing artifacts on the edges of the Capitol.

IV. PERFORMANCE OF SSIM AND VIF

The power of VIF and SSIM in predicting image quality was illustrated in the previous sections using example images. However, to test the performance of the quality assessment algorithm quantitatively, the Video Quality Experts Group (VQEG) Phase I FR-TV specifies four different metrics.[19] First, logistic functions are used in a fitting procedure to provide a nonlinear mapping between the objective and subjective scores. The performance of the algorithm is then tested with respect to the following aspects of their ability to predict quality:[20]

1. *Prediction Accuracy*: The ability to predict the subjective score with low error.
2. *Prediction Monotonicity*: The ability to accurately predict relative magnitudes of subjective scores.
3. *Prediction Consistency*: The robustness of the predictor in assigning accurate scores over a range of different images.

FIGURE 16.7 Illustrative example of VIF (a) Original 'Church and Capitol' image, (b) JPEG compressed image, (c) Image Information map, and (d) Absolute Error Map.

FIGURE 16.7	(Continued).

The first two metrics used are the correlation coefficient between the subjective and objective scores after variance-weighted and nonlinear regression analysis, respectively. These metrics characterize the prediction accuracy of the objective measure. The third metric is the Spearman rank-order correlation coefficient between the objective and subjective scores, which characterizes the prediction monotonicity. Finally, the outlier ratio measures the prediction consistency.

Performance of the SSIM algorithm was tested on images that were compressed using JPEG and JPEG2000 at different bit rates. The details of the experiments conducted to obtain subjective quality scores can be found in Ref. 8. Multiple subjects were asked to assign quality scores to the same image along a linear scale marked with adjectives ranging from "bad" to "good". However, not all subjects used the entire range of values in the numerical scale and this leads to variability.[21] The raw scores are hence converted to Z-scores. The Z-score, z_j, of a raw score, x_j, of a subject X is given by

$$z_j = \frac{x_j - \mu_x}{\sigma_x} \tag{16.24}$$

where

$$\mu_x = \frac{1}{N} \sum_{i=1}^{N} x_i$$

and

$$\sigma_x = \frac{1}{(N-1)} \sum_{i=1}^{N} (x_i - \mu_x)^2$$

where x_i, $i = 1, \ldots, N$ are the raw scores assigned by subject X to all images. Hence, the Z-scores tell us how many standard deviations from the mean the given score is. The Z-scores are then rescaled to fit the entire range of values from 1 to 100. The Mean Opinion Score (MOS) for each image is computed as the mean of the Z-scores for that image, after removing any outliers. The scatter plots of the MOS vs. the SSIM model prediction are shown in Figure 16.8. Each sample point represents one image. The best fitting logistic function is also plotted in the same graph. Also shown here is the scatter plot of the MOS vs. Peak Signal to Noise Ratio (PSNR). PSNR is defined by

$$\text{PSNR} = 10 \log_{10} \frac{255^2}{\text{MSE}} \tag{16.25}$$

for 8 bit images and is just a function of the MSE. The plots clearly illustrate that SSIM performs much better than PSNR in predicting image quality.

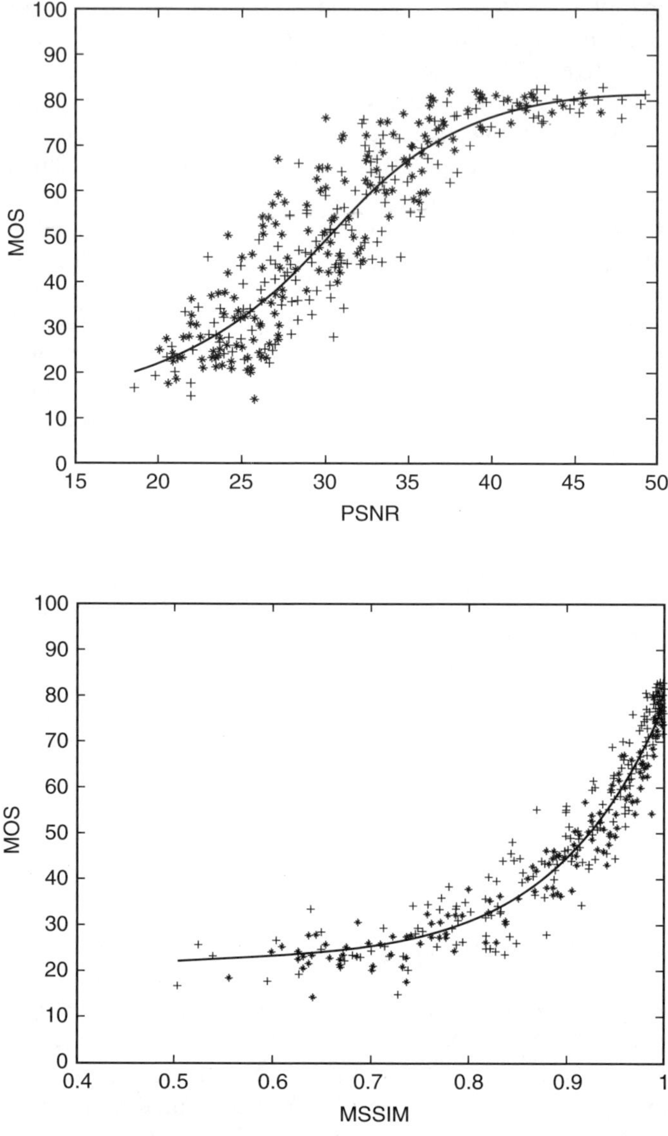

FIGURE 16.8 SSIM: Plot of MOS vs. model prediction and the best fitting logistic function to the subjective and objective scores (From Wang, Z., Bovik, A. C., Sheikh, H. R., and Simoncelli, E. P., Image quality assessment: from error visibility to structural similarity, *IEEE Trans. Image Process.*, 13(4), 1–14, 2004, April. With permission.)

TABLE 16.1
Validation of MOS for SSIM: The criteria Correlation Coefficient (CC), Outlier Ratio (OR), Spearman Rank-Order Correlation Coefficient (SROCC)

Model	CC (Variance Weighted Regression)	CC (Nonlinear Regression)	OR (Nonlinear Regression)	SROCC
PSNR	0.903	0.905	0.157	0.901
Sarnoff	0.956	0.956	0.064	0.947
MSSIM	0.967	0.967	0.041	0.963

Source: From Wang, Z., Bovik, A. C., Sheikh, H. R., and Simoncelli, E. P., Image quality assessment: from error visibility to structural similarity, *IEEE Trans. Image Process.*, 13(4), 1–14, 2004, April. With permission.

Table 16.1 shows the four metrics that were obtained for PSNR, the currently popular Sarnoff model (Sarnoff JND-Metrix 8.0)[22] and SSIM. Again, SSIM outperforms PSNR for every metric.

Performance of the VIF algorithm was tested on JPEG and JPEG2000 compressed images, blurred images, AWGN corrupted images and images reconstructed after transmission errors in a JPEG2000 bitstream while passing through a fast fading Rayleigh channel. The details of the experiments conducted to obtain subjective quality scores can be found in Ref. 7. Again, the raw scores were converted to Z-scores and then rescaled to fit the range of values from 1 to 100. The MOS for each image is computed. The raw scores were also converted to difference scores between the reference and test images and then converted to Z-scores and finally, a Difference Mean Opinion Score (DMOS). The scatter plots of the DMOS vs. the VIF model prediction, as well as PSNR are shown in Figure 16.9. The best fitting logistic function is also plotted in the same graph. The plots clearly illustrate that VIF performs much better than PSNR in predicting image quality. Table 16.2 shows the metrics that were obtained for PSNR, the Sarnoff model and VIF. Again, VIF outperforms PSNR by a sizeable margin for every metric.

Note that the databases used in testing the performance of VIF and SSIM are different, as indicated by the values of the correlation coefficients for both PSNR and the Sarnoff model. The correlation coefficients for both PSNR and SSIM are higher in Table 16.1 than the corresponding values in Table 16.2. This indicates that the database used to evaluate the performance of SSIM is in some sense easier than the one used to evaluate VIF. The metrics for SSIM and VIF presented here are, therefore, not comparable. Further comparisons can be found in Ref. 7.

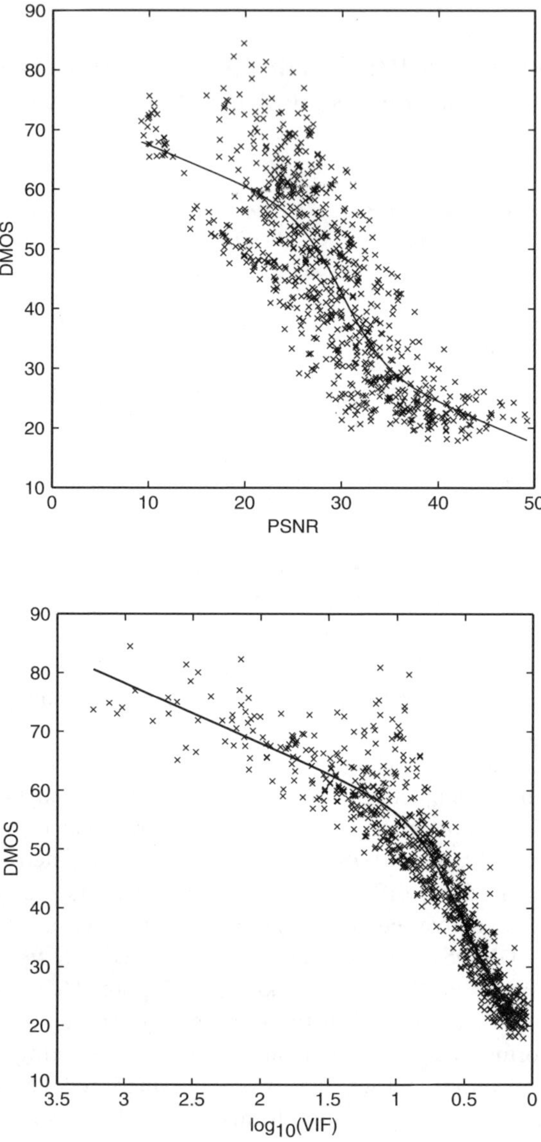

FIGURE 16.9 VIF: Plot of DMOS vs. model prediction and the best fitting logistic function to the subjective and objective scores (From Sheikh, H. R., and Bovik, A. C., Information theoretic approaches to image quality assessment, In *Handbook of Image and Video Processing*, Bovik, A. C. Ed., Academic Press, submitted for publication. With permission.)

TABLE 16.2
Validation of DMOS Scores for VIF: The criteria are Correlation Coefficient (CC), Outlier Ratio (OR), Spearman Rank-Order Correlation Coefficient (SROCC)

Model	CC	OR	SROCC
PSNR	0.826	0.114	0.820
Sarnoff	0.901	0.046	0.902
VIF	0.949	0.013	0.949

Source: Sheikh, H. R., and Bovik, A. C. Image information and visual quality. *IEEE Trans. Image Process.*, submitted for publication, 2003. With permission.

V. CONCLUSIONS

This chapter presented two different top–down approaches to image quality assessment. Both methods have been shown to out-perform several state-of-the-art quality assessment algorithms. We have presented only some of the results here and further details can be found in Refs. 7,8. Structural approaches to image quality assessment attempt to measure the closeness of two signals by measuring the amount of structural distortion present in the distorted signal. This approach can be thought of as complementary to traditional bottom–up HVS-based measures.[10] Information theoretic approaches, on the other hand, assume that the test image is the output of the channel through which the reference image passes and attempt to relate visual quality to the mutual information between the distorted and reference images. The equivalence of the information-theoretic setting to certain bottom–up HVS-based systems has also been shown.[23]

The success of both these methods in quality assessment as competitive state-of-the-art methods has been demonstrated beyond doubt. However, the question as to which role each of them plays in the future of quality assessment research is still unclear. It is even possible that the two paradigms will converge together in building a unified theory of quality assessment. Only further investigation into the structural and information-theoretic framework will answer these questions.

REFERENCES

1. Piella, G., New quality measures for image fusion, pp. 542–546. In *Proceedings of International Conference on Information Fusion*, 2004.
2. Ryan, D., and Tinkler, R., Night pilotage assessment of image fusion, pp. 50–65. In *Proceedings of SPIE*, Vol. 2465. Orlando, FL, 1995, April.
3. Rockinger, O., Image sequence fusion using a shift-invariant wavelet transform, pp. 288–291. In *Proceedings of IEEE International Conference on Image Processing*, Vol. 13, 1997.

4. Karam, L. J., Lossless coding, In *Handbook of Image and Video Processing*, Bovik, A. C., Ed., Academic Press, New York, pp. 461–474, 2000.

5. Sheikh, H. R., Wang, Z., Cormack, L., and Bovik, A. C., LIVE Image Quality Assessment Database, 2003; available: http://live.ece.utexas.edu/research/quality [Online].

6. Pappas, T. N., and Safranek, R. J., Perceptual criteria for image quality evaluation, In *Handbook of Image and Video Processing*, Bovik, A. C., Ed., Academic Press, New York, pp. 669–684, 2000.

7. Sheikh, H. R., and Bovik, A. C., Image information and visual quality. *IEEE Trans. Image Process.*, in press, 2005.

8. Wang, Z., Bovik, A. C., Sheikh, H. R., and Simoncelli, E. P., Image quality assessment: from error visibility to structural similarity, *IEEE Trans. Image Process.*, 13(4), 1–14, 2004, April.

9. Ruderman, D. L., The statistics of natural images, *Network: Computation in Neural Systems*, Vol. 5, pp. 517–548, 1994.

10. Wang, Z., and Bovik, A. C., Structural approaches to image quality assessment, In *Handbook of Image and Video Processing*, 2nd ed., Bovik, A. C. Ed., Academic Press, New York, submitted for publication.

11. Wang, Z., The SSIM index for image quality assessment; available: http://www.cns.nyu.edu/lcv/ssim [Online].

12. Sheikh, H. R., and Bovik, A. C., Information theoretic approaches to image quality assessment, In *Handbook of Image and Video Processing*, 2nd ed., Bovik, A. C. Ed., Academic Press, New York, submitted for publication.

13. Buccigrossi, R. W., and Simoncelli, E. P., Image compression via joint statistical characterization in the wavelet domain, *IEEE Trans. Image Process.*, 8(12), 1688–1701, 1999, December.

14. Portilla, J., Strela, V., Wainwright, M. J., and Simoncelli, E. P., Image denoising using scale mixtures of Gaussians in the wavelet domain, *IEEE Trans. Image Process.*, 12(11), 1338–1351, 2003, November.

15. Wainwright, M. J., and Simoncelli, E. P., Scale mixtures of Gaussians and the statistics of natural images, In *Adv. Neural Information Processing Systems*, Vol. 12, Solla, S. A., Leen, T. K., and Muller, K. R., Eds., MIT Press, Cambridge, MA, pp. 855–861, May, 2000.

16. Wainwright, M. J., Simoncelli, E. P., and Willsky, A. S., Random cascades on wavelet trees and their use in analyzing and modeling natural images, *Appl. Comput. Harmonic Anal.*, 11(1), 89–123, 2001, July.

17. Simoncelli, E. P., Freeman, W. T., Adelson, E. H., and Heeger, D. J., Shiftable multi-scale transforms, *IEEE Trans. Inform. Theory*, 38, 587–607, 1992, March.

18. Simoncelli, E. P., and Olshausen, B. A., Natural image statistics and neural representation, *Annu. Rev. Neural Sci.*, 24, 1193–1216, 2001, May.

19. VQEG, Final report from the video quality experts group on the validation of objective models of video quality assessment, 2000, March; available: http://www.vqeg.org [Online].

20. Rohaly, A., et. al., Video quality experts group: current results and future directions, 2000; available: citeseer.ist.psu.edu/rohaly00video.html [Online].

21. van Dijk, A., Martens, J. B., and Watson, A. B., Quality assessment of coded images using numerical category scaling, pp. 90–101. In *Proceedings of SPIE*, Vol. 2451, 1995, March.

22. Sarnoff Corporation. Jndmetrix technology, 2003; available:http://www.sarnoff. com/products_services/video_vision/jndmetrix/downloads.asp [Online].
23. Sheikh, H. R., Bovik, A. C., and de Veciana, G., An information theoretic criterion for image quality assessment using natural scene statistics, *IEEE Trans. Image Process.*, 2004.

Index

α-entropy, 189
α-geometric-arithmetic (α-GA) mean
divergence, 189
α-Jensen difference divergence, 189
α-mutual information (α-MI), 189

A

Acoustic emission, 376, 377
Active contour registration method, 75
Activity level measurement, 8, 9, 12, 16
Adaptive weight averaging (AWA), 18
Additive white Gausian noise (AWGN), 150,
478
Advanced very high resolution radiometer
(AVHRR) data, 444
Aircraft fuselage lap joints, 376
Algebraic reconstruction technique (ART), 152
Artifical neural networks, 19
multilayer perception neural networks
(MLP), 19
pulse-coupled neural networks (PCNN), 19

B

Backfitting algorithm, 396
Bayesian inference, 378, 388, 400
Biometric, 289–299

C

Canny edge detection, 25, 28
Chernoff bound, 193
Coefficient combining method, 17
Coefficient grouping method, 16
Coefficient-based measure, 16
Color fusion method, 92, 252, 260
Concealed weapon detection (CWD), 4, 5,
22–24, 267, 275, 277, 279
Conditional transition probability (CCTP), 307
Confocal microscopy, 409

Consistency verification, 9, 16, 17
Continuous quasiadditive functional, 197, 198
Contrast sensitive function (CSF), 476
Correlation ratio (CR), 111, 113, 114, 334
Corrosion damage, 376, 380, 383, 400, 401
Cross-correlation, 64, 68, 70, 71, 76, 334, 355,
359, 385
Cross-entropy, 97–100
Cross-power spectrum, 64, 70, 335

D

Dempster-Shafer theory, 314, 388
Diabetic retinopathy, 58–62, 84, 85, 89, 91, 100
Difference mean opinion score (*see also* mean
opinion score), 495
Difference moment, 315
inverse, 315
first-order, 315
second-order, 315
Direct volume rendering (DVR), 452
Discrete wavelet frame (DWF), 9, 12, 14, 15
Drawing web format (DWF), 457
Dual-band visual intensified (DII) camera, 248,
252–259

E

Eddy current (ET), 7, 349, 360, 362–366, 370,
371, 375–389, 397–401
Edge of Light (EOL), 401
Emission computed tomography (ECT), 158
Euler-Lagrange evolution equation, 120
Expectation-maximum (EM), 41, 270–279,
284–287
Eye-tracking technology, 439

F

FaceLAB, 463, 464, 467
Failure to enroll (FTE) rate, 292